Student Solutions Manual

College Algebra

SIXTH EDITION

James Stewart
McMaster University and University of Toronto

Lothar Redlin
The Pennsylvania State University

Saleem Watson
California State University, Long Beach

Prepared by

Andrew Bulman-Fleming

BROOKS/COLE
CENGAGE Learning

Australia • Brazil • Japan • Korea • Mexico • Singapore • Spain • United Kingdom • United States

ISBN-13: 978-1-111-99024-4
ISBN-10: 1-111-99024-7

Brooks/Cole
20 Davis Drive
Belmont, CA 94002-3098
USA

Cengage Learning is a leading provider of customized learning solutions with office locations around the globe, including Singapore, the United Kingdom, Australia, Mexico, Brazil, and Japan. Locate your local office at: **www.cengage.com/global**

Cengage Learning products are represented in Canada by Nelson Education, Ltd.

To learn more about Brooks/Cole, visit **www.cengage.com/brookscole**

Purchase any of our products at your local college store or at our preferred online store **www.cengagebrain.com**

Printed in the United States of America
1 2 3 4 5 6 7 16 15 14 13 12

CONTENTS

CHAPTER 2   FUNCTIONS   89

CHAPTER 3   POLYNOMIAL AND RATIONAL FUNCTIONS   135

CHAPTER 4   EXPONENTIAL AND LOGARITHMIC FUNCTIONS   197

CHAPTER 5   SYSTEMS OF EQUATIONS AND INEQUALITIES   227

CHAPTER 6   MATRICES AND DETERMINANTS   259

CHAPTER 7   CONIC SECTIONS   291

P PREREQUISITES

P.1 MODELING THE REAL WORLD WITH ALGEBRA

1. Using this model, we find that if $S = 12$, $L = 4S = 4(12) = 48$. Thus, 12 sheep have 48 legs.

3. If $x = \$120$ and $T = 0.06x$, then $T = 0.06(120) = 7.2$. The sales tax is \$7.20.

5. If $x = 32$ and $C = 50 + 1.25x$, then $C = 50 + 1.25(32) = 90$. The cost is \$90.

7. If $v = 70$, $t = 3.5$, and $d = vt$, then $d = 70 \cdot 3.5 = 245$. The car has traveled 245 miles.

9. (a) $M = \dfrac{N}{G} = \dfrac{240}{8} = 30$ miles/gallon

   (b) $25 = \dfrac{175}{G} \Leftrightarrow G = \dfrac{175}{25} = 7$ gallons

11. (a) $V = 9.5S = 9.5\left(4 \text{ km}^3\right) = 38 \text{ km}^3$

    (b) $19 \text{ km}^3 = 9.5S \Leftrightarrow S = 2 \text{ km}^3$

13. (a)

Depth (ft)	Pressure (lb/in^2)
0	$0.45(0) + 14.7 = 14.7$
10	$0.45(10) + 14.7 = 19.2$
20	$0.45(20) + 14.7 = 23.7$
30	$0.45(30) + 14.7 = 28.2$
40	$0.45(40) + 14.7 = 32.7$
50	$0.45(50) + 14.7 = 37.2$
60	$0.45(60) + 14.7 = 41.7$

   (b) We know that $P = 30$ and we want to find d, so we solve the equation $30 = 14.7 + 0.45d \Leftrightarrow 15.3 = 0.45d \Leftrightarrow$ $d = \dfrac{15.3}{0.45} = 34.0$. Thus, if the pressure is 30 lb/in^2, the depth is 34 ft.

15. The number N of days in w weeks is $N = 7w$.

17. The average A of two numbers, a and b, is $A = \dfrac{a+b}{2}$.

19. The cost C of purchasing x gallons of gas at \$3.50 a gallon is $C = 3.5x$.

21. The distance d in miles that a car travels in t hours at 60 mi/h is $d = 60t$.

23. (a) $\$12 + 3(\$1) = \$12 + \$3 = \$15$

    (b) The cost C, in dollars, of a pizza with n toppings is $C = 12 + n$.

    (c) Using the model $C = 12 + n$ with $C = 16$, we get $16 = 12 + n \Leftrightarrow n = 4$. So the pizza has four toppings.

25. (a) (i) For an all-electric car, the energy cost of driving x miles is $C_e = 0.04x$.

       (ii) For an average gasoline powered car, the energy cost of driving x miles is $C_g = 0.12x$.

    (b) (i) The cost of driving 10,000 miles with an all-electric car is $C_e = 0.04(10{,}000) = \$400$.

       (ii) The cost of driving 10,000 miles with a gasoline powered car is $C_g = 0.12(10{,}000) = \$1200$.

27. (a) A 10-minute call costs $1 + 0.1(10) = 1 + 1 = \$2$.

    (b) The cost is $\left(\begin{array}{c} \text{connect} \\ \text{fee} \end{array}\right) + \left(\begin{array}{c} \text{cost} \\ \text{per minute} \end{array}\right) \times (\text{minutes})$, so $C = 1 + 0.1t$.

    (c) When $C = 2.2$, we have $2.2 = 1 + 0.1t \Leftrightarrow 1.2 = 0.1t \Leftrightarrow t = 12$. The call lasted 12 minutes.

    (d) The cost is $\left(\begin{array}{c} \text{connect} \\ \text{fee} \end{array}\right) + \left(\begin{array}{c} \text{cost} \\ \text{per minute} \end{array}\right) \times (\text{minutes})$, so when the connection fee is F cents and the rate is r cents then the cost (in cents) is given by $C = F + rt$.

1

P.2   THE REAL NUMBERS

1. **(a)** The natural numbers are $\{1, 2, 3, \ldots\}$.

   (b) The numbers $\{\ldots, -3, -2, -1, 0\}$ are integers but not natural numbers.

   (c) Any irreducible fraction $\dfrac{p}{q}$ with $q \neq 1$ is rational but is not an integer. Examples: $\frac{3}{2}, -\frac{5}{12}, \frac{1729}{23}$.

   (d) Any number which cannot be expressed as a ratio $\dfrac{p}{q}$ of two integers is irrational. Examples are $\sqrt{2}, \sqrt{3}, \pi$, and e.

3. **(a)** In set-builder notation, the set of real numbers between but not including 2 and 7 is $\{x \mid 2 < x < 7\}$.

   (b) In interval notation, the set of real numbers between but not including 2 and 7 is $(2, 7)$.

5. The symbol $|x|$ stands for the *absolute value* of the number x. If x is not 0, then the sign of $|x|$ is always *positive*.

7. **(a)** Natural number: None

   (b) Integers: $-3, 0, -1000$

   (c) Rational numbers: $-3, 0, \frac{22}{7}, 3.14, 2.7\overline{6}, -1000, -\frac{2}{5}$

   (d) Irrational numbers: $\sqrt{7}, -\pi$

9. Commutative Property for addition

11. Associative Property for multiplication

13. Distributive Property

15. Associative Property for addition

17. Distributive Property

19. Commutative Property for multiplication

21. $x + 3 = 3 + x$

23. $4(A + B) = 4A + 4B$

25. $3(x + y) = 3x + 3y$

27. $4(2m) = (4 \cdot 2)m = 8m$

29. $-\frac{5}{2}(2x - 4y) = -\frac{5}{2}(2x) + \frac{5}{2}(4y) = -5x + 10y$

31. **(a)** $\frac{3}{10} + \frac{4}{15} = \frac{9}{30} + \frac{8}{30} = \frac{17}{30}$

   (b) $\frac{1}{4} + \frac{1}{5} = \frac{5}{20} + \frac{4}{20} = \frac{9}{20}$

33. **(a)** $\frac{2}{3}\left(6 - \frac{3}{2}\right) = \frac{2}{3} \cdot 6 - \frac{2}{3} \cdot \frac{3}{2} = 4 - 1 = 3$

   (b) $0.25\left(\frac{8}{9} + \frac{1}{2}\right) = \frac{1}{4}\left(\frac{16}{18} + \frac{9}{18}\right) = \frac{1}{4} \cdot \frac{25}{18} = \frac{25}{72}$

35. **(a)** $\dfrac{\frac{2}{3}}{\frac{2}{3}} - \dfrac{\frac{2}{3}}{2} = 2 \cdot \frac{3}{2} - \frac{2}{3} \cdot \frac{1}{2} = 3 - \frac{1}{3} = \frac{9}{3} - \frac{1}{3} = \frac{8}{3}$

   (b) $\dfrac{\frac{1}{12}}{\frac{1}{8} - \frac{1}{9}} = \dfrac{\frac{1}{12}}{\frac{1}{8} - \frac{1}{9}} \cdot \frac{72}{72} = \frac{6}{9 - 8} = \frac{6}{1} = 6$

37. **(a)** $2 \cdot 3 = 6$ and $2 \cdot \frac{7}{2} = 7$, so $3 < \frac{7}{2}$

   (b) $-6 > -7$

   (c) $3.5 = \frac{7}{2}$

37. **(a)** $x > 0$ **(b)** $t < 4$

   (c) $a \geq \pi$ **(d)** $-5 < x < \frac{1}{3}$

   (e) $|p - 3| \leq 5$

47. **(a)** $A \cup B = \{1, 2, 3, 4, 5, 6, 7, 8\}$

   (b) $A \cap B = \{2, 4, 6\}$

49. **(a)** $A \cup C = \{1, 2, 3, 4, 5, 6, 7, 8, 9, 10\}$

   (b) $A \cap C = \{7\}$

51. **(a)** $B \cup C = \{x \mid x \leq 5\}$

   (b) $B \cap C = \{x \mid -1 < x < 4\}$

53. $(-3, 0) = \{x \mid -3 < x < 0\}$

55. $[2, 8) = \{x \mid 2 \leq x < 8\}$

<!-- number line from 2 (filled) to 8 (open) -->

57. $[2, \infty) = \{x \mid x \geq 2\}$

<!-- number line from 2 (filled) to right -->

59. $x \leq 1 \Leftrightarrow x \in (-\infty, 1]$

<!-- number line to 1 (filled) -->

61. $-2 < x \leq 1 \Leftrightarrow x \in (-2, 1]$

<!-- number line from -2 (open) to 1 (filled) -->

63. $x > -1 \Leftrightarrow x \in (-1, \infty)$

<!-- number line from -1 (open) to right -->

65. (a) $[-3, 5]$ **(b)** $(-3, 5]$

67. (a) $[-4, -1)$ **(b)** $[1, 4)$

69. (a) $[-5, \infty)$ **(b)** $(-\infty, 5)$

71. $(-2, 0) \cup (-1, 1) = (-2, 1)$

<!-- number line from -2 (open) to 1 (open) -->

73. $[-4, 6] \cap [0, 8) = [0, 6]$

<!-- number line from 0 (filled) to 6 (filled) -->

75. $(-\infty, -4) \cup (4, \infty)$

<!-- number line -->

77. (a) $|100| = 100$

   (b) $|-73| = 73$

79. (a) $||-6| - |-4|| = |6 - 4| = |2| = 2$

   (b) $\frac{-1}{|-1|} = \frac{-1}{1} = -1$

81. (a) $|(-2) \cdot 6| = |-12| = 12$

   (b) $\left|\left(-\frac{1}{3}\right)(-15)\right| = |5| = 5$

83. $|(-2) - 3| = |-5| = 5$

85. (a) $|17 - 2| = 15$

   (b) $|21 - (-3)| = |21 + 3| = |24| = 24$

   (c) $\left|-\frac{3}{10} - \frac{11}{8}\right| = \left|-\frac{12}{40} - \frac{55}{40}\right| = \left|-\frac{67}{40}\right| = \frac{67}{40}$

87. (a) Let $x = 0.777\ldots$. So $10x = 7.7777\ldots \Leftrightarrow x = 0.7777\ldots \Leftrightarrow 9x = 7$. Thus, $x = \frac{7}{9}$.

   (b) Let $x = 0.2888\ldots$. So $100x = 28.8888\ldots \Leftrightarrow 10x = 2.8888\ldots \Leftrightarrow 90x = 26$. Thus, $x = \frac{26}{90} = \frac{13}{45}$.

   (c) Let $x = 0.575757\ldots$. So $100x = 57.5757\ldots \Leftrightarrow x = 0.5757\ldots \Leftrightarrow 99x = 57$. Thus, $x = \frac{57}{99} = \frac{19}{33}$.

89. Distributive Property

91. (a) When $L = 60$, $x = 8$, and $y = 6$, we have $L + 2(x + y) = 60 + 2(8 + 6) = 60 + 28 = 88$. Because $88 \leq 108$ the post office will accept this package.
   When $L = 48$, $x = 24$, and $y = 24$, we have $L + 2(x + y) = 48 + 2(24 + 24) = 48 + 96 = 144$, and since $144 \nleq 108$, the post office will *not* accept this package.

   (b) If $x = y = 9$, then $L + 2(9 + 9) \leq 108 \Leftrightarrow L + 36 \leq 108 \Leftrightarrow L \leq 72$. So the length can be as long as 72 in. = 6 ft.

93. Let $x = \dfrac{m_1}{n_1}$ and $y = \dfrac{m_2}{n_2}$ be rational numbers. Then $x + y = \dfrac{m_1}{n_1} + \dfrac{m_2}{n_2} = \dfrac{m_1 n_2 + m_2 n_1}{n_1 n_2}$,

$x - y = \dfrac{m_1}{n_1} - \dfrac{m_2}{n_2} = \dfrac{m_1 n_2 - m_2 n_1}{n_1 n_2}$, and $x \cdot y = \dfrac{m_1}{n_1} \cdot \dfrac{m_2}{n_2} = \dfrac{m_1 m_2}{n_1 n_2}$. This shows that the sum, difference, and product

of two rational numbers are again rational numbers. However the product of two irrational numbers is not necessarily

irrational; for example, $\sqrt{2} \cdot \sqrt{2} = 2$, which is rational. Also, the sum of two irrational numbers is not necessarily irrational;

for example, $\sqrt{2} + \left(-\sqrt{2}\right) = 0$ which is rational.

95.

x	1	2	10	100	1000
$\dfrac{1}{x}$	1	$\dfrac{1}{2}$	$\dfrac{1}{10}$	$\dfrac{1}{100}$	$\dfrac{1}{1000}$

As x gets large, the fraction $1/x$ gets small. Mathematically, we say that $1/x$ goes to zero.

x	1	0.5	0.1	0.01	0.001
$\dfrac{1}{x}$	1	$\dfrac{1}{0.5} = 2$	$\dfrac{1}{0.1} = 10$	$\dfrac{1}{0.01} = 100$	$\dfrac{1}{0.001} = 1000$

As x gets small, the fraction $1/x$ gets large. Mathematically, we say that $1/x$ goes to infinity.

97. **(a)** Subtraction is not commutative. For example, $5 - 1 \neq 1 - 5$.

(b) Division is not commutative. For example, $5 \div 1 \neq 1 \div 5$.

(c) Putting on your socks and putting on your shoes are not commutative. If you put on your socks first, then your shoes, the result is not the same as if you proceed the other way around.

(d) Putting on your hat and putting on your coat are commutative. They can be done in either order, with the same result.

(e) Washing laundry and drying it are not commutative.

(f) Answers will vary.

(g) Answers will vary.

P.3   INTEGER EXPONENTS AND SCIENTIFIC NOTATION

1. Using exponential notation we can write the product $5 \cdot 5 \cdot 5 \cdot 5 \cdot 5 \cdot 5$ as 5^6.

3. In the expression 3^4, the number 3 is called the *base* and the number 4 is called the *exponent*.

5. When we divide two powers with the same base, we *subtract* the exponents. So $\dfrac{3^5}{3^2} = 3^3$.

7. **(a)** $2^{-1} = \dfrac{1}{2}$ **(b)** $2^{-3} = \dfrac{1}{8}$ **(c)** $\left(\dfrac{1}{2}\right)^{-1} = 2$ **(d)** $\dfrac{1}{2^{-3}} = 2^3 = 8$

9. **(a)** $-3^2 = -9$ **(b)** $(-3)^2 = 9$ **(c)** $\left(\dfrac{2}{3}\right)^3 \cdot (-3)^3 = \dfrac{2^3 (-3)^3}{3^3} = -8$

11. **(a)** $\left(\dfrac{5}{3}\right)^0 \cdot 2^{-1} = \dfrac{1}{2}$ **(b)** $\dfrac{2^{-3}}{3^0} = \dfrac{1}{2^3} = \dfrac{1}{8}$ **(c)** $\left(\dfrac{1}{4}\right)^{-2} = 4^2 = 16$

13. **(a)** $5^3 \cdot 5 = 5^4 = 625$ **(b)** $3^2 \cdot 3^0 = 3^2 = 9$ **(c)** $\left(2^2\right)^3 = 2^6 = 64$

15. **(a)** $\left(2^8\right)^0 = 1$ **(b)** $(-5)^0 = 1$ **(c)** $-5^0 = -1$

17. **(a)** $5^4 \cdot 5^{-2} = 5^2 = 25$ **(b)** $\dfrac{10^7}{10^4} = 10^3 = 1000$ **(c)** $\dfrac{3^2}{3^4} = \dfrac{1}{3^2} = \dfrac{1}{9}$

19. **(a)** $x^3 \cdot x^4 = x^{3+4} = x^7$ **(b)** $\left(2y^2\right)^3 = 2^3 y^{2 \cdot 3} = 8y^6$ **(c)** $y^{-2} y^7 = y^{-2+7} = y^5$

21. (a) $x^{-5} \cdot x^3 = x^{-5+3} = x^{-2} = \dfrac{1}{x^2}$ **(b)** $w^{-2}w^{-4}w^5 = w^{-2-4+5} = w^{-1} = \dfrac{1}{w}$

(c) $\dfrac{y^{10}y^0}{y^7} = y^{10+0-7} = y^3$

23. (a) $\dfrac{a^9 a^{-2}}{a} = a^{9-2-1} = a^6$ **(b)** $\left(a^2 a^4\right)^3 = \left(a^{2+4}\right)^3 = \left(a^6\right)^3 = a^{6 \cdot 3} = a^{18}$

(c) $(2x)^2 \left(5x^6\right) = 2^2 x^2 \cdot 5x^6 = 20x^{2+6} = 20x^8$

25. (a) $\left(3x^3 y^2\right)\left(2y^3\right) = 3 \cdot 2x^3 y^{2+3} = 6x^3 y^5$ **(b)** $\left(6a^5 b\right)\left(\tfrac{1}{2}a^2 b^4\right) = 6\left(\tfrac{1}{2}\right)a^{5+2} b^{1+4} = 3a^7 b^5$

27. (a) $\left(5z^{-2}\right)^2 \left(z^3\right) = 5^2 z^{-2(2)} z^3 = 25z^{-4+3} = 25z^{-1} = \dfrac{25}{z}$

(b) $\left(3a^4 b^{-2}\right)^3 \left(a^2 b^{-1}\right) = 3^3 a^{4 \cdot 3} b^{-2 \cdot 3} a^2 b^{-1} = 27a^{12+2} b^{-6-1} = \dfrac{27a^{14}}{b^7}$

29. (a) $\dfrac{x^2 y^{-1}}{x^{-5}} = x^{2-(-5)} y^{-1} = \dfrac{x^7}{y}$ **(b)** $\dfrac{x^{-3} y^2}{x^{-2} y^{-1}} = x^{-3-(-2)} y^{2-(-1)} = x^{-1} y^3 = \dfrac{y^3}{x}$

31. (a) $\left(\dfrac{a^3}{2b^2}\right)^3 = \dfrac{a^{3 \cdot 3}}{2^3 b^{2 \cdot 3}} = \dfrac{a^9}{8b^6}$ **(b)** $\left(\dfrac{2x^4}{y^{-1}}\right)^{-1} = \dfrac{y^{-1}}{2x^4} = \dfrac{1}{2x^4 y}$

33. (a) $\left(\dfrac{a^2}{b}\right)^5 \left(\dfrac{a^3 b^2}{c^3}\right)^3 = a^{2 \cdot 5 + 3 \cdot 3} b^{-5+2 \cdot 3} c^{-3 \cdot 3} = \dfrac{a^{19} b}{c^9}$ **(b)** $\dfrac{\left(u^{-1} v^2\right)^2}{\left(u^3 v^{-2}\right)^3} = u^{-1 \cdot 2 - 3 \cdot 3} v^{2 \cdot 2 - (-2)3} = \dfrac{v^{10}}{u^{11}}$

35. (a) $\dfrac{8a^3 b^{-4}}{2a^{-5} b^5} = 4a^{3-(-5)} b^{-4-5} = \dfrac{4a^8}{b^9}$ **(b)** $\left(\dfrac{y}{5x^{-2}}\right)^{-3} = 5^{-1(-3)} x^{-(-2)(-3)} y^{-3} = \dfrac{125}{x^6 y^3}$

37. (a) $\left(\dfrac{3a}{b^3}\right)^{-1} = 3^{-1} a^{-1} b^{-3(-1)} = \dfrac{b^3}{3a}$

(b) $\left(\dfrac{q^{-1} r^{-1} s^{-2}}{r^{-5} s q^{-8}}\right)^{-1} = \dfrac{r^{-5} s q^{-8}}{q^{-1} r^{-1} s^{-2}} = q^{-8-(-1)} r^{-5-(-1)} s^{1-(-2)} = \dfrac{s^3}{q^7 r^4}$

39. (a) $69{,}300{,}000 = 6.93 \times 10^7$ **41. (a)** $3.19 \times 10^5 = 319{,}000$

(b) $7{,}200{,}000{,}000{,}000 = 7.2 \times 10^{12}$ **(b)** $2.721 \times 10^8 = 272{,}100{,}000$

(c) $0.000028536 = 2.8536 \times 10^{-5}$ **(c)** $2.670 \times 10^{-8} = 0.00000002670$

(d) $0.0001213 = 1.213 \times 10^{-4}$ **(d)** $9.999 \times 10^{-9} = 0.000000009999$

43. (a) $5{,}900{,}000{,}000{,}000 \text{ mi} = 5.9 \times 10^{12}$ mi

(b) $0.0000000000004 \text{ cm} = 4 \times 10^{-13}$ cm

(c) $33 \text{ billion billion molecules} = 33 \times 10^9 \times 10^9 = 3.3 \times 10^{19}$ molecules

45. $\left(7.2 \times 10^{-9}\right)\left(1.806 \times 10^{-12}\right) = 7.2 \cdot 1.806 \times 10^{-9} \times 10^{-12} \approx 13.0 \times 10^{-21} = 1.3 \times 10^{-20}$

47. $\dfrac{1.295643 \times 10^9}{\left(3.610 \times 10^{-17}\right)\left(2.511 \times 10^6\right)} = \dfrac{1.295643}{3.610 \cdot 2.511} \times 10^{9+17-6} \approx 0.1429 \times 10^{19} = 1.429 \times 10^{19}$

49. $\dfrac{(0.0000162)(0.01582)}{(594621000)(0.0058)} = \dfrac{\left(1.62 \times 10^{-5}\right)\left(1.582 \times 10^{-2}\right)}{\left(5.94621 \times 10^8\right)\left(5.8 \times 10^{-3}\right)} = \dfrac{1.62 \cdot 1.582}{5.94621 \cdot 5.8} \times 10^{-5-2-8+3} = 0.074 \times 10^{-12}$

$= 7.4 \times 10^{-14}$

51. (a) $\dfrac{a^m}{a^n} = \dfrac{a \cdot a \cdot a \cdot \cdots \cdot a \ (m \text{ factors})}{a \cdot a \cdot a \cdot \cdots \cdot a \ (n \text{ factors})} = a \cdot a \cdot a \cdot \cdots \cdot a \ (m - n \text{ factors}) = a^{m-n},\ a \neq 0$

(b) $\left(\dfrac{a}{b}\right)^n = \dfrac{a}{b} \cdot \dfrac{a}{b} \cdot \cdots \cdot \dfrac{a}{b}\ (n\ \text{factors}) = \dfrac{a \cdot a \cdot a \cdot \cdots \cdot a\ (n\ \text{factors})}{b \cdot b \cdot b \cdot \cdots \cdot b\ (n\ \text{factors})} = \dfrac{a^n}{b^n},\ b \neq 0$

53. Since one light year is 5.9×10^{12} miles, Centauri is about $4.3 \times 5.9 \times 10^{12} \approx 2.54 \times 10^{13}$ miles away or 25,400,000,000,000 miles away.

55. Volume = (average depth) (area) = $\left(3.7 \times 10^3\ \text{m}\right)\left(3.6 \times 10^{14}\ \text{m}^2\right)\left(\dfrac{10^3\ \text{liters}}{\text{m}^3}\right) \approx 1.33 \times 10^{21}$ liters

57. The number of molecules is equal to

$$(\text{volume}) \cdot \left(\dfrac{\text{liters}}{\text{m}^3}\right) \cdot \left(\dfrac{\text{molecules}}{22.4\ \text{liters}}\right) = (5 \cdot 10 \cdot 3) \cdot \left(10^3\right) \cdot \left(\dfrac{6.02 \times 10^{23}}{22.4}\right) \approx 4.03 \times 10^{27}$$

59.

Year	Total interest
1	\$152.08
2	308.79
3	470.26
4	636.64
5	808.08

61. (a) $\dfrac{18^5}{9^5} = \left(\dfrac{18}{9}\right)^5 = 2^5 = 32$

(b) $20^6 \cdot (0.5)^6 = (20 \cdot 0.5)^6 = 10^6 = 1{,}000{,}000$

63. (a) b^5 is negative since a negative number raised to an odd power is negative.

(b) b^{10} is positive since a negative number raised to an even power is positive.

(c) ab^2c^3 we have (positive) (negative)^2 (negative)^3 = (positive) (positive) (negative) which is negative.

(d) Since $b - a$ is negative, $(b - a)^3 = $ (negative)^3 which is negative.

(e) Since $b - a$ is negative, $(b - a)^4 = $ (negative)^4 which is positive.

(f) $\dfrac{a^3c^3}{b^6c^6} = \dfrac{(\text{positive})^3\ (\text{negative})^3}{(\text{negative})^6\ (\text{negative})^6} = \dfrac{(\text{positive})\ (\text{negative})}{(\text{positive})\ (\text{positive})} = \dfrac{\text{negative}}{\text{positive}}$ which is negative.

P.4   RATIONAL EXPONENTS AND RADICALS

1. Using exponential notation we can write $\sqrt[3]{5}$ as $5^{1/3}$.

3. No. $\sqrt{5^2} = \left(5^2\right)^{1/2} = 5^{2(1/2)} = 5$ and $\left(\sqrt{5}\right)^2 = \left(5^{1/2}\right)^2 = 5^{(1/2)2} = 5$.

5. Because the denominator is of the form $\sqrt{a}$, we multiply numerator and denominator by $\sqrt{a}$: $\dfrac{1}{\sqrt{3}} = \dfrac{1}{\sqrt{3}} \cdot \dfrac{\sqrt{3}}{\sqrt{3}} = \dfrac{\sqrt{3}}{3}$.

7. $\dfrac{1}{\sqrt{7}} = 7^{-1/2}$

9. $4^{2/3} = \sqrt[3]{4^2} = \sqrt[3]{16}$

11. $\sqrt[5]{5^3} = 5^{3/5}$

13. $a^{2/5} = \sqrt[5]{a^2}$

15. $\sqrt[3]{y^4} = y^{4/3}$

17. (a) $\sqrt{16} = \sqrt{4^2} = 4$

(b) $\sqrt[4]{16} = \sqrt[4]{2^4} = 2$

(c) $\sqrt[4]{\dfrac{1}{16}} = \sqrt[4]{\left(\dfrac{1}{2}\right)^4} = \dfrac{1}{2}$

19. (a) $3\sqrt[3]{16} = 3\sqrt[3]{2 \cdot 2^3} = 6\sqrt[3]{2}$

(b) $\dfrac{\sqrt{18}}{\sqrt{81}} = \sqrt{\dfrac{18}{81}} = \sqrt{\dfrac{2}{9}} = \dfrac{\sqrt{2}}{3}$

(c) $\sqrt{\dfrac{27}{4}} = \sqrt{\dfrac{3 \cdot 3^2}{2^2}} = \dfrac{3\sqrt{3}}{2}$

21. (a) $\sqrt{7}\sqrt{28} = \sqrt{7 \cdot 28} = \sqrt{196} = 14$

(b) $\dfrac{\sqrt{48}}{\sqrt{3}} = \sqrt{\dfrac{48}{3}} = \sqrt{16} = 4$

(c) $\sqrt[4]{24}\sqrt[4]{54} = \sqrt[4]{24 \cdot 54} = \sqrt[4]{1296} = 6$

23. (a) $\dfrac{\sqrt{216}}{\sqrt{6}} = \sqrt{\dfrac{216}{6}} = \sqrt{36} = 6$

(b) $\sqrt[3]{2}\sqrt[3]{32} = \sqrt[3]{64} = 4$

(c) $\sqrt[4]{\dfrac{1}{4}}\sqrt[4]{\dfrac{1}{64}} = \sqrt[4]{\dfrac{1}{256}} = \dfrac{1}{\sqrt[4]{256}} = \dfrac{1}{4}$

25. $\sqrt[4]{x^4} = |x|$

27. $\sqrt[5]{32y^6} = \sqrt[5]{2^5 y^6} = 2\sqrt[5]{y^6}$

29. $\sqrt[4]{16x^8} = \sqrt[4]{2^4 x^8} = 2x^2$

31. $\sqrt[3]{x^3 y} = \left(x^3\right)^{1/3} y^{1/3} = x\sqrt[3]{y}$

33. $\sqrt{36r^2 t^4} = \sqrt{\left(6rt^2\right)^2} = 6|r|t^2$

35. $\sqrt[5]{a^6 b^7} = a^{6/5} b^{7/5} = a \cdot a^{1/5} b \cdot b^{2/5} = ab\sqrt[5]{ab^2}$

37. $\sqrt[3]{\sqrt{64x^6}} = \left(8\left|x^3\right|\right)^{1/3} = 2|x|$

39. $\sqrt{32} + \sqrt{18} = \sqrt{16 \cdot 2} + \sqrt{9 \cdot 2} = \sqrt{4^2 \cdot 2} + \sqrt{3^2 \cdot 2} = 4\sqrt{2} + 3\sqrt{2} = 7\sqrt{2}$

41. $\sqrt{125} - \sqrt{45} = \sqrt{25 \cdot 5} - \sqrt{9 \cdot 5} = \sqrt{5^2 \cdot 5} - \sqrt{3^2 \cdot 5} = 5\sqrt{5} - 3\sqrt{5} = 2\sqrt{5}$

43. $\sqrt[3]{108} - \sqrt[3]{32} = 3\sqrt[3]{4} - 2\sqrt[3]{4} = \sqrt[3]{4}$

45. $\sqrt{9a^3} - \sqrt{a} = \sqrt{3^2 a^2 \cdot a} - \sqrt{a} = 3a\sqrt{a} - \sqrt{a} = (3a - 1)\sqrt{a}$

47. $\sqrt[3]{x^4} + \sqrt[3]{8x} = \sqrt[3]{x^3 x} + \sqrt[3]{2^3 x} = x\sqrt[3]{x} + 2\sqrt[3]{x} = (x + 2)\sqrt[3]{x}$

49. $\sqrt[3]{16a^5} - 3\sqrt[3]{2a^2} = \sqrt[3]{2a^2 \cdot 2^3 a^3} - 3\sqrt[3]{2a^2} = (2a - 3)\sqrt[3]{2a^2}$

51. (a) $16^{1/4} = 2$ **(b)** $-125^{1/3} = -5$ **(c)** $9^{-1/2} = \dfrac{1}{9^{1/2}} = \dfrac{1}{3}$

53. (a) $32^{2/5} = \left(32^{1/5}\right)^2 = 2^2 = 4$ **(b)** $\left(\dfrac{4}{9}\right)^{-1/2} = \left(\dfrac{9}{4}\right)^{1/2} = \dfrac{3}{2}$ **(c)** $\left(\dfrac{16}{81}\right)^{3/4} = \left(\dfrac{2}{3}\right)^3 = \dfrac{8}{27}$

55. (a) $5^{2/3} \cdot 5^{1/3} = 5^{2/3 + 1/3} = 5^1 = 5$ **(b)** $\dfrac{3^{3/5}}{3^{2/5}} = 3^{3/5 - 2/5} = \sqrt[5]{3}$ **(c)** $\left(\sqrt[3]{4}\right)^3 = 4^{(1/3)3} = 4$

57. When $x = 3$, $y = 4$, $z = -1$ we have $\sqrt{x^2 + y^2} = \sqrt{3^2 + 4^2} = \sqrt{9 + 16} = \sqrt{25} = 5$.

59. When $x = 3$, $y = 4$, $z = -1$ we have

$$(9x)^{2/3} + (2y)^{2/3} + z^{2/3} = (9 \cdot 3)^{2/3} + (2 \cdot 4)^{2/3} + (-1)^{2/3} = \left(3^3\right)^{2/3} + \left(2^3\right)^{2/3} + (1)^{1/3}$$

$$= 3^2 + 2^2 + 1 = 9 + 4 + 1 = 14.$$

61. (a) $x^{3/4} x^{5/4} = x^{3/4 + 5/4} = x^2$

(b) $y^{2/3} y^{4/3} = y^{2/3 + 4/3} = y^2$

63. (a) $(4b)^{1/2}\left(8b^{1/4}\right) = 4^{1/2} \cdot 8b^{1/2 + 1/4} = 16b^{3/4}$

(b) $\left(3a^{3/4}\right)^2\left(5a^{1/2}\right) = 3^2 \cdot 5a^{(3/4) \cdot 2 + (1/2)} = 45a^2$

65. (a) $\dfrac{w^{4/3} w^{2/3}}{w^{1/3}} = w^{4/3 + 2/3 - 1/3} = w^{5/3}$

(b) $\dfrac{s^{5/2}\left(2s^{5/4}\right)^2}{s^{1/2}} = 2^2 s^{(5/2) + (5/4)2 - 1/2} = 4s^{9/2}$

67. (a) $\left(8a^6 b^{3/2}\right)^{2/3} = 8^{2/3} a^{6(2/3)} b^{(3/2)(2/3)} = 4a^4 b$

(b) $(4a^6 b^8)^{3/2} = 4^{3/2} a^{6(3/2)} b^{8(3/2)} = 8a^9 b^{12}$

69. (a) $\left(8y^3\right)^{-2/3} = 8^{-2/3} y^{3(-2/3)} = \dfrac{1}{4y^2}$

(b) $\left(u^4 v^6\right)^{-1/3} = u^{4(-1/3)} v^{6(-1/3)} = \dfrac{1}{u^{4/3} v^2}$

71. (a) $\left(\dfrac{x^{-2/3}}{y^{1/2}}\right)\left(\dfrac{x^{-2}}{y^{-3}}\right)^{1/6} = x^{-2/3+-2(1/6)}y^{-1/2-(-3)(1/6)} = \dfrac{1}{x}$

(b) $\left(\dfrac{4y^3 z^{2/3}}{x^{1/2}}\right)^2 \left(\dfrac{x^{-3}y^6}{8z^4}\right)^{1/3} = 4^2 8^{-1/3} x^{-1/2(2)-3(1/3)} y^{3(2)+6(1/3)} z^{2/3(2)-4(1/3)} = \dfrac{8y^8}{x^2}$

73. $\sqrt[9]{x^5} = x^{5/9}$

75. $\left(\sqrt[6]{y^5}\right)\left(\sqrt[3]{y^2}\right) = y^{5/6} \cdot y^{2/3} = y^{5/6+2/3} = y^{3/2}$

77. $\left(5\sqrt[3]{x}\right)\left(2\sqrt[4]{x}\right) = 5 \cdot 2x^{1/3+1/4} = 10x^{7/12}$

79. $\sqrt{4st^3}\sqrt[6]{s^3 t^2} = 4^{1/2} s^{1/2+3/6} t^{3/2+2/6} = 2st^{11/6}$

81. $\dfrac{\sqrt[4]{x^7}}{\sqrt[4]{x^3}} = x^{7/4-3/4} = x$

83. $\dfrac{\sqrt{xy}}{\sqrt[4]{16xy}} = 16^{-1/4} x^{1/2-1/4} y^{1/2-1/4} = \dfrac{x^{1/4}y^{1/4}}{2}$

85. $\sqrt{\dfrac{16u^3 v}{uv^5}} = \sqrt{\dfrac{16u^2}{v^4}} = \dfrac{4u}{v^2}$

87. $\sqrt[3]{y\sqrt{y}} = \left(y^{1+1/2}\right)^{1/3} = y^{(3/2)(1/3)} = y^{1/2}$

89. (a) $\dfrac{1}{\sqrt{6}} = \dfrac{1}{\sqrt{6}} \cdot \dfrac{\sqrt{6}}{\sqrt{6}} = \dfrac{\sqrt{6}}{6}$

(b) $\dfrac{3}{\sqrt{2}} = \dfrac{3}{\sqrt{2}} \cdot \dfrac{\sqrt{2}}{\sqrt{2}} = \dfrac{3\sqrt{2}}{2}$

(c) $\dfrac{9}{\sqrt{3}} = \dfrac{9}{\sqrt{3}} \cdot \dfrac{\sqrt{3}}{\sqrt{3}} = \dfrac{9\sqrt{3}}{3} = 3\sqrt{3}$

91. (a) $\dfrac{1}{\sqrt[3]{4}} = \dfrac{1}{\sqrt[3]{2^2}} \cdot \dfrac{\sqrt[3]{2}}{\sqrt[3]{2}} = \dfrac{\sqrt[3]{2}}{2}$

(b) $\dfrac{1}{\sqrt[4]{3}} = \dfrac{1}{\sqrt[4]{3}} \cdot \dfrac{\sqrt[4]{3^3}}{\sqrt[4]{3^3}} = \dfrac{\sqrt[4]{3^3}}{3} = \dfrac{\sqrt[4]{27}}{3}$

(c) $\dfrac{8}{\sqrt[5]{2}} = \dfrac{8}{\sqrt[5]{2}} \cdot \dfrac{\sqrt[5]{2^4}}{\sqrt[5]{2^4}} = \dfrac{8\sqrt[5]{2^4}}{2} = 4\sqrt[5]{2^4} = 4\sqrt[5]{16}$

93. (a) $\dfrac{1}{\sqrt[3]{x}} = \dfrac{1}{\sqrt[3]{x}} \cdot \dfrac{\sqrt[3]{x^2}}{\sqrt[3]{x^2}} = \dfrac{\sqrt[3]{x^2}}{x}$

(b) $\dfrac{1}{\sqrt[6]{x^5}} = \dfrac{1}{\sqrt[6]{x^5}} \cdot \dfrac{\sqrt[6]{x}}{\sqrt[6]{x}} = \dfrac{\sqrt[6]{x}}{x}$

(c) $\dfrac{1}{\sqrt[7]{x^3}} = \dfrac{1}{\sqrt[7]{x^3}} \cdot \dfrac{\sqrt[7]{x^4}}{\sqrt[7]{x^4}} = \dfrac{\sqrt[7]{x^4}}{x}$

95. First convert 1135 feet to miles. This gives $1135 \text{ ft} = 1135 \cdot \dfrac{1 \text{ mile}}{5280 \text{ feet}} = 0.215 \text{ mi}$. Thus the distance you can see is given

by $D = \sqrt{2rh + h^2} = \sqrt{2\,(3960)\,(0.215) + (0.215)^2} \approx \sqrt{1702.8} \approx 41.3$ miles.

97. (a) Substituting, we get $0.30\,(60) + 0.38\,(3400)^{1/2} - 3\,(650)^{1/3} \approx 18 + 0.38\,(58.31) - 3\,(8.66) \approx 18 + 22.16 - 25.98 \approx 14.18$. Since this value is less than 16, the sailboat qualifies for the race.

(b) Solve for A when $L = 65$ and $V = 600$. Substituting, we get $0.30\,(65) + 0.38A^{1/2} - 3\,(600)^{1/3} \le 16 \Leftrightarrow$ $19.5 + 0.38A^{1/2} - 25.30 \le 16 \Leftrightarrow 0.38A^{1/2} - 5.80 \le 16 \Leftrightarrow 0.38A^{1/2} \le 21.80 \Leftrightarrow A^{1/2} \le 57.38 \Leftrightarrow A \le 3292.0$. Thus, the largest possible sail is 3292 ft^2.

99. (a)

n	1	2	5	10	100
$2^{1/n}$	$2^{1/1} = 2$	$2^{1/2} = 1.414$	$2^{1/5} = 1.149$	$2^{1/10} = 1.072$	$2^{1/100} = 1.007$

So when n gets large, $2^{1/n}$ decreases to 1.

(b)

n	1	2	5	10	100
$\left(\frac{1}{2}\right)^{1/n}$	$\left(\frac{1}{2}\right)^{1/1} = 0.5$	$\left(\frac{1}{2}\right)^{1/2} = 0.707$	$\left(\frac{1}{2}\right)^{1/5} = 0.871$	$\left(\frac{1}{2}\right)^{1/10} = 0.933$	$\left(\frac{1}{2}\right)^{1/100} = 0.993$

So when n gets large, $\left(\frac{1}{2}\right)^{1/n}$ increases to 1.

P.5 ALGEBRAIC EXPRESSIONS

1. (a) $2x^3 - \frac{1}{2}x + \sqrt{3}$ is a polynomial. (The constant term is not an integer, but all exponents are integers.)

(b) $x^2 - \frac{1}{2} - 3\sqrt{x} = x^2 - \frac{1}{2} - 3x^{1/2}$ is not a polynomial because the exponent $\frac{1}{2}$ is not an integer.

(c) $\dfrac{1}{x^2 + 4x + 7}$ is not a polynomial. (It is the reciprocal of the polynomial $x^2 + 4x + 7$.)

(d) $x^5 + 7x^2 - x + 100$ is a polynomial.

(e) $\sqrt[3]{8x^6 - 5x^3 + 7x - 3}$ is not a polynomial. (It is the cube root of the polynomial $8x^6 - 5x^3 + 7x - 3$.)

(f) $\sqrt{3}x^4 + \sqrt{5}x^2 - 15x$ is a polynomial. (Some coefficients are not integers, but all exponents are integers.)

3. To subtract polynomials we subtract *like* terms. So $\left(2x^3 + 9x^2 + x + 10\right) - \left(x^3 + x^2 + 6x + 8\right) = (2-1)x^3 +$
$(9-1)x^2 + (1-6)x + (10-8) = x^3 + 8x^2 - 5x + 2.$

5. The Special Product Formula for the "square of a sum" is $(A + B)^2 = A^2 + 2AB + B^2$. So
$(2x + 3)^2 = (2x)^2 + 2(2x)(3) + 3^2 = 4x^2 + 12x + 9.$

7. Trinomial, terms x^2, $-3x$, and 7, degree 2   **9.** Monomial, term -8, degree 0

11. Polynomial, terms x, $-x^2$, x^3, and $-x^4$, degree 4

13. $(6x - 3) + (3x + 7) = (6x + 3x) + (-3 + 7) = 9x + 4$

15. $\left(2x^2 - 5x\right) - \left(x^2 - 8x + 3\right) = \left(2x^2 - x^2\right) + [-5x - (-8x)] + (-3) = x^2 + 3x - 3$

17. $3(x - 1) + 4(x + 2) = 3x - 3 + 4x + 8 = 7x + 5$

19. $\left(x^3 + 6x^2 - 4x + 7\right) - \left(3x^2 + 2x - 4\right) = x^3 + 6x^2 - 4x + 7 - 3x^2 - 2x + 4 = x^3 + 3x^2 - 6x + 11$

21. $2x(x - 1) = 2x^2 - 2x$ **23.** $x^2(x + 3) = x^3 + 3x^2$

25. $2(2 - 5t) + t(t + 10) = 4 - 10t + t^2 + 10t = t^2 + 4$ **27.** $r\left(r^2 - 9\right) + 3r^2(2r - 1) = r^3 - 9r + 6r^3 - 3r^2$
$= 7r^3 - 3r^2 - 9r$

29. $x^2\left(2x^2 - x + 1\right) = 2x^4 - x^3 + x^2$ **31.** $(x - 3)(x + 5) = x^2 + 5x - 3x - 15 = x^2 + 2x - 15$

33. $(s + 6)(2s + 3) = 2s^2 + 3s + 12s + 18 = 2s^2 + 15s + 18$ **35.** $(3t - 2)(7t - 4) = 21t^2 - 12t - 14t + 8 = 21t^2 - 26t + 8$

37. $(3x + 5)(2x - 1) = 6x^2 + 10x - 3x - 5 = 6x^2 + 7x - 5$ **39.** $(x + 3y)(2x - y) = 2x^2 + 5xy - 3y^2$

41. $(2r - 5s)(3r - 2s) = 6r^2 - 19rs + 10s^2$ **43.** $(x + 5)^2 = x^2 + 2(x)(5) + 5^2 = x^2 + 10x + 25$

45. $(3y - 1)^2 = (3y)^2 - 2(3y)(1) + 1^2 = 9y^2 - 6y + 1$ **47.** $(2u + v)^2 = 4u^2 + 4uv + v^2$

49. $(2x + 3y)^2 = 4x^2 + 12xy + 9y^2$ **51.** $\left(x^2 + 1\right)^2 = x^4 + 2x^2 + 1$

53. $(x + 5)(x - 5) = x^2 - 5^2 = x^2 - 25$ **55.** $(3x - 4)(3x + 4) = (3x)^2 - 4^2 = 9x^2 - 16$

57. $(x + 3y)(x - 3y) = x^2 - (3y)^2 = x^2 - 9y^2$ **59.** $\left(\sqrt{x} + 2\right)\left(\sqrt{x} - 2\right) = x - 4$

61. $(y + 2)^3 = y^3 + 3y^2(2) + 3y\left(2^2\right) + 2^3 = y^3 + 6y^2 + 12y + 8$

63. $(1 - 2r)^3 = 1^3 - 3\left(1^2\right)(2r) + 3(1)(2r)^2 - (2r)^3 = -8r^3 + 12r^2 - 6r + 1$

65. $(x + 2)\left(x^2 + 2x + 3\right) = x^3 + 2x^2 + 3x + 2x^2 + 4x + 6 = x^3 + 4x^2 + 7x + 6$

67. $(2x - 5)\left(x^2 - x + 1\right) = 2x^3 - 2x^2 + 2x - 5x^2 + 5x - 5 = 2x^3 - 7x^2 + 7x - 5$

69. $\sqrt{x}\left(x - \sqrt{x}\right) = x\sqrt{x} - \left(\sqrt{x}\right)^2 = x\sqrt{x} - x$ **71.** $y^{1/3}\left(y^{2/3} + y^{5/3}\right) = y^{1/3+2/3} + y^{1/3+5/3} = y^2 + y$

73. $\left(x^2 + y^2\right)^2 = \left(x^2\right)^2 + \left(y^2\right)^2 + 2x^2y^2 = x^4 + y^4 + 2x^2y^2$

75. $\left(x^2 - a^2\right)\left(x^2 + a^2\right) = x^4 - a^4$ **77.** $\left(\sqrt{a} - b\right)\left(\sqrt{a} + b\right) = a - b^2$

79. $\left(1 + x^{2/3}\right)\left(1 - x^{2/3}\right) = 1 - x^{4/3}$

81. $\left((x - 1) + x^2\right)\left((x - 1) - x^2\right) = (x - 1)^2 - \left(x^2\right)^2 = x^2 - 2x + 1 - x^4 = -x^4 + x^2 - 2x + 1$

83. $(2x + y - 3)(2x + y + 3) = (2x + y)^2 - 3^2 = 4x^2 + 4xy + y^2 - 9$

85. (a) The height of the box is x, its width is $6 - 2x$, and its length is $10 - 2x$. Since Volume = height × width × length, we have $V = x(6 - 2x)(10 - 2x)$.

(b) $V = x\left(60 - 32x + 4x^2\right) = 60x - 32x^2 + 4x^3$, degree 3.

(c) When $x = 1$, the volume is $V = 60(1) - 32\left(1^2\right) + 4\left(1^3\right) = 32$, and when $x = 2$, the volume is $V = 60(2) - 32\left(2^2\right) + 4\left(2^3\right) = 24$.

87. (a) $A = 2000(1 + r)^3 = 2000\left(1 + 3r + 3r^2 + r^3\right) = 2000 + 6000r + 6000r^2 + 2000r^3$, degree 3.

(b) Remember that % means divide by 100, so 2% = 0.02.

Interest rate r	2%	3%	4.5%	6%	10%
Amount A	\$2122.42	\$2185.45	\$2282.33	\$2382.03	\$2662.00

89. (a) When $x = 1$, $(x + 5)^2 = (1 + 5)^2 = 36$ and $x^2 + 25 = 1^2 + 25 = 26$.

(b) $(x + 5)^2 = x^2 + 10x + 25$

P.6   FACTORING

1. (a) The polynomial $2x^5 + 6x^4 + 4x^3$ has three terms: $2x^5$, $6x^4$, and $4x^3$.

(b) The factor $2x^3$ is common to each term, so $2x^5 + 6x^4 + 4x^3 = 2x^3\left(x^2 + 3x + 2\right)$.

[In fact, the polynomial can be factored further as $2x^3(x + 2)(x + 1)$.]

3. The Special Factoring Formula for the "difference of squares" is $A^2 - B^2 = (A - B)(A + B)$. So $4x^2 - 25 = (2x - 5)(2x + 5)$.

5. $5a - 20 = 5(a - 4)$ **7.** $-2x^3 + 16x = -2x\left(x^2 - 8\right) = 2x\left(-x^2 + 8\right)$

9. $2x^2y - 6xy^2 + 3xy = xy(2x - 6y + 3)$ **11.** $y(y - 6) + 9(y - 6) = (y - 6)(y + 9)$

13. $x^2 + 2x - 3 = (x - 1)(x + 3)$ **15.** $x^2 + 2x - 15 = (x + 5)(x - 3)$

17. $3x^2 - 16x + 5 = (3x - 1)(x - 5)$

19. $(3x + 2)^2 + 8(3x + 2) + 12 = [(3x + 2) + 2][(3x + 2) + 6] = (3x + 4)(3x + 8)$

21. $x^2 - 25 = (x - 5)(x + 5)$ **23.** $49 - 4z^2 = (7 - 2z)(7 + 2z)$

25. $16y^2 - z^2 = (4y - z)(4y + z)$

27. $(x + 3)^2 - y^2 = [(x + 3) - y][(x + 3) + y] = (x - y + 3)(x + y + 3)$

29. $x^2 + 10x + 25 = (x + 5)^2$ 　　　　**31.** $z^2 - 12z + 36 = (z - 6)^2$

33. $4t^2 - 20t + 25 = (2t - 5)^2$ 　　　**35.** $9u^2 - 6uv + v^2 = (3u - v)^2$

37. $x^3 + 27 = (x + 3)(x^2 - 3x + 9)$ 　**39.** $8a^3 - 1 = (2a - 1)(4a^2 + 2a + 1)$

41. $27x^3 + y^3 = (3x + y)(9x^2 - 3xy + y^2)$ 　　**43.** $u^3 - v^6 = u^3 - (v^2)^3 = (u - v^2)(u^2 + uv^2 + v^4)$

45. $x^3 + 4x^2 + x + 4 = x^2(x + 4) + 1(x + 4) = (x + 4)(x^2 + 1)$

47. $2x^3 + x^2 - 6x - 3 = x^2(2x + 1) - 3(2x + 1) = (2x + 1)(x^2 - 3)$. If irrational coefficients are permitted, then this can

be further factored as $(2x + 1)(x - \sqrt{3})(x - \sqrt{3})$.

49. $x^3 + x^2 + x + 1 = x^2(x + 1) + 1(x + 1) = (x + 1)(x^2 + 1)$

51. $x^{5/2} - x^{1/2} = x^{1/2}(x^2 - 1) = \sqrt{x}\,(x - 1)(x + 1)$

53. Start by factoring out the power of x with the smallest exponent, that is, $x^{-3/2}$. So

$$x^{-3/2} + 2x^{-1/2} + x^{1/2} = x^{-3/2}(1 + 2x + x^2) = \frac{(1 + x)^2}{x^{3/2}}.$$

55. Start by factoring out the power of $(x^2 + 1)$ with the smallest exponent, that is, $(x^2 + 1)^{-1/2}$. So

$$(x^2 + 1)^{1/2} + 2(x^2 + 1)^{-1/2} = (x^2 + 1)^{-1/2}[(x^2 + 1) + 2] = \frac{x^2 + 3}{\sqrt{x^2 + 1}}.$$

57. $2x^{1/3}(x - 2)^{2/3} - 5x^{4/3}(x - 2)^{-1/3} = x^{1/3}(x - 2)^{-1/3}[2(x - 2) - 5x] = x^{1/3}(x - 2)^{-1/3}(2x - 4 - 5x)$

$$= x^{1/3}(x - 2)^{-1/3}(-3x - 4) = \frac{(-3x - 4)\sqrt[3]{x}}{\sqrt[3]{x - 2}}$$

59. $12x^3 + 18x = 6x(2x^2 + 3)$ 　　　　**61.** $6y^4 - 15y^3 = 3y^3(2y - 5)$

63. $x^2 - 2x - 8 = (x - 4)(x + 2)$ 　　　**65.** $y^2 - 8y + 15 = (y - 3)(y - 5)$

67. $2x^2 + 5x + 3 = (2x + 3)(x + 1)$ 　　**69.** $9x^2 - 36x - 45 = 9(x^2 - 4x - 5) = 9(x - 5)(x + 1)$

71. $6x^2 - 5x - 6 = (3x + 2)(2x - 3)$ 　　**73.** $x^2 - 36 = (x - 6)(x + 6)$

75. $49 - 4y^2 = (7 - 2y)(7 + 2y)$ 　　　**77.** $t^2 - 6t + 9 = (t - 3)^2$

79. $4x^2 + 4xy + y^2 = (2x + y)^2$ 　　　**81.** $t^3 + 1 = (t + 1)(t^2 - t + 1)$

83. $8x^3 - 125 = (2x)^3 - 5^3 = (2x - 5)[(2x)^2 + (2x)(5) + 5^2] = (2x - 5)(4x^2 + 10x + 25)$

85. $x^3 + 2x^2 + x = x(x^2 + 2x + 1) = x(x + 1)^2$

87. $x^4 + 2x^3 - 3x^2 = x^2(x^2 + 2x - 3) = x^2(x - 1)(x + 3)$

89. $x^4y^3 - x^2y^5 = x^2y^3(x^2 - y^2) = x^2y^3(x + y)(x - y)$

91. $x^6 - 8y^3 = \left(x^2\right)^3 - (2y)^3 = \left(x^2 - 2y\right)\left[\left(x^2\right)^2 + \left(x^2\right)(2y) + (2y)^2\right] = \left(x^2 - 2y\right)\left(x^4 + 2x^2y + 4y^2\right)$

93. $y^3 - 3y^2 - 4y + 12 = \left(y^3 - 3y^2\right) + (-4y + 12) = y^2(y - 3) + (-4)(y - 3) = (y - 3)\left(y^2 - 4\right)$

$= (y - 3)(y - 2)(y + 2)$ (factor by grouping)

95. $2x^3 + 4x^2 + x + 2 = \left(2x^3 + 4x^2\right) + (x + 2) = 2x^2(x + 2) + (1)(x + 2) = (x + 2)\left(2x^2 + 1\right)$ (factor by grouping)

97. $(a + b)^2 - (a - b)^2 = [(a + b) - (a - b)][(a + b) + (a - b)] = (2b)(2a) = 4ab$

99. $x^2\left(x^2 - 1\right) - 9\left(x^2 - 1\right) = \left(x^2 - 1\right)\left(x^2 - 9\right) = (x - 1)(x + 1)(x - 3)(x + 3)$

101. $(x - 1)(x + 2)^2 - (x - 1)^2(x + 2) = (x - 1)(x + 2)[(x + 2) - (x - 1)] = 3(x - 1)(x + 2)$

103. $y^4(y + 2)^3 + y^5(y + 2)^4 = y^4(y + 2)^3\left[(1) + y(y + 2)\right] = y^4(y + 2)^3\left(y^2 + 2y + 1\right) = y^4(y + 2)^3(y + 1)^2$

105. Start by factoring $y^2 - 7y + 10$, and then substitute $a^2 + 1$ for y. This gives

$$\left(a^2 + 1\right)^2 - 7\left(a^2 + 1\right) + 10 = \left[\left(a^2 + 1\right) - 2\right]\left[\left(a^2 + 1\right) - 5\right] = \left(a^2 - 1\right)\left(a^2 - 4\right) = (a - 1)(a + 1)(a - 2)(a + 2)$$

107. $3x^2(4x - 12)^2 + x^3(2)(4x - 12)(4) = x^2(4x - 12)[3(4x - 12) + x(2)(4)] = 4x^2(x - 3)(12x - 36 + 8x)$

$= 4x^2(x - 3)(20x - 36) = 16x^2(x - 3)(5x - 9)$

109. $3(2x - 1)^2(2)(x + 3)^{1/2} + (2x - 1)^3\left(\frac{1}{2}\right)(x + 3)^{-1/2} = (2x - 1)^2(x + 3)^{-1/2}\left[6(x + 3) + (2x - 1)\left(\frac{1}{2}\right)\right]$

$= (2x - 1)^2(x + 3)^{-1/2}\left(6x + 18 + x - \frac{1}{2}\right) = (2x - 1)^2(x + 3)^{-1/2}\left(7x + \frac{35}{2}\right)$

111. $\left(x^2 + 3\right)^{-1/3} - \frac{2}{3}x^2\left(x^2 + 3\right)^{-4/3} = \left(x^2 + 3\right)^{-4/3}\left[\left(x^2 + 3\right) - \frac{2}{3}x^2\right] = \left(x^2 + 3\right)^{-4/3}\left(\frac{1}{3}x^2 + 3\right) = \dfrac{\frac{1}{3}x^2 + 3}{(x^2 + 3)^{4/3}}$

113. **(a)** $\frac{1}{2}\left[(a + b)^2 - \left(a^2 + b^2\right)\right] = \frac{1}{2}\left[a^2 + 2ab + b^2 - a^2 - b^2\right] = \frac{1}{2}(2ab) = ab$.

(b) $\left(a^2 + b^2\right)^2 - \left(a^2 - b^2\right)^2 = \left[\left(a^2 + b^2\right) - \left(a^2 - b^2\right)\right]\left[\left(a^2 + b^2\right) + \left(a^2 - b^2\right)\right]$

$= \left(a^2 + b^2 - a^2 + b^2\right)\left(a^2 + b^2 + a^2 - b^2\right) = \left(2b^2\right)\left(2a^2\right) = 4a^2b^2$

(c) LHS $= \left(a^2 + b^2\right)\left(c^2 + d^2\right) = a^2c^2 + a^2d^2 + b^2c^2 + b^2d^2$.

RHS $= (ac + bd)^2 + (ad - bc)^2 = a^2c^2 + 2abcd + b^2d^2 + a^2d^2 - 2abcd + b^2c^2 = a^2c^2 + a^2d^2 + b^2c^2 + b^2d^2$.

So LHS $=$ RHS, that is, $\left(a^2 + b^2\right)\left(c^2 + d^2\right) = (ac + bd)^2 + (ad - bc)^2$.

(d) $4a^2c^2 - \left(c^2 - b^2 + a^2\right)^2 = (2ac)^2 - \left(c^2 - b^2 + a^2\right)$

$= \left[(2ac) - \left(c^2 - b^2 + a^2\right)\right]\left[(2ac) + \left(c^2 - b^2 + a^2\right)\right]$ (difference of squares)

$= \left(2ac - c^2 + b^2 - a^2\right)\left(2ac + c^2 - b^2 + a^2\right)$

$= \left[b^2 - \left(c^2 - 2ac + a^2\right)\right]\left[\left(c^2 + 2ac + a^2\right) - b^2\right]$ (regrouping)

$= \left[b^2 - (c - a)^2\right]\left[(c + a)^2 - b^2\right]$ (perfect squares)

$= [b - (c - a)][b + (c - a)][(c + a) - b][(c + a) + b]$ (each factor is a difference of squares)

$= (b - c + a)(b + c - a)(c + a - b)(c + a + b)$

$= (a + b - c)(-a + b + c)(a - b + c)(a + b + c)$

115. The volume of the shell is the difference between the volumes of the outside cylinder (with radius R) and the inside cylinder

(with radius r). Thus $V = \pi R^2 h - \pi r^2 h = \pi \left(R^2 - r^2\right) h = \pi (R - r)(R + r) h = 2\pi \cdot \dfrac{R + r}{2} \cdot h \cdot (R - r)$. The

average radius is $\dfrac{R + r}{2}$ and $2\pi \cdot \dfrac{R + r}{2}$ is the average circumference (length of the rectangular box), h is the height, and

$R - r$ is the thickness of the rectangular box. Thus $V = \pi R^2 h - \pi r^2 h = 2\pi \cdot \dfrac{R + r}{2} \cdot h \cdot (R - r) = 2\pi \cdot$ (average radius) $\cdot$

(height) $\cdot$ (thickness)

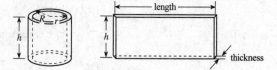

117. (a) $528^2 - 527^2 = (528 - 527)(528 + 527) = 1(1055) = 1055$

  (b) $122^2 - 120^2 = (122 - 120)(122 + 120) = 2(242) = 484$

  (c) $1020^2 - 1010^2 = (1020 - 1010)(1020 + 1010) = 10(2030) = 20{,}300$

  (d) $49 \cdot 51 = (50 - 1)(50 + 1) = 50^2 - 1 = 2500 - 1 = 2499$

  (e) $998 \cdot 1002 = (1000 - 2)(1000 + 2) = 1000^2 - 2^2 = 1{,}000{,}000 - 4 = 999{,}996$

119.

$$
\begin{array}{r}
A + 1 \\
\times \quad A - 1 \\
\hline
-A - 1 \\
A^2 + A \\
\hline
A^2 \qquad\ - 1
\end{array}
\qquad
\begin{array}{r}
A^2 + A + 1 \\
\times \quad A - 1 \\
\hline
-A^2 - A - 1 \\
A^3 + A^2 + A \\
\hline
A^3 \qquad\qquad - 1
\end{array}
\qquad
\begin{array}{r}
A^3 + A^2 + A + 1 \\
\times \quad A - 1 \\
\hline
-A^3 - A^2 - A - 1 \\
A^4 + A^3 + A^2 + A \\
\hline
A^4 \qquad\qquad\qquad - 1
\end{array}
$$

Based on the pattern, we suspect that $A^5 - 1 = (A - 1)\left(A^4 + A^3 + A^2 + A + 1\right)$. Check:

$$
\begin{array}{r}
A^4 + A^3 + A^2 + A + 1 \\
\times \qquad\qquad\qquad A - 1 \\
\hline
-A^4 - A^3 - A^2 - A - 1 \\
A^5 + A^4 + A^3 + A^2 + A \\
\hline
A^5 \qquad\qquad\qquad\qquad - 1
\end{array}
$$

The general pattern is $A^n - 1 = (A - 1)\left(A^{n-1} + A^{n-2} + \cdots + A^2 + A + 1\right)$, where n is a positive integer.

P.7 RATIONAL EXPRESSIONS

1. (a) $\dfrac{3x}{x^2 - 1}$ is a rational expression.

  (b) $\dfrac{\sqrt{x + 1}}{2x + 3}$ is not a rational expression. A rational expression must be a polynomial divided by a polynomial, and the

  numerator of the expression is $\sqrt{x + 1}$, which is not a polynomial.

  (c) $\dfrac{x(x^2 - 1)}{x + 3} = \dfrac{x^3 - x}{x + 3}$ is a rational expression.

3. (a) $\dfrac{x^2 + 3}{x^2 + 5}$ *does not simplify* to $\dfrac{3}{5}$. There is no common factor.

(b) $\dfrac{3x^2}{5x^2}$ *does* simplify to $\dfrac{3}{5}$. The common factor is x^2.

5. (a) $\dfrac{1}{x} - \dfrac{2}{(x+1)} - \dfrac{x}{(x+1)^2}$ has three terms.

(b) The least common denominator of all the terms is $x\,(x+1)^2$.

(c) $\dfrac{1}{x} - \dfrac{2}{(x+1)} - \dfrac{x}{(x+1)^2} = \dfrac{(x+1)^2}{x\,(x+1)^2} - \dfrac{2x\,(x+1)}{(x+1)} - \dfrac{x\,(x)}{(x+1)^2} = \dfrac{(x+1)^2 - 2x\,(x+1) - x^2}{x\,(x+1)^2}$

$\qquad = \dfrac{x^2 + 2x + 1 - 2x^2 - 2x - x^2}{x\,(x+1)^2} = \dfrac{1 - 2x^2}{x\,(x+1)^2}$

7. $-x^2 + 3x + \frac{1}{2}$ has domain $(-\infty, \infty)$.   **9.** $3x^{-2} = \dfrac{3}{x^2}$ has domain $\{x \mid x \neq 0\}$.

11. Since $x - 4 \neq 0$, we have $x \neq 4$, so the domain is $\{x \mid x \neq 4\}$.

13. $x^2 - x - 2 = (x+1)\,(x-2) \neq 0 \Leftrightarrow x \neq -1$ or 2, so the domain is $\{x \mid x \neq -1, 2\}$.

15. $x + 3 \geq 0 \Leftrightarrow x \geq -3$, so the domain is $\{x \mid x \geq -3\}$.

17. $\dfrac{12x}{6x^2} = \dfrac{6x \cdot 2}{6x \cdot x} = \dfrac{2}{x}$ **19.** $\dfrac{5y^2}{10y + y^2} = \dfrac{y \cdot 5y}{y \cdot (10 + y)} = \dfrac{5y}{10 + y}$

21. $\dfrac{3\,(x+2)\,(x-1)}{6\,(x-1)^2} = \dfrac{3\,(x-1) \cdot (x+2)}{3\,(x-1) \cdot 2\,(x-1)} = \dfrac{x+2}{2\,(x-1)}$ **23.** $\dfrac{x-2}{x^2 - 4} = \dfrac{x-2}{(x-2)\,(x+2)} = \dfrac{1}{x+2}$

25. $\dfrac{x^2 + 6x + 8}{x^2 + 5x + 4} = \dfrac{(x+2)\,(x+4)}{(x+1)\,(x+4)} = \dfrac{x+2}{x+1}$ **27.** $\dfrac{y^2 + y}{y^2 - 1} = \dfrac{y\,(y+1)}{(y-1)\,(y+1)} = \dfrac{y}{y-1}$

29. $\dfrac{2x^3 - x^2 - 6x}{2x^2 - 7x + 6} = \dfrac{x\left(2x^2 - x - 6\right)}{(2x-3)\,(x-2)} = \dfrac{x\,(2x+3)\,(x-2)}{(2x-3)\,(x-2)} = \dfrac{x\,(2x+3)}{2x-3}$

31. $\dfrac{4x}{x^2 - 4} \cdot \dfrac{x+2}{16x} = \dfrac{4x}{(x-2)\,(x+2)} \cdot \dfrac{x+2}{16x} = \dfrac{1}{4\,(x-2)}$

33. $\dfrac{x^2 - 2x - 15}{x^2 - 9} \cdot \dfrac{x+3}{x-5} = \dfrac{(x-5)\,(x+3)}{(x-3)\,(x+3)} \cdot \dfrac{x+3}{x-5} = \dfrac{x+3}{x-3}$

35. $\dfrac{t-3}{t^2 + 9} \cdot \dfrac{t+3}{t^2 - 9} = \dfrac{(t-3)\,(t+3)}{(t^2 + 9)\,(t-3)\,(t+3)} = \dfrac{1}{t^2 + 9}$

37. $\dfrac{x^2 + 7x + 12}{x^2 + 3x + 2} \cdot \dfrac{x^2 + 5x + 6}{x^2 + 6x + 9} = \dfrac{(x+3)\,(x+4)}{(x+1)\,(x+2)} \cdot \dfrac{(x+2)\,(x+3)}{(x+3)\,(x+3)} = \dfrac{x+4}{x+1}$

39. $\dfrac{x+3}{4x^2 - 9} \div \dfrac{x^2 + 7x + 12}{2x^2 + 7x - 15} = \dfrac{x+3}{4x^2 - 9} \cdot \dfrac{2x^2 + 7x - 15}{x^2 + 7x + 12} = \dfrac{x+3}{(2x-3)\,(2x+3)} \cdot \dfrac{(x+5)\,(2x-3)}{(x+3)\,(x+4)} = \dfrac{x+5}{(2x+3)\,(x+4)}$

41. $\dfrac{2x^2 + 3x + 1}{x^2 + 2x - 15} \div \dfrac{x^2 + 6x + 5}{2x^2 - 7x + 3} = \dfrac{2x^2 + 3x + 1}{x^2 + 2x - 15} \cdot \dfrac{2x^2 - 7x + 3}{x^2 + 6x + 5} = \dfrac{(2x+1)\,(x+1)}{(x-3)\,(x+5)} \cdot \dfrac{(2x-1)\,(x-3)}{(x+1)\,(x+5)}$

$\qquad = \dfrac{(2x+1)\,(2x-1)}{(x+5)\,(x+5)} = \dfrac{(2x+1)\,(2x-1)}{(x+5)^2}$

43. $\dfrac{\dfrac{x^3}{x+1}}{\dfrac{x}{x^2 + 2x + 1}} = \dfrac{x^3}{x+1} \cdot \dfrac{x^2 + 2x + 1}{x} = \dfrac{x^3\,(x+1)\,(x+1)}{(x+1)\,x} = x^2\,(x+1)$

45. $\dfrac{x/y}{z} = \dfrac{x}{y} \cdot \dfrac{1}{z} = \dfrac{x}{yz}$

47. $2 + \dfrac{x}{x+3} = \dfrac{2\,(x+3)}{x+3} + \dfrac{x}{x+3} = \dfrac{2x + 6 + x}{x+3} = \dfrac{3x + 6}{x+3} = \dfrac{3\,(x+2)}{(x+3)}$

49. $\dfrac{1}{x+5} + \dfrac{2}{x-3} = \dfrac{x-3}{(x+5)\,(x-3)} + \dfrac{2\,(x+5)}{(x+5)\,(x-3)} = \dfrac{x - 3 + 2x + 10}{(x+5)\,(x-3)} = \dfrac{3x + 7}{(x+5)\,(x-3)}$

51. $\dfrac{1}{x+1} - \dfrac{1}{x+2} = \dfrac{x+2}{(x+1)(x+2)} + \dfrac{-(x+1)}{(x+1)(x+2)} = \dfrac{x+2-x-1}{(x+1)(x+2)} = \dfrac{1}{(x+1)(x+2)}$

53. $\dfrac{x}{(x+1)^2} + \dfrac{2}{x+1} = \dfrac{x}{(x+1)^2} + \dfrac{2(x+1)}{(x+1)(x+1)} = \dfrac{x+2x+2}{(x+1)^2} = \dfrac{3x+2}{(x+1)^2}$

55. $u + 1 + \dfrac{u}{u+1} = \dfrac{(u+1)(u+1)}{u+1} + \dfrac{u}{u+1} = \dfrac{u^2+2u+1+u}{u+1} = \dfrac{u^2+3u+1}{u+1}$

57. $\dfrac{1}{x^2} + \dfrac{1}{x^2+x} = \dfrac{1}{x^2} + \dfrac{1}{x(x+1)} = \dfrac{x+1}{x^2(x+1)} + \dfrac{x}{x^2(x+1)} = \dfrac{2x+1}{x^2(x+1)}$

59. $\dfrac{2}{x+3} - \dfrac{1}{x^2+7x+12} = \dfrac{2}{x+3} - \dfrac{1}{(x+3)(x+4)} = \dfrac{2(x+4)}{(x+3)(x+4)} + \dfrac{-1}{(x+3)(x+4)}$

$= \dfrac{2x+8-1}{(x+3)(x+4)} = \dfrac{2x+7}{(x+3)(x+4)}$

61. $\dfrac{1}{x+3} + \dfrac{1}{x^2-9} = \dfrac{1}{x+3} + \dfrac{1}{(x-3)(x+3)} = \dfrac{x-3}{(x-3)(x+3)} + \dfrac{1}{(x-3)(x+3)} = \dfrac{x-2}{(x-3)(x+3)}$

63. $\dfrac{2}{x} + \dfrac{3}{x-1} - \dfrac{4}{x^2-x} = \dfrac{2}{x} + \dfrac{3}{x-1} - \dfrac{4}{x(x-1)} = \dfrac{2(x-1)}{x(x-1)} + \dfrac{3x}{x(x-1)} + \dfrac{-4}{x(x-1)} = \dfrac{2x-2+3x-4}{x(x-1)} = \dfrac{5x-6}{x(x-1)}$

65. $\dfrac{1}{x^2+3x+2} - \dfrac{1}{x^2-2x-3} = \dfrac{1}{(x+2)(x+1)} - \dfrac{1}{(x-3)(x+1)}$

$= \dfrac{x-3}{(x-3)(x+2)(x+1)} + \dfrac{-(x+2)}{(x-3)(x+2)(x+1)} = \dfrac{x-3-x-2}{(x-3)(x+2)(x+1)} = \dfrac{-5}{(x-3)(x+2)(x+1)}$

67. $\dfrac{2}{1-\dfrac{1}{x}} = \dfrac{x(2)}{x\left(1-\dfrac{1}{x}\right)} = \dfrac{2x}{x-1}$

69. $\dfrac{1-\dfrac{1}{x}}{1+\dfrac{1}{x}} = \dfrac{\dfrac{x}{x}-\dfrac{1}{x}}{\dfrac{x}{x}+\dfrac{1}{x}} = \dfrac{\dfrac{x-1}{x}}{\dfrac{x+1}{x}} = \dfrac{x-1}{x} \cdot \dfrac{x}{x+1} = \dfrac{x-1}{x+1}$. An alternative method is to multiply the

numerator and denominator by the common denominator of both the numerator and denominator, in this case x:

$\dfrac{1-\dfrac{1}{x}}{1+\dfrac{1}{x}} = \dfrac{1-\dfrac{1}{x}}{1+\dfrac{1}{x}} \cdot \dfrac{x}{x} = \dfrac{\left(1-\dfrac{1}{x}\right)x}{\left(1+\dfrac{1}{x}\right)x} = \dfrac{x-1}{x+1}$.

71. $\dfrac{x-\dfrac{x}{y}}{y-\dfrac{y}{x}} = \dfrac{xy\left(x-\dfrac{x}{y}\right)}{xy\left(y-\dfrac{y}{x}\right)} = \dfrac{x^2y-x^2}{xy^2-y^2} = \dfrac{x^2(y-1)}{y^2(x-1)}$

73. $\dfrac{x+\dfrac{1}{x+2}}{x-\dfrac{1}{x+2}} = \dfrac{x(x+2)+1}{x(x+2)-1} = \dfrac{x^2+2x+1}{x^2+2x-1} = \dfrac{(x+1)^2}{x^2+2x-1}$

75. $\dfrac{\dfrac{x+2}{x-1}-\dfrac{x-3}{x-2}}{x+2} = \dfrac{(x+2)(x-2)-(x-3)(x-1)}{(x-2)(x-1)(x+2)} = \dfrac{x^2-4-\left(x^2-4x+3\right)}{(x-2)(x-1)(x+2)} = \dfrac{4x-7}{(x-2)(x-1)(x+2)}$

77. $\dfrac{\dfrac{x}{y}-\dfrac{y}{x}}{\dfrac{1}{x^2}-\dfrac{1}{y^2}} = \dfrac{\dfrac{x^2-y^2}{xy}}{\dfrac{y^2-x^2}{x^2y^2}} = \dfrac{x^2-y^2}{xy} \cdot \dfrac{x^2y^2}{-\left(x^2-y^2\right)} = -xy$

79. $\dfrac{x^{-2}+y^{-2}}{x^{-1}+y^{-1}} = \dfrac{\frac{1}{x^2}+\frac{1}{y^2}}{\frac{1}{x}+\frac{1}{y}} = \dfrac{\frac{1}{x^2}+\frac{1}{y^2}}{\frac{1}{x}+\frac{1}{y}} \cdot \dfrac{x^2 y^2}{x^2 y^2} = \dfrac{y^2+x^2}{xy^2+yx^2} = \dfrac{x^2+y^2}{xy\,(x+y)}$

81. $1 - \dfrac{1}{1-\frac{1}{x}} = 1 - \dfrac{x}{x-1} = \dfrac{x-1-x}{x-1} = \dfrac{1}{1-x}$

83. $\dfrac{\frac{1}{1+x+h}-\frac{1}{1+x}}{h} = \dfrac{(1+x)-(1+x+h)}{h\,(1+x)\,(1+x+h)} = -\dfrac{1}{(1+x)\,(1+x+h)}$

85. $\dfrac{\frac{1}{(x+h)^2}-\frac{1}{x^2}}{h} = \dfrac{x^2-(x+h)^2}{hx^2\,(x+h)^2} = \dfrac{x^2-\left(x^2+2xh+h^2\right)}{hx^2\,(x+h)^2} = -\dfrac{2x+h}{x^2\,(x+h)^2}$

87. $\sqrt{1+\left(\dfrac{x}{\sqrt{1-x^2}}\right)^2} = \sqrt{1+\dfrac{x^2}{1-x^2}} = \sqrt{\dfrac{1-x^2}{1-x^2}+\dfrac{x^2}{1-x^2}} = \sqrt{\dfrac{1}{1-x^2}} = \dfrac{1}{\sqrt{1-x^2}}$

89. $\dfrac{3\,(x+2)^2\,(x-3)^2-(x+2)^3\,(2)\,(x-3)}{(x-3)^4} = \dfrac{(x+2)^2\,(x-3)\,[3\,(x-3)-(x+2)\,(2)]}{(x-3)^4}$

$= \dfrac{(x+2)^2\,(3x-9-2x-4)}{(x-3)^3} = \dfrac{(x+2)^2\,(x-13)}{(x-3)^3}$

91. $\dfrac{2\,(1+x)^{1/2}-x\,(1+x)^{-1/2}}{1+x} = \dfrac{(1+x)^{-1/2}\,[2\,(1+x)-x]}{1+x} = \dfrac{x+2}{(1+x)^{3/2}}$

93. $\dfrac{3\,(1+x)^{1/3}-x\,(1+x)^{-2/3}}{(1+x)^{2/3}} = \dfrac{(1+x)^{-2/3}\,[3\,(1+x)-x]}{(1+x)^{2/3}} = \dfrac{2x+3}{(1+x)^{4/3}}$

95. $\dfrac{1}{2-\sqrt{3}} = \dfrac{1}{2-\sqrt{3}} \cdot \dfrac{2+\sqrt{3}}{2+\sqrt{3}} = \dfrac{2+\sqrt{3}}{4-3} = \dfrac{2+\sqrt{3}}{1} = 2+\sqrt{3}$

97. $\dfrac{2}{\sqrt{2}+\sqrt{7}} = \dfrac{2}{\sqrt{2}+\sqrt{7}} \cdot \dfrac{\sqrt{2}-\sqrt{7}}{\sqrt{2}-\sqrt{7}} = \dfrac{2\left(\sqrt{2}-\sqrt{7}\right)}{2-7} = \dfrac{2\left(\sqrt{2}-\sqrt{7}\right)}{-5} = \dfrac{2\left(\sqrt{7}-\sqrt{2}\right)}{5}$

99. $\dfrac{y}{\sqrt{3}+\sqrt{y}} = \dfrac{y}{\sqrt{3}+\sqrt{y}} \cdot \dfrac{\sqrt{3}-\sqrt{y}}{\sqrt{3}-\sqrt{y}} = \dfrac{y\left(\sqrt{3}-\sqrt{y}\right)}{3-y} = \dfrac{y\sqrt{3}-y\sqrt{y}}{3-y}$

101. $\dfrac{1-\sqrt{5}}{3} = \dfrac{1-\sqrt{5}}{3} \cdot \dfrac{1+\sqrt{5}}{1+\sqrt{5}} = \dfrac{1-5}{3\left(1+\sqrt{5}\right)} = \dfrac{-4}{3\left(1+\sqrt{5}\right)}$

103. $\dfrac{\sqrt{r}+\sqrt{2}}{5} = \dfrac{\sqrt{r}+\sqrt{2}}{5} \cdot \dfrac{\sqrt{r}-\sqrt{2}}{\sqrt{r}-\sqrt{2}} = \dfrac{r-2}{5\left(\sqrt{r}-\sqrt{2}\right)}$

105. $\sqrt{x^2+1}-x = \dfrac{\sqrt{x^2+1}-x}{1} \cdot \dfrac{\sqrt{x^2+1}+x}{\sqrt{x^2+1}+x} = \dfrac{x^2+1-x^2}{\sqrt{x^2+1}+x} = \dfrac{1}{\sqrt{x^2+1}+x}$

107. $\dfrac{16+a}{16} = \dfrac{16}{16} + \dfrac{a}{16} = 1 + \dfrac{a}{16}$, so the statement is true.

109. This statement is false. For example, take $x = 2$. Then LHS $= \dfrac{2}{4+x} = \dfrac{2}{4+2} = \dfrac{2}{6} = \dfrac{1}{3}$, while

RHS $= \dfrac{1}{2} + \dfrac{2}{x} = \dfrac{1}{2} + \dfrac{2}{2} = \dfrac{3}{2}$, and $\dfrac{1}{3} \neq \dfrac{3}{2}$.

111. This statement is false. For example, take $x = 0$ and $y = 1$. Then LHS $= \dfrac{x}{x+y} = \dfrac{0}{0+1} = 0$, while

RHS $= \dfrac{1}{1+y} = \dfrac{1}{1+1} = \dfrac{1}{2}$, and $0 \neq \dfrac{1}{2}$.

113. This statement is true: $\dfrac{-a}{b} = (-a)\left(\dfrac{1}{b}\right) = (-1)(a)\left(\dfrac{1}{b}\right) = (-1)\left(\dfrac{a}{b}\right) = -\dfrac{a}{b}$.

115. (a) $R = \dfrac{1}{\dfrac{1}{R_1} + \dfrac{1}{R_2}} = \dfrac{1}{\dfrac{1}{R_1} + \dfrac{1}{R_2}} \cdot \dfrac{R_1 R_2}{R_1 R_2} = \dfrac{R_1 R_2}{R_2 + R_1}$

(b) Substituting $R_1 = 10$ ohms and $R_2 = 20$ ohms gives $R = \dfrac{(10)(20)}{(20) + (10)} = \dfrac{200}{30} \approx 6.7$ ohms.

117.

x	2.80	2.90	2.95	2.99	2.999	3	3.001	3.01	3.05	3.10	3.20
$\dfrac{x^2 - 9}{x - 3}$	5.80	5.90	5.95	5.99	5.999	?	6.001	6.01	6.05	6.10	6.20

From the table, we see that the expression $\dfrac{x^2 - 9}{x - 3}$ approaches 6 as x approaches 3. We simplify the expression:

$\dfrac{x^2 - 9}{x - 3} = \dfrac{(x - 3)(x + 3)}{x - 3} = x + 3, x \neq 3$. Clearly as x approaches 3, $x + 3$ approaches 6. This explains the result in the table.

119. Answers will vary.

Algebraic Error	**Counterexample**
$\dfrac{1}{a} + \dfrac{1}{b} \neq \dfrac{1}{a + b}$	$\dfrac{1}{2} + \dfrac{1}{2} \neq \dfrac{1}{2 + 2}$
$(a + b)^2 \neq a^2 + b^2$	$(1 + 3)^2 \neq 1^2 + 3^2$
$\sqrt{a^2 + b^2} \neq a + b$	$\sqrt{5^2 + 12^2} \neq 5 + 12$
$\dfrac{a + b}{a} \neq b$	$\dfrac{2 + 6}{2} \neq 6$
$\left(a^3 + b^3\right)^{1/3} \neq a + b$	$\left(2^3 + 2^3\right)^{1/3} \neq 2 + 2$
$\dfrac{a^m}{a^n} \neq a^{m/n}$	$\dfrac{3^5}{3^2} \neq 3^{5/2}$
$a^{-1/n} \neq \dfrac{1}{a^n}$	$64^{-1/3} \neq \dfrac{1}{64^3}$

P.8 SOLVING BASIC EQUATIONS

1. Substituting $x = 3$ in the equation $4x - 2 = 10$ makes the equation true, so the number 3 is a *solution* of the equation.

3. (a) $\dfrac{x}{2} + 2x = 10$ is equivalent to $\dfrac{5}{2}x - 10 = 0$, so it is a linear equation.

(b) $\dfrac{2}{x} - 2x = 1$ is not linear because it contains the term $\dfrac{2}{x}$, a multiple of the reciprocal of the variable.

(c) $x + 7 = 5 - 3x \Leftrightarrow 4x - 2 = 0$, so it is linear.

5. (a) This is true: If $a = b$, then $a + x = b + x$.

(b) This is false, because the number could be zero. However, it is true that multiplying each side of an equation by a *nonzero* number always gives an equivalent equation.

(c) This is false. For example, $-5 = 5$ is false, but $(-5)^2 = 5^2$ is true.

7. (a) When $x = -2$, LHS $= 4(-2) + 7 = -8 + 7 = -1$ and RHS $= 9(-2) - 3 = -18 - 3 = -21$. Since LHS $\neq$ RHS, $x = -2$ is not a solution.

(b) When $x = 2$, LHS $= 4(-2) + 7 = 8 + 7 = 15$ and RHS $= 9(2) - 3 = 18 - 3 = 15$. Since LHS $=$ RHS, $x = 2$ is a solution.

9. (a) When $x = 2$, LHS $= 1 - [2 - (3 - (2))] = 1 - [2 - 1] = 1 - 1 = 0$ and RHS $= 4(2) - (6 + (2)) = 8 - 8 = 0$. Since LHS $=$ RHS, $x = 2$ is a solution.

   (b) When $x = 4$ LHS $= 1 - [2 - (3 - (4))] = 1 - [2 - (-1)] = 1 - 3 = -2$ and RHS $= 4(4) - (6 + (4)) = 16 - 10 = 6$. Since LHS $\neq$ RHS, $x = 4$ is not a solution.

11. (a) When $x = -1$, LHS $= 2(-1)^{1/3} - 3 = 2(-1) - 3 = -2 - 3 = -5$. Since LHS $\neq 1$, $x = -1$ is not a solution.

    (b) When $x = 8$ LHS $= 2(8)^{1/3} - 3 = 2(2) - 3 = 4 - 3 = 1 =$ RHS. So $x = 8$ is a solution.

13. (a) When $x = 0$, LHS $= \dfrac{0 - a}{0 - b} = \dfrac{-a}{-b} = \dfrac{a}{b} =$ RHS. So $x = 0$ is a solution.

    (b) When $x = b$, LHS $= \dfrac{b - a}{b - b} = \dfrac{b - a}{0}$ is not defined, so $x = b$ is not a solution.

15. $3x + 7 = 0 \Leftrightarrow 3x = -7 \Leftrightarrow x = -\frac{7}{3}$ 17. $7 - 2x = 15 \Leftrightarrow 2x = -8 \Leftrightarrow x = -4$

19. $\frac{1}{2}x + 7 = 3 \Leftrightarrow \frac{1}{2}x = -4 \Leftrightarrow x = -8$ 21. $x - 3 = 2x + 6 \Leftrightarrow -9 = x$

23. $-7w = 15 - 2w \Leftrightarrow -5w = 15 \Leftrightarrow w = -3$

25. $\frac{1}{2}y - 2 = \frac{1}{3}y \Leftrightarrow 3y - 12 = 2y$ (multiply both sides by the LCD, 6) $\Leftrightarrow y = 12$

27. $2(1 - x) = 3(1 + 2x) + 5 \Leftrightarrow 2 - 2x = 3 + 6x + 5 \Leftrightarrow 2 - 2x = 8 + 6x \Leftrightarrow -6 = 8x \Leftrightarrow x = -\frac{3}{4}$

29. $4\left(y - \frac{1}{2}\right) - y = 6(5 - y) \Leftrightarrow 4y - 2 - y = 30 - 6y \Leftrightarrow 3y - 2 = 30 - 6y \Leftrightarrow 9y = 32 \Leftrightarrow y = \frac{32}{9}$

31. $x - \frac{1}{3}x - \frac{1}{2}x - 5 = 0 \Leftrightarrow 6x - 2x - 3x - 30 = 0$ (multiply both sides by 6) $\Leftrightarrow x = 30$

33. $2x - \dfrac{x}{2} + \dfrac{x + 1}{4} = 6x \Leftrightarrow 8x - 2x + x + 1 = 24x \Leftrightarrow 7x + 1 = 24x \Leftrightarrow 1 = 17x \Leftrightarrow x = \frac{1}{17}$

35. $(x - 1)(x + 2) = (x - 2)(x - 3) \Leftrightarrow x^2 + x - 2 = x^2 - 5x + 6 \Leftrightarrow x - 2 = -5x + 6 \Leftrightarrow 6x = 8 \Leftrightarrow x = \frac{4}{3}$

37. $(x - 1)(4x + 5) = (2x - 3)^2 \Leftrightarrow 4x^2 + x - 5 = 4x^2 - 12x + 9 \Leftrightarrow x - 5 = -12x + 9 \Leftrightarrow 13x = 14 \Leftrightarrow x = \frac{14}{13}$

39. $\dfrac{1}{x} = \dfrac{4}{3x} + 1 \Rightarrow 3 = 4 + 3x$ (multiply both sides by the LCD, $3x$) $\Leftrightarrow -1 = 3x \Leftrightarrow x = -\frac{1}{3}$

41. $\dfrac{2x - 1}{x + 2} = \dfrac{4}{5} \Rightarrow 5(2x - 1) = 4(x + 2) \Leftrightarrow 10x - 5 = 4x + 8 \Leftrightarrow 6x = 13 \Leftrightarrow x = \frac{13}{6}$

43. $\dfrac{2}{t + 6} = \dfrac{3}{t - 1} \Rightarrow 2(t - 1) = 3(t + 6)$ [multiply both sides by the LCD, $(t - 1)(t + 6)$] $\Leftrightarrow 2t - 2 = 3t + 18 \Leftrightarrow -20 = t$

45. $\dfrac{3}{x + 1} - \dfrac{1}{2} = \dfrac{1}{3x + 3} \Rightarrow 3(6) - (3x + 3) = 2$ [multiply both sides by $6(x + 1)$] $\Leftrightarrow 18 - 3x - 3 = 2 \Leftrightarrow -3x + 15 = 2 \Leftrightarrow$
   $-3x = -13 \Leftrightarrow x = \frac{13}{3}$

47. $\dfrac{1}{z} - \dfrac{1}{2z} - \dfrac{1}{5z} = \dfrac{10}{z + 1} \Rightarrow 10(z + 1) - 5(z + 1) - 2(z + 1) = 10(10z)$ [multiply both sides by $10z(z + 1)$] $\Leftrightarrow$
   $3(z + 1) = 100z \Leftrightarrow 3z + 3 = 100z \Leftrightarrow 3 = 97z \Leftrightarrow \frac{3}{97} = z$

49. $\dfrac{x}{2x - 4} - 2 = \dfrac{1}{x - 2} \Rightarrow x - 2(2x - 4) = 2$ [multiply both sides by $2(x - 2)$] $\Leftrightarrow x - 4x + 8 = 2 \Leftrightarrow -3x = -6 \Leftrightarrow x = 2$.
   But substituting $x = 2$ into the original equation does not work, since we cannot divide by 0. Thus there is no solution.

51. $\dfrac{3}{x + 4} = \dfrac{1}{x} + \dfrac{6x + 12}{x^2 + 4x} \Rightarrow 3(x) = (x + 4) + 6x + 12$ (multiply both sides by $x(x + 4)$] $\Leftrightarrow 3x = 7x + 16 \Leftrightarrow -4x = 16$
   $\Leftrightarrow x = -4$. But substituting $x = -4$ into the original equation does not work, since we cannot divide by 0. Thus, there is no solution.

53. $x^2 = 49 \Rightarrow x = \pm 7$

55. $x^2 - 24 = 0 \Leftrightarrow x^2 = 24 \Rightarrow x = \pm\sqrt{24} = \pm 2\sqrt{6}$

57. $8x^2 - 64 = 0 \Leftrightarrow x^2 - 8 = 0 \Leftrightarrow x^2 = 8 \Rightarrow x = \pm\sqrt{8} = \pm 2\sqrt{2}$

59. $x^2 + 16 = 0 \Leftrightarrow x^2 = -16$ which has no real solution.

61. $(x + 2)^2 = 4 \Leftrightarrow (x + 2)^2 = 4 \Rightarrow x + 2 = \pm 2$. If $x + 2 = 2$, then $x = 0$. If $x + 2 = -2$, then $x = -4$. The solutions are -4 and 0.

63. $x^3 = 27 \Leftrightarrow x = 27^{1/3} = 3$

65. $0 = x^4 - 16 = \left(x^2 + 4\right)\left(x^2 - 4\right) = \left(x^2 + 4\right)(x - 2)(x + 2)$. $x^2 + 4 = 0$ has no real solution. If $x - 2 = 0$, then $x = 2$. If $x + 2 = 0$, then $x = -2$. The solutions are ± 2.

67. $x^4 + 64 = 0 \Leftrightarrow x^4 = -64$ which has no real solution.

69. $(x + 2)^4 - 81 = 0 \Leftrightarrow (x + 2)^4 = 81 \Leftrightarrow \left[(x + 2)^4\right]^{1/4} = \pm 81^{1/4} \Leftrightarrow x + 2 = \pm 3$. So $x + 2 = 3$, then $x = 1$. If $x + 2 = -3$, then $x = -5$. The solutions are -5 and 1.

71. $3(x - 3)^3 = 375 \Leftrightarrow (x - 3)^3 = 125 \Leftrightarrow (x - 3) = 125^{1/3} = 5 \Leftrightarrow x = 3 + 5 = 8$

73. $\sqrt[3]{x} = 5 \Leftrightarrow x = 5^3 = 125$

75. $2x^{5/3} + 64 = 0 \Leftrightarrow 2x^{5/3} = -64 \Leftrightarrow x^{5/3} = -32 \Leftrightarrow x = (-32)^{3/5} = \left(-2^5\right)^{1/5} = (-2)^3 = -8$

77. $3.02x + 1.48 = 10.92 \Leftrightarrow 3.02x = 9.44 \Leftrightarrow x = \dfrac{9.44}{3.02} \approx 3.13$

79. $2.15x - 4.63 = x + 1.19 \Leftrightarrow 1.15x = 5.82 \Leftrightarrow x = \dfrac{5.82}{1.19} \approx 5.06$

81. $3.16(x + 4.63) = 4.19(x - 7.24) \Leftrightarrow 3.16x + 14.63 = 4.19x - 30.34 \Leftrightarrow 44.97 = 1.03x \Leftrightarrow x = \dfrac{44.97}{1.03} \approx 43.66$

83. $\dfrac{0.26x - 1.94}{3.03 - 2.44x} = 1.76 \Rightarrow 0.26x - 1.94 = 1.76(3.03 - 2.44x) \Leftrightarrow 0.26x - 1.94 = 5.33 - 4.29x \Leftrightarrow 4.55x = 7.27 \Leftrightarrow$

$x = \dfrac{7.27}{4.55} \approx 1.60$

85. $r = \dfrac{12}{M} \Leftrightarrow M = \dfrac{12}{r}$
      87. $PV = nRT \Leftrightarrow R = \dfrac{PV}{nT}$

89. $P = 2l + 2w \Leftrightarrow 2w = P - 2l \Leftrightarrow w = \dfrac{P - 2l}{2}$

91. $V = \frac{1}{3}\pi r^2 h \Leftrightarrow r^2 = \dfrac{3V}{\pi h} \Rightarrow r = \pm\sqrt{\dfrac{3V}{\pi h}}$

93. $V = \frac{4}{3}\pi r^3 \Leftrightarrow r^3 = \dfrac{3V}{4\pi} \Leftrightarrow r = \sqrt[3]{\dfrac{3V}{4\pi}}$

95. $A = P\left(1 + \dfrac{i}{100}\right)^2 \Leftrightarrow \dfrac{A}{P} = \left(1 + \dfrac{i}{100}\right)^2 \Rightarrow 1 + \dfrac{i}{100} = \pm\sqrt{\dfrac{A}{P}} \Leftrightarrow \dfrac{i}{100} = -1 \pm \sqrt{\dfrac{A}{P}} \Leftrightarrow i = -100 \pm 100\sqrt{\dfrac{A}{P}}$

97. $\dfrac{ax + b}{cx + d} = 2 \Leftrightarrow ax + b = 2(cx + d) \Leftrightarrow ax + b = 2cx + 2d \Leftrightarrow ax - 2cx = 2d - b \Leftrightarrow (a - 2c)x = 2d - b \Leftrightarrow x = \dfrac{2d - b}{a - 2c}$

99. (a) The shrinkage factor when $w = 250$ is $S = \dfrac{0.032(250) - 2.5}{10,000} = \dfrac{8 - 2.5}{10,000} = 0.00055$. So the beam shrinks

$0.00055 \times 12.025 \approx 0.007$ m, so when it dries it will be $12.025 - 0.007 = 12.018$ m long.

(b) Substituting $S = 0.00050$ we get $0.00050 = \dfrac{0.032w - 2.5}{10,000} \Leftrightarrow 5 = 0.032w - 2.5 \Leftrightarrow 7.5 = 0.032w \Leftrightarrow$

$w = \dfrac{7.5}{0.032} \approx 234.375$. So the water content should be 234.375 kg/m^3.

101. (a) Solving for v when $P = 10,000$ we get $10,000 = 15.6v^3 \Leftrightarrow v^3 \approx 641.02 \Leftrightarrow v \approx 8.6$ km/h.

(b) Solving for v when $P = 50,000$ we get $50,000 = 15.6v^3 \Leftrightarrow v^3 \approx 3205.13 \Leftrightarrow v \approx 14.7$ km/h.

103. (a) $3(0) + k - 5 = k(0) - k + 1 \Leftrightarrow k - 5 = -k + 1 \Leftrightarrow 2k = 6 \Leftrightarrow k = 3$

   (b) $3(1) + k - 5 = k(1) - k + 1 \Leftrightarrow 3 + k - 5 = k - k + 1 \Leftrightarrow k - 2 = 1 \Leftrightarrow k = 3$

   (c) $3(2) + k - 5 = k(2) - k + 1 \Leftrightarrow 6 + k - 5 = 2k - k + 1 \Leftrightarrow k + 1 = k + 1$. $x = 2$ is a solution for every value of k.
   That is, $x = 2$ is a solution to every member of this family of equations.

CHAPTER P REVIEW

1. (a) Since there are initially 250 tablets and she takes 2 tablets per day, the number of tablets T that are left in the bottle after she has been taking the tablets for x days is $T = 250 - 2x$.

   (b) After 30 days, there are $250 - 2(30) = 190$ tablets left.

   (c) We set $T = 0$ and solve: $T = 250 - 2x = 0 \Leftrightarrow 250 = 2x \Leftrightarrow x = 125$. She will run out after 125 days.

3. (a) 16 is rational. It is an integer, and more precisely, a natural number.

   (b) -16 is rational. It is an integer, but because it is negative, it is not a natural number.

   (c) $\sqrt{16} = 4$ is rational. It is an integer, and more precisely, a natural number.

   (d) $\sqrt{2}$ is irrational.

   (e) $\frac{8}{3}$ is rational, but is neither a natural number nor an integer.

   (f) $-\frac{8}{2} = -4$ is rational. It is an integer, but because it is negative, it is not a natural number.

5. Commutative Property for addition.

7. Distributive Property.

9. (a) $\dfrac{5}{6} + \dfrac{2}{3} = \dfrac{5}{6} + \dfrac{4}{6} = \dfrac{9}{6} = \dfrac{3}{2}$

   (b) $\dfrac{5}{6} - \dfrac{2}{3} = \dfrac{5}{6} - \dfrac{4}{6} = \dfrac{1}{6}$

11. (a) $\dfrac{15}{8} \cdot \dfrac{12}{5} = \dfrac{15 \cdot 12}{8 \cdot 5} = \dfrac{3 \cdot 3}{2 \cdot 1} = \dfrac{9}{2}$

   (b) $\dfrac{15}{8} \div \dfrac{12}{5} = \dfrac{15 \cdot 5}{8 \cdot 12} = \dfrac{5 \cdot 5}{8 \cdot 4} = \dfrac{25}{32}$

13. $x \in [-2, 6) \Leftrightarrow -2 \le x < 6$

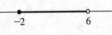

15. $x \in (-\infty, 4] \Leftrightarrow x \le 4$

17. $x \ge 5 \Leftrightarrow x \in [5, \infty)$

19. $-1 < x \le 5 \Leftrightarrow x \in (-1, 5]$

21. (a) $A \cup B = \left\{ -1, 0, \tfrac{1}{2}, 1, 2, 3, 4 \right\}$

   (b) $A \cap B = \{1\}$

23. (a) $A \cap C = \{1, 2\}$

   (b) $B \cap D = \left\{ \tfrac{1}{2}, 1 \right\}$

25. $|7 - 10| = |-3| = 3$

27. $|3 - |-9|| = |3 - 9| = |-6| = 6$

29. $2^{-3} - 3^{-2} = \dfrac{1}{8} - \dfrac{1}{9} = \dfrac{9}{72} - \dfrac{8}{72} = \dfrac{1}{72}$

31. $216^{-1/3} = \dfrac{1}{216^{1/3}} = \dfrac{1}{\sqrt[3]{216}} = \dfrac{1}{6}$

33. $\dfrac{\sqrt{242}}{\sqrt{2}} = \sqrt{\dfrac{242}{2}} = \sqrt{121} = 11$

35. $\sqrt[3]{-125} = \sqrt[3]{(-5)^3} = -5$

37. (a) $|5 - 3| = |2| = 2$

   (b) $|-5 - 3| = |-8| = 8$

39. (a) $\sqrt[3]{7} = 7^{1/3}$

   (b) $\sqrt[5]{7^4} = 7^{4/5}$

41. (a) $\sqrt[6]{x^5} = x^{5/6}$

    (b) $\left(\sqrt{x}\right)^9 = \left(x^{1/2}\right)^9 = x^{9/2}$

43. $\left(2x^3y\right)^2 \left(3x^{-1}y^2\right) = 4x^6y^2 \cdot 3x^{-1}y^2 = 4 \cdot 3x^{6-1}y^{2+2}$

                                                  $= 12x^5y^4$

45. $\dfrac{x^4 (3x)^2}{x^3} = \dfrac{x^4 \cdot 9x^2}{x^3} = 9x^{4+2-3} = 9x^3$

47. $\sqrt[3]{\left(x^3y\right)^2 y^4} = \sqrt[3]{x^6 y^4 y^2} = \sqrt[3]{x^6 y^6} = x^2 y^2$

49. $\dfrac{8r^{1/2}s^{-3}}{2r^{-2}s^4} = 4r^{(1/2)-(-2)}s^{-3-4} = 4r^{5/2}s^{-7} = \dfrac{4r^{5/2}}{s^7}$

51. $78{,}250{,}000{,}000 = 7.825 \times 10^{10}$

53. $\dfrac{ab}{c} \approx \dfrac{(0.00000293)\left(1.582 \times 10^{-14}\right)}{2.8064 \times 10^{12}} = \dfrac{\left(2.93 \times 10^{-6}\right)\left(1.582 \times 10^{-14}\right)}{2.8064 \times 10^{12}} = \dfrac{2.93 \cdot 1.582}{2.8064} \times 10^{-6-14-12}$

        $\approx 1.65 \times 10^{-32}$

55. $2x^2y - 6xy^2 = 2xy\,(x - 3y)$

57. $x^2 - 9x + 18 = (x - 6)\,(x - 3)$

59. $3x^2 - 2x - 1 = (3x + 1)\,(x - 1)$

61. $4t^2 - 13t - 12 = (4t + 3)\,(t - 4)$

63. $25 - 16t^2 = (5 - 4t)\,(5 + 4t)$

65. $a^4b^2 + ab^5 = ab^2\left(a^3 + b^3\right) = ab^2\,(a + b)\left(a^2 - ab + b^2\right)$

67. $8x^3 + y^6 = (2x)^3 + \left(y^2\right)^3 = \left(2x + y^2\right)\left(4x^2 - 2xy^2 + y^4\right)$

69. $4x^3 - 8x^2 + 3x - 6 = 4x^2\,(x - 2) + 3\,(x - 2) = \left(4x^2 + 3\right)(x - 2)$

71. $(x + 1)^2 - 2\,(x + 1) + 1 = [(x + 1) - 1]^2 = x^2.$ You can also obtain this result by expanding each term and then simplifying.

73. $(2x + 1)\,(3x - 2) - 5\,(4x - 1) = 6x^2 - 4x + 3x - 2 - 20x + 5 = 6x^2 - 21x + 3$

75. $\left(2a^2 - b\right)^2 = \left(2a^2\right)^2 - 2\left(2a^2\right)(b) + (b)^2 = 4a^4 - 4a^2b + b^2$

77. $(2x + 1)^3 = (2x)^3 + 3\,(2x)^2\,(1) + 3\,(2x)\,(1)^2 + (1)^3 = 8x^3 + 12x^2 + 6x + 1$

79. $x^2\,(x - 2) + x\,(x - 2)^2 = x^3 - 2x^2 + x\left(x^2 - 4x + 4\right) = x^3 - 2x^2 + x^3 - 4x^2 + 4x = 2x^3 - 6x^2 + 4x$

81. $\dfrac{x^2 - 2x - 3}{2x^2 + 5x + 3} = \dfrac{(x - 3)\,(x + 1)}{(2x + 3)\,(x + 1)} = \dfrac{x - 3}{2x + 3}$

83. $\dfrac{x^2 + 2x - 3}{x^2 + 8x + 16} \cdot \dfrac{3x + 12}{x - 1} = \dfrac{(x + 3)\,(x - 1)}{(x + 4)\,(x + 4)} \cdot \dfrac{3\,(x + 4)}{(x - 1)} = \dfrac{3\,(x + 3)}{x + 4}$

85. $\dfrac{x^2 - 2x - 15}{x^2 - 6x + 5} \div \dfrac{x^2 - x - 12}{x^2 - 1} = \dfrac{(x - 5)\,(x + 3)}{(x - 5)\,(x - 1)} \cdot \dfrac{(x - 1)\,(x + 1)}{(x - 4)\,(x + 3)} = \dfrac{x + 1}{x - 4}$

87. $\dfrac{1}{x - 1} - \dfrac{x}{x^2 + 1} = \dfrac{x^2 + 1}{(x - 1)\left(x^2 + 1\right)} - \dfrac{x\,(x - 1)}{(x - 1)\left(x^2 + 1\right)} = \dfrac{x^2 + 1 - x^2 + x}{(x - 1)\left(x^2 + 1\right)} = \dfrac{x + 1}{(x - 1)\left(x^2 + 1\right)}$

89. $\dfrac{1}{x - 1} - \dfrac{2}{x^2 - 1} = \dfrac{1}{x - 1} - \dfrac{2}{(x - 1)\,(x + 1)} = \dfrac{x + 1}{(x - 1)\,(x + 1)} - \dfrac{2}{(x - 1)\,(x + 1)}$

                  $= \dfrac{x + 1 - 2}{(x - 1)\,(x + 1)} = \dfrac{x - 1}{(x - 1)\,(x + 1)} = \dfrac{1}{x + 1}$

91. $\dfrac{\dfrac{1}{x} - \dfrac{1}{2}}{x - 2} = \dfrac{\dfrac{2}{2x} - \dfrac{x}{2x}}{x - 2} = \dfrac{2 - x}{2x} \cdot \dfrac{1}{x - 2} = \dfrac{-1\,(x - 2)}{2x} \cdot \dfrac{1}{x - 2} = \dfrac{-1}{2x}$

93.
$$\frac{3(x+h)^2 - 5(x+h) - \left(3x^2 - 5x\right)}{h} = \frac{3x^2 + 6xh + 3h^2 - 5x - 5h - 3x^2 + 5x}{h} = \frac{6xh + 3h^2 - 5h}{h}$$
$$= \frac{h(6x + 3h - 5)}{h} = 6x + 3h - 5$$

95. $\dfrac{1}{\sqrt{7}} = \dfrac{1}{\sqrt{7}} \cdot \dfrac{\sqrt{7}}{\sqrt{7}} = \dfrac{\sqrt{7}}{7}$

97. $\dfrac{12}{\sqrt{3}-1} = \dfrac{12}{\sqrt{3}-1} \cdot \dfrac{\sqrt{3}+1}{\sqrt{3}+1} = \dfrac{12\sqrt{3}+12}{\left(\sqrt{3}\right)^2 - 1^2} = \dfrac{12 + 12\sqrt{3}}{2} = 6 + 6\sqrt{3}$

99. $\dfrac{x}{2+\sqrt{x}} = \dfrac{x}{2+\sqrt{x}} \cdot \dfrac{2-\sqrt{x}}{2-\sqrt{x}} = \dfrac{2x - x\sqrt{x}}{2^2 - \left(\sqrt{x}\right)^2} = \dfrac{x\left(\sqrt{x}-2\right)}{x-4}$

101. $\dfrac{x+5}{x+10}$ is defined whenever $x + 10 \neq 0 \Leftrightarrow x \neq -10$, so its domain is $\{x \mid x \neq -10\}$.

103. $\dfrac{\sqrt{x}}{x^2 - 3x - 4}$ is defined whenever $x \geq 0$ (so that $\sqrt{x}$ is defined) and $x^2 - 3x - 4 = (x+1)(x-4) \neq 0 \Leftrightarrow x \neq -1$ and $x \neq 4$. Thus, its domain is $\{x \mid x \geq 0 \text{ and } x \neq 4\}$.

105. This statement is false. For example, take $x = 1$ and $y = 1$. Then LHS $= (x+y)^3 = (1+1)^3 = 2^3 = 8$, while RHS $= x^3 + y^3 = 1^3 + 1^3 = 1 + 1 = 2$, and $8 \neq 2$.

107. This statement is true: $\dfrac{12+y}{y} = \dfrac{12}{y} + \dfrac{y}{y} = \dfrac{12}{y} + 1$.

109. This statement is false. For example, take $a = -1$. Then LHS $= \sqrt{a^2} = \sqrt{(-1)^2} = \sqrt{1} = 1$, which does not equal $a = -1$. The true statement is $\sqrt{a^2} = |a|$.

111. $3x + 12 = 24 \Leftrightarrow 3x = 12 \Leftrightarrow x = 4$ **113.** $7x - 6 = 4x + 9 \Leftrightarrow 3x = 15 \Leftrightarrow x = 5$

115. $\frac{1}{3}x - \frac{1}{2} = 2 \Leftrightarrow 2x - 3 = 12 \Leftrightarrow 2x = 15 \Leftrightarrow x = \frac{15}{2}$

117. $2(x+3) - 4(x-5) = 8 - 5x \Leftrightarrow 2x + 6 - 4x + 20 = 8 - 5x \Leftrightarrow -2x + 26 = 8 - 5x \Leftrightarrow 3x = -18 \Leftrightarrow x = -6$

119. $\dfrac{x+1}{x-1} = \dfrac{2x-1}{2x+1} \Leftrightarrow (x+1)(2x+1) = (2x-1)(x-1) \Leftrightarrow 2x^2 + 3x + 1 = 2x^2 - 3x + 1 \Leftrightarrow 6x = 0 \Leftrightarrow x = 0$

121. $\dfrac{x+1}{x-1} = \dfrac{3x}{3x-6} = \dfrac{3x}{3(x-2)} = \dfrac{x}{x-2} \Leftrightarrow (x+1)(x-2) = x(x-1) \Leftrightarrow x^2 - x - 2 = x^2 - x \Leftrightarrow -2 = 0$. Since this last equation is never true, there is no real solution to the original equation.

123. $x^2 = 144 \Rightarrow x = \pm 12$

125. $x^3 - 27 = 0 \Leftrightarrow x^3 = 27 \Rightarrow x = 3$.

127. $(x+1)^3 = -64 \Leftrightarrow x + 1 = -4 \Leftrightarrow x = -1 - 4 = -5$.

129. $\sqrt[3]{x} = -3 \Leftrightarrow x = (-3)^3 = -27$.

131. $4x^{3/4} - 500 = 0 \Leftrightarrow 4x^{3/4} = 500 \Leftrightarrow x^{3/4} = 125 \Leftrightarrow x = 125^{4/3} = 5^4 = 625$.

133. $A = \dfrac{x+y}{2} \Leftrightarrow 2A = x + y \Leftrightarrow x = 2A - y$.

135. Multiply through by t: $J = \dfrac{1}{t} + \dfrac{1}{2t} + \dfrac{1}{3t} \Leftrightarrow tJ = 1 + \dfrac{1}{2} + \dfrac{1}{3} = \dfrac{11}{6} \Leftrightarrow t = \dfrac{11}{6J}, J \neq 0$.

CHAPTER P TEST

1. (a) The cost is $C = 9 + 1.5x$.

   (b) There are four extra toppings, so $x = 4$ and $C = 9 + 1.5\,(4) = \$15$.

3. (a) $A \cap B = \{0, 1, 5\}$

   (b) $A \cup B = \left\{-2, 0, \frac{1}{2}, 1, 3, 5, 7\right\}$

5. (a) $-2^6 = -64$ (b) $(-2)^6 = 64$ (c) $2^{-6} = \dfrac{1}{2^6} = \dfrac{1}{64}$ (d) $\dfrac{7^{10}}{7^{12}} = 7^{-2} = \dfrac{1}{49}$

   (e) $\left(\dfrac{3}{2}\right)^{-2} = \left(\dfrac{2}{3}\right)^2 = \dfrac{4}{9}$ (f) $\dfrac{\sqrt[5]{32}}{\sqrt{16}} = \dfrac{2}{4} = \dfrac{1}{2}$ (g) $\sqrt[4]{\dfrac{3^8}{2^{16}}} = \dfrac{3^2}{2^4} = \dfrac{9}{16}$ (h) $81^{-3/4} = \left(3^4\right)^{-3/4} = 3^{-3} = \dfrac{1}{27}$

7. (a) $\dfrac{a^2 b}{a^{-1} b^5} = \dfrac{a^3}{b^4}$ (b) $\left(3x^2 y^{1/2} x^{-2}\right)^3 = \left(3x^{2-2} y^{1/2}\right)^3 = \left(3y^{1/2}\right)^3 = 27 y^{3/2}$

   (c) $\left(3a^3 b^3\right)\left(4ab^2\right)^2 = 3a^3 b^3 \cdot 4^2 a^2 b^4 = 48 a^5 b^7$ (d) $\sqrt{200} - \sqrt{32} = 10\sqrt{2} - 4\sqrt{2} = 6\sqrt{2}$

   (e) $\sqrt{48 x^4 y^5} = \sqrt{3y \cdot \left(4x^2 y^2\right)^2} = 4x^2 y^2 \sqrt{3y}$ (f) $\sqrt[3]{\dfrac{125}{x^{-9}}} = \sqrt[3]{5^3 \cdot x^9} = 5x^3$

   (g) $\left(\dfrac{3x^{3/2} y^3}{x^2 y^{-1/2}}\right)^{-2} = \dfrac{3^{-2} x^{-3} y^{-6}}{x^{-4} y^1} = \tfrac{1}{9} x^{-3-(-4)} y^{-6-1} = \tfrac{1}{9} x y^{-7} = \dfrac{x}{9y^7}$

9. (a) $4x^2 - 25 = (2x - 5)(2x + 5)$

   (b) $2x^2 + 5x - 12 = (2x - 3)(x + 4)$

   (c) $x^3 - 3x^2 - 4x + 12 = x^2\,(x - 3) - 4\,(x - 3) = (x - 3)\left(x^2 - 4\right) = (x - 3)(x - 2)(x + 2)$

   (d) $x^4 + 27x = x\left(x^3 + 27\right) = x\,(x + 3)\left(x^2 - 3x + 9\right)$

   (e) $(2x - y)^2 - 10\,(2x - y) + 25 = (2x - y)^2 - 2\,(5)\,(2x - y) + 5^2 = (2x - y - 5)^2$

   (f) $x^3 y - 4xy = xy\left(x^2 - 4\right) = xy\,(x - 2)(x + 2)$

11. (a) $\dfrac{6}{\sqrt[3]{4}} = \dfrac{6}{\sqrt[3]{2^2}} = \dfrac{6}{\sqrt[3]{2^2}} \cdot \dfrac{\sqrt[3]{2}}{\sqrt[3]{2}} = \dfrac{6\sqrt[3]{2}}{2} = 3\sqrt[3]{2}$

    (b) $\dfrac{\sqrt{6}}{2 + \sqrt{3}} = \dfrac{\sqrt{6}}{2 + \sqrt{3}} \cdot \dfrac{2 - \sqrt{3}}{2 - \sqrt{3}} = \dfrac{2\sqrt{6} - \sqrt{18}}{4 - 3} = \dfrac{2\sqrt{6} - \sqrt{9 \cdot 2}}{1} = 2\sqrt{6} - 3\sqrt{2}$

13. $E = mc^2 \Leftrightarrow \dfrac{E}{m} = c^2 \Leftrightarrow c = \sqrt{\dfrac{E}{m}}$. (We take the positive root because c represents the speed of light, which is positive.)

FOCUS ON MODELING   Making the Best Decisions

1. (a) The total cost is $\begin{pmatrix} \text{cost of} \\ \text{copier} \end{pmatrix} + \begin{pmatrix} \text{maintenance} \\ \text{cost} \end{pmatrix} \begin{pmatrix} \text{number} \\ \text{of months} \end{pmatrix} + \begin{pmatrix} \text{copy} \\ \text{cost} \end{pmatrix} \begin{pmatrix} \text{number} \\ \text{of months} \end{pmatrix}$. Each month

the copy cost is $8000 \cdot 0.03 = 240$. Thus we get $C_1 = 5800 + 25n + 240n = 5800 + 265n$.

(b) In this case the cost is $\begin{pmatrix} \text{rental} \\ \text{cost} \end{pmatrix} \begin{pmatrix} \text{number} \\ \text{of months} \end{pmatrix} + \begin{pmatrix} \text{copy} \\ \text{cost} \end{pmatrix} \begin{pmatrix} \text{number} \\ \text{of months} \end{pmatrix}$. Each month the copy cost is

$8000 \cdot 0.06 = 480$. Thus we get $C_2 = 95n + 480n = 575n$.

(c)

Years	n	Purchase	Rental
1	12	8,980	6,900
2	24	12,160	13,800
3	36	15,340	20,700
4	48	18,520	27,600
5	60	21,700	34,500
6	72	24,880	41,400

(d) The cost is the same when $C_1 = C_2$ are equal. So $5800 + 265n = 575n \Leftrightarrow 5800 = 310n \Leftrightarrow n \approx 18.71$ months.

3. (a) The total cost is $\begin{pmatrix} \text{setup} \\ \text{cost} \end{pmatrix} + \begin{pmatrix} \text{cost per} \\ \text{tire} \end{pmatrix} \begin{pmatrix} \text{number} \\ \text{of tires} \end{pmatrix}$. So $C = 8000 + 22x$.

(b) The revenue is $\begin{pmatrix} \text{price per} \\ \text{tire} \end{pmatrix} \begin{pmatrix} \text{number} \\ \text{of tires} \end{pmatrix}$. So $R = 49x$.

(c) Profit = Revenue − Cost. So $P = R - C = 49x - (8000 + 22x) = 27x - 8000$.

(d) Break even is when profit is zero. Thus $27x - 8000 = 0 \Leftrightarrow 27x = 8000 \Leftrightarrow x \approx 296.3$. So they need to sell at least 297 tires to break even.

5. (a) Design 1 is a square and the perimeter of a square is four times the length of a side. $24 = 4x$, so each side is $x = 6$ feet long. Thus the area is $6^2 = 36$ ft^2.

Design 2 is a circle with perimeter $2\pi r$ and area πr^2. Thus we must solve $2\pi r = 24 \Leftrightarrow r = \dfrac{12}{\pi}$. Thus, the area is

$\pi \left(\dfrac{12}{\pi} \right)^2 = \dfrac{144}{\pi} \approx 45.8$ ft^2. Design 2 gives the largest area.

(b) In Design 1, the cost is \$3 times the perimeter p, so $120 = 3p$ and the perimeter is 40 feet. By part (a), each side is then $\frac{40}{4} = 10$ feet long. So the area is $10^2 = 100$ ft^2.

In Design 2, the cost is \$4 times the perimeter p. Because the perimeter is $2\pi r$, we get $120 = 4(2\pi r)$ so

$r = \dfrac{120}{8\pi} = \dfrac{15}{\pi}$. The area is $\pi r^2 = \pi \left(\dfrac{15}{\pi} \right)^2 = \dfrac{225}{\pi} \approx 71.6$ ft^2. Design 1 gives the largest area.

7. (a)

Minutes	Plan A	Plan B	Plan C
500	$ 30	$ 40	$ 60
600	$30 + 100\,(0.50) = \$ 80$	$40 + 100\,(0.30) = \$ 70$	$60 + 100\,(0.10) = \$ 70$
700	$30 + 200\,(0.50) = \$130$	$40 + 200\,(0.30) = \$100$	$60 + 200\,(0.10) = \$ 80$
800	$30 + 300\,(0.50) = \$180$	$40 + 300\,(0.30) = \$130$	$60 + 300\,(0.10) = \$ 90$
900	$30 + 400\,(0.50) = \$230$	$40 + 400\,(0.30) = \$160$	$60 + 400\,(0.10) = \$100$
1000	$30 + 500\,(0.50) = \$280$	$40 + 500\,(0.30) = \$190$	$60 + 500\,(0.10) = \$110$
1100	$30 + 600\,(0.50) = \$330$	$40 + 600\,(0.30) = \$220$	$60 + 600\,(0.10) = \$120$

(b) For Plan A: $C_A = 30 + 0.50\,(x - 500) = 0.5x - 220$. For Plan B: $C_B = 40 + 0.30\,(x - 500) = 0.3x - 110$. For Plan C: $C_C = 60 + 0.10\,(x - 500) = 0.1x + 10$. Note that these equations are valid only for $x \geq 500$.

(c) If Genevieve uses 550 minutes, $C_A = 0.5\,(550) - 220 = \$55$, $C_B = 0.3\,(550) - 110 = \$55$, and $C_C = 0.1\,(550) + 10 = \$65$. If she uses 975 minutes, $C_A = 0.5\,(975) - 220 = \$267.50$, $C_B = 0.3\,(975) - 110 = \$182.50$, and $C_C = 0.1\,(975) + 10 = \$107.50$. If she uses 1200 minutes, $C_A = 0.5\,(1200) - 220 = \$380$, $C_B = 0.3\,(1200) - 110 = \$250$, and $C_C = 0.1\,(1200) + 10 = \$130$.

(d) (i) We set $C_A = C_B \Leftrightarrow 0.5x - 220 = 0.3x - 110 \Leftrightarrow 0.2x = 110 \Leftrightarrow x = \dfrac{110}{0.2} = 550$. Plans A and B cost the same when 550 minutes are used.

(ii) We set $C_A = C_C \Leftrightarrow 0.5x - 220 = 0.1x + 10 \Leftrightarrow 0.4x = 230 \Leftrightarrow x = \dfrac{230}{0.4} = 575$. Plans A and C cost the same when 575 minutes are used.

1 EQUATIONS AND GRAPHS

1.1 THE COORDINATE PLANE

1. The point that is 2 units to the left of the y-axis and 4 units above the x-axis has coordinates $(-2, 4)$.

3. The distance between the points (a, b) and (c, d) is $\sqrt{(c - a)^2 + (d - b)^2}$. So the distance between $(1, 2)$ and $(7, 10)$ is
$\sqrt{(7 - 1)^2 + (10 - 2)^2} = \sqrt{6^2 + 8^2} = \sqrt{36 + 64} = \sqrt{100} = 10$.

5. $A\,(5, 1)$, $B\,(1, 2)$, $C\,(-2, 6)$, $D\,(-6, 2)$, $E\,(-4, -1)$, $F\,(-2, 0)$, $G\,(-1, -3)$, $H\,(2, -2)$

7. $(0, 5)$, $(-1, 0)$, $(-1, -2)$, and $\left(\frac{1}{2}, \frac{2}{3}\right)$

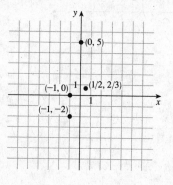

9. $\{(x, y) \mid y \geq 0\}$

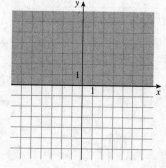

11. $\{(x, y) \mid y = 5\}$

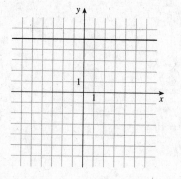

13. $\{(x, y) \mid -1 < x < 1\}$

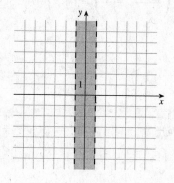

27

15. $\{(x, y) \mid xy < 0\}$
$= \{(x, y) \mid x < 0 \text{ and } y > 0 \text{ or } x > 0 \text{ and } y < 0\}$

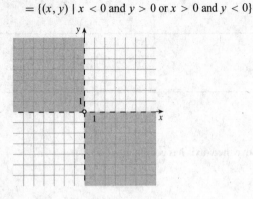

17. $\{(x, y) \mid x \geq 1 \text{ and } y < 3\}$

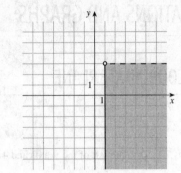

19. $\{(x, y) \mid x < 2 \text{ and } y \geq 1\}$

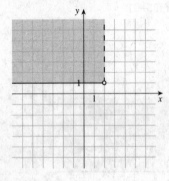

21. $\{(x, y) \mid -1 < x < 1 \text{ and } -2 < y < 2\}$

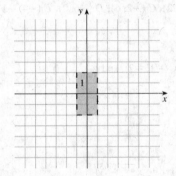

23. The two points are $(0, 2)$ and $(3, 0)$.

    (a) $d = \sqrt{(3 - 0)^2 + (0 - (-2))^2} = \sqrt{3^2 + 2^2} = \sqrt{9 + 4} = \sqrt{13}$

    (b) midpoint: $\left(\dfrac{3 + 0}{2}, \dfrac{0 + 2}{2}\right) = \left(\dfrac{3}{2}, 1\right)$

25. The two points are $(-3, 3)$ and $(5, -3)$.

    (a) $d = \sqrt{(-3 - 5)^2 + (3 - (-3))^2} = \sqrt{(-8)^2 + 6^2} = \sqrt{64 + 36} = \sqrt{100} = 10$

    (b) midpoint: $\left(\dfrac{-3 + 5}{2}, \dfrac{3 + (-3)}{2}\right) = (1, 0)$

27. (a)

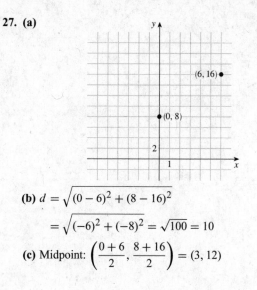

(b) $d = \sqrt{(0-6)^2 + (8-16)^2}$

$= \sqrt{(-6)^2 + (-8)^2} = \sqrt{100} = 10$

(c) Midpoint: $\left(\dfrac{0+6}{2}, \dfrac{8+16}{2}\right) = (3, 12)$

29. (a)

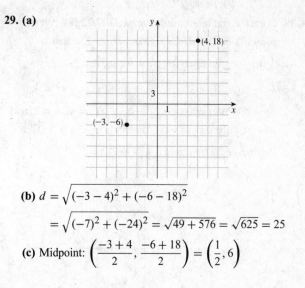

(b) $d = \sqrt{(-3-4)^2 + (-6-18)^2}$

$= \sqrt{(-7)^2 + (-24)^2} = \sqrt{49+576} = \sqrt{625} = 25$

(c) Midpoint: $\left(\dfrac{-3+4}{2}, \dfrac{-6+18}{2}\right) = \left(\dfrac{1}{2}, 6\right)$

31. (a)

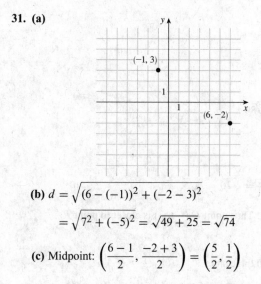

(b) $d = \sqrt{(6-(-1))^2 + (-2-3)^2}$

$= \sqrt{7^2 + (-5)^2} = \sqrt{49+25} = \sqrt{74}$

(c) Midpoint: $\left(\dfrac{6-1}{2}, \dfrac{-2+3}{2}\right) = \left(\dfrac{5}{2}, \dfrac{1}{2}\right)$

33. (a)

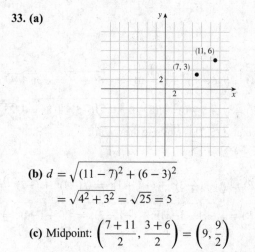

(b) $d = \sqrt{(11-7)^2 + (6-3)^2}$

$= \sqrt{4^2 + 3^2} = \sqrt{25} = 5$

(c) Midpoint: $\left(\dfrac{7+11}{2}, \dfrac{3+6}{2}\right) = \left(9, \dfrac{9}{2}\right)$

35. (a)

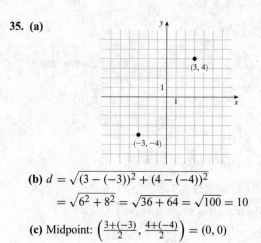

(b) $d = \sqrt{(3-(-3))^2 + (4-(-4))^2}$

$= \sqrt{6^2 + 8^2} = \sqrt{36+64} = \sqrt{100} = 10$

(c) Midpoint: $\left(\dfrac{3+(-3)}{2}, \dfrac{4+(-4)}{2}\right) = (0, 0)$

37. $d(A, B) = \sqrt{(1-5)^2 + (3-3)^2} = \sqrt{(-4)^2} = 4$.

$d(A, C) = \sqrt{(1-1)^2 + (3-(-3))^2} = \sqrt{(6)^2} = 6$. So the area is $4 \cdot 6 = 24$.

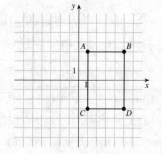

39. From the graph, the quadrilateral $ABCD$ has a pair of parallel sides, so $ABCD$ is a trapezoid. The area is $\frac{1}{2}(b_1 + b_2)h$. From the graph we see that

$b_1 = d(A, B) = \sqrt{(1-5)^2 + (0-0)^2} = \sqrt{4^2} = 4;$

$b_2 = d(C, D) = \sqrt{(4-2)^2 + (3-3)^2} = \sqrt{2^2} = 2;$ and

h is the difference in y-coordinates is $|3 - 0| = 3$. Thus the area of the trapezoid is $\frac{1}{2}(4+2)\,3 = 9$.

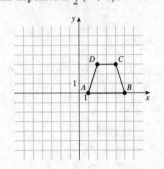

41. $d(0, A) = \sqrt{(6-0)^2 + (7-0)^2} = \sqrt{6^2 + 7^2} = \sqrt{36+49} = \sqrt{85}.$

$d(0, B) = \sqrt{(-5-0)^2 + (8-0)^2} = \sqrt{(-5)^2 + 8^2} = \sqrt{25+64} = \sqrt{89}.$

Thus point $A(6, 7)$ is closer to the origin.

43. $d(P, R) = \sqrt{(-1-3)^2 + (-1-1)^2} = \sqrt{(-4)^2 + (-2)^2} = \sqrt{16+4} = \sqrt{20} = 2\sqrt{5}.$

$d(Q, R) = \sqrt{(-1-(-1))^2 + (-1-3)^2} = \sqrt{0 + (-4)^2} = \sqrt{16} = 4.$ Thus point $Q(-1, 3)$ is closer to point R.

45. Since we do not know which pair are isosceles, we find the length of all three sides.

$d(A, B) = \sqrt{(-3-0)^2 + (-1-2)^2} = \sqrt{(-3)^2 + (-3)^2} = \sqrt{9+9} = \sqrt{18} = 3\sqrt{2}.$

$d(C, B) = \sqrt{(-3-(-4))^2 + (-1-3)^2} = \sqrt{1^2 + (-4)^2} = \sqrt{1+16} = \sqrt{17}.$

$d(A, C) = \sqrt{(0-(-4))^2 + (2-3)^2} = \sqrt{4^2 + (-1)^2} = \sqrt{16+1} = \sqrt{17}.$ So sides AC and CB have the same length.

47. (a) Here we have $A = (2, 2)$, $B = (3, -1)$, and $C = (-3, -3)$. So

$d(A, B) = \sqrt{(3-2)^2 + (-1-2)^2} = \sqrt{1^2 + (-3)^2} = \sqrt{1+9} = \sqrt{10};$

$d(C, B) = \sqrt{(3-(-3))^2 + (-1-(-3))^2} = \sqrt{6^2 + 2^2} = \sqrt{36+4} = \sqrt{40} = 2\sqrt{10};$

$d(A, C) = \sqrt{(-3-2)^2 + (-3-2)^2} = \sqrt{(-5)^2 + (-5)^2} = \sqrt{25+25} = \sqrt{50} = 5\sqrt{2}.$

Since $[d(A, B)]^2 + [d(C, B)]^2 = [d(A, C)]^2$, we conclude that the triangle is a right triangle.

(b) The area of the triangle is $\frac{1}{2} \cdot d(C, B) \cdot d(A, B) = \frac{1}{2} \cdot \sqrt{10} \cdot 2\sqrt{10} = 10.$

49. We show that all sides are the same length (its a rhombus) and then show that the diagonals are equal. Here we have
$A = (-2, 9)$, $B = (4, 6)$, $C = (1, 0)$, and $D = (-5, 3)$. So

$$d(A, B) = \sqrt{(4 - (-2))^2 + (6 - 9)^2} = \sqrt{6^2 + (-3)^2} = \sqrt{36 + 9} = \sqrt{45};$$

$$d(B, C) = \sqrt{(1 - 4)^2 + (0 - 6)^2} = \sqrt{(-3)^2 + (-6)^2} = \sqrt{9 + 36} = \sqrt{45};$$

$$d(C, D) = \sqrt{(-5 - 1)^2 + (3 - 0)^2} = \sqrt{(-6)^2 + (-3)^2} = \sqrt{36 + 9} = \sqrt{45};$$

$$d(D, A) = \sqrt{(-2 - (-5))^2 + (9 - 3)^2} = \sqrt{3^2 + 6^2} = \sqrt{9 + 36} = \sqrt{45}.$$ So the points form a

rhombus. Also $d(A, C) = \sqrt{(1 - (-2))^2 + (0 - 9)^2} = \sqrt{3^2 + (-9)^2} = \sqrt{9 + 81} = \sqrt{90} = 3\sqrt{10}$,

and $d(B, D) = \sqrt{(-5 - 4)^2 + (3 - 6)^2} = \sqrt{(-9)^2 + (-3)^2} = \sqrt{81 + 9} = \sqrt{90} = 3\sqrt{10}$. Since the diagonals are equal,
the rhombus is a square.

51. Let $P = (0, y)$ be such a point. Setting the distances equal we get

$$\sqrt{(0 - 5)^2 + (y - (-5))^2} = \sqrt{(0 - 1)^2 + (y - 1)^2} \Leftrightarrow$$

$$\sqrt{25 + y^2 + 10y + 25} = \sqrt{1 + y^2 - 2y + 1} \Rightarrow y^2 + 10y + 50 = y^2 - 2y + 2 \Leftrightarrow 12y = -48 \Leftrightarrow y = -4.$$ Thus, the point
is $P = (0, -4)$. Check:

$$\sqrt{(0 - 5)^2 + (-4 - (-5))^2} = \sqrt{(-5)^2 + 1^2} = \sqrt{25 + 1} = \sqrt{26};$$

$$\sqrt{(0 - 1)^2 + (-4 - 1)^2} = \sqrt{(-1)^2 + (-5)^2} = \sqrt{25 + 1} = \sqrt{26}.$$

53. We find the midpoint M of PQ, and then the midpoint of PM. Now $M = \left(\frac{-1+7}{2}, \frac{3+5}{2}\right) = (3, 4)$, and the midpoint of PM
is thus $\left(\frac{-1+3}{2}, \frac{3+4}{2}\right) = \left(1, \frac{7}{2}\right)$.

55. As indicated by Example 5, we must find a point $S(x_1, y_1)$ such that the midpoints of PR

and QS are the same. Thus $\left(\frac{4 + (-1)}{2}, \frac{2 + (-4)}{2}\right) = \left(\frac{x_1 + 1}{2}, \frac{y_1 + 1}{2}\right)$. Setting the

x-coordinates equal, we get $\frac{4 + (-1)}{2} = \frac{x_1 + 1}{2} \Leftrightarrow 4 - 1 = x_1 + 1 \Leftrightarrow x_1 = 2$. Setting the

y-coordinates equal, we get $\frac{2 + (-4)}{2} = \frac{y_1 + 1}{2} \Leftrightarrow 2 - 4 = y_1 + 1 \Leftrightarrow y_1 = -3$. Thus

$S = (2, -3)$.

57. (a)

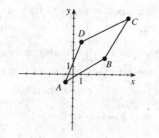

(b) The midpoint of AC is $\left(\frac{-2 + 7}{2}, \frac{-1 + 7}{2}\right) = \left(\frac{5}{2}, 3\right)$, the

midpoint of BD is $\left(\frac{4 + 1}{2}, \frac{2 + 4}{2}\right) = \left(\frac{5}{2}, 3\right)$.

(c) Since the they have the same midpoint, we conclude that the
diagonals bisect each other.

59. (a) $d(A, B) = \sqrt{3^2 + 4^2} = \sqrt{25} = 5$.

(b) We want the distances from $C = (4, 2)$ to $D = (11, 26)$. The walking distance is

$$|4 - 11| + |2 - 26| = 7 + 24 = 31 \text{ blocks. Straight-line distance is}$$

$$\sqrt{(4 - 11)^2 + (2 - 26)^2} = \sqrt{7^2 + 24^2} = \sqrt{625} = 25 \text{ blocks.}$$

(c) The two points are on the same avenue or the same street.

61. The midpoint of the line segment is $(66, 45)$. The pressure experienced by an ocean diver at a depth of 66 feet is 45 lb/in^2.

63. (a) The point $(3, 7)$ is reflected to the point $(-3, 7)$.

(b) The point (a, b) is reflected to the point $(-a, b)$.

(c) Since the point $(-a, b)$ is the reflection of (a, b), the point $(-4, -1)$ is the reflection of $(4, -1)$.

(d) $A = (3, 3)$, so $A' = (-3, 3)$; $B = (6, 1)$, so $B' = (-6, 1)$; and $C = (1, -4)$, so $C' = (-1, -4)$.

65. We need to find a point $S(x_1, y_1)$ such that $PQRS$ is a parallelogram. As indicated by Example 3, this will be the case if the diagonals PR and QS bisect each other. So the midpoints of PR and QS are the same. Thus

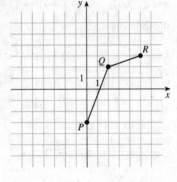

$$\left(\frac{0 + 5}{2}, \frac{-3 + 3}{2}\right) = \left(\frac{x_1 + 2}{2}, \frac{y_1 + 2}{2}\right). \text{ Setting the } x\text{-coordinates equal, we get}$$

$$\frac{0 + 5}{2} = \frac{x_1 + 2}{2} \Leftrightarrow 0 + 5 = x_1 + 2 \Leftrightarrow x_1 = 3.$$

Setting the y-coordinates equal, we get $\dfrac{-3 + 3}{2} = \dfrac{y_1 + 2}{2} \Leftrightarrow -3 + 3 = y_1 + 2 \Leftrightarrow$

$y_1 = -2$. Thus $S = (3, -2)$.

1.2 GRAPHS OF EQUATIONS IN TWO VARIABLES

1. If the point $(2, 3)$ is on the graph of an equation in x and y, then the equation is satisfied when we replace x by 2 and y by 3.

We check whether $2(3) \overset{?}{=} 2 + 1 \Leftrightarrow 6 \overset{?}{=} 3$. This is false, so the point $(2, 3)$ is not on the graph of the equation $2y = x + 1$.

To complete the table, we express y in terms of x: $2y = x + 1 \Leftrightarrow y = \frac{1}{2}(x + 1) = \frac{1}{2}x + \frac{1}{2}$.

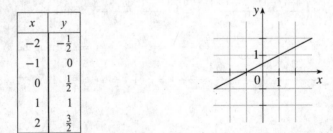

x	y
-2	$-\frac{1}{2}$
-1	0
0	$\frac{1}{2}$
1	1
2	$\frac{3}{2}$

3. To find the y-intercept(s) of the graph of an equation we set x equal to 0 in the equation and solve for y: $2y = 0 + 1 \Leftrightarrow$

$y = \frac{1}{2}$, so the y-intercept of $2y = x + 1$ is $\frac{1}{2}$.

5. (a) If a graph is symmetric with respect to the x-axis and (a, b) is on the graph, then $(a, -b)$ is also on the graph.

(b) If a graph is symmetric with respect to the y-axis and (a, b) is on the graph, then $(-a, b)$ is also on the graph.

(c) If a graph is symmetric about the origin and (a, b) is on the graph, then $(-a, -b)$ is also on the graph.

7. $y = 3 - 4x$. For the point $(0, 3)$: $3 \overset{?}{=} 3 - 4(0) \Leftrightarrow 3 = 3$. Yes. For $(4, 0)$: $0 \overset{?}{=} 3 - 4(4) \Leftrightarrow 0 \overset{?}{=} -13$. No. For $(1, -1)$:

$-1 \overset{?}{=} 3 - 4(1) \Leftrightarrow -1 \overset{?}{=} -1$. Yes.

So the points $(0, 3)$ and $(1, -1)$ are on the graph of this equation.

9. $x - 2y - 1 = 0$. For the point $(0, 0)$: $0 - 2(0) - 1 \overset{?}{=} 0 \Leftrightarrow -1 \overset{?}{=} 0$. No. For $(1, 0)$: $1 - 2(0) - 1 \overset{?}{=} 0 \Leftrightarrow -1 + 1 \overset{?}{=} 0$. Yes.

For $(-1, -1)$: $(-1) - 2(-1) - 1 \overset{?}{=} 0 \Leftrightarrow -1 + 2 - 1 \overset{?}{=} 0$. Yes.

So the points $(1, 0)$ and $(-1, -1)$ are on the graph of this equation.

11. $x^2 + 2xy + y^2 = 1$. For the point $(0, 1)$: $0^2 + 2(0)(1) + 1^2 \overset{?}{=} 1 \Leftrightarrow 1 \overset{?}{=} 1$. Yes. For $(2, -1)$: $2^2 + 2(2)(-1) + (-1)^2 \overset{?}{=} 1$

$\Leftrightarrow 4 - 4 + 1 \overset{?}{=} 1 \Leftrightarrow 1 = 1$. Yes. For $(-2, 3)$: $(-2)^2 + 2(-2)(3) + 3^2 \overset{?}{=} 1 \Leftrightarrow 4 - 12 + 9 \overset{?}{=} 1 \Leftrightarrow 1 \overset{?}{=} 1$. Yes.

So the points $(0, 1)$, $(2, -1)$, and $(-2, 3)$ are on the graph of this equation.

13. $y = -x$.

x	y
-4	4
-2	2
0	0
1	-1
2	-2
3	-3
4	-4

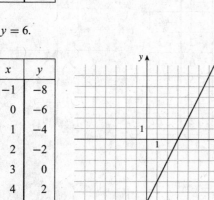

15. $y = x - 3$.

x	y
-4	-7
-2	-5
0	-3
2	-1
4	1

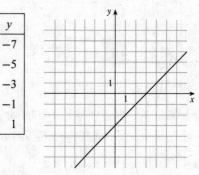

17. $2x - y = 6$.

x	y
-1	-8
0	-6
1	-4
2	-2
3	0
4	2
5	4

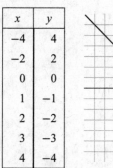

19. $y = 1 - x^2$.

x	y
-3	-8
-2	-3
-1	0
0	1
1	0
2	-3
3	-8

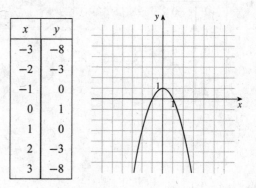

21. $y = 2x^2 - 1$.

x	y
-3	17
-2	8
-1	1
0	-1
1	1
2	8
4	17

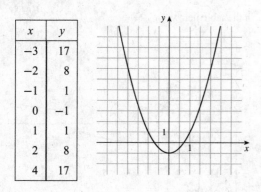

23. $9y = x^2$. To make a table, we rewrite the equation as

$$y = \tfrac{1}{9}x^2.$$

x	y
-9	9
-3	1
0	0
3	1
9	9

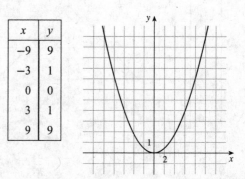

25. $x + y^2 = 4$.

x	y
-12	-4
-5	-3
0	-2
3	-1
4	0
3	1
0	2
-5	3
-12	4

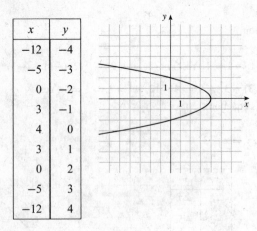

27. $y = \sqrt{x}$.

x	y
0	0
$\frac{1}{4}$	$\frac{1}{2}$
1	1
2	$\sqrt{2}$
4	2
9	3
16	4

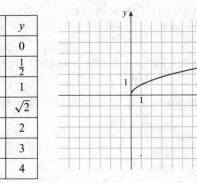

29. $y = \sqrt{4 - x^2}$. Since the radicand (the expression inside the square root) cannot be negative, we must have

$$4 - x^2 \geq 0 \Leftrightarrow x^2 \leq 4 \Leftrightarrow |x| \leq 2.$$

x	y
-2	0
-1	$\sqrt{3}$
0	4
1	$\sqrt{3}$
2	0

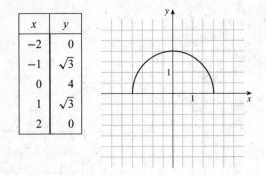

31. $y = -|x|$.

x	y
-6	-6
-4	-4
-2	-2
0	0
2	-2
4	-4
6	-6

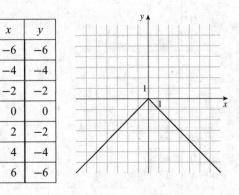

33. $y = 4 - |x|$.

x	y
-6	-2
-4	0
-2	2
0	4
2	2
4	0
6	-2

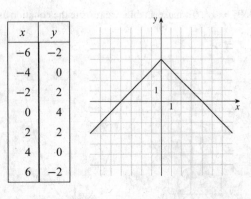

35. $x = y^3$. Since $x = y^3$ is solved for x in terms of y, we insert values for y and find the corresponding values of x in the table below.

x	y
-27	-3
-8	-2
-1	-1
0	0
1	1
8	2
27	3

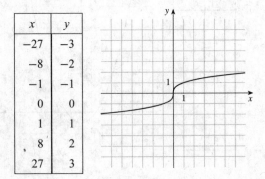

37. $y = x^4$.

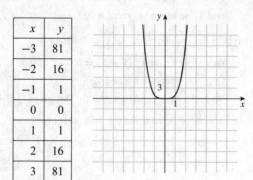

x	y
-3	81
-2	16
-1	1
0	0
1	1
2	16
3	81

39. $y = 0.01x^3 - x^2 + 5$; $[-100, 150]$ by $[-2000, 2000]$

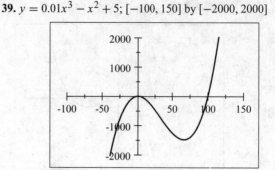

41. $y = \sqrt{12x - 17}$; $[-1, 10]$ by $[-1, 20]$

43. $y = \dfrac{x}{x^2 + 25}$; $[-50, 50]$ by $[-0.2, 0.2]$

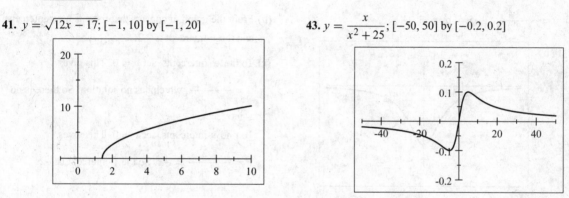

45. $y = x - 3$. To find x-intercepts, set $y = 0$. This gives $0 = x - 3 \Leftrightarrow x = 3$, so the x-intercept is 3.
To find y-intercepts, set $x = 0$. This gives $y = 0 - 3 \Leftrightarrow y = -3$, so the y-intercept is -3.

47. $y = x^2 - 9$. To find x-intercepts, set $y = 0$. This gives $0 = x^2 - 9 \Leftrightarrow x^2 = 9 \Rightarrow x = \pm 3$, so the x-intercepts are ± 3.
To find y-intercepts, set $x = 0$. This gives $y = (0)^2 - 9 \Leftrightarrow y = -9$, so the y-intercept is -9.

49. $y - 2xy + 2x = 1$. To find x-intercepts, set $y = 0$. This gives $0 - 2x(0) + 2x = 1 \Leftrightarrow 2x = 1 \Leftrightarrow x = \dfrac{1}{2}$, so the x-intercept is $\dfrac{1}{2}$.
To find y-intercepts, set $x = 0$. This gives $y - 2(0)y + 2(0) = 1 \Leftrightarrow y = 1$, so the y-intercept is 1.

51. $y = \sqrt{x + 1}$. To find x-intercepts, set $y = 0$. This gives $0 = \sqrt{x + 1} \Leftrightarrow 0 = x + 1 \Leftrightarrow x = -1$, so the x-intercept is -1.
To find y-intercepts, set $x = 0$. This gives $y = \sqrt{0 + 1} \Leftrightarrow y = 1$, so the y-intercept is 1.

53. $25x^2 + 4y^2 = 100$. To find x-intercepts, set $y = 0$. This gives $25x^2 + 4(0)^2 = 100 \Leftrightarrow x^2 = 4 \Leftrightarrow x = \pm 2$, so the x-intercepts are -2 and 2.
To find y-intercepts, set $x = 0$. This gives $25(0)^2 + 4y^2 = 100 \Leftrightarrow y^2 = 25 \Leftrightarrow y = \pm 5$, so the y-intercepts are -5 and 5.

55. $y = 4x - x^2$. To find x-intercepts, set $y = 0$. This gives $0 = 4x - x^2 \Leftrightarrow 0 = x(4 - x) \Leftrightarrow 0 = x$ or $x = 4$, so the x-intercepts are 0 and 4.
To find y-intercepts, set $x = 0$. This gives $y = 4(0) - 0^2 \Leftrightarrow y = 0$, so the y-intercept is 0.

57. $x^4 + y^2 - xy = 16$. To find x-intercepts, set $y = 0$. This gives $x^4 + 0^2 - x(0) = 16 \Leftrightarrow x^4 = 16 \Leftrightarrow x = \pm 2$. So the x-intercepts are -2 and 2.
To find y-intercepts, set $x = 0$. This gives $0^4 + y^2 - (0)y = 16 \Leftrightarrow y^2 = 16 \Leftrightarrow y = \pm 4$. So the y-intercepts are -4 and 4.

59. (a) $y = x^3 - x^2$; $[-2, 2]$ by $[-1, 1]$

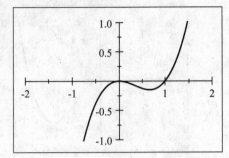

(b) From the graph, it appears that the x-intercepts are 0 and 1 and the y-intercept is 0.

(c) To find x-intercepts, set $y = 0$. This gives
$0 = x^3 - x^2 \Leftrightarrow x^2 (x - 1) = 0 \Leftrightarrow x = 0$ or 1. So the x-intercepts are 0 and 1.
To find y-intercepts, set $x = 0$. This gives
$y = 0^3 - 0^2 = 0$. So the y-intercept is 0.

61. (a) $y = -\dfrac{2}{x^2 + 1}$; $[-5, 5]$ by $[-3, 1]$

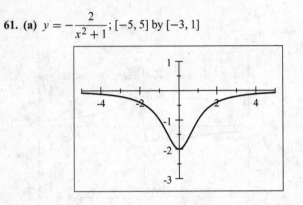

(b) From the graph, it appears that there is no x-intercept and the y-intercept is -2.

(c) To find x-intercepts, set $y = 0$. This gives
$0 = -\dfrac{2}{x^2 + 1}$, which has no solution. So there is no x-intercept.
To find y-intercepts, set $x = 0$. This gives
$y = -\dfrac{2}{0^2 + 1} = -2$. So the y-intercept is -2.

63. (a) $y = \sqrt[3]{x}$; $[-5, 5]$ by $[-2, 2]$

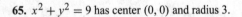
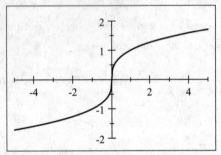

(b) From the graph, it appears that and the x- and y-intercepts are 0.

(c) To find x-intercepts, set $y = 0$. This gives $0 = \sqrt[3]{x}$ $\Leftrightarrow x = 0$. So the x-intercept is 0.
To find y-intercepts, set $x = 0$. This gives
$y = \sqrt[3]{0} = 0$. So the y-intercept is 0.

65. $x^2 + y^2 = 9$ has center $(0, 0)$ and radius 3.

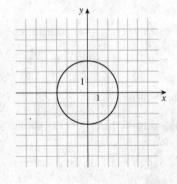

67. $(x - 3)^2 + y^2 = 16$ has center $(3, 0)$ and radius 4.

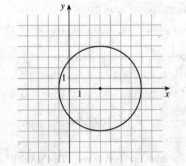

69. $(x + 3)^2 + (y - 4)^2 = 25$ has center $(-3, 4)$ and radius 5.

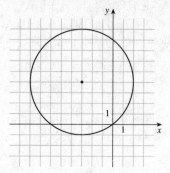

71. Using $h = 2$, $k = -1$, and $r = 3$, we get $(x - 2)^2 + (y - (-1))^2 = 3^2 \Leftrightarrow (x - 2)^2 + (y + 1)^2 = 9$.

73. The equation of a circle centered at the origin is $x^2 + y^2 = r^2$. Using the point $(4, 7)$ we solve for r^2. This gives $(4)^2 + (7)^2 = r^2 \Leftrightarrow 16 + 49 = 65 = r^2$. Thus, the equation of the circle is $x^2 + y^2 = 65$.

75. The center is at the midpoint of the line segment, which is $\left(\dfrac{-1 + 5}{2}, \dfrac{1 + 9}{2} \right) = (2, 5)$. The radius is one half the diameter, so $r = \frac{1}{2}\sqrt{(-1 - 5)^2 + (1 - 9)^2} = \frac{1}{2}\sqrt{36 + 64} = \frac{1}{2}\sqrt{100} = 5$. Thus, an equation of the circle is $(x - 2)^2 + (y - 5)^2 = 5^2$ $\Leftrightarrow (x - 2)^2 + (y - 5)^2 = 25$.

77. Since the circle is tangent to the x-axis, it must contain the point $(7, 0)$, so the radius is the change in the y-coordinates. That is, $r = |-3 - 0| = 3$. So the equation of the circle is $(x - 7)^2 + (y - (-3))^2 = 3^2$, which is $(x - 7)^2 + (y + 3)^2 = 9$.

79. From the figure, the center of the circle is at $(-2, 2)$. The radius is the change in the y-coordinates, so $r = |2 - 0| = 2$. Thus the equation of the circle is $(x - (-2))^2 + (y - 2)^2 = 2^2$, which is $(x + 2)^2 + (y - 2)^2 = 4$.

81. Completing the square gives $x^2 + y^2 - 2x + 4y + 1 = 0 \Leftrightarrow x^2 - 2x + \left(\dfrac{-2}{2}\right)^2 + y^2 + 4y + \left(\dfrac{4}{2}\right)^2 = -1 + \left(\dfrac{-2}{2}\right)^2 + \left(\dfrac{4}{2}\right)^2$ $\Leftrightarrow x^2 - 2x + 1 + y^2 + 4y + 4 = -1 + 1 + 4 \Leftrightarrow (x - 1)^2 + (y + 2)^2 = 4$. Thus, the center is $(1, -2)$, and the radius is 2.

83. Completing the square gives $x^2 + y^2 - 4x + 10y + 13 = 0 \Leftrightarrow x^2 - 4x + \left(\dfrac{-4}{2}\right)^2 + y^2 + 10y + \left(\dfrac{10}{2}\right)^2 = -13 + \left(\dfrac{4}{2}\right)^2 + \left(\dfrac{10}{2}\right)^2$ $\Leftrightarrow x^2 - 4x + 4 + y^2 + 10y + 25 = -13 + 4 + 25 \Leftrightarrow (x - 2)^2 + (y + 5)^2 = 16$.
Thus, the center is $(2, -5)$, and the radius is 4.

85. Completing the square gives $x^2 + y^2 + x = 0 \Leftrightarrow x^2 + x + \left(\frac{1}{2}\right)^2 + y^2 = \left(\frac{1}{2}\right)^2 \Leftrightarrow x^2 + x + \frac{1}{4} + y^2 = \frac{1}{4} \Leftrightarrow$ $\left(x + \frac{1}{2}\right)^2 + y^2 = \frac{1}{4}$. Thus, the circle has center $\left(-\frac{1}{2}, 0\right)$ and radius $\frac{1}{2}$.

87. Completing the square gives $x^2 + y^2 - \frac{1}{2}x + \frac{1}{2}y = \frac{1}{8} \Leftrightarrow x^2 - \frac{1}{2}x + \left(\dfrac{-1/2}{2}\right)^2 + y^2 + \frac{1}{2}y + \left(\dfrac{1/2}{2}\right)^2 = \frac{1}{8} + \left(\dfrac{-1/2}{2}\right)^2 + \left(\dfrac{1/2}{2}\right)^2$ $\Leftrightarrow x^2 - \frac{1}{2}x + \frac{1}{16} + y^2 + \frac{1}{2}y + \frac{1}{16} = \frac{1}{8} + \frac{1}{16} + \frac{1}{16} = \frac{2}{8} = \frac{1}{4} \Leftrightarrow \left(x - \frac{1}{4}\right)^2 + \left(y + \frac{1}{4}\right)^2 = \frac{1}{4}$. Thus, the circle has center $\left(\frac{1}{4}, -\frac{1}{4}\right)$ and radius $\frac{1}{2}$.

89. Completing the square gives $x^2 + y^2 + 4x - 10y = 21 \Leftrightarrow$

$$x^2 + 4x + \left(\frac{4}{2}\right)^2 + y^2 - 10y + \left(\frac{-10}{2}\right)^2 = 21 + \left(\frac{4}{2}\right)^2 +$$

$$\left(\frac{-10}{2}\right)^2 \Leftrightarrow (x+2)^2 + (y-5)^2 = 21 + 4 + 25 = 50.$$

Thus, the circle has center $(-2, 5)$ and radius $\sqrt{50} = 5\sqrt{2}$.

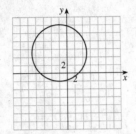

91. Completing the square gives $x^2 + y^2 + 6x - 12y + 45 = 0$
$\Leftrightarrow (x+3)^2 + (y-6)^2 = -45 + 9 + 36 = 0$. Thus, the
center is $(-3, 6)$, and the radius is 0. This is a degenerate
circle whose graph consists only of the point $(-3, 6)$.

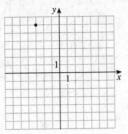

93. x-axis symmetry: $(-y) = x^4 + x^2 \Leftrightarrow y = -x^4 - x^2$, which is not the same as $y = x^4 + x^2$, so the graph is not symmetric
with respect to the x-axis.

y-axis symmetry: $y = (-x)^4 + (-x)^2 = x^4 + x^2$, so the graph is symmetric with respect to the y-axis.

Origin symmetry: $(-y) = (-x)^4 + (-x)^2 \Leftrightarrow -y = x^4 + x^2$, which is not the same as $y = x^4 + x^2$, so the graph is not
symmetric with respect to the origin.

95. x-axis symmetry: $(-y) = x^3 + 10x \Leftrightarrow y = -x^3 - 10x$, which is not the same as $y = x^3 + 10x$, so the graph is not
symmetric with respect to the x-axis.

y-axis symmetry: $y = (-x)^3 + 10(-x) \Leftrightarrow y = -x^3 - 10x$, which is not the same as $y = x^3 + 10x$, so the graph is not
symmetric with respect to the y-axis.

Origin symmetry: $(-y) = (-x)^3 + 10(-x) \Leftrightarrow -y = -x^3 - 10x \Leftrightarrow y = x^3 + 10x$, so the graph is symmetric with respect
to the origin.

97. x-axis symmetry: $x^4(-y)^4 + x^2(-y)^2 = 1 \Leftrightarrow x^4y^4 + x^2y^2 = 1$, so the graph is symmetric with respect to the x-axis.

y-axis symmetry: $(-x)^4 y^4 + (-x)^2 y^2 = 1 \Leftrightarrow x^4y^4 + x^2y^2 = 1$, so the graph is symmetric with respect to the y-axis.

Origin symmetry: $(-x)^4(-y)^4 + (-x)^2(-y)^2 = 1 \Leftrightarrow x^4y^4 + x^2y^2 = 1$, so the graph is symmetric with respect to the
origin.

99. Symmetric with respect to the y-axis.

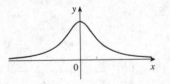

101. Symmetric with respect to the origin.

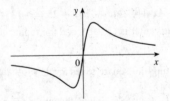

103. (a) In 1980 inflation was 14%; in 1990, it was 6%; in 1999, it was 2%.

(b) Inflation exceeded 6% from 1975 to 1976 and from 1978 to 1982.

(c) Between 1980 and 1985 the inflation rate generally decreased. Between 1987 and 1992 the inflation rate generally
increased.

(d) The highest rate was about 14% in 1980. The lowest was about 1% in 2002.

105. Completing the square gives $x^2 + y^2 + ax + by + c = 0 \Leftrightarrow x^2 + ax + \left(\dfrac{a}{2}\right)^2 + y^2 + by + \left(\dfrac{b}{2}\right)^2 = -c + \left(\dfrac{a}{2}\right)^2 + \left(\dfrac{b}{2}\right)^2$

$\Leftrightarrow \left(x + \dfrac{a}{2}\right)^2 + \left(y + \dfrac{b}{2}\right)^2 = -c + \dfrac{a^2 + b^2}{4}$. This equation represents a circle only when $-c + \dfrac{a^2 + b^2}{4} > 0$. This

equation represents a point when $-c + \dfrac{a^2 + b^2}{4} = 0$, and this equation represents the empty set when $-c + \dfrac{a^2 + b^2}{4} < 0$.

When the equation represents a circle, the center is $\left(-\dfrac{a}{2}, -\dfrac{b}{2}\right)$, and the radius is $\sqrt{-c + \dfrac{a^2 + b^2}{4}} = \frac{1}{2}\sqrt{a^2 + b^2 - 4ac}$.

107. (a) Symmetric about the x-axis.   **(b)** Symmetric about the y-axis.   **(c)** Symmetric about the origin.

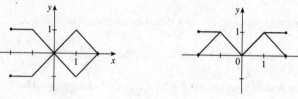
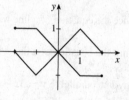

1.3 LINES

1. We find the "steepness" or slope of a line passing through two points by dividing the difference in the y-coordinates of these points by the difference in the x-coordinates. So the line passing through the points $(0, 1)$ and $(2, 5)$ has slope $\dfrac{5 - 1}{2 - 0} = 2$.

3. The point-slope form of the equation of the line with slope 3 passing through the point $(1, 2)$ is $y - 2 = 3(x - 1)$.

5. The slope of a vertical line is undefined. The equation of the vertical line passing through $(2, 3)$ is $x = 2$.

7. $m = \dfrac{y_2 - y_1}{x_2 - x_1} = \dfrac{4 - 0}{0 - 3} = \dfrac{4}{-3} = -\dfrac{4}{3}$ **9.** $m = \dfrac{y_2 - y_1}{x_2 - x_1} = \dfrac{2 - 4}{5 - (-1)} = \dfrac{-2}{6} = -\dfrac{1}{3}$

11. $m = \dfrac{y_2 - y_1}{x_2 - x_1} = \dfrac{4 - 3}{2 - 4} = \dfrac{1}{-2} = -\dfrac{1}{2}$ **13.** $m = \dfrac{y_2 - y_1}{x_2 - x_1} = \dfrac{6 - (-3)}{-1 - 1} = \dfrac{9}{-2} = -\dfrac{9}{2}$

15. For ℓ_1, we find two points, $(-1, 2)$ and $(0, 0)$ that lie on the line. Thus the slope of ℓ_1 is $m = \dfrac{y_2 - y_1}{x_2 - x_1} = \dfrac{2 - 0}{-1 - 0} = -2$.

For ℓ_2, we find two points $(0, 2)$ and $(2, 3)$. Thus, the slope of ℓ_2 is $m = \dfrac{y_2 - y_1}{x_2 - x_1} = \dfrac{3 - 2}{2 - 0} = \dfrac{1}{2}$. For ℓ_3 we find the points

$(2, -2)$ and $(3, 1)$. Thus, the slope of ℓ_3 is $m = \dfrac{y_2 - y_1}{x_2 - x_1} = \dfrac{1 - (-2)}{3 - 2} = 3$. For ℓ_4, we find the points $(-2, -1)$ and

$(2, -2)$. Thus, the slope of ℓ_4 is $m = \dfrac{y_2 - y_1}{x_2 - x_1} = \dfrac{-2 - (-1)}{2 - (-2)} = \dfrac{-1}{4} = -\dfrac{1}{4}$.

17. First we find two points $(0, 4)$ and $(4, 0)$ that lie on the line. So the slope is $m = \dfrac{0 - 4}{4 - 0} = -1$. Since the y-intercept is 4, the equation of the line is $y = mx + b = -1x + 4$. So $y = -x + 4$, or $x + y - 4 = 0$.

19. We choose the two intercepts as points, $(0, -3)$ and $(2, 0)$. So the slope is $m = \dfrac{0 - (-3)}{2 - 0} = \dfrac{3}{2}$. Since the y-intercept is -3, the equation of the line is $y = mx + b = \frac{3}{2}x - 3$, or $3x - 2y - 6 = 0$.

21. Using $y = mx + b$, we have $y = 3x + (-2)$ or $3x - y - 2 = 0$.

23. Using the equation $y - y_1 = m(x - x_1)$, we get $y - 3 = 5(x - 2) \Leftrightarrow -5x + y = -7 \Leftrightarrow 5x - y - 7 = 0$.

25. Using the equation $y - y_1 = m(x - x_1)$, we get $y - 7 = \frac{2}{3}(x - 1) \Leftrightarrow 3y - 21 = 2x - 2 \Leftrightarrow -2x + 3y = 19 \Leftrightarrow 2x - 3y + 19 = 0$.

27. First we find the slope, which is $m = \dfrac{y_2 - y_1}{x_2 - x_1} = \dfrac{6 - 1}{1 - 2} = \dfrac{5}{-1} = -5$. Substituting into $y - y_1 = m(x - x_1)$, we get $y - 6 = -5(x - 1) \Leftrightarrow y - 6 = -5x + 5 \Leftrightarrow 5x + y - 11 = 0$.

29. We are given two points, $(-2, 5)$ and $(-1, -3)$. Thus, the slope is $m = \dfrac{y_2 - y_1}{x_2 - x_1} = \dfrac{-3 - 5}{-1 - (-2)} = \dfrac{-8}{1} = -8$. Substituting into $y - y_1 = m(x - x_1)$, we get $y - 5 = -8[x - (-2)] \Leftrightarrow y = -8x - 11$ or $8x + y + 11 = 0$.

31. We are given two points, $(1, 0)$ and $(0, -3)$. Thus, the slope is $m = \dfrac{y_2 - y_1}{x_2 - x_1} = \dfrac{-3 - 0}{0 - 1} = \dfrac{-3}{-1} = 3$. Using the y-intercept, we have $y = 3x + (-3)$ or $y = 3x - 3$ or $3x - y - 3 = 0$.

33. Since the equation of a line with slope 0 passing through (a, b) is $y = b$, the equation of this line is $y = 3$.

35. Since the equation of a line with undefined slope passing through (a, b) is $x = a$, the equation of this line is $x = 2$.

37. Any line parallel to $y = 3x - 5$ has slope 3. The desired line passes through $(1, 2)$, so substituting into $y - y_1 = m(x - x_1)$, we get $y - 2 = 3(x - 1) \Leftrightarrow y = 3x - 1$ or $3x - y - 1 = 0$.

39. Since the equation of a horizontal line passing through (a, b) is $y = b$, the equation of the horizontal line passing through $(4, 5)$ is $y = 5$.

41. Since $x + 2y = 6 \Leftrightarrow 2y = -x + 6 \Leftrightarrow y = -\frac{1}{2}x + 3$, the slope of this line is $-\frac{1}{2}$. Thus, the line we seek is given by $y - (-6) = -\frac{1}{2}(x - 1) \Leftrightarrow 2y + 12 = -x + 1 \Leftrightarrow x + 2y + 11 = 0$.

43. Any line parallel to $x = 5$ has undefined slope and an equation of the form $x = a$. Thus, an equation of the line is $x = -1$.

45. First find the slope of $2x + 5y + 8 = 0$. This gives $2x + 5y + 8 = 0 \Leftrightarrow 5y = -2x - 8 \Leftrightarrow y = -\frac{2}{5}x - \frac{8}{5}$. So the slope of the line that is perpendicular to $2x + 5y + 8 = 0$ is $m = -\dfrac{1}{-2/5} = \dfrac{5}{2}$. The equation of the line we seek is $y - (-2) = \frac{5}{2}(x - (-1)) \Leftrightarrow 2y + 4 = 5x + 5 \Leftrightarrow 5x - 2y + 1 = 0$.

47. First find the slope of the line passing through $(2, 5)$ and $(-2, 1)$. This gives $m = \dfrac{1 - 5}{-2 - 2} = \dfrac{-4}{-4} = 1$, and so the equation of the line we seek is $y - 7 = 1(x - 1) \Leftrightarrow x - y + 6 = 0$.

49. (a)

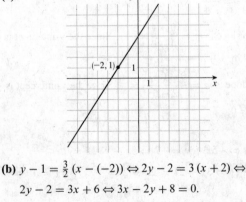

(b) $y - 1 = \frac{3}{2}(x - (-2)) \Leftrightarrow 2y - 2 = 3(x + 2) \Leftrightarrow 2y - 2 = 3x + 6 \Leftrightarrow 3x - 2y + 8 = 0$.

51.

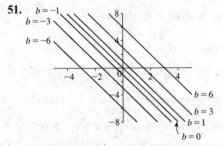

$y = -2x + b$, $b = 0, \pm 1, \pm 3, \pm 6$. They have the same slope, so they are parallel.

53.

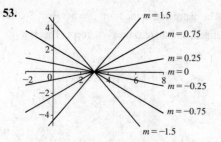

$y = m(x - 3)$, $m = 0, \pm 0.25, \pm 0.75, \pm 1.5$. Each of the lines contains the point $(3, 0)$ because the point $(3, 0)$ satisfies each equation $y = m(x - 3)$. Since $(3, 0)$ is on the x-axis, we could also say that they all have the same x-intercept.

55. $y = 3 - x = -x + 3$. So the slope is -1 and the y-intercept is 3.

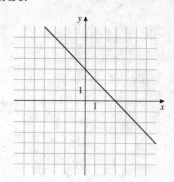

57. $-2x + y = 7 \Leftrightarrow y = 2x + 7$. So the slope is 2 and the y-intercept is 7.

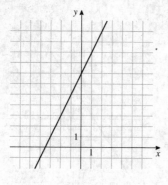

59. $x + 3y = 0 \Leftrightarrow 3y = -x \Leftrightarrow y = -\frac{1}{3}x$. So the slope is $-\frac{1}{3}$ and the y-intercept is 0.

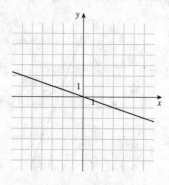

61. $4x + 5y = 10 \Leftrightarrow 5y = -4x + 10 \Leftrightarrow y = -\frac{4}{5}x + 2$. So the slope is $-\frac{4}{5}$ and the y-intercept is 2.

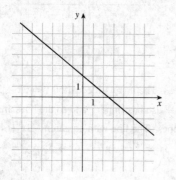

63. $-3x - 5y + 30 = 0 \Leftrightarrow -5y = 3x - 30 \Leftrightarrow y = -\frac{3}{5}x + 6$. So the slope is $-\frac{3}{5}$ and the y-intercept is 6.

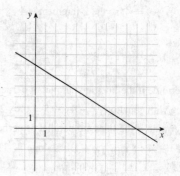

65. $y = 4$ can also be expressed as $y = 0x + 4$. So the slope is 0 and the y-intercept is 4.

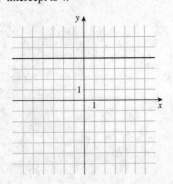

67. $x = 3$ cannot be expressed in the form $y = mx + b$. So the slope is undefined, and there is no y-intercept. This is a vertical line.

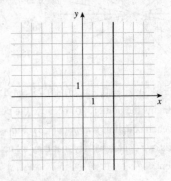

69. $5x + 2y - 10 = 0$. To find x-intercepts, we set $y = 0$ and solve for x: $5x + 2(0) - 10 = 0 \Leftrightarrow 5x = 10 \Leftrightarrow x = 2$, so the x-intercept is 2.
To find y-intercepts, we set $x = 0$ and solve for y:
$5(0) + 2y - 10 = 0 \Leftrightarrow 2y = 10 \Leftrightarrow y = 5$, so the y-intercept is 5.

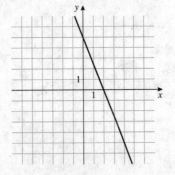

71. $\frac{1}{2}x - \frac{1}{3}y + 1 = 0$. To find x-intercepts, we set $y = 0$ and solve for x: $\frac{1}{2}x - \frac{1}{3}(0) + 1 = 0 \Leftrightarrow \frac{1}{2}x = -1 \Leftrightarrow x = -2$, so the x-intercept is -2.
To find y-intercepts, we set $x = 0$ and solve for y:
$\frac{1}{2}(0) - \frac{1}{3}y + 1 = 0 \Leftrightarrow \frac{1}{3}y = 1 \Leftrightarrow y = 3$, so the y-intercept is 3.

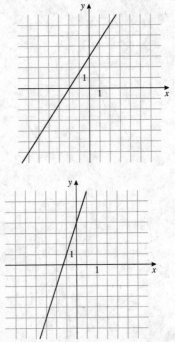

73. $y = 6x + 4$. To find x-intercepts, we set $y = 0$ and solve for x: $0 = 6x + 4 \Leftrightarrow 6x = -4 \Leftrightarrow x = -\frac{2}{3}$, so the x-intercept is $-\frac{2}{3}$.
To find y-intercepts, we set $x = 0$ and solve for y:
$y = 6(0) + 4 = 4$, so the y-intercept is 4.

75. To determine if the lines are parallel or perpendicular, we find their slopes. The line with equation $y = 2x + 3$ has slope 2. The line with equation $2y - 4x - 5 = 0 \Leftrightarrow 2y = 4x + 5 \Leftrightarrow y = 2x + \frac{5}{2}$ also has slope 2, and so the lines are parallel.

77. To determine if the lines are parallel or perpendicular, we find their slopes. The line with equation $-3x + 4y = 4 \Leftrightarrow 4y = 3x + 4 \Leftrightarrow y = \frac{3}{4}x + 1$ has slope $\frac{3}{4}$. The line with equation $4x + 3y = 5 \Leftrightarrow 3y = -4x + 5 \Leftrightarrow y = -\frac{4}{3}x + \frac{5}{3}$ has slope $-\frac{4}{3} = -\frac{1}{3/4}$, and so the lines are perpendicular.

79. To determine if the lines are parallel or perpendicular, we find their slopes. The line with equation $7x - 3y = 2 \Leftrightarrow$ $3y = 7x - 2 \Leftrightarrow y = \frac{7}{3}x - \frac{2}{3}$ has slope $\frac{7}{3}$. The line with equation $9y + 21x = 1 \Leftrightarrow 9y = -21x - 1 \Leftrightarrow y = -\frac{7}{3} - \frac{1}{9}$ has slope $-\frac{7}{3} \neq -\frac{1}{7/3}$, and so the lines are neither parallel nor perpendicular.

81. (a) We plot the points $P_1\ (0, 10)$, $P_2\ (2, 16)$, $P_3\ (4, 22)$, and $P_4\ (6, 28)$.

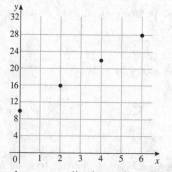

The points do appear to lie along a line. We find the slopes of the lines joining the first point with each of the other points: $m_2 = \dfrac{y_2 - y_1}{x_2 - x_1} = \dfrac{16 - 10}{2 - 0} = 3$, $m_3 = \frac{22-10}{4-0} = 3$, and $m_4 = \frac{28-10}{6-0} = 3$. Therefore, the points are collinear.

(b) From part (a), the slope of the line is $m = 3$. Using the formula $y - y_1 = m\ (x - x_1)$, we find its equation: $y - 10 = 3\ (x - 0) \Leftrightarrow y = 3x + 10$.

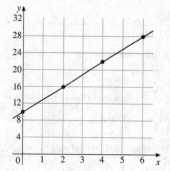

83. We first plot the points to find the pairs of points that determine each side. Next we find the slopes of opposite sides. The slope of AB is $\dfrac{4 - 1}{7 - 1} = \dfrac{3}{6} = \dfrac{1}{2}$, and the slope of DC is $\dfrac{10 - 7}{5 - (-1)} = \dfrac{3}{6} = \dfrac{1}{2}$. Since these slope are equal, these two sides are parallel. The slope of AD is $\dfrac{7 - 1}{-1 - 1} = \dfrac{6}{-2} = -3$, and the slope of BC is $\dfrac{10 - 4}{5 - 7} = \dfrac{6}{-2} = -3$. Since these slope are equal, these two sides are parallel. Hence $ABCD$ is a parallelogram.

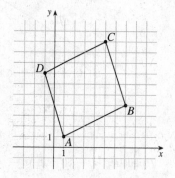

85. We first plot the points to find the pairs of points that determine each side. Next we find the slopes of opposite sides. The slope of AB is $\dfrac{3 - 1}{11 - 1} = \dfrac{2}{10} = \dfrac{1}{5}$ and the slope of DC is $\dfrac{6 - 8}{0 - 10} = \dfrac{-2}{-10} = \dfrac{1}{5}$. Since these slope are equal, these two sides are parallel. Slope of AD is $\dfrac{6 - 1}{0 - 1} = \dfrac{5}{-1} = -5$, and the slope of BC is $\dfrac{3 - 8}{11 - 10} = \dfrac{-5}{1} = -5$. Since these slope are equal, these two sides are parallel. Since (slope of AB) $\times$ (slope of AD) $= \frac{1}{5} \times (-5) = -1$, the first two sides are each perpendicular to the second two sides. So the sides form a rectangle.

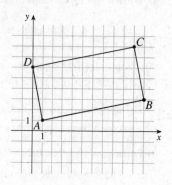

87. We need the slope and the midpoint of the line AB. The midpoint of AB is $\left(\dfrac{1+7}{2}, \dfrac{4-2}{2}\right) = (4, 1)$, and the slope of

AB is $m = \dfrac{-2-4}{7-1} = \dfrac{-6}{6} = -1$. The slope of the perpendicular bisector will have slope $\dfrac{-1}{m} = \dfrac{-1}{-1} = 1$. Using the

point-slope form, the equation of the perpendicular bisector is $y - 1 = 1\,(x - 4)$ or $x - y - 3 = 0$.

89. (a) We start with the two points $(a, 0)$ and $(0, b)$. The slope of the line that contains them is $\dfrac{b-0}{0-a} = -\dfrac{b}{a}$. So the equation

of the line containing them is $y = -\dfrac{b}{a}x + b$ (using the slope-intercept form). Dividing by b (since $b \neq 0$) gives

$\dfrac{y}{b} = -\dfrac{x}{a} + 1 \Leftrightarrow \dfrac{x}{a} + \dfrac{y}{b} = 1$.

(b) Setting $a = 6$ and $b = -8$, we get $\dfrac{x}{6} + \dfrac{y}{-8} = 1 \Leftrightarrow 4x - 3y = 24 \Leftrightarrow 4x - 3y - 24 = 0$.

91. Let h be the change in your horizontal distance, in feet. Then $-\dfrac{6}{100} = \dfrac{-1000}{h} \Leftrightarrow h = \dfrac{100,000}{6} \approx 16{,}667$. So the change

in your horizontal distance is about 16,667 feet.

93. (a) The slope is $0.0417D = 0.0417\,(200) = 8.34$. It represents the increase in dosage for each one-year increase in the

child's age.

(b) When $a = 0$, $c = 8.34\,(0 + 1) = 8.34$ mg.

95. (a)

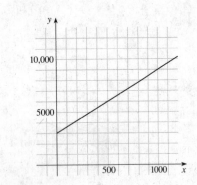

(b) The slope is the cost per toaster oven, \$6. The y-intercept, \$3000, is
the monthly fixed cost — the cost that is incurred no matter how
many toaster ovens are produced.

97. (a) Using n in place of x and t in place of y, we find that the slope is $\dfrac{t_2 - t_1}{n_2 - n_1} = \dfrac{80 - 70}{168 - 120} = \dfrac{10}{48} = \dfrac{5}{24}$. So the linear

equation is $t - 80 = \dfrac{5}{24}\,(n - 168) \Leftrightarrow t - 80 = \dfrac{5}{24}n - 35 \Leftrightarrow t = \dfrac{5}{24}n + 45$.

(b) When $n = 150$, the temperature is approximately given by $t = \dfrac{5}{24}\,(150) + 45 = 76.25°$ F $\approx 76°$ F.

99. (a) We are given $\dfrac{\text{change in pressure}}{10\ \text{feet change in depth}} = \dfrac{4.34}{10} = 0.434$. Using P for

pressure and d for depth, and using the point $P = 15$ when $d = 0$, we

have $P - 15 = 0.434\,(d - 0) \Leftrightarrow P = 0.434d + 15$.

(b)

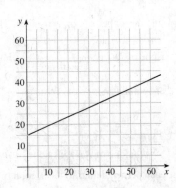

(c) The slope represents the increase in pressure per foot of descent. The
y-intercept represents the pressure at the surface.

(d) When $P = 100$, then $100 = 0.434d + 15 \Leftrightarrow 0.434d = 85 \Leftrightarrow$
$d = 195.9$ ft. Thus the pressure is 100 lb/in^3 at a depth of
approximately 196 ft.

101. (a) Using d in place of x and C in place of y, we find the slope to be

$$\frac{C_2 - C_1}{d_2 - d_1} = \frac{460 - 380}{800 - 480} = \frac{80}{320} = \frac{1}{4}.$$ So the linear equation is

$$C - 460 = \frac{1}{4}(d - 800) \Leftrightarrow C - 460 = \frac{1}{4}d - 200 \Leftrightarrow C = \frac{1}{4}d + 260.$$

(b) Substituting $d = 1500$ we get $C = \frac{1}{4}(1500) + 260 = 635$. Thus, the cost of driving 1500 miles is $635.

(d) The y-intercept represents the fixed cost, $260.

(e) It is a suitable model because you have fixed monthly costs such as insurance and car payments, as well as costs that occur as you drive, such as gasoline, oil, tires, etc., and the cost of these for each additional mile driven is a constant.

(c)

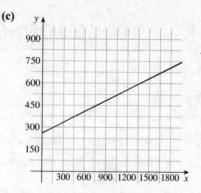

The slope of the line represents the cost per mile, $0.25.

103. (a) We plot the points $P_0\,(0, 212.0)$, $P_1\,(1, 210.2)$, $P_2\,(2, 208.4)$, $P_3\,(3, 206.6)$, and $P_4\,(4, 204.8)$. The points do appear to lie along a line. We find the slopes of the lines joining the first point with each of the other points:

$m_1 = \frac{T_1 - T_0}{h_1 - h_0} = \frac{210.2 - 212.0}{1 - 0} = -1.8$, and similarly

$m_2 = \frac{208.4 - 212.0}{2 - 0} = -1.8$, $m_3 = \frac{206.6 - 212.0}{3 - 0} = -1.8$, and

$m_4 = \frac{204.8 - 212.0}{4 - 0} = -1.8$. Therefore, the points are collinear.

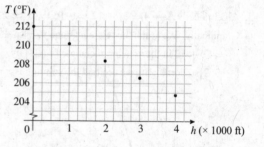

(b) From part (a), the slope of the line is $m = -1.8$. Using the formula $T - T_0 = m(h - h_0)$, we find its equation: $T - 212.0 = -1.8(h - 0) \Leftrightarrow$ $T = -1.8h + 212$.

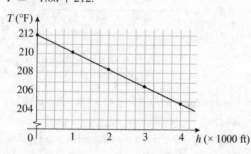

(c) Substituting $h = 19.34$ into the equation found in part (a), we have $T = -1.8(19.34) + 212 = 177.19$. So we expect the boiling point of water at the peak of Mount Kilimanjaro to be approximately $177.2°$F.

(d) The slope of the line represents the rate at which the boiling point of water changes as the elevation above sea level increases.

105. Slope is the rate of change of one variable per unit change in another variable. So if the slope is positive, then the temperature is rising. Likewise, if the slope is negative then the temperature is decreasing. If the slope is 0, then the temperature is not changing.

1.4 SOLVING EQUATIONS GRAPHICALLY

1. (a) The solutions of the equation $x^4 - 3x^3 - x^2 + 3x = 0$ are the x-intercepts of the graph of $y = x^4 - 3x^3 - x^2 + 3x$.

   (b) From the graph, it appears that the solutions of the equation $x^4 - 3x^3 - x^2 + 3x = 0$ are $x = -1$, $x = 0$, $x = 1$, and $x = 3$.

3. Algebraically: $x - 4 = 5x + 12 \Leftrightarrow -16 = 4x \Leftrightarrow x = -4$.

Graphically: We graph the two equations $y_1 = x - 4$ and $y_2 = 5x + 12$ in the viewing rectangle $[-6, 4]$ by $[-10, 2]$. Zooming in, we see that the solution is $x = -4$.

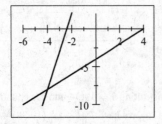

5. Algebraically: $\dfrac{2}{x} + \dfrac{1}{2x} = 7 \Leftrightarrow 2x\left(\dfrac{2}{x} + \dfrac{1}{2x}\right) = 2x\,(7)$

$\Leftrightarrow 4 + 1 = 14x \Leftrightarrow x = \dfrac{5}{14}$.

Graphically: We graph the two equations $y_1 = \dfrac{2}{x} + \dfrac{1}{2x}$ and $y_2 = 7$ in the viewing rectangle $[-2, 2]$ by $[-2, 8]$. Zooming in, we see that the solution is $x \approx 0.36$.

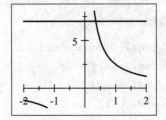

7. Algebraically: $x^2 - 32 = 0 \Leftrightarrow x^2 = 32 \Rightarrow$

$x = \pm\sqrt{32} = \pm 4\sqrt{2}$.

Graphically: We graph the equation $y_1 = x^2 - 32$ and determine where this curve intersects the x-axis. We use the viewing rectangle $[-10, 10]$ by $[-5, 5]$. Zooming in, we see that solutions are $x \approx 5.66$ and $x \approx -5.66$.

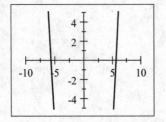

9. Algebraically: $x^2 + 9 = 0 \Leftrightarrow x^2 = -9$, which has no real solution.

Graphically: We graph the equation $y = x^2 + 9$ and see that this curve does not intersect the x-axis. We use the viewing rectangle $[-5, 5]$ by $[-5, 30]$.

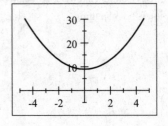

11. Algebraically: $16x^4 = 625 \Leftrightarrow x^4 = \frac{625}{16} \Rightarrow$

$x = \pm\frac{5}{2} = \pm2.5$.

Graphically: We graph the two equations $y_1 = 16x^4$ and $y_2 = 625$ in the viewing rectangle $[-5, 5]$ by $[610, 640]$. Zooming in, we see that solutions are $x = \pm2.5$.

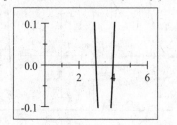

13. Algebraically: $(x-5)^4 - 80 = 0 \Leftrightarrow (x-5)^4 = 80 \Rightarrow$

$x - 5 = \pm\sqrt[4]{80} = \pm2\sqrt[4]{5} \Leftrightarrow x = 5 \pm 2\sqrt[4]{5}$.

Graphically: We graph the equation $y_1 = (x-5)^4 - 80$ and determine where this curve intersects the x-axis. We use the viewing rectangle $[-1, 9]$ by $[-5, 5]$. Zooming in, we see that solutions are $x \approx 2.01$ and $x \approx 7.99$.

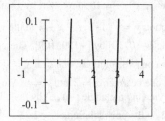

15. We graph $y = x^2 - 7x + 12$ in the viewing rectangle $[0, 6]$ by $[-0.1, 0.1]$. The solutions appear to be exactly $x = 3$ and $x = 4$. [In fact $x^2 - 7x + 12 = (x-3)(x-4)$.]

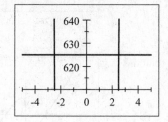

17. We graph $y = x^3 - 6x^2 + 11x - 6$ in the viewing rectangle $[-1, 4]$ by $[-0.1, 0.1]$. The solutions are $x = 1.00$, $x = 2.00$, and $x = 3.00$.

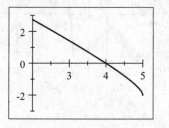

19. $\sqrt{5-x} + 1 = x - 2 \Leftrightarrow \sqrt{5-x} - x + 3 = 0$, so we graph $\sqrt{5-x} - x + 3$ in the viewing rectangle $[2, 5]$ by $[-3, 3]$. The only solution is $x = 4$.

21. We first graph $y = x - \sqrt{x+1}$ in the viewing rectangle $[-1, 5]$ by $[-0.1, 0.1]$ and find that the solution is near 1.6. Zooming in, we see that the solution is $x \approx 1.62$.

23. We graph $x - 5\sqrt{x} + 6$ in the viewing rectangle $[2, 11]$ by $[-1, 1]$ and see that the solutions of the equation $x - 5\sqrt{x} + 6 = 0$ in the interval $[2, 11]$ are $x = 4$ and $x = 9$. You can verify by substitution that these solutions are exact.

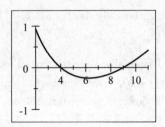

25. We graph $y = x^{1/3} - x$ in the viewing rectangle $[-3, 3]$ by $[-1, 1]$. The solutions are $x = -1$, $x = 0$, and $x = 1$, as can be verified by substitution.

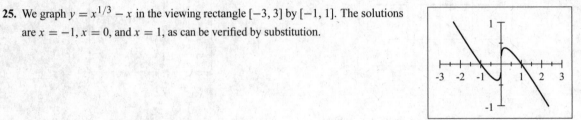

27. $x^3 - 2x^2 - x - 1 = 0$, so we start by graphing the function $y = x^3 - 2x^2 - x - 1$ in the viewing rectangle $[-10, 10]$ by $[-100, 100]$. There appear to be two solutions, one near $x = 0$ and another one between $x = 2$ and $x = 3$. We then use the viewing rectangle $[-1, 5]$ by $[-1, 1]$ and zoom in on the only solution, $x \approx 2.55$.

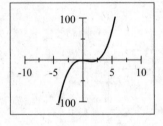
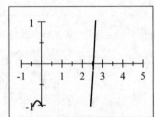

29. $x(x - 1)(x + 2) = \frac{1}{6}x \Leftrightarrow$

$x(x - 1)(x + 2) - \frac{1}{6}x = 0$. We start by graphing the function $y = x(x - 1)(x + 2) - \frac{1}{6}x$ in the viewing rectangle $[-5, 5]$ by $[-10, 10]$. There appear to be three solutions. We then use the viewing rectangle $[-2.5, 2.5]$ by $[-1, 1]$ and zoom into the solutions at $x \approx -2.05$, $x = 0.00$, and $x \approx 1.05$.

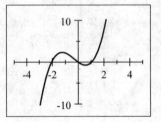
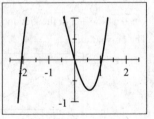

31. $5 - 3x = 8x - 20 \Leftrightarrow 11x - 25 = 0$, so we graph the equation $y = 11x - 25$ in the viewing rectangle $[1, 3]$ by $[-5, 5]$. Using a zoom or trace function, we find that the solution is $x \approx 2.27$, as found in Example 2.

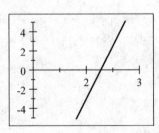

33. (a) We graph the equation

$$y = 10x + 0.5x^2 - 0.001x^3 - 5000 \text{ in the viewing}$$

rectangle $[0, 600]$ by $[-30000, 20000]$.

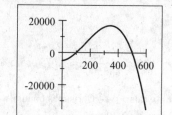

(b) From the graph it appears that

$0 < 10x + 0.05x^2 - 0.001x^3 - 5000$ for

$100 < x < 500$, and so 101 cooktops must be produced

to *begin* to make a profit.

35. Answers will vary.

1.5   MODELING WITH EQUATIONS

1. An equation modeling a real-world situation can be used to help us understand a real-world problem using mathematical methods. We translate real-world ideas into the language of algebra to construct our model, and translate our mathematical results back into real-world ideas in order to interpret our findings.

3. (a) A square of side x has area $A = x^2$.

(b) A rectangle of length l and width w has area $A = lw$.

(c) A circle of radius r has area $A = \pi r^2$.

5. A painter paints a wall in x hours, so the fraction of the wall she paints in one hour is $\dfrac{1 \text{ wall}}{x \text{ hours}} = \dfrac{1}{x}$.

7. If n is the first integer, then $n + 1$ is the middle integer and $n + 2$ is the third integer. So the sum of the three consecutive integers is $n + (n + 1) + (n + 2) = 3n + 3$.

9. If n is the first even integer, then $n + 2$ is the second and $n + 4$ is the third. Thus, the sum of the three integers is $n + (n + 2) + (n + 4) = 3n + 6$.

11. If s is the third test score, then since the other test scores are 78 and 82, the average of the three test scores is $\dfrac{78 + 82 + s}{3} = \dfrac{160 + s}{3}$.

13. If x dollars are invested at $2\frac{1}{2}\%$ simple interest, then the first year you will receive $0.025x$ dollars in interest.

15. Since w is the width of the rectangle, the length is five times the width, or $5w$. Then area $= \text{length} \times \text{width} = 5w \times w = 5w^2 \text{ ft}^2$.

17. If d is the given distance, in miles, and distance $= \text{rate} \times \text{time}$, we have time $= \dfrac{\text{distance}}{\text{rate}} = \dfrac{d}{55}$ min.

19. If x is the quantity of pure water added, the mixture will contain 25 oz of salt and $3 + x$ gallons of water. Thus the concentration is $\dfrac{25}{3 + x}$.

21. If d is the number of days and m the number of miles, then the cost of a rental is $C = 65d + 0.20m$. In this case, $d = 3$ and $C = 275$, so we solve for m: $275 = 65 \cdot 3 + 0.20m \Leftrightarrow 275 = 195 + 0.2m \Leftrightarrow 0.2m = 80 \Leftrightarrow m = \dfrac{80}{0.2} = 400$. Thus, Michael drove 400 miles.

23. Let x be Linh's final exam score. Because the final counts twice as much as each midterm, the weighted average of the three midterms and the final exam is $\dfrac{82 + 75 + 71 + 2x}{5} = \dfrac{228 + 2x}{5}$. In order for her average to be 80, we must have $\dfrac{228 + 2x}{5} = 80 \Leftrightarrow 228 + 2x = 400 \Leftrightarrow 2x = 172 \Leftrightarrow x = 86$. Thus, Linh must score 86% on the final exam in order to get an average of 80%.

25. Let m be the amount invested at $4\frac{1}{2}\%$. Then $12{,}000 - m$ is the amount invested at 4%.

Since the total interest is equal to the interest earned at $4\frac{1}{2}\%$ plus the interest earned at 4%, we have

$525 = 0.045m + 0.04\,(12{,}000 - m) \Leftrightarrow 525 = 0.045m + 480 - 0.04m \Leftrightarrow 45 = 0.005m \Leftrightarrow m = \dfrac{45}{0.005} = 9000$. Thus

$9000 is invested at $4\frac{1}{2}\%$, and $12{,}000 - 9000 = \$3000$ is invested at 4%.

27. Using the formula $I = Prt$ and solving for r, we get $262.50 = 3500 \cdot r \cdot 1 \Leftrightarrow r = \dfrac{262.5}{3500} = 0.075$ or 7.5%.

29. Let x be her monthly salary. Since her annual salary $= 12 \times$ (monthly salary) $+$ (Christmas bonus) we have
$97{,}300 = 12x + 8{,}500 \Leftrightarrow 88{,}800 = 12x \Leftrightarrow x \approx 7{,}400$. Her monthly salary is $7,400.

31. Let x be the overtime hours Helen works. Since gross pay $=$ regular salary $+$ overtime pay, we obtain the equation
$352.50 = 7.50 \times 35 + 7.50 \times 1.5 \times x \Leftrightarrow 352.50 = 262.50 + 11.25x \Leftrightarrow 90 = 11.25x \Leftrightarrow x = \dfrac{90}{11.25} = 8$. Thus Helen
worked 8 hours of overtime.

33. All ages are in terms of the daughter's age 7 years ago. Let y be age of the daughter 7 years ago. Then $11y$ is the age of
the movie star 7 years ago. Today, the daughter is $y + 7$, and the movie star is $11y + 7$. But the movie star is also 4 times
his daughter's age today. So $4\,(y + 7) = 11y + 7 \Leftrightarrow 4y + 28 = 11y + 7 \Leftrightarrow 21 = 7y \Leftrightarrow y = 3$. Thus the movie star's age
today is $11\,(3) + 7 = 40$ years.

35. Let p be the number of pennies. Then p is the number of nickels and p is the number of dimes. So the value of
the coins in the purse is the value of the pennies plus the value of the nickels plus the value of the dimes. Thus
$1.44 = 0.01p + 0.05p + 0.10p \Leftrightarrow 1.44 = 0.16p \Leftrightarrow p = \frac{1.44}{0.16} = 9$. So the purse contains 9 pennies, 9 nickels, and 9 dimes.

37. Let l be the length of the garden. Since area $=$ width $\cdot$ length, we obtain the equation $1125 = 25l \Leftrightarrow l = \frac{1125}{25} = 45$ ft. So
the garden is 45 feet long.

39. Let x be the length of a side of the square plot. As shown in the figure,
area of the plot $=$ area of the building $+$ area of the parking lot. Thus,
$x^2 = 60\,(40) + 12{,}000 = 2{,}400 + 12{,}000 = 14{,}400 \Rightarrow x = \pm 120$. So the plot of
land measures 120 feet by 120 feet.

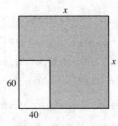

41. The figure is a trapezoid, so its area is $\dfrac{\text{base}_1 + \text{base}_2}{2}$ (height). Putting in the known quantities, we have

$120 = \dfrac{y + 2y}{2}\,(y) = \frac{3}{2}y^2 \Leftrightarrow y^2 = 80 \Rightarrow y = \pm\sqrt{80} = \pm 4\sqrt{5}$. Since length is positive, $y = 4\sqrt{5} \approx 8.94$ inches.

43. Let x be the width of the strip. Then the length of the mat is $20 + 2x$, and the width of the mat is $15 + 2x$. Now the
perimeter is twice the length plus twice the width, so $102 = 2\,(20 + 2x) + 2\,(15 + 2x) \Leftrightarrow 102 = 40 + 4x + 30 + 4x \Leftrightarrow$
$102 = 70 + 8x \Leftrightarrow 32 = 8x \Leftrightarrow x = 4$. Thus the strip of mat is 4 inches wide.

45. Let x be the length of the man's shadow, in meters. Using similar triangles, $\dfrac{10 + x}{6} = \dfrac{x}{2} \Leftrightarrow 20 + 2x = 6x \Leftrightarrow 4x = 20 \Leftrightarrow$
$x = 5$. Thus the man's shadow is 5 meters long.

47. Let x be the amount (in mL) of 60% acid solution to be used. Then $300 - x$ mL of 30% solution would have to be used to yield a total of 300 mL of solution.

	60% acid	30% acid	Mixture
mL	x	$300 - x$	300
Rate (% acid)	0.60	0.30	0.50
Value	$0.60x$	$0.30(300 - x)$	$0.50(300)$

Thus the total amount of pure acid used is $0.60x + 0.30(300 - x) = 0.50(300) \Leftrightarrow 0.3x + 90 = 150 \Leftrightarrow x = \dfrac{60}{0.3} = 200$. So 200 mL of 60% acid solution must be mixed with 100 mL of 30% solution to get 300 mL of 50% acid solution.

49. Let x be the number of grams of silver added. The weight of the rings is 5×18 g $= 90$ g.

	5 rings	Pure silver	Mixture
Grams	90	x	$90 + x$
Rate (% gold)	0.90	0	0.75
Value	$0.90(90)$	$0x$	$0.75(90 + x)$

So $0.90(90) + 0x = 0.75(90 + x) \Leftrightarrow 81 = 67.5 + 0.75x \Leftrightarrow 0.75x = 13.5 \Leftrightarrow x = \dfrac{13.5}{0.75} = 18$. Thus 18 grams of silver must be added to get the required mixture.

51. Let x be the number of liters of coolant removed and replaced by water.

	60% antifreeze	60% antifreeze (removed)	Water	Mixture
Liters	3.6	x	x	3.6
Rate (% antifreeze)	0.60	0.60	0	0.50
Value	$0.60(3.6)$	$-0.60x$	$0x$	$0.50(3.6)$

So $0.60(3.6) - 0.60x + 0x = 0.50(3.6) \Leftrightarrow 2.16 - 0.6x = 1.8 \Leftrightarrow -0.6x = -0.36 \Leftrightarrow x = \dfrac{-0.36}{-0.6} = 0.6$. Thus 0.6 liters must be removed and replaced by water.

53. Let c be the concentration of fruit juice in the cheaper brand. The new mixture that Jill makes will consist of 650 mL of the original fruit punch and 100 mL of the cheaper fruit punch.

	Original Fruit Punch	Cheaper Fruit Punch	Mixture
mL	650	100	750
Concentration	0.50	c	0.48
Juice	$0.50 \cdot 650$	$100c$	$0.48 \cdot 750$

So $0.50 \cdot 650 + 100c = 0.48 \cdot 750 \Leftrightarrow 325 + 100c = 360 \Leftrightarrow 100c = 35 \Leftrightarrow c = 0.35$. Thus the cheaper brand is only 35% fruit juice.

55. Let t be the time in minutes it would take Candy and Tim if they work together. Candy delivers the papers at a rate of $\frac{1}{70}$ of the job per minute, while Tim delivers the paper at a rate of $\frac{1}{80}$ of the job per minute. The sum of the fractions of the job that each can do individually in one minute equals the fraction of the job they can do working together. So we have $\dfrac{1}{t} = \dfrac{1}{70} + \dfrac{1}{80} \Leftrightarrow 560 = 8t + 7t \Leftrightarrow 560 = 15t \Leftrightarrow t = 37\frac{1}{3}$ minutes. Since $\frac{1}{3}$ of a minute is 20 seconds, it would take them 37 minutes 20 seconds if they worked together.

57. Let t be the time, in hours, it takes Karen to paint a house alone. Then working together, Karen and Betty can paint a house in $\frac{2}{3}t$ hours. The sum of their individual rates equals their rate working together, so $\frac{1}{t} + \frac{1}{6} = \frac{1}{\frac{2}{3}t} \Leftrightarrow \frac{1}{t} + \frac{1}{6} = \frac{3}{2t} \Leftrightarrow$ $6 + t = 9 \Leftrightarrow t = 3$. Thus it would take Karen 3 hours to paint a house alone.

59. Let t be the time in hours that Wendy spent on the train. Then $\frac{11}{2} - t$ is the time in hours that Wendy spent on the bus. We construct a table:

	Rate	Time	Distance
By train	40	t	$40t$
By bus	60	$\frac{11}{2} - t$	$60\left(\frac{11}{2} - t\right)$

The total distance traveled is the sum of the distances traveled by bus and by train, so $300 = 40t + 60\left(\frac{11}{2} - t\right) \Leftrightarrow$ $300 = 40t + 330 - 60t \Leftrightarrow -30 = -20t \Leftrightarrow t = \frac{-30}{-20} = 1.5$ hours. So the time spent on the train is $5.5 - 1.5 = 4$ hours.

61. Let r be the speed of the plane from Montreal to Los Angeles. Then $r + 0.20r = 1.20r$ is the speed of the plane from Los Angeles to Montreal.

	Rate	Time	Distance
Montreal to L.A.	r	$\dfrac{2500}{r}$	2500
L.A. to Montreal	$1.2r$	$\dfrac{2500}{1.2r}$	2500

The total time is the sum of the times each way, so $9\frac{1}{6} = \frac{2500}{r} + \frac{2500}{1.2r} \Leftrightarrow \frac{55}{6} = \frac{2500}{r} + \frac{2500}{1.2r} \Leftrightarrow$ $55 \cdot 1.2r = 2500 \cdot 6 \cdot 1.2 + 2500 \cdot 6 \Leftrightarrow 66r = 18,000 + 15,000 \Leftrightarrow 66r = 33,000 \Leftrightarrow r = \frac{33,000}{66} = 500$. Thus the plane flew at a speed of 500 mi/h on the trip from Montreal to Los Angeles.

63. Let x be the distance from the fulcrum to where the mother sits. Then substituting the known values into the formula given, we have $100\,(8) = 125x \quad \Leftrightarrow \quad 800 = 125x \Leftrightarrow x = 6.4$. So the mother should sit 6.4 feet from the fulcrum.

65. Let l be the length of the lot in feet. Then the length of the diagonal is $l + 10$. We apply the Pythagorean Theorem with the hypotenuse as the diagonal. So $l^2 + 50^2 = (l + 10)^2 \Leftrightarrow l^2 + 2500 = l^2 + 20l + 100 \Leftrightarrow 20l = 2400 \Leftrightarrow l = 120$. Thus the length of the lot is 120 feet.

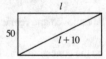

67. Let h be the height in feet of the structure. The structure is composed of a right cylinder with radius 10 and height $\frac{2}{3}h$ and a cone with base radius 10 and height $\frac{1}{3}h$. Using the formulas for the volume of a cylinder and that of a cone, we obtain the equation $1400\pi = \pi\,(10)^2\left(\frac{2}{3}h\right) + \frac{1}{3}\pi\,(10)^2\left(\frac{1}{3}h\right) \Leftrightarrow 1400\pi = \frac{200\pi}{3}h + \frac{100\pi}{9}h \Leftrightarrow 126 = 6h + h$ (multiply both sides by $\frac{9}{100\pi}$) $\Leftrightarrow 126 = 7h \Leftrightarrow h = 18$. Thus the height of the structure is 18 feet.

69. Pythagoras was born about 569 BC in Samos, Ionia and died about 475 BC.

Euclid was born about 325 BC and died about 265 BC in Alexandria, Egypt.

Archimedes was born in 287 BC in Syracuse, Sicily and died in 212 BC in Syracuse.

1.6   SOLVING QUADRATIC EQUATIONS

1. (a) The Quadratic Formula states that $x = \dfrac{-b \pm \sqrt{b^2 - 4ac}}{2a}$.

(b) In the equation $\frac{1}{2}x^2 - x - 4 = 0$, $a = \frac{1}{2}$, $b = -1$, and $c = -4$. So, the solution of the equation is

$$x = \frac{-(-1) \pm \sqrt{(-1)^2 - 4\left(\frac{1}{2}\right)(-4)}}{2\left(\frac{1}{2}\right)} = \frac{1 \pm 3}{1} = -2 \text{ or } 4.$$

3. For the quadratic equation $ax^2 + bx + c = 0$ the discriminant is $D = b^2 - 4ac$. If $D > 0$, the equation has two real solutions; if $D = 0$, the equation has one real solution; and if $D < 0$, the equation has no real solution.

5. $x^2 - x = 12 \Leftrightarrow x^2 - x - 12 = 0 \Leftrightarrow (x - 4)(x + 3) = 0 \Leftrightarrow x - 4 = 0 \text{ or } x + 3 = 0$. Thus, $x = 4$ or $x = -3$.

7. $x^2 - 7x + 12 = 0 \Leftrightarrow (x - 4)(x - 3) = 0 \Leftrightarrow x - 4 = 0 \text{ or } x - 3 = 0$. Thus, $x = 4$ or $x = 3$.

9. $3x^2 - 5x - 2 = 0 \Leftrightarrow (3x + 1)(x - 2) = 0 \Leftrightarrow 3x + 1 = 0 \text{ or } x - 2 = 0$. Thus, $x = -\frac{1}{3}$ or $x = 2$.

11. $2s^2 = 5s + 3 \Leftrightarrow 2s^2 - 5s - 3 = 0 \Leftrightarrow (2s + 1)(s - 3) = 0 \Leftrightarrow 2s + 1 = 0 \text{ or } s - 3 = 0$. Thus, $s = -\frac{1}{2}$ or $s = 3$.

13. $12z^2 - 44z = 45 \Leftrightarrow 12z^2 - 44z - 45 = 0 \Leftrightarrow (6z + 5)(2z - 9) = 0 \Leftrightarrow 6z + 5 = 0 \text{ or } 2z - 9 = 0$. Thus, $z = -\frac{5}{6}$ or $z = \frac{9}{2}$.

15. $6x^2 + 5x = 4 \Leftrightarrow 6x^2 + 5x - 4 = 0 \Leftrightarrow (2x - 1)(3x + 4) = 0 \Leftrightarrow 2x - 1 = 0 \text{ or } 3x + 4 = 0$. If $2x - 1 = 0$, then $x = \frac{1}{2}$; if $3x + 4 = 0$, then $x = -\frac{4}{3}$.

17. $x^2 = 5(x + 100) \Leftrightarrow x^2 = 5x + 500 \Leftrightarrow x^2 - 5x - 500 = 0 \Leftrightarrow (x - 25)(x + 20) = 0 \Leftrightarrow x - 25 = 0 \text{ or } x + 20 = 0$. Thus, $x = 25$ or $x = -20$.

19. $x^2 + 2x - 5 = 0 \Leftrightarrow x^2 + 2x = 5 \Leftrightarrow x^2 + 2x + 1 = 5 + 1 \Leftrightarrow (x + 1)^2 = 6 \Rightarrow x + 1 = \pm\sqrt{6} \Leftrightarrow x = -1 \pm \sqrt{6}$.

21. $x^2 - 6x - 11 = 0 \Leftrightarrow x^2 - 6x = 11 \Leftrightarrow x^2 - 6x + 9 = 11 + 9 \Leftrightarrow (x - 3)^2 = 20 \Rightarrow x - 3 = \pm 2\sqrt{5} \Leftrightarrow x = 3 \pm 2\sqrt{5}$.

23. $x^2 + x - \frac{3}{4} = 0 \Leftrightarrow x^2 + x = \frac{3}{4} \Leftrightarrow x^2 + x + \frac{1}{4} = \frac{3}{4} + \frac{1}{4} \Leftrightarrow \left(x + \frac{1}{2}\right)^2 = 1 \Rightarrow x + \frac{1}{2} = \pm 1 \Leftrightarrow x = -\frac{1}{2} \pm 1$. So $x = -\frac{1}{2} - 1 = -\frac{3}{2}$ or $x = -\frac{1}{2} + 1 = \frac{1}{2}$.

25. $x^2 + 22x + 21 = 0 \Leftrightarrow x^2 + 22x = -21 \Leftrightarrow x^2 + 22x + 11^2 = -21 + 11^2 = -21 + 121 \Leftrightarrow (x + 11)^2 = 100 \Rightarrow x + 11 = \pm 10 \Leftrightarrow x = -11 \pm 10$. Thus, $x = -1$ or $x = -21$.

27. $2x^2 + 8x + 1 = 0 \Leftrightarrow x^2 + 4x + \frac{1}{2} = 0 \Leftrightarrow x^2 + 4x = -\frac{1}{2} \Leftrightarrow x^2 + 4x + 4 = -\frac{1}{2} + 4 \Leftrightarrow (x + 2)^2 = \frac{7}{2} \Rightarrow x + 2 = \pm\sqrt{\frac{7}{2}}$ $\Leftrightarrow x = -2 \pm \frac{\sqrt{14}}{2}$.

29. $2x^2 + 7x + 4 = 0 \Leftrightarrow x^2 + \frac{7}{2}x + 2 = 0 \Leftrightarrow x^2 + \frac{7}{2}x = -2 \Leftrightarrow x^2 + \frac{7}{2}x + \frac{49}{16} = -2 + \frac{49}{16} \Leftrightarrow \left(x + \frac{7}{4}\right)^2 = \frac{17}{16} \Leftrightarrow$ $x + \frac{7}{4} = \pm\sqrt{\frac{17}{16}} \Leftrightarrow x = -\frac{7}{4} \pm \frac{\sqrt{17}}{4}$.

31. $x^2 - 2x - 15 = 0 \Leftrightarrow (x + 3)(x - 5) = 0 \Leftrightarrow x + 3 = 0 \text{ or } x - 5 = 0$. Thus, $x = -3$ or $x = 5$.

33. $x^2 - 7x + 10 = 0 \Leftrightarrow (x - 5)(x - 2) = 0 \Leftrightarrow x - 5 = 0 \text{ or } x - 2 = 0$. Thus, $x = 5$ or $x = 2$.

35. $2x^2 + x - 3 = 0 \Leftrightarrow (x - 1)(2x + 3) = 0 \Leftrightarrow x - 1 = 0 \text{ or } 2x + 3 = 0$. If $x - 1 = 0$, then $x = 1$; if $2x + 3 = 0$, then $x = -\frac{3}{2}$.

37. $x^2 + 12x - 27 = 0 \Leftrightarrow x^2 + 12x = 27 \Leftrightarrow x^2 + 12x + 36 = 27 + 36 \Leftrightarrow (x + 6)^2 = 63 \Rightarrow x + 6 = \pm 3\sqrt{7} \Leftrightarrow x = -6 \pm 3\sqrt{7}$.

39. $3x^2 + 6x - 5 = 0 \Leftrightarrow x^2 + 2x - \frac{5}{3} = 0 \Leftrightarrow x^2 + 2x = \frac{5}{3} \Leftrightarrow x^2 + 2x + 1 = \frac{5}{3} + 1 \Leftrightarrow (x + 1)^2 = \frac{8}{3} \Rightarrow x + 1 = \pm\sqrt{\frac{8}{3}} \Leftrightarrow$ $x = -1 \pm \frac{2\sqrt{6}}{3}$.

41. $z^2 - \frac{3}{2}z + \frac{9}{16} = 0 \Rightarrow \left(z - \frac{3}{4}\right)^2 = 0 \Leftrightarrow z = \frac{3}{4}$.

43. $4x^2 + 16x - 9 = 0 \Leftrightarrow (2x - 1)(2x + 9) = 0 \Leftrightarrow 2x - 1 = 0$ or $2x + 9 = 0$. If $2x - 1 = 0$, then $x = \frac{1}{2}$; if $2x + 9 = 0$, then $x = -\frac{9}{2}$.

45. $w^2 = 3(w - 1) \Leftrightarrow w^2 - 3w + 3 = 0 \Rightarrow w = \dfrac{-(-3) \pm \sqrt{(-3)^2 - 4(1)(3)}}{2(1)} = \dfrac{3 \pm \sqrt{9 - 12}}{2} = \dfrac{3 \pm \sqrt{-3}}{2}$. Since the discriminant is less than 0, the equation has no real solution.

47. $x^2 - \sqrt{5}x + 1 = 0 \Rightarrow x = \dfrac{-b \pm \sqrt{b^2 - 4ac}}{2a} = \dfrac{-\left(-\sqrt{5}\right) \pm \sqrt{\left(-\sqrt{5}\right)^2 - 4(1)(1)}}{2(1)} = \dfrac{\sqrt{5} \pm \sqrt{5 - 4}}{2} = \dfrac{\sqrt{5} \pm 1}{2}$.

49. $10y^2 - 16y + 5 = 0 \Rightarrow$

$x = \dfrac{-b \pm \sqrt{b^2 - 4ac}}{2a} = \dfrac{-(-16) \pm \sqrt{(-16)^2 - 4(10)(5)}}{2(10)} = \dfrac{16 \pm \sqrt{256 - 200}}{20} = \dfrac{16 \pm \sqrt{56}}{20} = \dfrac{8 \pm \sqrt{14}}{10}$.

51. $3x^2 + 2x + 2 = 0 \Rightarrow x = \dfrac{-b \pm \sqrt{b^2 - 4ac}}{2a} = \dfrac{-(2) \pm \sqrt{(2)^2 - 4(3)(2)}}{2(3)} = \dfrac{-2 \pm \sqrt{4 - 24}}{6} = \dfrac{-2 \pm \sqrt{-20}}{6}$. Since the discriminant is less than 0, the equation has no real solution.

53. $x^2 - 0.011x - 0.064 = 0 \Rightarrow$

$x = \dfrac{-(-0.011) \pm \sqrt{(-0.011)^2 - 4(1)(-0.064)}}{2(1)} = \dfrac{0.011 \pm \sqrt{0.000121 + 0.256}}{2} \approx \dfrac{0.011 \pm 0.506}{2}$.

Thus, $x \approx \dfrac{0.011 + 0.506}{2} = 0.259$ or $x \approx \dfrac{0.011 - 0.506}{2} = -0.248$.

55. $x^2 - 2.450x + 1.501 = 0 \Rightarrow$

$x = \dfrac{-(-2.450) \pm \sqrt{(-2.450)^2 - 4(1)(1.501)}}{2(1)} = \dfrac{2.450 \pm \sqrt{6.0025 - 6.004}}{2} = \dfrac{2.450 \pm \sqrt{-0.0015}}{2}$.

Thus, there is no real solution.

57. $h = \frac{1}{2}gt^2 + v_0 t \Leftrightarrow \frac{1}{2}gt^2 + v_0 t - h = 0$. Using the Quadratic Formula,

$t = \dfrac{-(v_0) \pm \sqrt{(v_0)^2 - 4\left(\frac{1}{2}g\right)(-h)}}{2\left(\frac{1}{2}g\right)} = \dfrac{-v_0 \pm \sqrt{v_0^2 + 2gh}}{g}$.

59. $A = 2x^2 + 4xh \Leftrightarrow 2x^2 + 4xh - A = 0$. Using the Quadratic Formula,

$x = \dfrac{-(4h) \pm \sqrt{(4h)^2 - 4(2)(-A)}}{2(2)} = \dfrac{-4h \pm \sqrt{16h^2 + 8A}}{4} = \dfrac{-4h \pm \sqrt{4(4h^2 + 2A)}}{4} = \dfrac{-4h \pm 2\sqrt{4h^2 + 2A}}{4}$

$= \dfrac{2\left(-2h \pm \sqrt{4h^2 + 2A}\right)}{4} = \dfrac{-2h \pm \sqrt{4h^2 + 2A}}{2}$

61. $\dfrac{1}{s+a} + \dfrac{1}{s+b} = \dfrac{1}{c} \Leftrightarrow c(s+b) + c(s+a) = (s+a)(s+b) \Leftrightarrow cs + bc + cs + ac = s^2 + as + bs + ab \Leftrightarrow$
$s^2 + (a+b-2c)s + (ab - ac - bc) = 0$. Using the Quadratic Formula,

$$s = \dfrac{-(a+b-2c) \pm \sqrt{(a+b-2c)^2 - 4(1)(ab - ac - bc)}}{2(1)}$$

$$= \dfrac{-(a+b-2c) \pm \sqrt{a^2 + b^2 + 4c^2 + 2ab - 4ac - 4bc - 4ab + 4ac + 4bc}}{2}$$

$$= \dfrac{-(a+b-2c) \pm \sqrt{a^2 + b^2 + 4c^2 - 2ab}}{2}$$

63. $D = b^2 - 4ac = (-6)^2 - 4(1)(1) = 32$. Since D is positive, this equation has two real solutions.

65. $D = b^2 - 4ac = (2.20)^2 - 4(1)(1.21) = 4.84 - 4.84 = 0$. Since $D = 0$, this equation has one real solution.

67. $D = b^2 - 4ac = (5)^2 - 4(4)\left(\dfrac{13}{8}\right) = 25 - 26 = -1$. Since D is negative, this equation has no real solution.

69. $a^2 x^2 + 2ax + 1 = 0 \Leftrightarrow (ax+1)^2 = 0 \Leftrightarrow ax + 1 = 0$. So $ax + 1 = 0$ then $ax = -1 \Leftrightarrow x = -\dfrac{1}{a}$.

71. Algebraically: We use the Quadratic Formula to solve the equation

$x^2 - x - 6 = 0$. $x = \dfrac{-(-1) \pm \sqrt{(-1)^2 - 4(1)(-6)}}{2(1)} = \dfrac{1 \pm \sqrt{25}}{2}$. There are two

solutions, $x = -2$ and $x = 3$.

Graphically: We graph $y = x^2 - x - 6$ in the viewing rectangle $[-5, 5]$ by
$[-7, 3]$, and find that the x-intercepts are -2 and 3.

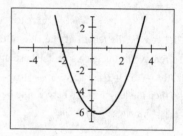

73. Algebraically: We use the Quadratic Formula to solve the equation

$x^2 + 6x + 9 = 0$. $x = \dfrac{-6 \pm \sqrt{6^2 - 4(1)(9)}}{2(1)} = \dfrac{-6 \pm \sqrt{0}}{2}$. There is one solution,

$x = -3$.

Graphically: We graph $y = x^2 + 6x + 9$ in the viewing rectangle $[-6, 0]$ by
$[-5, 5]$, and find that the x-intercept is -3.

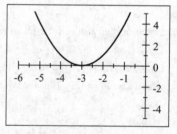

75. Algebraically: We use the Quadratic Formula to solve the equation

$x^2 - 6x + 14 = 0$. $x = \dfrac{-(-6) \pm \sqrt{(-6)^2 - 4(1)(14)}}{2(1)} = \dfrac{6 \pm \sqrt{-20}}{2}$. There is

no real solution.

Graphically: We graph $y = x^2 - 6x + 14$ in the viewing rectangle $[-2, 8]$ by
$[-2, 12]$, and find that there is no x-intercept.

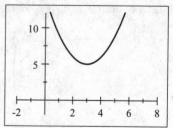

77. We want to find the values of k that make the discriminant 0. Thus $k^2 - 4(4)(25) = 0 \Leftrightarrow k^2 = 400 \Leftrightarrow k = \pm 20$.

79. Let n be one number. Then the other number must be $55 - n$, since $n + (55 - n) = 55$. Because
the product is 684, we have $(n)(55 - n) = 684 \Leftrightarrow 55n - n^2 = 684 \Leftrightarrow n^2 - 55n + 684 = 0 \Rightarrow$

$n = \dfrac{-(-55) \pm \sqrt{(-55)^2 - 4(1)(684)}}{2(1)} = \dfrac{55 \pm \sqrt{3025 - 2736}}{2} = \dfrac{55 \pm \sqrt{289}}{2} = \dfrac{55 \pm 17}{2}$. So $n = \dfrac{55+17}{2} = \dfrac{72}{2} = 36$ or

$n = \dfrac{55-17}{2} = \dfrac{38}{2} = 19$. In either case, the two numbers are 19 and 36.

81. Let w be the width of the garden in feet. Then the length is $w + 10$. Thus $875 = w(w+10) \Leftrightarrow w^2 + 10w - 875 = 0 \Leftrightarrow (w+35)(w-25) = 0$. So $w + 35 = 0$ in which case $w = -35$, which is not possible, or $w - 25 = 0$ and so $w = 25$. Thus the width is 25 feet and the length is 35 feet.

83. Let w be the width of the garden in feet. We use the perimeter to express the length l of the garden in terms of width. Since the perimeter is twice the width plus twice the length, we have $200 = 2w + 2l \Leftrightarrow 2l = 200 - 2w \Leftrightarrow l = 100 - w$. Using the formula for area, we have $2400 = w(100-w) = 100w - w^2 \Leftrightarrow w^2 - 100w + 2400 = 0 \Leftrightarrow (w-40)(w-60) = 0$. So $w - 40 = 0 \Leftrightarrow w = 40$, or $w - 60 = 0 \Leftrightarrow w = 60$. If $w = 40$, then $l = 100 - 40 = 60$. And if $w = 60$, then $l = 100 - 60 = 40$. So the length is 60 feet and the width is 40 feet.

85. The shaded area is the sum of the area of a rectangle and the area of a triangle. So $A = y(1) + \frac{1}{2}(y)(y) = \frac{1}{2}y^2 + y$. We are given that the area is 1200 cm^2, so $1200 = \frac{1}{2}y^2 + y \Leftrightarrow y^2 + 2y - 2400 = 0 \Leftrightarrow (y+50)(y-48) = 0$. y is positive, so $y = 48$ cm.

87. Let x be the length of one side of the cardboard, so we start with a piece of cardboard x by x. When 4 inches are removed from each side, the base of the box is $x - 8$ by $x - 8$. Since the volume is 100 in^3, we get $4(x-8)^2 = 100 \Leftrightarrow x^2 - 16x + 64 = 25 \Leftrightarrow x^2 - 16x + 39 = 0 \Leftrightarrow (x-3)(x-13) = 0$. So $x = 3$ or $x = 13$. But $x = 3$ is not possible, since then the length of the base would be $3 - 8 = -5$, and all lengths must be positive. Thus $x = 13$, and the piece of cardboard is 13 inches by 13 inches.

89. Let w be the width of the lot in feet. Then the length is $w + 6$. Using the Pythagorean Theorem, we have $w^2 + (w+6)^2 = (174)^2 \Leftrightarrow w^2 + w^2 + 12w + 36 = 30{,}276 \Leftrightarrow 2w^2 + 12w - 30240 = 0 \Leftrightarrow w^2 + 6w - 15120 = 0 \Leftrightarrow (w+126)(w-120) = 0$. So either $w + 126 = 0$ in which case $w = -126$, which is not possible, or $w - 120 = 0$ in which case $w = 120$. Thus the width is 120 feet and the length is 126 feet.

91. Let x be the rate, in mi/h, at which the salesman drove between Ajax and Barrington.

Direction	Distance	Rate	Time
Ajax $\rightarrow$ Barrington	120	x	$\dfrac{120}{x}$
Barrington $\rightarrow$ Collins	150	$x + 10$	$\dfrac{150}{x+10}$

We have used the equation time $= \dfrac{\text{distance}}{\text{rate}}$ to fill in the "Time" column of the table. Since the second part of the trip took 6 minutes (or $\frac{1}{10}$ hour) more than the first, we can use the time column to get the equation $\dfrac{120}{x} + \dfrac{1}{10} = \dfrac{150}{x+10} \Rightarrow$ $120(10)(x+10) + x(x+10) = 150(10x) \Leftrightarrow 1200x + 12{,}000 + x^2 + 10x = 1500x \Leftrightarrow x^2 - 290x + 12{,}000 = 0 \Leftrightarrow$ $x = \dfrac{-(-290) \pm \sqrt{(-290)^2 - 4(1)(12{,}000)}}{2} = \dfrac{290 \pm \sqrt{84{,}100 - 48{,}000}}{2} = \dfrac{290 \pm \sqrt{36{,}100}}{2} = \dfrac{290 \pm 190}{2} = 145 \pm 95$. Hence, the salesman drove either 50 mi/h or 240 mi/h between Ajax and Barrington. (The first choice seems more likely!) •

93. Let r be the rowing rate in km/h of the crew in still water. Then their rate upstream was $r - 3$ km/h, and their rate downstream was $r + 3$ km/h.

Direction	Distance	Rate	Time
Upstream	6	$r - 3$	$\dfrac{6}{r-3}$
Downstream	6	$r + 3$	$\dfrac{6}{r+3}$

Since the time to row upstream plus the time to row downstream was 2 hours 40 minutes $= \frac{8}{3}$ hour, we get the equation

$$\frac{6}{r-3} + \frac{6}{r+3} = \frac{8}{3} \Leftrightarrow 6\,(3)\,(r+3) + 6\,(3)\,(r-3) = 8\,(r-3)\,(r+3) \Leftrightarrow 18r + 54 + 18r - 54 = 8r^2 - 72 \Leftrightarrow$$

$0 = 8r^2 - 36r - 72 = 4\left(2r^2 - 9r - 18\right) = 4\,(2r+3)\,(r-6) \Leftrightarrow 2r + 3 = 0$ or $r - 6 = 0$. If $2r + 3 = 0$, then $r = -\frac{3}{2}$, which is impossible because the rowing rate is positive. If $r - 6 = 0$, then $r = 6$. So the rate of the rowing crew in still water is 6 km/h.

95. Using $h_0 = 288$, we solve $0 = -16t^2 + 288$, for $t \geq 0$. So $0 = -16t^2 + 288 \Leftrightarrow 16t^2 = 288 \Leftrightarrow t^2 = 18 \Rightarrow$ $t = \pm\sqrt{18} = \pm 3\sqrt{2}$. Thus it takes $3\sqrt{2} \approx 4.24$ seconds for the ball the hit the ground.

97. We are given $v_o = 40$ ft/s.

(a) Setting $h = 24$, we have $24 = -16t^2 + 40t \Leftrightarrow 16t^2 - 40t + 24 = 0 \Leftrightarrow 8\left(2t^2 - 5t + 3\right) = 0 \Leftrightarrow 8\,(2t-3)\,(t-1) = 0$ $\Leftrightarrow t = 1$ or $t = 1\frac{1}{2}$. Therefore, the ball reaches 24 feet in 1 second (ascending) and again after $1\frac{1}{2}$ seconds (descending).

(b) Setting $h = 48$, we have $48 = -16t^2 + 40t \Leftrightarrow 16t^2 - 40t + 48 = 0 \Leftrightarrow 2t^2 - 5t + 6 = 0 \Leftrightarrow$ $t = \dfrac{5 \pm \sqrt{25 - 48}}{4} = \dfrac{5 \pm \sqrt{-23}}{4}$. However, since the discriminant $D < 0$, there is no real solution, and hence the ball never reaches a height of 48 feet.

(c) The greatest height h is reached only once. So $h = -16t^2 + 40t \Leftrightarrow 16t^2 - 40t + h = 0$ has only one solution. Thus $D = (-40)^2 - 4\,(16)\,(h) = 0 \Leftrightarrow 1600 - 64h = 0 \Leftrightarrow h = 25$. So the greatest height reached by the ball is 25 feet.

(d) Setting $h = 25$, we have $25 = -16t^2 + 40t \Leftrightarrow 16t^2 - 40t + 25 = 0 \Leftrightarrow (4t - 5)^2 = 0 \Leftrightarrow t = 1\frac{1}{4}$. Thus the ball reaches the highest point of its path after $1\frac{1}{4}$ seconds.

(e) Setting $h = 0$ (ground level), we have $0 = -16t^2 + 40t \Leftrightarrow 2t^2 - 5t = 0 \Leftrightarrow t\,(2t-5) = 0 \Leftrightarrow t = 0$ (start) or $t = 2\frac{1}{2}$. So the ball hits the ground in $2\frac{1}{2}$ s.

99. (a) The fish population on January 1, 2002 corresponds to $t = 0$, so $F = 1000\left(30 + 17\,(0) - (0)^2\right) = 30,000$. To find when the population will again reach this value, we set $F = 30,000$, giving

$30000 = 1000\left(30 + 17t - t^2\right) = 30000 + 17000t - 1000t^2 \Leftrightarrow 0 = 17000t - 1000t^2 = 1000t\,(17 - t) \Leftrightarrow t = 0$ or $t = 17$. Thus the fish population will again be the same 17 years later, that is, on January 1, 2019.

(b) Setting $F = 0$, we have $0 = 1000\left(30 + 17t - t^2\right) \Leftrightarrow t^2 - 17t - 30 = 0 \Leftrightarrow$ $t = \dfrac{17 \pm \sqrt{289 + 120}}{-2} = \dfrac{17 \pm \sqrt{409}}{-2} = \dfrac{17 \pm 20.22}{2}$. Thus $t \approx -1.612$ or $t \approx 18.612$. Since $t < 0$ is inadmissible, it follows that the fish in the lake will have died out 18.612 years after January 1, 2002, that is on August 12, 2020.

101. Let w be the uniform width of the lawn. With w cut off each end, the area of the factory is $(240 - 2w)(180 - 2w)$. Since the lawn and the factory are equal in size this area, is $\frac{1}{2} \cdot 240 \cdot 180$. So $21,600 = 43,200 - 480w - 360w + 4w^2 \Leftrightarrow$ $0 = 4w^2 - 840w + 21,600 = 4\left(w^2 - 210w + 5400\right) = 4(w - 30)(w - 180) \Rightarrow w = 30$ or $w = 180$. Since 180 ft is too wide, the width of the lawn is 30 ft, and the factory is 120 ft by 180 ft.

103. Let t be the time, in hours it takes Irene to wash all the windows. Then it takes Henry $t + \frac{3}{2}$ hours to wash all the windows, and the sum of the fraction of the job per hour they can do individually equals the fraction of the job they can do together. Since 1 hour 48 minutes $= 1 + \frac{48}{60} = 1 + \frac{4}{5} = \frac{9}{5}$, we have $\dfrac{1}{t} + \dfrac{1}{t + \frac{3}{2}} = \dfrac{1}{\frac{9}{5}} \Leftrightarrow$

$\dfrac{1}{t} + \dfrac{2}{2t + 3} = \dfrac{5}{9} \Rightarrow 9(2t + 3) + 2(9t) = 5t(2t + 3) \Leftrightarrow 18t + 27 + 18t = 10t^2 + 15t \Leftrightarrow 10t^2 - 21t - 27 = 0$

$\Leftrightarrow t = \dfrac{-(-21) \pm \sqrt{(-21)^2 - 4(10)(-27)}}{2(10)} = \dfrac{21 \pm \sqrt{441 + 1080}}{20} = \dfrac{21 \pm 39}{20}$. So $t = \dfrac{21 - 39}{20} = -\dfrac{9}{10}$

or $t = \dfrac{21 + 39}{20} = 3$. Since $t < 0$ is impossible, all the windows are washed by Irene alone in 3 hours and by Henry alone in $3 + \frac{3}{2} = 4\frac{1}{2}$ hours.

105. Let x be the distance from the center of the earth to the dead spot (in thousands of miles). Now setting $F = 0$, we have $0 = -\dfrac{K}{x^2} + \dfrac{0.012K}{(239 - x)^2} \Leftrightarrow \dfrac{K}{x^2} = \dfrac{0.012K}{(239 - x)^2} \Leftrightarrow K(239 - x)^2 = 0.012Kx^2 \Leftrightarrow$ $57121 - 478x + x^2 = 0.012x^2 \Leftrightarrow 0.988x^2 - 478x + 57121 = 0$. Using the Quadratic Formula, we obtain $x = \dfrac{-(-478) \pm \sqrt{(-478)^2 - 4(0.988)(57121)}}{2(0.988)} = \dfrac{478 \pm \sqrt{228484 - 225742.192}}{1.976} = \dfrac{478 \pm \sqrt{2741.808}}{1.976} \approx \dfrac{478 \pm 52.362}{1.976} \approx 241.903 \pm 26.499$. So either $x \approx 241.903 + 26.499 \approx 268$ or $x \approx 241.903 - 26.499 \approx 215$. Since 268 is greater than the distance from the earth to the moon, we reject it; thus $x \approx 215,000$ miles.

1.7   SOLVING OTHER TYPES OF EQUATIONS

Note: In cases where both sides of an equation are squared, the implication symbols $\Rightarrow$ and $\Leftrightarrow$ are sometimes used loosely. For example, $\sqrt{x} = x - 1$ "$\Leftrightarrow$" $\left(\sqrt{x}\right)^2 = (x - 1)^2$ is valid only for positive x. In these cases, inadmissible solutions are identified later in the solution.

1. (a) To solve the equation $x^3 - 4x^2 = 0$ we *factor* the left-hand side: $x^2(x - 4) = 0$, as above.

   (b) The solutions of the equation $x^2(x - 4) = 0$ are $x = 0$ and $x = 4$.

3. The equation $(x + 1)^2 - 5(x + 1) + 6 = 0$ is of *quadratic* type. To solve the equation we set $W = x + 1$. The resulting quadratic equation is $W^2 - 5W + 6 = 0 \Leftrightarrow (W - 3)(W - 2) = 0 \Leftrightarrow W = 2$ or $W = 3 \Leftrightarrow x + 1 = 2$ or $x + 1 = 3 \Leftrightarrow x = 1$ or $x = 2$. You can verify that these are both solutions to the original equation.

5. $x^3 = 64x \Leftrightarrow x^3 - 64x = 0 \Leftrightarrow x\left(x^2 - 64\right) = 0 \Leftrightarrow x(x - 8)(x + 8) = 0 \Leftrightarrow x = 0$, $x - 8 = 0$, or $x + 8 = 0$. Thus, the three real solutions are -8, 0, and 8.

7. $0 = x^6 - 81x^2 = x^2\left(x^4 - 81\right) = x^2\left(x^2 - 9\right)\left(x^2 + 9\right) = x^2(x - 3)(x + 3)\left(x^2 + 9\right)$. So $x^2 = 0 \Leftrightarrow x = 0$, or $x - 3 = 0 \Leftrightarrow x = 3$, or $x + 3 = 0 \Leftrightarrow x = -3$. $x^2 + 9 = 0 \Leftrightarrow x^2 = -9$, which has no real solution. The solutions are 0 and ± 3.

9. $0 = 4z^5 - 10z^2 = 2z^2\left(2z^3 - 5\right)$. If $2z^2 = 0$, then $z = 0$. If $2z^3 - 5 = 0$, then $2z^3 = 5 \Leftrightarrow z = \sqrt[3]{\frac{5}{2}}$. The solutions are 0 and $\sqrt[3]{\frac{5}{2}}$.

11. $0 = x^5 + 8x^2 = x^2\left(x^3 + 8\right) = x^2\left(x + 2\right)\left(x^2 - 2x + 4\right) \Leftrightarrow x^2 = 0, x + 2 = 0,$ or $x^2 - 2x + 4 = 0.$ If $x^2 = 0,$ then $x = 0;$ if $x + 2 = 0,$ then $x = -2,$ and $x^2 - 2x + 4 = 0$ has no real solution. Thus the solutions are $x = 0$ and $x = -2.$

13. $0 = x^3 - 5x^2 + 6x = x\left(x^2 - 5x + 6\right) = x\left(x - 2\right)\left(x - 3\right) \Leftrightarrow x = 0, x - 2 = 0,$ or $x - 3 = 0.$ Thus $x = 0,$ or $x = 2,$ or $x = 3.$ The solutions are $x = 0, x = 2,$ and $x = 3.$

15. $0 = x^4 + 4x^3 + 2x^2 = x^2\left(x^2 + 4x + 2\right).$ So either $x^2 = 0 \Leftrightarrow x = 0,$ or using the Quadratic Formula on $x^2 + 4x + 2 = 0,$ we have $x = \dfrac{-4 \pm \sqrt{4^2 - 4(1)(2)}}{2(1)} = \dfrac{-4 \pm \sqrt{16 - 8}}{2} = \dfrac{-4 \pm \sqrt{8}}{2} = \dfrac{-4 \pm 2\sqrt{2}}{2} = -2 \pm \sqrt{2}.$ The solutions are $0, -2 - \sqrt{2},$ and $-2 + \sqrt{2}.$

17. $(2r + 1)^6 - 16(2r + 1)^4 = 0.$ Let $x = 2r + 1.$ The equation becomes $x^6 - 16x^4 = 0 \Leftrightarrow x^4\left(x^2 - 16\right) = 0 \Leftrightarrow x^4\left(x + 4\right)\left(x - 4\right) = 0.$ If $x^4 = 0,$ then $x = 2r + 1 = 0 \Leftrightarrow r = -\frac{1}{2}.$ If $x + 4 = 0,$ then $2r + 1 + 4 = 0 \Leftrightarrow r = -\frac{5}{2}.$ If $x - 4 = 0,$ then $2r + 1 - 4 = 0 \Leftrightarrow r = \frac{3}{2}.$ Thus, the three solutions are $-\frac{5}{2}, -\frac{1}{2},$ and $\frac{3}{2}.$

19. $0 = x^3 - 5x^2 - 2x + 10 = x^2\left(x - 5\right) - 2\left(x - 5\right) = \left(x - 5\right)\left(x^2 - 2\right).$ If $x - 5 = 0,$ then $x = 5.$ If $x^2 - 2 = 0,$ then $x^2 = 2 \Leftrightarrow x = \pm\sqrt{2}.$ The solutions are 5 and $\pm\sqrt{2}.$

21. $x^3 - x^2 + x - 1 = x^2 + 1 \Leftrightarrow 0 = x^3 - 2x^2 + x - 2 = x^2\left(x - 2\right) + \left(x - 2\right) = \left(x - 2\right)\left(x^2 + 1\right).$ Since $x^2 + 1 = 0$ has no real solution, the only solution comes from $x - 2 = 0 \Leftrightarrow x = 2.$

23. $z + \dfrac{4}{z + 1} = 3 \Leftrightarrow (z + 1)\left(z + \dfrac{4}{z + 1}\right) = (z + 1)(3) \Leftrightarrow z^2 + z + 4 = 3z + 3 \Leftrightarrow z^2 - 2z + 1 = 0 \Leftrightarrow (z - 1)^2 = 0.$ The solution is $z = 1.$ We must check the original equation to make sure this value of z does not result in a zero denominator.

25. $\dfrac{1}{x - 1} + \dfrac{1}{x + 2} = \dfrac{5}{4} \Leftrightarrow 4\left(x - 1\right)\left(x + 2\right)\left(\dfrac{1}{x - 1} + \dfrac{1}{x + 2}\right) = 4\left(x - 1\right)\left(x + 2\right)\left(\dfrac{5}{4}\right) \Leftrightarrow$
$4\left(x + 2\right) + 4\left(x - 1\right) = 5\left(x - 1\right)\left(x + 2\right) \Leftrightarrow 4x + 8 + 4x - 4 = 5x^2 + 5x - 10 \Leftrightarrow 5x^2 - 3x - 14 = 0 \Leftrightarrow$
$(5x + 7)\left(x - 2\right) = 0.$ If $5x + 7 = 0,$ then $x = -\frac{7}{5};$ if $x - 2 = 0,$ then $x = 2.$ The solutions are $-\dfrac{7}{5}$ and $2.$

27. $\dfrac{x^2}{x + 100} = 50 \Leftrightarrow x^2 = 50\left(x + 100\right) = 50x + 5000 \Leftrightarrow x^2 - 50x - 5000 = 0 \Leftrightarrow \left(x - 100\right)\left(x + 50\right) = 0 \Leftrightarrow x - 100 = 0$ or $x + 50 = 0.$ Thus $x = 100$ or $x = -50.$ The solutions are 100 and $-50.$

29. $\dfrac{x + 5}{x - 2} = \dfrac{5}{x + 2} + \dfrac{28}{x^2 - 4} \Leftrightarrow \left(x + 2\right)\left(x + 5\right) = 5\left(x - 2\right) + 28 \Leftrightarrow x^2 + 7x + 10 = 5x - 10 + 28 \Leftrightarrow x^2 + 2x - 8 = 0 \Leftrightarrow$
$\left(x - 2\right)\left(x + 4\right) = 0 \Leftrightarrow x - 2 = 0$ or $x + 4 = 0 \Leftrightarrow x = 2$ or $x = -4.$ However, $x = 2$ is inadmissible because it results in a zero denominator in the original equation, so the only solution is $-4.$

31. $\dfrac{x}{2x + 7} - \dfrac{x + 1}{x + 3} = 1 \Leftrightarrow x\left(x + 3\right) - \left(x + 1\right)\left(2x + 7\right) = \left(2x + 7\right)\left(x + 3\right) \Leftrightarrow x^2 + 3x - 2x^2 - 9x - 7 = 2x^2 + 13x + 21$
$\Leftrightarrow 3x^2 + 19x + 28 = 0 \Leftrightarrow \left(3x + 7\right)\left(x + 4\right) = 0.$ Thus either $3x + 7 = 0,$ so $x = -\frac{7}{3},$ or $x = -4.$ The solutions are $-\dfrac{7}{3}$ and $-4.$

33. $\dfrac{x + \frac{2}{x}}{3 + \frac{4}{x}} = 5x \Leftrightarrow \left(\dfrac{x + \frac{2}{x}}{3 + \frac{4}{x}}\right) \cdot \dfrac{x}{x} = \dfrac{x^2 + 2}{3x + 4} = 5x \Leftrightarrow x^2 + 2 = 5x\left(3x + 4\right) \Leftrightarrow x^2 + 2 = 15x^2 + 20x \Leftrightarrow 0 = 14x^2 + 20x - 2$
$\Leftrightarrow x = \dfrac{-(20) \pm \sqrt{(20)^2 - 4(14)(-2)}}{2(14)} = \dfrac{-20 \pm \sqrt{400 + 112}}{28} = \dfrac{-20 \pm \sqrt{512}}{28} = \dfrac{-20 \pm 16\sqrt{2}}{28} = \dfrac{-5 \pm 4\sqrt{2}}{7}.$ The solutions are $\dfrac{-5 \pm 4\sqrt{2}}{7}.$

35. $5 = \sqrt{4x - 3} \Leftrightarrow 5^2 = \left(\sqrt{4x - 3}\right)^2 \Leftrightarrow 25 = 4x - 3 \Leftrightarrow 4x = 28 \Leftrightarrow x = 7$ is a potential solution. Substituting into the original equation, we get $5 = \sqrt{4(7) - 3} \Leftrightarrow 5 = \sqrt{25},$ which is true, so the solution is $x = 7.$

37. $\sqrt{2x-1} = \sqrt{3x-5} \Leftrightarrow \left(\sqrt{2x-1}\right)^2 = \left(\sqrt{3x-5}\right)^2 \Leftrightarrow 2x-1 = 3x-5 \Leftrightarrow x = 4$. Substituting into the original equation, we get $\sqrt{2(4)-1} = \sqrt{3(4)-5} \Leftrightarrow \sqrt{7} = \sqrt{7}$, which is true, so the solution is $x = 4$.

39. $\sqrt{x+2} = x \Leftrightarrow \left(\sqrt{x+2}\right)^2 = x^2 \Leftrightarrow x+2 = x^2 \Leftrightarrow x^2 - x - 2 = (x+1)(x-2) = 0 \Leftrightarrow x = -1$ or $x = 2$. Substituting into the original equation, we get $\sqrt{(-1)+2} = -1 \Leftrightarrow \sqrt{1} = -1$, which is false, and $\sqrt{2+2} = 2 \Leftrightarrow \sqrt{4} = 2$, which is true. So $x = 2$ is the only real solution.

41. $\sqrt{2x+1} + 1 = x \Leftrightarrow \sqrt{2x+1} = x - 1 \Leftrightarrow 2x+1 = (x-1)^2 \Leftrightarrow 2x+1 = x^2 - 2x + 1 \Leftrightarrow 0 = x^2 - 4x = x(x-4)$. Potential solutions are $x = 0$ and $x - 4 \Leftrightarrow x = 4$. These are only potential solutions since squaring is not a reversible operation. We must check each potential solution in the original equation.
Checking $x = 0$: $\sqrt{2(0)+1} + 1 = (0) \Leftrightarrow \sqrt{1} + 1 = 0$ is false.
Checking $x = 4$: $\sqrt{2(4)+1} + 1 = (4) \Leftrightarrow \sqrt{9} + 1 = 4 \Leftrightarrow 3 + 1 = 4$ is true. The only solution is $x = 4$.

43. $x - \sqrt{x-1} = 3 \Leftrightarrow x - 3 = \sqrt{x-1} \Leftrightarrow (x-3)^2 = \left(\sqrt{x-1}\right)^2 \Leftrightarrow x^2 - 6x + 9 = x - 1 \Leftrightarrow x^2 - 7x + 10 = 0 \Leftrightarrow (x-2)(x-5) = 0$. Potential solutions are $x = 2$ and $x = 5$. We must check each potential solution in the original equation. Checking $x = 2$: $2 - \sqrt{2-1} = 3$, which is false, so $x = 2$ is not a solution. Checking $x = 5$: $5 - \sqrt{5-1} = 3 \Leftrightarrow 5 - 2 = 3$, which is true, so $x = 5$ is the only solution.

45. $x - \sqrt{x+3} = \frac{x}{2} \Leftrightarrow \frac{x}{2} = \sqrt{x+3} \Leftrightarrow \left(\frac{1}{2}x\right)^2 = x+3 \Leftrightarrow \frac{1}{4}x^2 = x+3 \Leftrightarrow x^2 = 4x+12 \Leftrightarrow 0 = x^2 - 4x - 12 = (x-6)(x+2)$. Potential solutions are $x = 6$ and $x = -2$. We must check each potential solution in the original equation. Checking $x = 6$: $6 - \sqrt{6+3} = \frac{6}{2} \Leftrightarrow 6 - \sqrt{6+3} = 6 - \sqrt{9} = 6 - 3$, which is true. Checking $x = -2$: $-2 - \sqrt{-2+3} = -\frac{2}{2} \Leftrightarrow -2 - \sqrt{-2+3} = -2 - \sqrt{1} = -2 - 1 = -3 = -\frac{2}{2} \Leftrightarrow -3 = -1$, hence this not a solution. The only solution is $x = 6$.

47. $0 = (x+5)^2 - 3(x+5) - 10 = [(x+5)-5][(x+5)+2] = x(x+7) \Leftrightarrow x = 0$ or $x = -7$. The solutions are 0 and -7.

49. Let $w = \dfrac{1}{x+1}$. Then $\left(\dfrac{1}{x+1}\right)^2 - 2\left(\dfrac{1}{x+1}\right) - 8 = 0$ becomes $w^2 - 2w - 8 = 0 \Leftrightarrow (w-4)(w+2) = 0$. So $w - 4 = 0 \Leftrightarrow w = 4$, and $w + 2 = 0 \Leftrightarrow w = -2$. When $w = 4$, we have $\dfrac{1}{x+1} = 4 \Leftrightarrow 1 = 4x + 4 \Leftrightarrow -3 = 4x \Leftrightarrow x = -\frac{3}{4}$. When $w = -2$, we have $\dfrac{1}{x+1} = -2 \Leftrightarrow 1 = -2x - 2 \Leftrightarrow 3 = -2x \Leftrightarrow x = -\frac{3}{2}$. Solutions are $-\frac{3}{4}$ and $-\frac{3}{2}$.

51. Let $w = x^2$. Then $x^4 - 13x^2 + 40 = \left(x^2\right)^2 - 13x^2 + 40 = 0$ becomes $w^2 - 13w + 40 = 0 \Leftrightarrow (w-5)(w-8) = 0$. So $w - 5 = 0 \Leftrightarrow w = 5$, and $w - 8 = 0 \Leftrightarrow w = 8$. When $w = 5$, we have $x^2 = 5 \Leftrightarrow x = \pm\sqrt{5}$. When $w = 8$, we have $x^2 = 8 \Leftrightarrow x = \pm\sqrt{8} = \pm 2\sqrt{2}$. The solutions are $\pm\sqrt{5}$ and $\pm 2\sqrt{2}$.

53. $2x^4 + 4x^2 + 1 = 0$. The LHS is the sum of two nonnegative numbers and a positive number, so $2x^4 + 4x^2 + 1 \geq 1 \neq 0$. This equation has no real solution.

55. $0 = x^6 - 26x^3 - 27 = \left(x^3 - 27\right)\left(x^3 + 1\right)$. If $x^3 - 27 = 0 \Leftrightarrow x^3 = 27$, so $x = 3$. If $x^3 + 1 = 0 \Leftrightarrow x^3 = -1$, so $x = -1$. The solutions are 3 and -1.

57. Let $u = x^{2/3}$. Then $0 = x^{4/3} - 5x^{2/3} + 6$ becomes $u^2 - 5u + 6 = 0 \Leftrightarrow (u-3)(u-2) = 0 \Leftrightarrow u - 3 = 0$ or $u - 2 = 0$. If $u - 3 = 0$, then $x^{2/3} - 3 = 0 \Leftrightarrow x^{2/3} = 3 \Leftrightarrow x = \pm 3^{3/2} = \pm 3\sqrt{3}$. If $u - 2 = 0$, then $x^{2/3} - 2 = 0 \Leftrightarrow x^{2/3} = 2 \Leftrightarrow x = \pm 2^{3/2} = 2\sqrt{2}$. The solutions are $\pm 3\sqrt{3}$ and $\pm 2\sqrt{2}$.

59. $4(x+1)^{1/2} - 5(x+1)^{3/2} + (x+1)^{5/2} = 0 \Leftrightarrow \sqrt{x+1}\left[4 - 5(x+1) + (x+1)^2\right] = 0 \Leftrightarrow$
$\sqrt{x+1}\left(4 - 5x - 5 + x^2 + 2x + 1\right) = 0 \Leftrightarrow \sqrt{x+1}\left(x^2 - 3x\right) = 0 \Leftrightarrow \sqrt{x+1} \cdot x(x-3) = 0 \Leftrightarrow x = -1$ or $x = 0$ or $x = 3$. The solutions are -1, 0, and 3.

61. $0 = x^{3/2} + 8x^{1/2} + 16x^{-1/2} = x^{-1/2}\left(x^2 + 8x + 16\right) = x^{-1/2}(x+4)^2$. Now $x^{-1/2} = \dfrac{1}{\sqrt{x}}$ so $x \neq 0$. So $x + 4 = 0 \Leftrightarrow$

$x = -4$. But $\sqrt{-4}$ is not a real number, so this equation has no real solution. Alternatively, we see that this is the sum of three positive real numbers (remember $x \neq 0$), so it never equals zero.

63. Let $u = x^{1/6}$. (We choose the exponent $\frac{1}{6}$ because the LCD of 2, 3, and 6 is 6.) Then $x^{1/2} - 3x^{1/3} = 3x^{1/6} - 9 \Leftrightarrow$

$x^{3/6} - 3x^{2/6} = 3x^{1/6} - 9 \Leftrightarrow u^3 - 3u^2 = 3u - 9 \Leftrightarrow 0 = u^3 - 3u^2 - 3u + 9 = u^2(u-3) - 3(u-3) = (u-3)\left(u^2 - 3\right)$.

So $u - 3 = 0$ or $u^2 - 3 = 0$. If $u - 3 = 0$, then $x^{1/6} - 3 = 0 \Leftrightarrow x^{1/6} = 3 \Leftrightarrow x = 3^6 = 729$. If $u^2 - 3 = 0$, then

$x^{1/3} - 3 = 0 \Leftrightarrow x^{1/3} = 3 \Leftrightarrow x = 3^3 = 27$. The solutions are 729 and 27.

65. $\dfrac{1}{x^3} + \dfrac{4}{x^2} + \dfrac{4}{x} = 0 \Leftrightarrow 1 + 4x + 4x^2 = 0 \Leftrightarrow (1 + 2x)^2 = 0 \Leftrightarrow 1 + 2x = 0 \Leftrightarrow 2x = -1 \Leftrightarrow x = -\frac{1}{2}$. The solution is $-\frac{1}{2}$.

67. $\sqrt{\sqrt{x+5} + x} = 5$. Squaring both sides, we get $\sqrt{x+5} + x = 25 \Leftrightarrow \sqrt{x+5} = 25 - x$. Squaring both sides again, we

get $x + 5 = (25 - x)^2 \Leftrightarrow x + 5 = 625 - 50x + x^2 \Leftrightarrow 0 = x^2 - 51x + 620 = (x - 20)(x - 31)$. Potential solutions are

$x = 20$ and $x = 31$. We must check each potential solution in the original equation.

Checking $x = 20$: $\sqrt{\sqrt{20+5} + 20} = 5 \Leftrightarrow \sqrt{\sqrt{25} + 20} = 5 \Leftrightarrow \sqrt{5 + 20} = 5$, which is true, and hence $x = 20$ is a

solution.

Checking $x = 31$: $\sqrt{\sqrt{(31)+5} + 31} = 5 \Leftrightarrow \sqrt{\sqrt{36} + 31} = 5 \Leftrightarrow \sqrt{37} = 5$, which is false, and hence $x = 31$ is not a

solution. The only real solution is $x = 20$.

69. $x^2\sqrt{x+3} = (x+3)^{3/2} \Leftrightarrow 0 = x^2\sqrt{x+3} - (x+3)^{3/2} \Leftrightarrow 0 = \sqrt{x+3}\left[\left(x^2\right) - (x+3)\right] \Leftrightarrow 0 = \sqrt{x+3}\left(x^2 - x - 3\right)$.

If $(x+3)^{1/2} = 0$, then $x + 3 = 0 \Leftrightarrow x = -3$. If $x^2 - x - 3 = 0$, then using the Quadratic Formula $x = \dfrac{1 \pm \sqrt{13}}{2}$. The

solutions are -3 and $\dfrac{1 \pm \sqrt{13}}{2}$.

71. $\sqrt{x + \sqrt{x+2}} = 2$. Squaring both sides, we get $x + \sqrt{x+2} = 4 \Leftrightarrow \sqrt{x+2} = 4 - x$. Squaring both sides again, we get

$x + 2 = (4 - x)^2 = 16 - 8x + x^2 \Leftrightarrow 0 = x^2 - 9x + 14 \Leftrightarrow 0 = (x-7)(x-2)$. If $x - 7 = 0$, then $x = 7$. If $x - 2 = 0$,

then $x = 2$. So $x = 2$ is a solution but $x = 7$ is not, since it does not satisfy the original equation.

73. We graph $y = x^3 - 5x^2 - 2x + 10$ in the viewing rectangle $[-4, 7]$ by $[-20, 15]$.

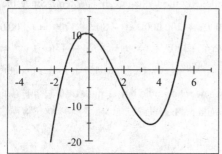

From the graph, we see that the roots are 5 and approximately -1.41 and 1.41. These agree with the values found in Exercise 19.

75. We graph $y = x^{4/3} - 5x^{2/3} + 6$ in the viewing rectangle $[-8, 8]$ by $[-1, 6]$. Note that some graphing devices have trouble with this type of expression; we can rewrite

$x^{4/3} - 5x^{2/3} + 6$ as $(-x)^{4/3} - 5(-x)^{2/3} + 6$ for $x \leq 0$ if necessary.

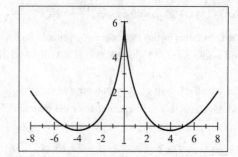

From the graph, we see that the roots are approximately -5.20, -2.83, 2.83, and 5.20. These agree with the values found in Exercise 57.

77. $0 = x^4 - 5ax^2 + 4a^2 = \left(a - x^2\right)\left(4a - x^2\right)$. Since a is positive, $a - x^2 = 0 \Leftrightarrow x^2 = a \Leftrightarrow x = \sqrt{a}$. Again, since a is

positive, $4a - x^2 = 0 \Leftrightarrow x^2 = 4a \Leftrightarrow x = \pm 2\sqrt{a}$. Thus the four solutions are $\pm\sqrt{a}$ and $\pm 2\sqrt{a}$.

79. $\sqrt{x+a} + \sqrt{x-a} = \sqrt{2}\sqrt{x+6}$. Squaring both sides, we have

$x + a + 2\left(\sqrt{x+a}\right)\left(\sqrt{x-a}\right) + x - a = 2(x+6) \Leftrightarrow 2x + 2\left(\sqrt{x+a}\right)\left(\sqrt{x-a}\right) = 2x + 12 \Leftrightarrow 2\left(\sqrt{x+a}\right)\left(\sqrt{x-a}\right) = 12$

$\Leftrightarrow \left(\sqrt{x+a}\right)\left(\sqrt{x-a}\right) = 6$. Squaring both sides again we have $(x+a)(x-a) = 36 \Leftrightarrow x^2 - a^2 = 36 \Leftrightarrow x^2 = a^2 + 36$

$\Leftrightarrow x = \pm\sqrt{a^2+36}$. Checking these answers, we see that $x = -\sqrt{a^2+36}$ is not a solution (for example, try substituting $a = 8$), but $x = \sqrt{a^2+36}$ is a solution.

81. Let x be the number of people originally intended to take the trip. Then originally, the cost of the trip is $\dfrac{900}{x}$. After 5 people

cancel, there are now $x - 5$ people, each paying $\dfrac{900}{x} + 2$. Thus $900 = (x-5)\left(\dfrac{900}{x} + 2\right) \Leftrightarrow 900 = 900 + 2x - \dfrac{4500}{x} - 10$

$\Leftrightarrow 0 = 2x - 10 - \dfrac{4500}{x} \Leftrightarrow 0 = 2x^2 - 10x - 4500 = (2x - 100)(x + 45)$. Thus either $2x - 100 = 0$, so $x = 50$, or

$x + 45 = 0$, $x = -45$. Since the number of people on the trip must be positive, originally 50 people intended to take the trip.

83. We want to solve for t when $P = 500$. Letting $u = \sqrt{t}$ and substituting, we have $500 = 3t + 10\sqrt{t} + 140 \Leftrightarrow$

$500 = 3u^2 + 10u + 140 \Leftrightarrow 0 = 3u^2 + 10u - 360 \Leftrightarrow u = \dfrac{-5 \pm \sqrt{1105}}{3}$. Since $u = \sqrt{t}$, we must have $u \geq 0$. So

$\sqrt{t} = u = \dfrac{-5 + \sqrt{1105}}{3} \approx 9.414 \Leftrightarrow t = \approx 88.62$. So it will take 89 days for the fish population to reach 500.

85. Let x be the height of the pile in feet. Then the diameter is $3x$ and the radius is $\frac{3}{2}x$ feet. Since the volume of the cone is

1000 ft^3, we have $\dfrac{\pi}{3}\left(\dfrac{3x}{2}\right)^2 x = 1000 \Leftrightarrow \dfrac{3\pi x^3}{4} = 1000 \Leftrightarrow x^3 = \dfrac{4000}{3\pi} \Leftrightarrow x = \sqrt[3]{\dfrac{4000}{3\pi}} \approx 7.52$ feet.

87. Let r be the radius of the larger sphere, in mm. Equating the volumes, we have $\frac{4}{3}\pi r^3 = \frac{4}{3}\pi\left(2^3 + 3^3 + 4^3\right) \Leftrightarrow$

$r^3 = 2^3 + 3^3 + 4^4 \Leftrightarrow r^3 = 99 \Leftrightarrow r = \sqrt[3]{99} \approx 4.63$. Therefore, the radius of the larger sphere is about 4.63 mm.

89. Let x be the length, in miles, of the abandoned road to be used. Then the length of the abandoned road not used

is $40 - x$, and the length of the new road is $\sqrt{10^2 + (40-x)^2}$ miles, by the Pythagorean Theorem. Since the

cost of the road is cost per mile $\times$ number of miles, we have $100{,}000x + 200{,}000\sqrt{x^2 - 80x + 1700} = 6{,}800{,}000$

$\Leftrightarrow 2\sqrt{x^2 - 80x + 1700} = 68 - x$. Squaring both sides, we get $4x^2 - 320x + 6800 = 4624 - 136x + x^2 \Leftrightarrow$

$3x^2 - 184x + 2176 = 0 \Leftrightarrow x = \frac{184 \pm \sqrt{33856 - 26112}}{6} = \frac{184 \pm 88}{6} \Leftrightarrow x = \frac{136}{3}$ or $x = 16$. Since $45\frac{1}{3}$ is longer than the existing

road, 16 miles of the abandoned road should be used. A completely new road would have length $\sqrt{10^2 + 40^2}$ (let $x = 0$)

and would cost $\sqrt{1700} \times 200{,}000 \approx 8.3$ million dollars. So no, it would not be cheaper.

91. Let x be the length of the hypotenuse of the triangle, in feet. Then one of the other

sides has length $x - 7$ feet, and since the perimeter is 392 feet, the remaining side

must have length $392 - x - (x - 7) = 399 - 2x$. From the Pythagorean Theorem,

we get $(x-7)^2 + (399-2x)^2 = x^2 \Leftrightarrow 4x^2 - 1610x + 159250 = 0$. Using the

Quadratic Formula, we get

$x = \frac{1610 \pm \sqrt{1610^2 - 4(4)(159250)}}{2(4)} = \frac{1610 \pm \sqrt{44100}}{8} = \frac{1610 \pm 210}{8}$, and so $x = 227.5$ or $x = 175$. But if $x = 227.5$, then the

side of length $x - 7$ combined with the hypotenuse already exceeds the perimeter of 392 feet, and so we must have $x = 175$.

Thus the other sides have length $175 - 7 = 168$ and $399 - 2(175) = 49$. The lot has sides of length 49 feet, 168 feet, and

175 feet.

93. Since the total time is 3 s, we have $3 = \dfrac{\sqrt{d}}{4} + \dfrac{d}{1090}$. Letting $w = \sqrt{d}$, we have $3 = \frac{1}{4}w + \frac{1}{1090}w^2 \Leftrightarrow \frac{1}{1090}w^2 + \frac{1}{4}w - 3 = 0$

$\Leftrightarrow 2w^2 + 545w - 6540 = 0 \Leftrightarrow w = \dfrac{-545 \pm 591.054}{4}$. Since $w \geq 0$, we have $\sqrt{d} = w \approx 11.51$, so $d = 132.56$. The well is 132.6 ft deep.

1.8   SOLVING INEQUALITIES

1. (a) If $x < 5$, then $x - 3 < 5 - 3 \Rightarrow x - 3 < 2$.

(b) If $x \leq 5$, then $3 \cdot x \leq 3 \cdot 5 \Rightarrow 3x \leq 15$.

(c) If $x \geq 2$, then $-3 \cdot x \leq -3 \cdot 2 \Rightarrow -3x \leq -6$.

(d) If $x < -2$, then $-x > 2$.

3. From the graph, we see that where $-1 \leq x \leq 0$ or $1 \leq x \leq 3$, the graph lies below the x-axis. Thus, the inequality $x^4 - 3x^3 - x^2 + 3x \leq 0$ is satisfied for $\{x \mid -1 \leq x \leq 0 \text{ or } 1 \leq x \leq 3\} = [-1, 0] \cup [1, 3]$.

5. $x = -2$: $-2 - 3 \overset{?}{>} 0$. No, $-5 \not> 0$. $x = -1$: $-1 - 3 \overset{?}{>} 0$. No, $-4 \not> 0$. $x = 0$: $0 - 3 \overset{?}{>} 0$. No, $-3 \not> 0$.

$x = \frac{1}{2}$: $\frac{1}{2} - 3 \overset{?}{>} 0$. No, $-\frac{5}{2} \not> 0$. $x = 1$: $1 - 3 \overset{?}{>} 0$. No, $-2 \not> 0$. $x = \sqrt{2}$: $\sqrt{2} - 3 \overset{?}{>} 0$. No, $\sqrt{2} - 3 \not> 0$.

$x = 2$: $2 - 3 \overset{?}{>} 0$. No, $-1 \not> 0$. $x = 4$: $4 - 3 \overset{?}{>} 0$. Yes, $1 > 0$. Only 4 satisfies the inequality.

7. $x = -2$: $3 - 2(-2) \overset{?}{\leq} \frac{1}{2}$. No, $7 \not\leq \frac{1}{2}$. $x = -1$: $3 - 2(-1) \overset{?}{\leq} \frac{1}{2}$. No, $6 \not\leq \frac{1}{2}$. $x = 0$: $3 - 2(0) \overset{?}{\leq} \frac{1}{2}$. No, $3 \not\leq \frac{1}{2}$.

$x = \frac{1}{2}$: $3 - 2\left(\frac{1}{2}\right) \overset{?}{\leq} \frac{1}{2}$. No, $2 \not\leq \frac{1}{2}$. $x = 1$: $3 - 2(1) \overset{?}{\leq} \frac{1}{2}$. No, $1 \not\leq \frac{1}{2}$.

$x = \sqrt{2}$: $3 - 2\left(\sqrt{2}\right) \overset{?}{\leq} \frac{1}{2}$. Yes, $3 - 2\sqrt{2} \leq \frac{1}{2}$. $x = 2$: $3 - 2(2) \overset{?}{\leq} \frac{1}{2}$. Yes, $-1 \leq \frac{1}{2}$.

$x = 4$: $3 - 2(4) \overset{?}{\leq} \frac{1}{2}$. Yes, $5 \leq \frac{1}{2}$. The elements $\sqrt{2}$, 2, and 4 satisfy the inequality.

9. $x = -2$: $1 \overset{?}{<} 2(-2) - 4 \overset{?}{\leq} 7$. No, $2(-2) - 4 = -8$ and $1 \not< -8$.

$x = -1$: $1 \overset{?}{<} 2(-1) - 4 \overset{?}{\leq} 7$. No, $2(-1) - 4 = -6$ and $1 \not< -6$.

$x = 0$: $1 \overset{?}{<} 2(0) - 4 \overset{?}{\leq} 7$. No, $2(0) - 4 = -4$ and $1 \not< -4$.

$x = \frac{1}{2}$: $1 \overset{?}{<} 2\left(\frac{1}{2}\right) - 4 \overset{?}{\leq} 7$. No, $2\left(\frac{1}{2}\right) - 4 = -3$ and $1 \not< -3$.

$x = 1$: $1 \overset{?}{<} 2(1) - 4 \overset{?}{\leq} 7$. No, $2(1) - 4 = -2$ and $1 \not< -2$.

$x = \sqrt{2}$: $1 \overset{?}{<} 2\left(\sqrt{2}\right) - 4 \overset{?}{\leq} 7$. No, $2\sqrt{2} - 4 < 0$ and $1 \not< 0$.

$x = 2$: $1 \overset{?}{<} 2(2) - 4 \overset{?}{\leq} 7$. No, $2(1) - 4 = -2$ and $1 \not< -2$.

$x = 4$: $1 \overset{?}{<} 2(4) - 4 \overset{?}{\leq} 7$. Yes, $2(4) - 4 = 4$ and $1 < 4 \leq 7$. Only 4 satisfies the inequality.

11. $x = -2$: $\frac{1}{(-2)} \overset{?}{\leq} \frac{1}{2}$. Yes, $-\frac{1}{2} \leq \frac{1}{2}$. $x = -1$: $\frac{1}{(-1)} \overset{?}{\leq} \frac{1}{2}$. Yes, $-1 \leq \frac{1}{2}$. $x = 0$: $\frac{1}{0} \overset{?}{\leq} \frac{1}{2}$. No, $\frac{1}{0}$ is not defined.

$x = \frac{1}{2}$: $\frac{1}{1/2} \overset{?}{\leq} \frac{1}{2}$. No, $2 \not\leq \frac{1}{2}$. $x = 1$: $\frac{1}{1} \overset{?}{\leq} \frac{1}{2}$. No, $1 \not\leq \frac{1}{2}$. $x = \sqrt{2}$: $\frac{1}{\sqrt{2}} \overset{?}{\leq} \frac{1}{2}$. No, $2 \not\leq \sqrt{2}$.

$x = 2$: $\frac{1}{2} \overset{?}{\leq} \frac{1}{2}$. Yes, $\frac{1}{2} \leq \frac{1}{2}$. $x = 4$: $\frac{1}{4} \overset{?}{\leq} \frac{1}{2}$. Yes, $\frac{1}{4} \leq \frac{1}{2}$. The elements -2, -1, 2, and 4 satisfy the inequality.

13. $x = -2$: $1 - 2^2 \overset{?}{\leq} -1$. Yes, $-3 \leq -1$. $x = -1$: $1 - (-1)^2 \overset{?}{\leq} -1$. No, $0 \not\leq -1$. $x = 0$: $1 - 0^2 \overset{?}{\leq} -1$. No, $1 \not\leq -1$.

$x = \frac{1}{2}$: $1 - \left(\frac{1}{2}\right)^2 \overset{?}{\leq} -1$. No, $\frac{3}{4} \not\leq -1$. $x = 1$: $1 - 1^2 \overset{?}{\leq} -1$. No, $0 \not\leq -1$. $x = \sqrt{2}$: $1 - \left(\sqrt{2}\right)^2 \overset{?}{\leq} -1$. Yes, $-1 \leq -1$.

$x = 2$: $1 - 2^2 \overset{?}{\leq} -1$. Yes, $-3 \leq -1$. $x = 4$: $1 - 4^2 \overset{?}{\leq} -1$. Yes, $-15 \leq -1$. The elements -2, $\sqrt{2}$, 2, and 4 satisfy the inequality.

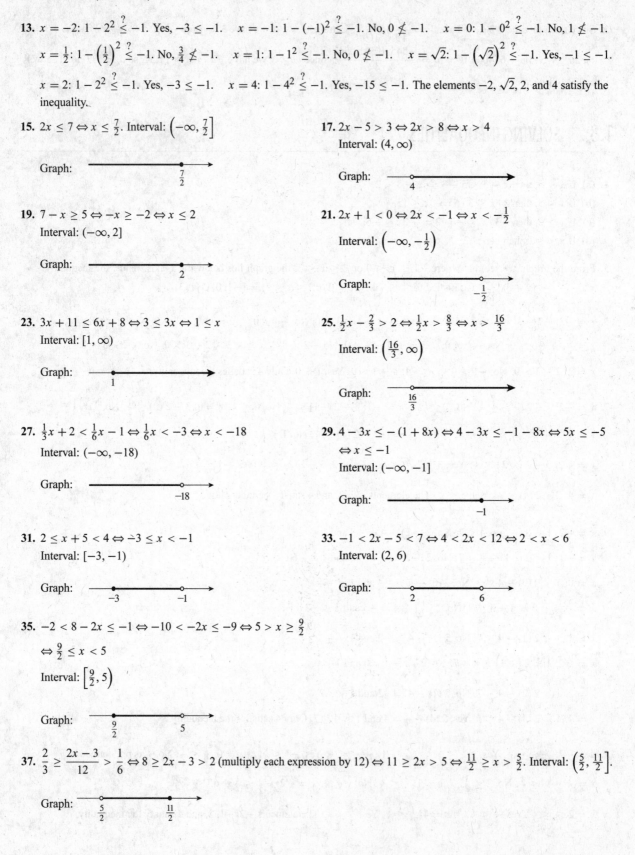

15. $2x \leq 7 \Leftrightarrow x \leq \frac{7}{2}$. Interval: $\left(-\infty, \frac{7}{2}\right]$

Graph:

17. $2x - 5 > 3 \Leftrightarrow 2x > 8 \Leftrightarrow x > 4$

Interval: $(4, \infty)$

Graph:

19. $7 - x \geq 5 \Leftrightarrow -x \geq -2 \Leftrightarrow x \leq 2$

Interval: $(-\infty, 2]$

Graph:

21. $2x + 1 < 0 \Leftrightarrow 2x < -1 \Leftrightarrow x < -\frac{1}{2}$

Interval: $\left(-\infty, -\frac{1}{2}\right)$

Graph:

23. $3x + 11 \leq 6x + 8 \Leftrightarrow 3 \leq 3x \Leftrightarrow 1 \leq x$

Interval: $[1, \infty)$

Graph:

25. $\frac{1}{2}x - \frac{2}{3} > 2 \Leftrightarrow \frac{1}{2}x > \frac{8}{3} \Leftrightarrow x > \frac{16}{3}$

Interval: $\left(\frac{16}{3}, \infty\right)$

Graph:

27. $\frac{1}{3}x + 2 < \frac{1}{6}x - 1 \Leftrightarrow \frac{1}{6}x < -3 \Leftrightarrow x < -18$

Interval: $(-\infty, -18)$

Graph:

29. $4 - 3x \leq -(1 + 8x) \Leftrightarrow 4 - 3x \leq -1 - 8x \Leftrightarrow 5x \leq -5$

$\Leftrightarrow x \leq -1$

Interval: $(-\infty, -1]$

Graph:

31. $2 \leq x + 5 < 4 \Leftrightarrow -3 \leq x < -1$

Interval: $[-3, -1)$

Graph:

33. $-1 < 2x - 5 < 7 \Leftrightarrow 4 < 2x < 12 \Leftrightarrow 2 < x < 6$

Interval: $(2, 6)$

Graph:

35. $-2 < 8 - 2x \leq -1 \Leftrightarrow -10 < -2x \leq -9 \Leftrightarrow 5 > x \geq \frac{9}{2}$

$\Leftrightarrow \frac{9}{2} \leq x < 5$

Interval: $\left[\frac{9}{2}, 5\right)$

Graph:

37. $\frac{2}{3} \geq \frac{2x - 3}{12} > \frac{1}{6} \Leftrightarrow 8 \geq 2x - 3 > 2$ (multiply each expression by 12) $\Leftrightarrow 11 \geq 2x > 5 \Leftrightarrow \frac{11}{2} \geq x > \frac{5}{2}$. Interval: $\left(\frac{5}{2}, \frac{11}{2}\right]$.

Graph:

39. $(x+2)(x-3) < 0$. The expression on the left of the inequality changes sign where $x = -2$ and where $x = 3$. Thus we must check the intervals in the following table.

Interval	$(-\infty, -2)$	$(-2, 3)$	$(3, \infty)$
Sign of $x + 2$	$-$	$+$	$+$
Sign of $x - 3$	$-$	$-$	$+$
Sign of $(x + 2)(x - 3)$	$+$	$-$	$+$

From the table, the solution set is $\{x \mid -2 < x < 3\}$. Interval: $(-2, 3)$.

Graph:

41. $x(2x + 7) \geq 0$. The expression on the left of the inequality changes sign where $x = 0$ and where $x = -\frac{7}{2}$. Thus we must check the intervals in the following table.

Interval	$\left(-\infty, -\frac{7}{2}\right)$	$\left(-\frac{7}{2}, 0\right)$	$(0, \infty)$
Sign of x	$-$	$-$	$+$
Sign of $2x + 7$	$-$	$+$	$+$
Sign of $x(2x + 7)$	$+$	$-$	$+$

From the table, the solution set is $\left\{x \mid x \leq -\frac{7}{2} \text{ or } 0 \leq x\right\}$.
Interval: $\left(-\infty, -\frac{7}{2}\right] \cup [0, \infty)$.

Graph:

43. $x^2 - 3x - 18 \leq 0 \Leftrightarrow (x + 3)(x - 6) \leq 0$. The expression on the left of the inequality changes sign where $x = 6$ and where $x = -3$. Thus we must check the intervals in the following table.

Interval	$(-\infty, -3)$	$(-3, 6)$	$(6, \infty)$
Sign of $x + 3$	$-$	$+$	$+$
Sign of $x - 6$	$-$	$-$	$+$
Sign of $(x + 3)(x - 6)$	$+$	$-$	$+$

From the table, the solution set is $\{x \mid -3 \leq x \leq 6\}$. Interval: $[-3, 6]$.

Graph:

45. $2x^2 + x \geq 1 \Leftrightarrow 2x^2 + x - 1 \geq 0 \Leftrightarrow (x + 1)(2x - 1) \geq 0$. The expression on the left of the inequality changes sign where $x = -1$ and where $x = \frac{1}{2}$. Thus we must check the intervals in the following table.

Interval	$(-\infty, -1)$	$\left(-1, \frac{1}{2}\right)$	$\left(\frac{1}{2}, \infty\right)$
Sign of $x + 1$	$-$	$+$	$+$
Sign of $2x - 1$	$-$	$-$	$+$
Sign of $(x + 1)(2x - 1)$	$+$	$-$	$+$

From the table, the solution set is $\left\{x \mid x \leq -1 \text{ or } \frac{1}{2} \leq x\right\}$.
Interval: $(-\infty, -1] \cup \left[\frac{1}{2}, \infty\right)$.

Graph:

47. $3x^2 - 3x < 2x^2 + 4 \Leftrightarrow x^2 - 3x - 4 < 0 \Leftrightarrow (x + 1)(x - 4) < 0$. The expression on the left of the inequality changes sign where $x = -1$ and where $x = 4$. Thus we must check the intervals in the following table.

Interval	$(-\infty, -1)$	$(-1, 4)$	$(4, \infty)$
Sign of $x + 1$	$-$	$+$	$+$
Sign of $x - 4$	$-$	$-$	$+$
Sign of $(x + 1)(x - 4)$	$+$	$-$	$+$

From the table, the solution set is $\{x \mid -1 < x < 4\}$. Interval: $(-1, 4)$.

Graph:

49. $x^2 > 3(x+6) \Leftrightarrow x^2 - 3x - 18 > 0 \Leftrightarrow (x+3)(x-6) > 0$. The expression on the left of the inequality changes sign where $x = 6$ and where $x = -3$. Thus we must check the intervals in the following table.

Interval	$(-\infty, -3)$	$(-3, 6)$	$(6, \infty)$
Sign of $x + 3$	$-$	$+$	$+$
Sign of $x - 6$	$-$	$-$	$+$
Sign of $(x+3)(x-6)$	$+$	$-$	$+$

From the table, the solution set is $\{x \mid x < -3 \text{ or } 6 < x\}$.
Interval: $(-\infty, -3) \cup (6, \infty)$.

Graph: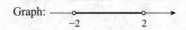

51. $x^2 < 4 \Leftrightarrow x^2 - 4 < 0 \Leftrightarrow (x+2)(x-2) < 0$. The expression on the left of the inequality changes sign where $x = -2$ and where $x = 2$. Thus we must check the intervals in the following table.

Interval	$(-\infty, -2)$	$(-2, 2)$	$(2, \infty)$
Sign of $x + 2$	$-$	$+$	$+$
Sign of $x - 2$	$-$	$-$	$+$
Sign of $(x+2)(x-2)$	$+$	$-$	$+$

From the table, the solution set is $\{x \mid -2 < x < 2\}$. Interval: $(-2, 2)$.

Graph:

53. $(x+2)(x-1)(x-3) \le 0$. The expression on the left of the inequality changes sign when $x = -2$, $x = 1$, and $x = 3$. Thus we must check the intervals in the following table.

Interval	$(-\infty, -2)$	$(-2, 1)$	$(1, 3)$	$(3, \infty)$
Sign of $x + 2$	$-$	$+$	$+$	$+$
Sign of $x - 1$	$-$	$-$	$+$	$+$
Sign of $x - 3$	$-$	$-$	$-$	$+$
Sign of $(x+2)(x-1)(x-3)$	$-$	$+$	$-$	$+$

From the table, the solution set is $\{x \mid x \le -2 \text{ or } 1 \le x \le 3\}$. Interval: $(-\infty, -2] \cup [1, 3]$. Graph:

55. $(x-4)(x+2)^2 < 0$. Note that $(x+2)^2 > 0$ for all $x \ne -2$, so the expression on the left of the original inequality changes sign only when $x = 4$. We check the intervals in the following table.

Interval	$(-\infty, -2)$	$(-2, 4)$	$(4, \infty)$
Sign of $x - 4$	$-$	$-$	$+$
Sign of $(x+2)^2$	$+$	$+$	$+$
Sign of $(x-4)(x+2)^2$	$-$	$-$	$+$

From the table, the solution set is $\{x \mid x \ne -2 \text{ and } x < 4\}$. We exclude the endpoint -2 since the original expression cannot be 0. Interval: $(-\infty, -2) \cup (-2, 4)$.

Graph:

57. $(x-2)^2(x-3)(x+1) \le 0$. Note that $(x-2)^2 \ge 0$ for all x, so the expression on the left of the original inequality changes sign only when $x = -1$ and $x = 3$. We check the intervals in the following table.

Interval	$(-\infty, -1)$	$(-1, 2)$	$(2, 3)$	$(3, \infty)$
Sign of $(x-2)^2$	+	+	+	+
Sign of $x - 3$	−	−	−	+
Sign of $x + 1$	−	+	+	+
Sign of $(x-2)^2(x-3)(x+1)$	+	−	−	+

From the table, the solution set is $\{x \mid -1 \le x \le 3\}$. Interval: $[-1, 3]$. Graph:

59. $x^3 - 4x > 0 \Leftrightarrow x\left(x^2 - 4\right) > 0 \Leftrightarrow x(x+2)(x-2) > 0$. The expression on the left of the inequality changes sign where $x = 0$, $x = -2$ and where $x = 4$. Thus we must check the intervals in the following table.

Interval	$(-\infty, -2)$	$(-2, 0)$	$(0, 2)$	$(2, \infty)$
Sign of x	−	−	+	+
Sign of $x + 2$	−	+	+	+
Sign of $x - 2$	−	−	−	+
Sign of $x(x+2)(x-2)$	−	+	−	+

From the table, the solution set is $\{x \mid -2 < x < 0 \text{ or } x > 2\}$. Interval: $(-2, 0) \cup (2, \infty)$. Graph:

61. $\dfrac{x-3}{x+1} \ge 0$. The expression on the left of the inequality changes sign where $x = -1$ and where $x = 3$. Thus we must check the intervals in the following table.

Interval	$(-\infty, -1)$	$(-1, 3)$	$(3, \infty)$
Sign of $x + 1$	−	+	+
Sign of $x - 3$	−	−	+
Sign of $\dfrac{x-3}{x+1}$	+	−	+

From the table, the solution set is $\{x \mid x < -1 \text{ or } x \le 3\}$. Since the denominator cannot equal 0 we must have $x \ne -1$. Interval: $(-\infty, -1) \cup [3, \infty)$.

Graph:

63. $\dfrac{4x}{2x+3} > 2 \Leftrightarrow \dfrac{4x}{2x+3} - 2 > 0 \Leftrightarrow \dfrac{4x}{2x+3} - \dfrac{2(2x+3)}{2x+3} > 0 \Leftrightarrow \dfrac{-6}{2x+3} > 0$. The expression on the left of the inequality changes sign where $x = -\frac{3}{2}$. Thus we must check the intervals in the following table.

Interval	$\left(-\infty, -\frac{3}{2}\right)$	$\left(-\frac{3}{2}, \infty\right)$
Sign of -6	−	−
Sign of $2x + 3$	−	+
Sign of $\dfrac{-6}{2x+3}$	+	−

From the table, the solution set is $\left\{x \mid x < -\frac{3}{2}\right\}$. Interval: $\left(-\infty, -\frac{3}{2}\right)$.

Graph:

65. $\dfrac{2x+1}{x-5} \le 3 \Leftrightarrow \dfrac{2x+1}{x-5} - 3 \le 0 \Leftrightarrow \dfrac{2x+1}{x-5} - \dfrac{3(x-5)}{x-5} \le 0 \Leftrightarrow \dfrac{-x+16}{x-5} \le 0$. The expression on the left of the inequality changes sign where $x = 16$ and where $x = 5$. Thus we must check the intervals in the following table.

Interval	$(-\infty, 5)$	$(5, 16)$	$(16, \infty)$
Sign of $-x + 16$	$+$	$+$	$-$
Sign of $x - 5$	$-$	$+$	$+$
Sign of $\dfrac{-x+16}{x-5}$	$-$	$+$	$-$

From the table, the solution set is $\{x \mid x < 5 \text{ or } x \ge 16\}$. Since the denominator cannot equal 0, we must have $x \ne 5$.
Interval: $(-\infty, 5) \cup [16, \infty)$.

Graph: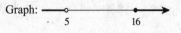

67. $\dfrac{4}{x} < x \Leftrightarrow \dfrac{4}{x} - x < 0 \Leftrightarrow \dfrac{4}{x} - \dfrac{x \cdot x}{x} < 0 \Leftrightarrow \dfrac{4 - x^2}{x} < 0 \Leftrightarrow \dfrac{(2-x)(2+x)}{x} < 0$. The expression on the left of the inequality changes sign where $x = 0$, where $x = -2$, and where $x = 2$. Thus we must check the intervals in the following table.

Interval	$(-\infty, -2)$	$(-2, 0)$	$(0, 2)$	$(2, \infty)$
Sign of $2 + x$	$-$	$+$	$+$	$+$
Sign of x	$-$	$-$	$+$	$+$
Sign of $2 - x$	$+$	$+$	$+$	$-$
Sign of $\dfrac{(2-x)(2+x)}{x}$	$+$	$-$	$+$	$-$

From the table, the solution set is $\{x \mid -2 < x < 0 \text{ or } 2 < x\}$. Interval: $(-2, 0) \cup (2, \infty)$. Graph:

69. $1 + \dfrac{2}{x+1} \le \dfrac{2}{x} \Leftrightarrow 1 + \dfrac{2}{x+1} - \dfrac{2}{x} \le 0 \Leftrightarrow \dfrac{x(x+1)}{x(x+1)} + \dfrac{2x}{x(x+1)} - \dfrac{2(x+1)}{x(x+1)} \le 0 \Leftrightarrow \dfrac{x^2 + x + 2x - 2x - 2}{x(x+1)} \le 0 \Leftrightarrow$

$\dfrac{x^2 + x - 2}{x(x+1)} \le 0 \Leftrightarrow \dfrac{(x+2)(x-1)}{x(x+1)} \le 0$. The expression on the left of the inequality changes sign where $x = -2$, where $x = -1$, where $x = 0$, and where $x = 1$. Thus we must check the intervals in the following table.

Interval	$(-\infty, -2)$	$(-2, -1)$	$(-1, 0)$	$(0, 1)$	$(1, \infty)$
Sign of $x + 2$	$-$	$+$	$+$	$+$	$+$
Sign of $x - 1$	$-$	$-$	$-$	$-$	$+$
Sign of x	$-$	$-$	$-$	$+$	$+$
Sign of $x + 1$	$-$	$-$	$+$	$+$	$+$
Sign of $\dfrac{(x+2)(x-1)}{x(x+1)}$	$+$	$-$	$+$	$-$	$+$

Since $x = -1$ and $x = 0$ yield undefined expressions, we cannot include them in the solution. From the table, the solution set is $\{x \mid -2 \le x < -1 \text{ or } 0 < x \le 1\}$. Interval: $[-2, -1) \cup (0, 1]$. Graph:

71. $\dfrac{6}{x-1} - \dfrac{6}{x} \geq 1 \Leftrightarrow \dfrac{6}{x-1} - \dfrac{6}{x} - 1 \geq 0 \Leftrightarrow \dfrac{6x}{x(x-1)} - \dfrac{6(x-1)}{x(x-1)} - \dfrac{x(x-1)}{x(x-1)} \geq 0 \Leftrightarrow$

$\dfrac{6x - 6x + 6 - x^2 + x}{x(x-1)} \geq 0 \Leftrightarrow \dfrac{-x^2 + x + 6}{x(x-1)} \geq 0 \Leftrightarrow \dfrac{(-x+3)(x+2)}{x(x-1)} \geq 0.$ The

expression on the left of the inequality changes sign where $x = 3$, where $x = -2$, where $x = 0$, and where $x = 1$. Thus we

must check the intervals in the following table.

Interval	$(-\infty, -2)$	$(-2, 0)$	$(0, 1)$	$(1, 3)$	$(3, \infty)$
Sign of $-x + 3$	+	+	+	+	−
Sign of $x + 2$	−	+	+	+	+
Sign of x	−	−	+	+	+
Sign of $x - 1$	−	−	−	+	+
Sign of $\dfrac{(-x+3)(x+2)}{x(x-1)}$	−	+	−	+	−

From the table, the solution set is $\{x \mid -2 \leq x < 0 \text{ or } 1 < x \leq 3\}$. The points $x = 0$ and $x = 1$ are excluded from the

solution set because they make the denominator zero. Interval: $[-2, 0) \cup (1, 3]$. Graph:

73. $\dfrac{x+2}{x+3} < \dfrac{x-1}{x-2} \Leftrightarrow \dfrac{x+2}{x+3} - \dfrac{x-1}{x-2} < 0 \Leftrightarrow \dfrac{(x+2)(x-2)}{(x+3)(x-2)} - \dfrac{(x-1)(x+3)}{(x-2)(x+3)} < 0 \Leftrightarrow$

$\dfrac{x^2 - 4 - x^2 - 2x + 3}{(x+3)(x-2)} < 0 \Leftrightarrow \dfrac{-2x - 1}{(x+3)(x-2)} < 0.$ The expression on the left of the inequality

changes sign where $x = -\frac{1}{2}$, where $x = -3$, and where $x = 2$. Thus we must check the intervals in the following table.

Interval	$(-\infty, -3)$	$\left(-3, -\frac{1}{2}\right)$	$\left(-\frac{1}{2}, 2\right)$	$(2, \infty)$
Sign of $-2x - 1$	+	+	−	−
Sign of $x + 3$	−	+	+	+
Sign of $x - 2$	−	−	−	+
Sign of $\dfrac{-2x - 1}{(x+3)(x-2)}$	+	−	+	−

From the table, the solution set is $\left\{x \mid -3 < x < -\frac{1}{2} \text{ or } 2 < x\right\}$. Interval: $\left(-3, -\frac{1}{2}\right) \cup (2, \infty)$.

Graph:

75. $\dfrac{(x-1)(x+2)}{(x-2)^2} \geq 0$. Note that $(x-2)^2 \geq 0$ for all x. The expression on the left of the original inequality changes sign when $x = -2$ and $x = 1$. We check the intervals in the following table.

Interval	$(-\infty, -2)$	$(-2, 1)$	$(1, 2)$	$(2, \infty)$
Sign of $x - 1$	$-$	$-$	$+$	$+$
Sign of $x + 2$	$-$	$+$	$+$	$+$
Sign of $(x - 2)^2$	$+$	$+$	$+$	$+$
Sign of $\dfrac{(x-1)(x+2)}{(x-2)^2}$	$+$	$-$	$+$	$+$

From the table, and recalling that the point $x = 2$ is excluded from the solution because the expression is undefined at those values, the solution set is $\{x \mid x \leq -2 \text{ or } x \geq 1 \text{ and } x \neq 2\}$. Interval: $(-\infty, -2] \cup [1, 2) \cup (2, \infty)$.

Graph:

77. $x^4 > x^2 \Leftrightarrow x^4 - x^2 > 0 \Leftrightarrow x^2\left(x^2 - 1\right) > 0 \Leftrightarrow x^2(x-1)(x+1) > 0$. The expression on the left of the inequality changes sign where $x = 0$, where $x = 1$, and where $x = -1$. Thus we must check the intervals in the following table.

Interval	$(-\infty, -1)$	$(-1, 0)$	$(0, 1)$	$(1, \infty)$
Sign of x^2	$+$	$+$	$+$	$+$
Sign of $x - 1$	$-$	$-$	$-$	$+$
Sign of $x + 1$	$-$	$+$	$+$	$+$
Sign of $x^2(x-1)(x+1)$	$+$	$-$	$-$	$+$

From the table, the solution set is $\{x \mid x < -1 \text{ or } 1 < x\}$. Interval: $(-\infty, -1) \cup (1, \infty)$. Graph:

79. We graph $y_1 = x^2$ and $y_2 = 3x + 10$ in the viewing rectangle $[-5, 8]$ by $[-2, 30]$.

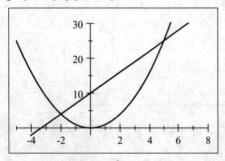

From the graph, we see that $x^2 \leq 3x + 10$ on $[-2, 5]$.

81. We graph $y_1 = x^3 + 11x$ and $y_2 = 6x^2 + 6$ in the viewing rectangles $[-3, 3]$ by $[-10, 40]$ and $[1.5, 3, 5]$ by $[20, 75]$.

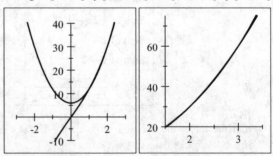

Zooming in or using a trace function, we see that $x^3 + 11x \leq 6x^2 + 6$ on $(-\infty, 1]$ and $[2, 3]$.

83. We graph $y_1 = x^{1/3}$ and $y_2 = x$ in the viewing rectangle $[-3, 3]$ by $[-2, 2]$.

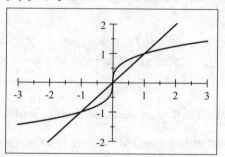

From the graph, we see that $x^{1/3} < x$ on $(-1, 0)$ and $(1, \infty)$.

85. We graph $y_1 = (x + 1)^2$ and $y_2 = (x - 1)^2$ in the viewing rectangle $[-3, 3]$ by $[-1, 10]$.

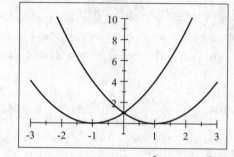

From the graph, we see that $(x + 1)^2 < (x - 1)^2$ on $(-\infty, 0)$.

87. For $\sqrt{16 - 9x^2}$ to be defined as a real number we must have $16 - 9x^2 \geq 0 \Leftrightarrow (4 - 3x)(4 + 3x) \geq 0$. The expression in the inequality changes sign at $x = \frac{4}{3}$ and $x = -\frac{4}{3}$.

Interval	$\left(-\infty, -\frac{4}{3}\right)$	$\left(-\frac{4}{3}, \frac{4}{3}\right)$	$\left(\frac{4}{3}, \infty\right)$
Sign of $4 - 3x$	+	+	−
Sign of $4 + 3x$	−	+	+
Sign of $(4 - 3x)(4 + 3x)$	−	+	−

Thus $-\frac{4}{3} \leq x \leq \frac{4}{3}$.

89. For $\left(\dfrac{1}{x^2 - 5x - 14}\right)^{1/2}$ to be defined as a real number we must have $x^2 - 5x - 14 > 0 \Leftrightarrow (x - 7)(x + 2) > 0$. The expression in the inequality changes sign at $x = 7$ and $x = -2$.

Interval	$(-\infty, -2)$	$(-2, 7)$	$(7, \infty)$
Sign of $x - 7$	−	−	+
Sign of $x + 2$	−	+	+
Sign of $(x - 7)(x + 2)$	+	−	+

Thus $x < -2$ or $7 < x$, and the solution set is $(-\infty, -2) \cup (7, \infty)$.

91. (a) $a(bx - c) \geq bc$ (where $a, b, c > 0$) $\Leftrightarrow bx - c \geq \dfrac{bc}{a} \Leftrightarrow bx \geq \dfrac{bc}{a} + c \Leftrightarrow x \geq \dfrac{1}{b}\left(\dfrac{bc}{a} + c\right) = \dfrac{c}{a} + \dfrac{c}{b} \Leftrightarrow x \geq \dfrac{c}{a} + \dfrac{c}{b}$.

(b) We have $a \leq bx + c < 2a$, where $a, b, c > 0 \Leftrightarrow a - c \leq bx < 2a - c \Leftrightarrow \dfrac{a - c}{b} \leq x < \dfrac{2a - c}{b}$.

93. Inserting the relationship $C = \frac{5}{9}(F - 32)$, we have $20 \leq C \leq 30 \Leftrightarrow 20 \leq \frac{5}{9}(F - 32) \leq 30 \Leftrightarrow 36 \leq F - 32 \leq 54 \Leftrightarrow 68 \leq F \leq 86$.

95. Let x be the average number of miles driven per day. Each day the cost of Plan A is $30 + 0.10x$, and the cost of Plan B is 50. Plan B saves money when $50 < 30 + 0.10x \Leftrightarrow 20 < 0.1x \Leftrightarrow 200 < x$. So Plan B saves money when you average more than 200 miles a day.

97. We need to solve $6400 \leq 0.35m + 2200 \leq 7100$ for m. So $6400 \leq 0.35m + 2200 \leq 7100 \Leftrightarrow 4200 \leq 0.35m \leq 4900 \Leftrightarrow 12{,}000 \leq m \leq 14{,}000$. She plans on driving between 12,000 and 14,000 miles.

99. (a) Let x be the number of \$3 increases. Then the number of seats sold is $120 - x$. So $P = 200 + 3x$
$\Leftrightarrow 3x = P - 200 \Leftrightarrow x = \frac{1}{3}(P - 200)$. Substituting for x we have that the number of seats sold is
$120 - x = 120 - \frac{1}{3}(P - 200) = -\frac{1}{3}P + \frac{560}{3}$.

(b) $90 \le -\frac{1}{3}P + \frac{560}{3} \le 115 \Leftrightarrow 270 \le 360 - P + 200 \le 345 \Leftrightarrow 270 \le -P + 560 \le 345 \Leftrightarrow -290 \le -P \le -215 \Leftrightarrow$
$290 \ge P \ge 215$. Putting this into standard order, we have $215 \le P \le 290$. So the ticket prices are between \$215 and
\$290.

101. $0.0004 \le \dfrac{4{,}000{,}000}{d^2} \le 0.01$. Since $d^2 \ge 0$ and $d \ne 0$, we can multiply each expression by d^2 to obtain
$0.0004d^2 \le 4{,}000{,}000 \le 0.01d^2$. Solving each pair, we have $0.0004d^2 \le 4{,}000{,}000 \Leftrightarrow d^2 \le 10{,}000{,}000{,}000$
$\Rightarrow d \le 100{,}000$ (recall that d represents distance, so it is always nonnegative). Solving $4{,}000{,}000 \le 0.01d^2 \Leftrightarrow$
$400{,}000{,}000 \le d^2 \Rightarrow 20{,}000 \le d$. Putting these together, we have $20{,}000 \le d \le 100{,}000$.

103. $128 + 16t - 16t^2 \ge 32 \Leftrightarrow -16t^2 + 16t + 96 \ge 0 \Leftrightarrow -16\left(t^2 - t - 6\right) \ge 0 \Leftrightarrow -16(t - 3)(t + 2) \ge 0$. The expression on
the left of the inequality changes sign at $x = -2$, at $t = 3$, and at $t = -2$. However, $t \ge 0$, so the only endpoint is $t = 3$.

Interval	$(0, 3)$	$(3, \infty)$
Sign of -16	$-$	$-$
Sign of $t - 3$	$-$	$+$
Sign of $t + 2$	$+$	$+$
Sign of $-16(t - 3)(t + 2)$	$+$	$-$

So $0 \le t \le 3$.

105. $240 \ge v + \dfrac{v^2}{20} \Leftrightarrow \frac{1}{20}v^2 + v - 240 \le 0 \Leftrightarrow \left(\frac{1}{20}v - 3\right)(v + 80) \le 0$. The expression in the inequality changes sign at
$v = 60$ and $v = -80$. However, since v represents the speed, we must have $v \ge 0$.

Interval	$(0, 60)$	$(60, \infty)$
Sign of $\frac{1}{20}v - 3$	$-$	$+$
Sign of $v + 80$	$+$	$+$
Sign of $\left(\frac{1}{20}v - 3\right)(v + 80)$	$-$	$+$

So Kerry must drive between 0 and 60 mi/h.

107. Let x be the length of the garden and w its width. Using the fact that the perimeter is 120 ft, we must have $2x + 2w = 120$
$\Leftrightarrow w = 60 - x$. Now since the area must be at least 800 ft^2, we have $800 < x(60 - x) \Leftrightarrow 800 < 60x - x^2 \Leftrightarrow$
$x^2 - 60x + 800 < 0 \Leftrightarrow (x - 20)(x - 40) < 0$. The expression in the inequality changes sign at $x = 20$ and $x = 40$.
However, since x represents length, we must have $x > 0$.

Interval	$(0, 20)$	$(20, 40)$	$(40, \infty)$
Sign of $x - 20$	$-$	$+$	$+$
Sign of $x - 40$	$-$	$-$	$+$
Sign of $(x - 20)(x - 40)$	$+$	$-$	$+$

The length of the garden should be between 20 and 40 feet.

109. The rule we want to apply here is "$a < b \Rightarrow ac < bc$ if $c > 0$ and $a < b \Rightarrow ac > bc$ if $c < 0$". Thus we cannot simply multiply by x, since we don't yet know if x is positive or negative, so in solving $1 < \dfrac{3}{x}$, we must consider two cases.

Case 1: $x > 0$ Multiplying both sides by x, we have $x < 3$. Together with our initial condition, we have $0 < x < 3$.

Case 2: $x < 0$ Multiplying both sides by x, we have $x > 3$. But $x < 0$ and $x > 3$ have no elements in common, so this gives no additional solution.

Hence, the only solutions are $0 < x < 3$.

1.9   SOLVING ABSOLUTE VALUE EQUATIONS AND INEQUALITIES

1. The equation $|x| = 3$ has the two solutions -3 and 3.

3. The solution of the inequality $|x| \geq 3$ is a union of two intervals $(-\infty, -3] \cup [3, \infty)$.

5. $|4x| = 24 \Leftrightarrow 4x = \pm 24 \Leftrightarrow x = \pm 6$.

7. $5|x| + 3 = 28 \Leftrightarrow 5|x| = 25 \Leftrightarrow |x| = 5 \Leftrightarrow x = \pm 5$.

9. $|x - 3| = 2$ is equivalent to $x - 3 = \pm 2 \Leftrightarrow x = 3 \pm 2 \Leftrightarrow x = 1$ or $x = 5$.

11. $|x + 4| = 0.5$ is equivalent to $x + 4 = \pm 0.5 \Leftrightarrow x = -4 \pm 0.5 \Leftrightarrow x = -4.5$ or $x = -3.5$.

13. $|4x + 7| = 9$ is equivalent to either $4x + 7 = 9 \Leftrightarrow 4x = 2 \Leftrightarrow x = \dfrac{1}{2}$; or $4x + 7 = -9 \Leftrightarrow 4x = -16 \Leftrightarrow x = -4$. The two solutions are $x = \frac{1}{2}$ and $x = -4$.

15. $4 - |3x + 6| = 1 \Leftrightarrow -|3x + 6| = -3 \Leftrightarrow |3x + 6| = 3$, which is equivalent to either $3x + 6 = 3 \Leftrightarrow 3x = -3 \Leftrightarrow x = -1$; or $3x + 6 = -3 \Leftrightarrow 3x = -9 \Leftrightarrow x = -3$. The two solutions are $x = -1$ and $x = -3$.

17. $3|x + 5| + 6 = 15 \Leftrightarrow 3|x + 5| = 9 \Leftrightarrow |x + 5| = 3$, which is equivalent to either $x + 5 = 3 \Leftrightarrow x = -2$; or $x + 5 = -3 \Leftrightarrow x = -8$. The two solutions are $x = -2$ and $x = -8$.

19. $8 + 5\left|\frac{1}{3}x - \frac{5}{6}\right| = 33 \Leftrightarrow 5\left|\frac{1}{3}x - \frac{5}{6}\right| = 25 \Leftrightarrow \left|\frac{1}{3}x - \frac{5}{6}\right| = 5$, which is equivalent to either $\frac{1}{3}x - \frac{5}{6} = 5 \Leftrightarrow \frac{1}{3}x = \frac{35}{6} \Leftrightarrow x = \frac{35}{2}$; or $\frac{1}{3}x - \frac{5}{6} = -5 \Leftrightarrow \frac{1}{3}x = -\frac{25}{6} \Leftrightarrow x = -\frac{25}{2}$. The two solutions are $x = -\frac{25}{2}$ and $x = \frac{35}{2}$.

21. $|x - 1| = |3x + 2|$, which is equivalent to either $x - 1 = 3x + 2 \Leftrightarrow -2x = 3 \Leftrightarrow x = -\frac{3}{2}$; or $x - 1 = -(3x + 2) \Leftrightarrow x - 1 = -3x - 2 \Leftrightarrow 4x = -1 \Leftrightarrow x = -\frac{1}{4}$. The two solutions are $x = -\frac{3}{2}$ and $x = -\frac{1}{4}$.

23. $|x| \leq 4 \Leftrightarrow -4 \leq x \leq 4$. Interval: $[-4, 4]$.

25. $|2x| > 7$ is equivalent to $2x > 7 \Leftrightarrow x > \frac{7}{2}$; or $2x < -7 \Leftrightarrow x < -\frac{7}{2}$. Interval: $\left(-\infty, -\frac{7}{2}\right) \cup \left(\frac{7}{2}, \infty\right)$.

27. $|x - 5| \leq 3 \Leftrightarrow -3 \leq x - 5 \leq 3 \Leftrightarrow 2 \leq x \leq 8$. Interval: $[2, 8]$.

29. $|x + 1| \geq 1$ is equivalent to $x + 1 \geq 1 \Leftrightarrow x \geq 0$; or $x + 1 \leq -1 \Leftrightarrow x \leq -2$. Interval: $(-\infty, -2] \cup [0, \infty)$.

31. $|x + 5| \geq 2$ is equivalent to $x + 5 \geq 2 \Leftrightarrow x \geq -3$; or $x + 5 \leq -2 \Leftrightarrow x \leq -7$. Interval: $(-\infty, -7] \cup [-3, \infty)$.

33. $|2x - 3| \leq 0.4 \Leftrightarrow -0.4 \leq 2x - 3 \leq 0.4 \Leftrightarrow 2.6 \leq 2x \leq 3.4 \Leftrightarrow 1.3 \leq x \leq 1.7$. Interval: $[1.3, 1.7]$.

35. $\left|\dfrac{x - 2}{3}\right| < 2 \Leftrightarrow -2 < \dfrac{x - 2}{3} < 2 \Leftrightarrow -6 < x - 2 < 6 \Leftrightarrow -4 < x < 8$. Interval: $(-4, 8)$.

37. $|x + 6| < 0.001 \Leftrightarrow -0.001 < x + 6 < 0.001 \Leftrightarrow -6.001 < x < -5.999$. Interval: $(-6.001, -5.999)$.

39. $4|x + 2| - 3 < 13 \Leftrightarrow 4|x + 2| < 16 \Leftrightarrow |x + 2| < 4 \Leftrightarrow -4 < x + 2 < 4 \Leftrightarrow -6 < x < 2$. Interval: $(-6, 2)$.

41. $8 - |2x - 1| \geq 6 \Leftrightarrow -|2x - 1| \geq -2 \Leftrightarrow |2x - 1| \leq 2 \Leftrightarrow -2 \leq 2x - 1 \leq 2 \Leftrightarrow -1 \leq 2x \leq 3 \Leftrightarrow -\frac{1}{2} \leq x \leq \frac{3}{2}$. Interval: $\left[-\frac{1}{2}, \frac{3}{2}\right]$.

43. $\frac{1}{2}\left|4x + \frac{1}{3}\right| > \frac{5}{6} \Leftrightarrow \left|4x + \frac{1}{3}\right| > \frac{5}{3}$, which is equivalent to either $4x + \frac{1}{3} > \frac{5}{3} \Leftrightarrow 4x > \frac{4}{3} \Leftrightarrow x > \frac{1}{3}$; or $4x + \frac{1}{3} < -\frac{5}{3} \Leftrightarrow 4x < -2 \Leftrightarrow x < -\frac{1}{2}$. Interval: $\left(-\infty, -\frac{1}{2}\right) \cup \left(\frac{1}{3}, \infty\right)$.

45. $1 \le |x| \le 4$. If $x \ge 0$, then this is equivalent to $1 \le x \le 4$. If $x < 0$, then this is equivalent to $1 \le -x \le 4 \Leftrightarrow -1 \ge x \ge -4$ $\Leftrightarrow -4 \le x \le -1$. Interval: $[-4, -1] \cup [1, 4]$.

47. $\dfrac{1}{|x+7|} > 2 \Leftrightarrow 1 > 2|x+7| \; (x \ne -7) \Leftrightarrow |x+7| < \frac{1}{2} \Leftrightarrow -\frac{1}{2} < x+7 < \frac{1}{2} \Leftrightarrow -\frac{15}{2} < x < -\frac{13}{2}$ and $x \ne -7$.

Interval: $\left(-\frac{15}{2}, -7\right) \cup \left(-7, -\frac{13}{2}\right)$.

49. $|x| < 3$ **51.** $|x - 7| \ge 5$ **53.** $|x| \le 2$ **55.** $|x| > 3$

57. (a) Let x be the thickness of the laminate. Then $|x - 0.020| \le 0.003$.

  (b) $|x - 0.020| \le 0.003 \Leftrightarrow -0.003 \le x - 0.020 \le 0.003 \Leftrightarrow 0.017 \le x \le 0.023$.

59. $|x - 1|$ is the distance between x and 1; $|x - 3|$ is the distance between x and 3. So $|x - 1| < |x - 3|$ represents those points closer to 1 than to 3, and the solution is $x < 2$, since 2 is the point halfway between 1 and 3. If $a < b$, then the solution to $|x - a| < |x - b|$ is $x < \dfrac{a + b}{2}$.

CHAPTER 1 REVIEW

1. (a)

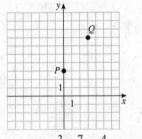

  (b) The distance from P to Q is
$$d(P, Q) = \sqrt{(0-3)^2 + (3-7)^2}$$
$$= \sqrt{9 + 16} = \sqrt{25} = 5$$

  (c) The midpoint is $\left(\dfrac{0+3}{2}, \dfrac{3+7}{2}\right) = \left(\dfrac{3}{2}, 5\right)$.

  (d) The line has slope $m = \dfrac{3-7}{0-3} = \dfrac{4}{3}$ and equation
$$y - 3 = \tfrac{4}{3}(x - 0) \Leftrightarrow y - 3 = \tfrac{4}{3}x \Leftrightarrow y = \tfrac{4}{3}x + 3.$$

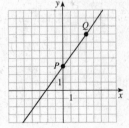

  (e) The radius of this circle was found in part (b). It is $r = d(P, Q) = 5$. So an equation is
$$(x - 0)^2 + (y - 3)^2 = (5)^2 \Leftrightarrow x^2 + (y - 3)^2 = 25.$$

3. (a)

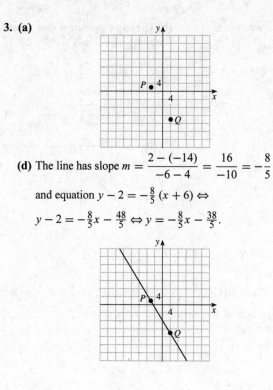

(b) The distance from P to Q is

$$d(P, Q) = \sqrt{(-6-4)^2 + [2 - (-14)]^2}$$
$$= \sqrt{100 + 256} = \sqrt{356} = 2\sqrt{89}$$

(c) The midpoint is $\left(\dfrac{-6+4}{2}, \dfrac{2 + (-14)}{2}\right) = (-1, -6)$.

(d) The line has slope $m = \dfrac{2 - (-14)}{-6-4} = \dfrac{16}{-10} = -\dfrac{8}{5}$

and equation $y - 2 = -\dfrac{8}{5}(x+6) \Leftrightarrow$

$y - 2 = -\dfrac{8}{5}x - \dfrac{48}{5} \Leftrightarrow y = -\dfrac{8}{5}x - \dfrac{38}{5}$.

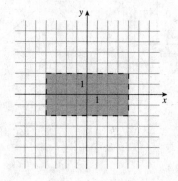

(e) The radius of this circle was found in part (b). It is

$r = d(P, Q) = 2\sqrt{89}$. So an equation is

$[x - (-6)]^2 + (y-2)^2 = \left(2\sqrt{89}\right)^2 \Leftrightarrow$

$(x+6)^2 + (y-2)^2 = 356$.

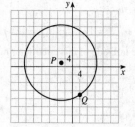

5. $\{(x, y) \mid -4 < x < 4 \text{ and } -2 < y < 2\}$

7. $d(A, C) = \sqrt{(4 - (-1))^2 + (4 - (-3))^2} = \sqrt{(4+1)^2 + (4+3)^2} = \sqrt{74}$ and

$d(B, C) = \sqrt{(5 - (-1))^2 + (3 - (-3))^2} = \sqrt{(5+1)^2 + (3+3)^2} = \sqrt{72}$. Therefore, B is closer to C.

9. The center is $C = (-5, -1)$, and the point $P = (0, 0)$ is on the circle. The radius of the circle is

$r = d(P, C) = \sqrt{(0 - (-5))^2 + (0 - (-1))^2} = \sqrt{(0+5)^2 + (0+1)^2} = \sqrt{26}$. Thus, an equation of the circle is

$(x+5)^2 + (y+1)^2 = 26$.

11. (a) $x^2 + y^2 + 2x - 6y + 9 = 0 \Leftrightarrow \left(x^2 + 2x\right) + \left(y^2 - 6y\right) = -9 \Leftrightarrow$

$\left(x^2 + 2x + 1\right) + \left(y^2 - 6y + 9\right) = -9 + 1 + 9 \Leftrightarrow$

$(x + 1)^2 + (y - 3)^2 = 1$, an equation of a circle.

(b) The circle has center $(-1, 3)$ and radius 1.

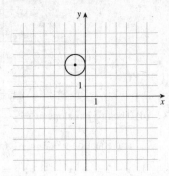

13. (a) $x^2 + y^2 + 72 = 12x \Leftrightarrow \left(x^2 - 12x\right) + y^2 = -72 \Leftrightarrow \left(x^2 - 12x + 36\right) + y^2 = -72 + 36 \Leftrightarrow (x - 6)^2 + y^2 = -36.$

Since the left side of this equation must be greater than or equal to zero, this equation has no graph.

15. $y = 2 - 3x$

x	y
-2	8
0	2
$\frac{2}{3}$	0

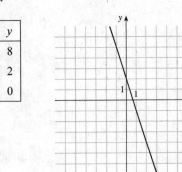

17. $x + 3y = 21 \Leftrightarrow y = -\frac{1}{3}x + 7$

x	y
-3	8
0	7
21	0

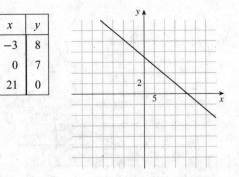

19. $\frac{x}{2} - \frac{y}{7} = 1 \Leftrightarrow y = \frac{7}{2}x - 7$

x	y
-2	-14
0	-7
2	0

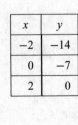

21. $y = 16 - x^2$

x	y
-3	7
-1	15
0	16
1	15
3	7

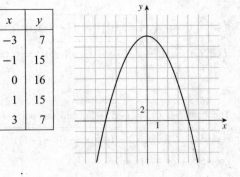

23. $x = \sqrt{y}$

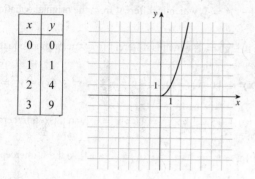

x	y
0	0
1	1
2	4
3	9

25. $y = 9 - x^2$

    (a) x-axis symmetry: replacing y by $-y$ gives $-y = 9 - x^2$, which is not the same as the original equation, so the graph is not symmetric about the x-axis.

    y-axis symmetry: replacing x by $-x$ gives $y = 9 - (-x)^2 = 9 - x^2$, which is the same as the original equation, so the graph is symmetric about the y-axis.

    Origin symmetry: replacing x by $-x$ and y by $-y$ gives $-y = 9 - (-x)^2 \Leftrightarrow y = -9 + x^2$, which is not the same as the original equation, so the graph is not symmetric about the origin.

    (b) To find x-intercepts, we set $y = 0$ and solve for x: $0 = 9 - x^2 \Leftrightarrow x^2 = 9 \Leftrightarrow x = \pm 3$, so the x-intercepts are -3 and 3.
    To find y-intercepts, we set $x = 0$ and solve for y: $y = 9 - 0^2 = 9$, so the y-intercept is 9.

27. $x^2 + (y - 1)^2 = 1$

    (a) x-axis symmetry: replacing y by $-y$ gives $x^2 + \left[(-y) - 1\right]^2 = 1 \Leftrightarrow x^2 + (y + 1)^2 = 1$, so the graph is not symmetric about the x-axis.

    y-axis symmetry: replacing x by $-x$ gives $(-x)^2 + (y - 1)^2 = 1 \Leftrightarrow x^2 + (y - 1)^2 = 1$, so the graph is symmetric about the y-axis.

    Origin symmetry: replacing x by $-x$ and y by $-y$ gives $(-x)^2 + \left[(-y) - 1\right]^2 = 1 \Leftrightarrow x^2 + (y + 1)^2 = 1$, so the graph is not symmetric about the origin.

    (b) To find x-intercepts, we set $y = 0$ and solve for x: $x^2 + (0 - 1)^2 = 1 \Leftrightarrow x^2 = 0$, so the x-intercept is 0.
    To find y-intercepts, we set $x = 0$ and solve for y: $0^2 + (y - 1)^2 = 1 \Leftrightarrow y - 1 = \pm 1 \Leftrightarrow y = 0$ or 2, so the y-intercepts are 0 and 2.

29. $9x^2 - 16y^2 = 144$

    (a) x-axis symmetry: replacing y by $-y$ gives $9x^2 - 16(-y)^2 = 144 \Leftrightarrow 9x^2 - 16y^2 = 144$, so the graph is symmetric about the x-axis.

    y-axis symmetry: replacing x by $-x$ gives $9(-x)^2 - 16y^2 = 144 \Leftrightarrow 9x^2 - 16y^2 = 144$, so the graph is symmetric about the y-axis.

    Origin symmetry: replacing x by $-x$ and y by $-y$ gives $9(-x)^2 - 16(-y)^2 = 144 \Leftrightarrow 9x^2 - 16y^2 = 144$, so the graph is symmetric about the origin.

    (b) To find x-intercepts, we set $y = 0$ and solve for x: $9x^2 - 16(0)^2 = 144 \Leftrightarrow 9x^2 = 144 \Leftrightarrow x = \pm 4$, so the x-intercepts are -4 and 4.
    To find y-intercepts, we set $x = 0$ and solve for y: $9(0)^2 - 16y^2 = 144 \Leftrightarrow 16y^2 = -144$, so there is no y-intercept.

31. $x^2 + 4xy + y^2 = 1$

    (a) x-axis symmetry: replacing y by $-y$ gives $x^2 + 4x(-y) + (-y)^2 = 1$, which is different from the original equation, so the graph is not symmetric about the x-axis.

    y-axis symmetry: replacing x by $-x$ gives $(-x)^2 + 4(-x)y + y^2 = 1$, which is different from the original equation, so the graph is not symmetric about the y-axis.

    Origin symmetry: replacing x by $-x$ and y by $-y$ gives $(-x)^2 + 4(-x)(-y) + (-y)^2 = 1 \Leftrightarrow x^2 + 4xy + y^2 = 1$, so the graph is symmetric about the origin.

    (b) To find x-intercepts, we set $y = 0$ and solve for x: $x^2 + 4x(0) + 0^2 = 1 \Leftrightarrow x^2 = 1 \Leftrightarrow x = \pm 1$, so the x-intercepts are -1 and 1.

    To find y-intercepts, we set $x = 0$ and solve for y: $0^2 + 4(0)y + y^2 = 1 \Leftrightarrow y^2 = 1 \Leftrightarrow y = \pm 1$, so the y-intercepts are -1 and 1.

33. (a) We graph $y = x^2 - 6x$ in the viewing rectangle $[-10, 10]$ by $[-10, 10]$.

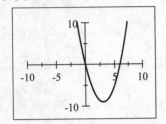

    (b) From the graph, we see that the x-intercepts are 0 and 6 and the y-intercept is 0.

35. (a) We graph $y = x^3 - 4x^2 - 5x$ in the viewing rectangle $[-4, 8]$ by $[-30, 20]$.

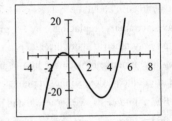

    (b) From the graph, we see that the x-intercepts are -1, 0, and 5 and the y-intercept is 0.

37. (a) The line that has slope 2 and y-intercept 6 has the slope-intercept equation $y = 2x + 6$.

    (b) An equation of the line in general form is $2x - y + 6 = 0$.

(c)

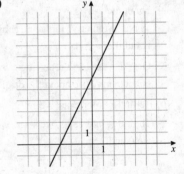

39. (a) The line that passes through the points $(-1, -6)$ and $(2, -4)$ has slope

$$m = \frac{-4 - (-6)}{2 - (-1)} = \frac{2}{3}, \text{ so } y - (-6) = \frac{2}{3}[x - (-1)] \Leftrightarrow y + 6 = \frac{2}{3}x + \frac{2}{3}$$

$$\Leftrightarrow y = \frac{2}{3}x - \frac{16}{3}.$$

    (b) $y = \frac{2}{3}x - \frac{16}{3} \Leftrightarrow 3y = 2x - 16 \Leftrightarrow 2x - 3y - 16 = 0.$

(c)

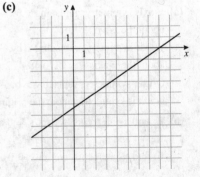

41. (a) The vertical line that passes through the point $(3, -2)$ has equation $x = 3$.   **(c)**

(b) $x = 3 \Leftrightarrow x - 3 = 0$.

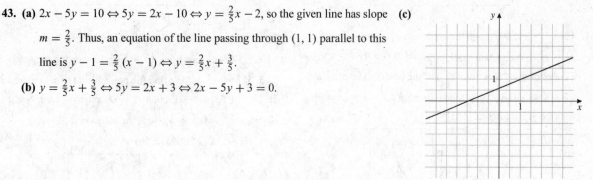

43. (a) $2x - 5y = 10 \Leftrightarrow 5y = 2x - 10 \Leftrightarrow y = \frac{2}{5}x - 2$, so the given line has slope   **(c)**

$m = \frac{2}{5}$. Thus, an equation of the line passing through $(1, 1)$ parallel to this

line is $y - 1 = \frac{2}{5}(x - 1) \Leftrightarrow y = \frac{2}{5}x + \frac{3}{5}$.

(b) $y = \frac{2}{5}x + \frac{3}{5} \Leftrightarrow 5y = 2x + 3 \Leftrightarrow 2x - 5y + 3 = 0$.

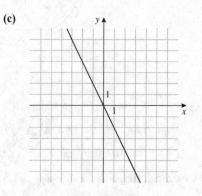

45. (a) The line $y = \frac{1}{2}x - 10$ has slope $\frac{1}{2}$, so a line perpendicular to this one has   **(c)**

slope $-\dfrac{1}{1/2} = -2$. In particular, the line passing through the origin

perpendicular to the given line has equation $y = -2x$.

(b) $y = -2x \Leftrightarrow 2x + y = 0$.

47. The line with equation $y = -\frac{1}{3}x - 1$ has slope $-\frac{1}{3}$. The line with equation $9y + 3x + 3 = 0 \Leftrightarrow 9y = -3x - 3 \Leftrightarrow$
$y = -\frac{1}{3}x - \frac{1}{3}$ also has slope $-\frac{1}{3}$, so the lines are parallel.

49. (a) The slope, 0.3, represents the increase in length of the spring for each unit increase in weight w. The S-intercept is the resting or natural length of the spring.

(b) When $w = 5$, $S = 0.3(5) + 2.5 = 1.5 + 2.5 = 4.0$ inches.

51. From the graph, we see that the graphs of $y = x^2 - 4x$ and $y = x + 6$ intersect at $x = -1$ and $x = 6$, so these are the solutions of the equation $x^2 - 4x = x + 6$.

53. From the graph, we see that the graph of $y = x^2 - 4x$ lies below the graph of $y = x + 6$ for $-1 < x < 6$, so the inequality $x^2 - 4x \le x + 6$ is satisfied on the interval $[-1, 6]$.

55. From the graph, we see that the graph of $y = x^2 - 4x$ lies above the x-axis for $x < 0$ and for $x > 4$, so the inequality $x^2 - 4x \ge 0$ is satisfied on the intervals $(-\infty, 0]$ and $[4, \infty)$.

57. $x^2 - 4x = 2x + 7$. We graph the equations $y_1 = x^2 - 4x$ and $y_2 = 2x + 7$ in the viewing rectangle $[-10, 10]$ by $[-5, 25]$. Using a zoom or trace function, we get the solutions $x = -1$ and $x = 7$.

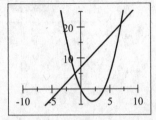

59. $x^4 - 9x^2 = x - 9$. We graph the equations $y_1 = x^4 - 9x^2$ and $y_2 = x - 9$ in the viewing rectangle $[-5, 5]$ by $[-25, 10]$. Using a zoom or trace function, we get the solutions $x \approx -2.72$, $x \approx -1.15$, $x = 1.00$, and $x \approx 2.87$.

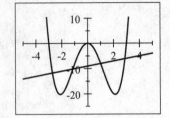

61. $4x - 3 \geq x^2$. We graph the equations $y_1 = 4x - 3$ and $y_2 = x^2$ in the viewing rectangle $[-5, 5]$ by $[0, 15]$. Using a zoom or trace function, we find the points of intersection are at $x = 1$ and $x = 3$. Since we want $4x - 3 \geq x^2$, the solution is the interval $[1, 3]$.

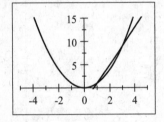

63. $x^4 - 4x^2 < \frac{1}{2}x - 1$. We graph the equations $y_1 = x^4 - 4x^2$ and $y_2 = \frac{1}{2}x - 1$ in the viewing rectangle $[-5, 5]$ by $[-5, 5]$. We find the points of intersection are at $x \approx -1.85$, $x \approx -0.60$, $x \approx 0.45$, and $x = 2.00$. Since we want $x^4 - 4x^2 < \frac{1}{2}x - 1$, the solution is $(-1.85, -0.60) \cup (0.45, 2.00)$.

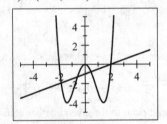

65. Here the center is at $(0, 0)$, and the circle passes through the point $(-5, 12)$, so the radius is

$r = \sqrt{(-5-0)^2 + (12-0)^2} = \sqrt{25 + 144} = \sqrt{169} = 13$. The equation of the circle is $x^2 + y^2 = 13^2 \Leftrightarrow$

$x^2 + y^2 = 169$. The line shown is the tangent that passes through the point $(-5, 12)$, so it is perpendicular to the line

through the points $(0, 0)$ and $(-5, 12)$. This line has slope $m_1 = \dfrac{12 - 0}{-5 - 0} = -\dfrac{12}{5}$. The slope of the line we seek is

$m_2 = -\dfrac{1}{m_1} = -\dfrac{1}{-12/5} = \dfrac{5}{12}$. Thus, an equation of the tangent line is $y - 12 = \frac{5}{12}(x + 5) \Leftrightarrow y - 12 = \frac{5}{12}x + \frac{25}{12} \Leftrightarrow$

$y = \frac{5}{12}x + \frac{169}{12} \Leftrightarrow 5x - 12y + 169 = 0$.

67. $x^2 - 9x + 14 = 0 \Leftrightarrow (x - 7)(x - 2) = 0 \Leftrightarrow x = 7$ or $x = 2$.

69. $2x^2 + x = 1 \Leftrightarrow 2x^2 + x - 1 = 0 \Leftrightarrow (2x - 1)(x + 1) = 0$. So either $2x - 1 = 0$ or $x + 1 = 0$. If $2x - 1 = 0$, then $2x = 1$ $\Leftrightarrow x = \frac{1}{2}$. If $x - 1 = 0$, then $x = -1$. Thus $x = -1$ or $x = \frac{1}{2}$.

71. $0 = 4x^3 - 25x = x\left(4x^2 - 25\right) = x(2x - 5)(2x + 5) = 0$. So either $x = 0$, $2x - 5 = 0$, or $2x + 5 = 0$. If $2x - 5 = 0$, then $2x = 5 \Leftrightarrow x = \frac{5}{2}$. If $2x + 5 = 0$, then $2x = -5 \Leftrightarrow x = -\frac{5}{2}$. Thus $x = -\frac{5}{2}$, $x = \frac{5}{2}$, or $x = 0$.

73. $3x^2 + 4x - 1 = 0 \Rightarrow$

$x = \dfrac{-b \pm \sqrt{b^2 - 4ac}}{2a} = \dfrac{-(4) \pm \sqrt{(4)^2 - 4(3)(-1)}}{2(3)} = \dfrac{-4 \pm \sqrt{16 + 12}}{6} = \dfrac{-4 \pm \sqrt{28}}{6} = \dfrac{-4 \pm 2\sqrt{7}}{6} = \dfrac{2\left(-2 \pm \sqrt{7}\right)}{6} = \dfrac{-2 \pm \sqrt{7}}{3}$.

75. $\frac{1}{x} + \frac{2}{x-1} = 3 \Leftrightarrow (x-1) + 2(x) = 3(x)(x-1) \Leftrightarrow x - 1 + 2x = 3x^2 - 3x \Leftrightarrow 0 = 3x^2 - 6x + 1 \Rightarrow$

$x = \frac{-b \pm \sqrt{b^2 - 4ac}}{2a} = \frac{-(-6) \pm \sqrt{(-6)^2 - 4(3)(1)}}{2(3)} = \frac{6 \pm \sqrt{36 - 12}}{6} = \frac{6 \pm \sqrt{24}}{6} = \frac{6 \pm 2\sqrt{6}}{6} = \frac{2(3 \pm \sqrt{6})}{6} = \frac{3 \pm \sqrt{6}}{3}.$

77. $x^4 - 8x^2 - 9 = 0 \Leftrightarrow (x^2 - 9)(x^2 + 1) = 0 \Leftrightarrow (x-3)(x+3)(x^2+1) = 0 \Rightarrow x - 3 = 0, x + 3 = 0, \text{ or } x^2 + 1 = 0.$

If $x - 3 = 0$, then $x = 3$. If $x + 3 = 0$, then $x = -3$. If $x^2 + 1 = 0$, then $x^2 = -1$, which has no real solution. So the solutions are $x = \pm 3$.

79. $x^{-1/2} - 2x^{1/2} + x^{3/2} = 0 \Leftrightarrow x^{-1/2}(1 - 2x + x^2) = 0 \Leftrightarrow x^{-1/2}(1-x)^2 = 0$. Since $x^{-1/2} - 1/\sqrt{x}$ is never 0, the only solution comes from $(1-x)^2 = 0 \Leftrightarrow 1 - x = 0 \Leftrightarrow x = 1$.

81. $|x - 7| = 4 \Leftrightarrow x - 7 = \pm 4 \Leftrightarrow x = 7 \pm 4$, so $x = 11$ or $x = 3$.

83. $|2x - 5| = 9$ is equivalent to $2x - 5 = \pm 9 \Leftrightarrow 2x = 5 \pm 9 \Leftrightarrow x = \frac{5 \pm 9}{2}$. So $x = -2$ or $x = 7$.

85. Let x be the number of pounds of raisins. Then the number of pounds of nuts is $50 - x$.

	Raisins	Nuts	Mixture
Pounds	x	$50 - x$	50
Rate (cost per pound)	3.20	2.40	2.72

So $3.20x + 2.40(50 - x) = 2.72(50) \Leftrightarrow 3.20x + 120 - 2.40x = 136 \Leftrightarrow 0.8x = 16 \Leftrightarrow x = 20$. Thus the mixture uses 20 pounds of raisins and $50 - 20 = 30$ pounds of nuts.

87. Let r be the rate the woman runs in mi/h. Then she cycles at $r + 8$ mi/h.

	Rate	Time	Distance
Cycle	$r + 8$	$\frac{4}{r+8}$	4
Run	r	$\frac{2.5}{r}$	2.5

Since the total time of the workout is 1 hour, we have $\frac{4}{r+8} + \frac{2.5}{r} = 1$. Multiplying by $2r(r+8)$, we

get $4(2r) + 2.5(2)(r+8) = 2r(r+8) \Leftrightarrow 8r + 5r + 40 = 2r^2 + 16r \Leftrightarrow 0 = 2r^2 + 3r - 40 \Rightarrow$

$r = \frac{-3 \pm \sqrt{(3)^2 - 4(2)(-40)}}{2(2)} = \frac{-3 \pm \sqrt{9 + 320}}{4} = \frac{-3 \pm \sqrt{329}}{4}$. Since $r \geq 0$, we reject the negative value. She runs at

$r = \frac{-3 + \sqrt{329}}{4} \approx 3.78$ mi/h.

89. Let x be the amount invested in the account earning 1.5% interest. Then the amount invested in the account earning 2.5% is $7000 - x$.

	1.5% Account	2.5% Account	Total
Amount invested	x	$7000 - x$	7000
Interest earned	$0.015x$	$0.025(7000 - x)$	120.25

From the table, we see that $0.015x + 0.025(7000 - x) = 120.25 \Leftrightarrow 0.015x + 175 - 0.025x = 120.25 \Leftrightarrow 54.75 = 0.01x \Leftrightarrow x = 5475$. Thus, Luc invested \$5475 in the account earning 1.5% interest and \$1525 in the account earning 2.5% interest.

91. Let x be the length of one side in cm. Then $28 - x$ is the length of the other side. Using the Pythagorean Theorem, we have $x^2 + (28 - x)^2 = 20^2 \Leftrightarrow x^2 + 784 - 56x + x^2 = 400 \Leftrightarrow 2x^2 - 56x + 384 = 0 \Leftrightarrow 2(x^2 - 28x + 192) = 0 \Leftrightarrow$ $2(x - 12)(x - 16) = 0$. So $x = 12$ or $x = 16$. If $x = 12$, then the other side is $28 - 12 = 16$. Similarly, if $x = 16$, then the other side is 12. The sides are 12 cm and 16 cm.

93. Let w be width of the pool. Then the length of the pool is $2w$, and its volume is $8(w)(2w) = 8464 \Leftrightarrow 16w^2 = 8464 \Leftrightarrow$ $w^2 = 529 \Rightarrow w = \pm 23$. Since $w > 0$, we reject the negative value. The pool is 23 feet wide, $2(23) = 46$ feet long, and 8 feet deep.

95. $3x - 2 > -11 \Leftrightarrow 3x > -9 \Leftrightarrow x > -3$.
Interval: $(-3, \infty)$

Graph: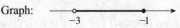

97. $-1 < 2x + 5 \leq 3 \Leftrightarrow -6 < 2x \leq -2 \Leftrightarrow -3 < x \leq -1$
Interval: $(-3, -1]$

Graph: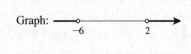

99. $x^2 + 4x - 12 > 0 \Leftrightarrow (x - 2)(x + 6) > 0$. The expression on the left of the inequality changes sign where $x = 2$ and where $x = -6$. Thus we must check the intervals in the following table.

Interval	$(-\infty, -6)$	$(-6, 2)$	$(2, \infty)$
Sign of $x - 2$	$-$	$-$	$+$
Sign of $x + 6$	$-$	$+$	$+$
Sign of $(x - 2)(x + 6)$	$+$	$-$	$+$

Interval: $(-\infty, -6) \cup (2, \infty)$.

Graph:

101. $\dfrac{2x + 5}{x + 1} \leq 1 \Leftrightarrow \dfrac{2x + 5}{x + 1} - 1 \leq 0 \Leftrightarrow \dfrac{2x + 5}{x + 1} - \dfrac{x + 1}{x + 1} \leq 0 \Leftrightarrow \dfrac{x + 4}{x + 1} \leq 0$. The expression on the left of the inequality changes sign where $x = -1$ and where $x = -4$. Thus we must check the intervals in the following table.

Interval	$(-\infty, -4)$	$(-4, -1)$	$(-1, \infty)$
Sign of $x + 4$	$-$	$+$	$+$
Sign of $x + 1$	$-$	$-$	$+$
Sign of $\dfrac{x + 4}{x + 1}$	$+$	$-$	$+$

We exclude $x = -1$, since the expression is not defined at this value. Thus the solution is $[-4, -1)$.

Graph:

103. $\dfrac{x - 4}{x^2 - 4} \leq 0 \Leftrightarrow \dfrac{x - 4}{(x - 2)(x + 2)} \leq 0$. The expression on the left of the inequality changes sign where $x = -2$, where $x = 2$, and where $x = 4$. Thus we must check the intervals in the following table.

Interval	$(-\infty, -2)$	$(-2, 2)$	$(2, 4)$	$(4, \infty)$
Sign of $x - 4$	$-$	$-$	$-$	$+$
Sign of $x - 2$	$-$	$-$	$+$	$+$
Sign of $x + 2$	$-$	$+$	$+$	$+$
Sign of $\dfrac{x - 4}{(x - 2)(x + 2)}$	$-$	$+$	$-$	$+$

Since the expression is not defined when $x = \pm 2$, we exclude these values and the solution is $(-\infty, -2) \cup (2, 4]$.

Graph:

105. $|x - 5| \leq 3 \Leftrightarrow -3 \leq x - 5 \leq 3 \Leftrightarrow 2 \leq x \leq 8$. Interval: $[2, 8]$. Graph:

107. $|2x + 1| \geq 1$ is equivalent to $2x + 1 \geq 1$ or $2x + 1 \leq -1$. *Case 1:* $2x + 1 \geq 1 \Leftrightarrow 2x \geq 0 \Leftrightarrow x \geq 0$. *Case 2:* $2x + 1 \leq -1$

$\Leftrightarrow 2x \leq -2 \Leftrightarrow x \leq -1$. Interval: $(-\infty, -1] \cup [0, \infty)$. Graph:

109. (a) For $\sqrt{24 - x - 3x^2}$ to define a real number, we must have $24 - x - 3x^2 \geq 0 \Leftrightarrow (8 - 3x)(3 + x) \geq 0$. The expression on the left of the inequality changes sign where $8 - 3x = 0 \Leftrightarrow -3x = -8 \Leftrightarrow x = \frac{8}{3}$; or where $x = -3$. Thus we must check the intervals in the following table.

Interval	$(-\infty, -3)$	$\left(-3, \frac{8}{3}\right)$	$\left(\frac{8}{3}, \infty\right)$
Sign of $8 - 3x$	+	+	−
Sign of $3 + x$	−	+	+
Sign of $(8 - 3x)(3 + x)$	−	+	−

Interval: $\left[-3, \frac{8}{3}\right]$.

Graph:

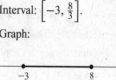

(b) For $\dfrac{1}{\sqrt[4]{x - x^4}}$ to define a real number we must have $x - x^4 > 0 \Leftrightarrow x\left(1 - x^3\right) > 0 \Leftrightarrow x(1 - x)\left(1 + x + x^2\right) > 0$.

The expression on the left of the inequality changes sign where $x = 0$; or where $x = 1$; or where $1 + x + x^2 = 0 \Rightarrow$ $x = \dfrac{-1 \pm \sqrt{1^2 - 4(1)(1)}}{2(1)} = \dfrac{1 \pm \sqrt{1 - 4}}{2}$ which is imaginary. We check the intervals in the following table.

Interval	$(-\infty, 0)$	$(0, 1)$	$(1, \infty)$
Sign of x	−	+	+
Sign of $1 - x$	+	+	−
Sign of $1 + x + x^2$	+	+	+
Sign of $x(1 - x)\left(1 + x + x^2\right)$	−	+	−

Interval: $(0, 1)$.

Graph:

CHAPTER 1 TEST

1. (a)

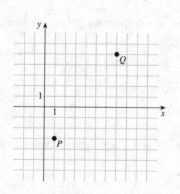

(b) The distance between P and Q is
$$d(P, Q) = \sqrt{(7 - 1)^2 + [5 - (-3)]^2} = \sqrt{36 + 64} = \sqrt{100} = 10.$$

(c) The midpoint is $\left(\dfrac{1 + 7}{2}, \dfrac{-3 + 5}{2}\right) = (4, 1)$.

(d) The slope of the line is $\dfrac{5 - (-3)}{7 - 1} = \dfrac{8}{6} = \dfrac{4}{3}$.

(e) The perpendicular bisector of PQ contains the midpoint, $(4, 1)$, and its slope is the negative reciprocal of $\frac{4}{3}$. Thus the slope is $-\dfrac{1}{4/3} = -\dfrac{3}{4}$. Hence the equation is $y - 1 = -\dfrac{3}{4}(x - 4) \Leftrightarrow$ $y = -\dfrac{3}{4}x + 3 + 1 = -\dfrac{3}{4}x + 4$. That is, $y = -\dfrac{3}{4}x + 4$.

(f) The center of the circle is the midpoint, $(4, 1)$, and the length of the radius is $\frac{1}{2} \cdot 10 = 5$. Thus an equation of the circle whose diameter is PQ is $(x - 4)^2 + (y - 1)^2 = 25$.

3. (a) $x = 4 - y^2$. To test for symmetry about the x-axis, we replace y with $-y$:

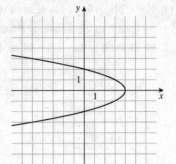

$x = 4 - (-y)^2 \Leftrightarrow x = 4 - y^2$, so the graph is symmetric about the x-axis.

To test for symmetry about the y-axis, we replace x with $-x$:

$-x = 4 - y^2$ is different from the original equation, so the graph is not symmetric about the y-axis.

For symmetry about the origin, we replace x with $-x$ and y with $-y$:

$-x = 4 - (-y)^2 \Leftrightarrow -x = 4 - y^2$, which is different from the original equation, so the graph is not symmetric about the origin.

To find x-intercepts, we set $y = 0$ and solve for x: $x = 4 - 0^2 = 4$, so the x-intercept is 4.

To find y-intercepts, we set $x = 0$ and solve for y:: $0 = 4 - y^2 \Leftrightarrow y^2 = 4$ $\Leftrightarrow y = \pm 2$, so the y-intercepts are -2 and 2.

(b) $y = |x - 2|$. To test for symmetry about the x-axis, we replace y with $-y$:

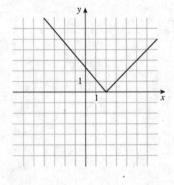

$-y = |x - 2|$ is different from the original equation, so the graph is not symmetric about the x-axis.

To test for symmetry about the y-axis, we replace x with $-x$:

$y = |-x - 2| = |x + 2|$ is different from the original equation, so the graph is not symmetric about the y-axis.

To test for symmetry about the origin, we replace x with $-x$ and y with $-y$: $-y = |-x - 2| \Leftrightarrow y = -|x + 2|$, which is different from the original equation, so the graph is not symmetric about the origin.

To find x-intercepts, we set $y = 0$ and solve for x: $0 = |x - 2| \Leftrightarrow$ $x - 2 = 0 \Leftrightarrow x = -2$, so the x-intercept is 2.

To find y-intercepts, we set $x = 0$ and solve for y:

$y = |0 - 2| = |-2| = 2$, so the y-intercept is 2.

5. (a) $3x + y - 10 = 0 \Leftrightarrow y = -3x + 10$, so the slope of the line we seek is -3. Using the point-slope form,

$\qquad y - (-6) = -3(x - 3) \Leftrightarrow y + 6 = -3x + 9 \Leftrightarrow 3x + y - 3 = 0$.

(b) Using the intercept form we get $\dfrac{x}{6} + \dfrac{y}{4} = 1 \Leftrightarrow 2x + 3y = 12 \Leftrightarrow 2x + 3y - 12 = 0$.

7. (a) We see that the graphs of $y = x^2 - 4x$ and $y = 2x - x^2$ intersect at $x = 0$ and $x = 3$, so these are the solutions of the equation $x^2 - 4x = 2x - x^2$.

(b) We see that the graph of $y = x^2 - 4x$ lies above that of $y = 2x - x^2$ for $-\infty < x < 0$ and $3 < x < \infty$, so $x^2 - 4x > 2x - x^2$ on the intervals $(-\infty, 0)$ and $(3, \infty)$.

(c) We see that the graph of $y = 2x - x^2$ crosses the x-axis at $x = 0$ and $x = 2$, so these are the solutions of the equation $2x - x^2 = 0$.

(d) We see that the graph of $y = x^2 - 4x$ lies below the x-axis for $0 < x < 4$, so $x^2 - 4x \leq 0$ on the interval $[0, 4]$.

9. Let d be the distance in km, between Bedingfield and Portsmouth.

Direction	Distance	Rate	Time
Bedingfield $\rightarrow$ Portsmouth	d	100	$\dfrac{d}{100}$
Portsmouth $\rightarrow$ Bedingfield	d	75	$\dfrac{d}{75}$

We have used time $= \dfrac{\text{distance}}{\text{rate}}$ to fill in the time column of the table. We are given that the sum of the times is 3.5 hours.

Thus we get the equation $\dfrac{d}{100} + \dfrac{d}{75} = 3.5 \Leftrightarrow 300\left(\dfrac{d}{100} + \dfrac{d}{75}\right) = 300\,(3.5) \Leftrightarrow 3d + 4d = 1050 \Leftrightarrow d = \dfrac{1050}{7} = 150$ km.

11. Let w be the width of the parcel of land. Then $w + 70$ is the length of the parcel of land. Then $w^2 + (w + 70)^2 = 130^2 \Leftrightarrow$ $w^2 + w^2 + 140w + 4900 = 16{,}900 \Leftrightarrow 2w^2 + 140w - 12{,}000 = 0 \Leftrightarrow w^2 + 70w - 6000 = 0 \Leftrightarrow (w - 50)\,(w + 120) = 0$. So $w = 50$ or $w = -120$. Since $w \geq 0$, the width is $w = 50$ ft and the length is $w + 70 = 120$ ft.

13. $5 \leq \dfrac{5}{9}\,(F - 32) \leq 10 \Leftrightarrow 9 \leq F - 32 \leq 18 \Leftrightarrow 41 \leq F \leq 50$. Thus the medicine is to be stored at a temperature between $41°$ F and $50°$ F.

FOCUS ON MODELING   Fitting Lines to Data

1. (a)

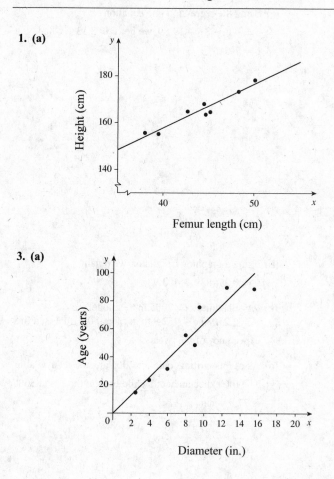

Femur length (cm)

(b) Using a graphing calculator, we obtain the regression line $y = 1.8807x + 82.65$.

(c) Using $x = 58$ in the equation $y = 1.8807x + 82.65$, we get $y = 1.8807\,(58) + 82.65 \approx 191.7$ cm.

3. (a)

Diameter (in.)

(b) Using a graphing calculator, we obtain the regression line $y = 6.451x - 0.1523$.

(c) Using $x = 18$ in the equation $y = 6.451x - 0.1523$, we get $y = 6.451\,(18) - 0.1523 \approx 116$ years.

5. (a)

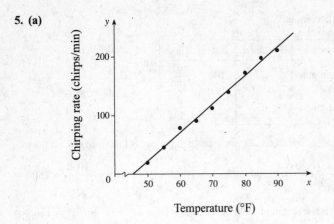

Temperature (°F)

(b) Using a graphing calculator, we obtain the regression line $y = 4.857x - 220.97$.

(c) Using $x = 100°$ F in the equation $y = 4.857x - 220.97$, we get $y \approx 265$ chirps per minute.

7. (a)

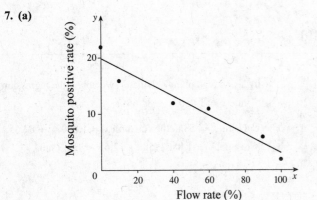

Flow rate (%)

(b) Using a graphing calculator, we obtain the regression line $y = -0.168x + 19.89$.

(c) Using the regression line equation $y = -0.168x + 19.89$, we get $y \approx 8.13\%$ when $x = 70\%$.

9. (a)

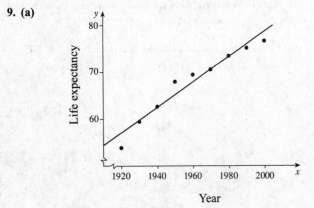

Year

(b) Using a graphing calculator, we obtain $y = 0.27083x - 462.9$.

(c) We substitute $x = 2006$ in the model $y = 0.27083x - 462.9$ to get $y = 80.4$, that is, a life expectancy of 80.4 years.

(d) As of this writing, data for 2006 are not yet available. The life expectancy of a child born in the US in 2005 is 77.9 years.

11. (a) If we take $x = 0$ in 1900 for both men and women, then the regression equation for the men's data is $y = -0.170x + 64.61$ and the regression equation for the women's data is $y = -0.260x + 78.27$.

(b)

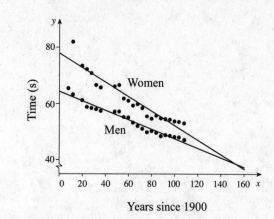

Years since 1900

These lines predict that the women will overtake the men in this event when $-0.170x + 64.61 = -0.260x + 78.27 \Leftrightarrow 0.09x = 13.66 \Leftrightarrow$ $x = 151.78$, or in 2052. This seems unlikely, but who knows?

13. Results will depend on student surveys in each class.

2 FUNCTIONS

2.1 FUNCTIONS

1. If $f(x) = x^3 + 1$, then

    (a) the value of f at $x = -1$ is $f(-1) = (-1)^3 + 1 = 0$.

    (b) the value of f at $x = 2$ is $f(2) = 2^3 + 1 = 9$.

    (c) the net change in the value of f between $x = -1$ and $x = 2$ is $f(2) - f(-1) = 9 - 0 = 9$.

3. (a) $f(x) = x^2 - 3x$ and $g(x) = \dfrac{x-5}{x}$ have 5 in their domain because they are defined when $x = 5$. However,

    $h(x) = \sqrt{x - 10}$ is undefined when $x = 5$ because $\sqrt{5 - 10} = \sqrt{-5}$, so 5 is not in the domain of h.

    (b) $f(5) = 5^2 - 3(5) = 25 - 15 = 10$ and $g(5) = \dfrac{5-5}{5} = \dfrac{0}{5} = 0$.

5. $f(x) = \dfrac{x-2}{5}$ **7.** $f(x) = 4x - 1$

9. Square, then add 2.   **11.** Subtract 4, then divide by 3.

13. Machine diagram for $f(x) = \sqrt{x - 1}$.   **15.** $f(x) = 2(x - 1)^2$

$1 \rightarrow$ [subtract 1, then take square root] $\rightarrow 0$

$2 \rightarrow$ [subtract 1, then take square root] $\rightarrow 1$

$5 \rightarrow$ [subtract 1, then take square root] $\rightarrow 2$

x	$f(x)$
-1	$2(-1-1)^2 = 8$
0	$2(-1)^2 = 2$
1	$2(1-1)^2 = 0$
2	$2(2-1)^2 = 2$
3	$2(3-1)^2 = 8$

17. $f(x) = x^2 - 6$; $f(-3) = (-3)^2 - 6 = 9 - 6 = 3$; $f(3) = 3^2 - 6 = 9 - 6 = 3$; $f(0) = 0^2 - 6 = -6$;

    $f\left(\frac{1}{2}\right) = \left(\frac{1}{2}\right)^2 - 6 = \frac{1}{4} - 6 = -\frac{23}{4}$.

19. $f(x) = 2x + 1$; $f(1) = 2(1) + 1 = 3$; $f(-2) = 2(-2) + 1 = -3$; $f\left(\frac{1}{2}\right) = 2\left(\frac{1}{2}\right) + 1 = 2$; $f(a) = 2(a) + 1 = 2a + 1$;

    $f(-a) = 2(-a) + 1 = -2a + 1$; $f(a - 1) = 2(a - 1) + 1 = 2a - 1$.

21. $f(x) = x^2 + 2x$; $f(0) = 0^2 + 2(0) = 0$; $f(3) = 3^2 + 2(3) = 9 + 6 = 15$; $f(-3) = (-3)^2 + 2(-3) = 9 - 6 = 3$;

    $f(a) = a^2 + 2(a) = a^2 + 2a$; $f(-x) = (-x)^2 + 2(-x) = x^2 - 2x$; $f\left(\frac{1}{a}\right) = \left(\frac{1}{a}\right)^2 + 2\left(\frac{1}{a}\right) = \frac{1}{a^2} + \frac{2}{a}$.

23. $g(x) = \dfrac{1-x}{1+x}$; $g(2) = \dfrac{1-(2)}{1+(2)} = \dfrac{-1}{3} = -\dfrac{1}{3}$; $g(-1) = \dfrac{1-(-1)}{1+(-1)}$, which is undefined; $g\left(\frac{1}{2}\right) = \dfrac{1-\left(\frac{1}{2}\right)}{1+\left(\frac{1}{2}\right)} = \dfrac{\frac{1}{2}}{\frac{3}{2}} = \dfrac{1}{3}$;

    $g(a) = \dfrac{1-(a)}{1+(a)} = \dfrac{1-a}{1+a}$; $g(a-1) = \dfrac{1-(a-1)}{1+(a-1)} = \dfrac{1-a+1}{1+a-1} = \dfrac{2-a}{a}$; $g\left(x^2 - 1\right) = \dfrac{1-\left(x^2-1\right)}{1+\left(x^2-1\right)} = \dfrac{2-x^2}{x^2}$.

89

25. $f(x) = 2x^2 + 3x - 4$; $f(0) = 2(0)^2 + 3(0) - 4 = -4$; $f(2) = 2(2)^2 + 3(2) - 4 = 8 + 6 - 4 = 10$; $f\left(\sqrt{2}\right) = 2\left(\sqrt{2}\right)^2 + 3\left(\sqrt{2}\right) - 4 = 4 + 3\sqrt{2} - 4 = 3\sqrt{2}$; $f(x+1) = 2(x+1)^2 + 3(x+1) - 4 = 2x^2 + 4x + 2 + 3x + 3 - 4 = 2x^2 + 7x + 1$; $f(-x) = 2(-x)^2 + 3(-x) - 4 = 2x^2 - 3x - 4$; $f\left(x^3\right) = 2\left(x^3\right)^2 + 3x^3 - 4 = 2x^6 + 3x^3 - 4$.

27. $f(x) = 2|x - 1|$; $f(-2) = 2|-2 - 1| = 2(3) = 6$; $f(0) = 2|0 - 1| = 2(1) = 2$; $f\left(\frac{1}{2}\right) = 2\left|\frac{1}{2} - 1\right| = 2\left(\frac{1}{2}\right) = 1$; $f(2) = 2|2 - 1| = 2(1) = 2$; $f(x+1) = 2|(x+1) - 1| = 2|x|$; $f\left(x^2 + 2\right) = 2\left|\left(x^2 + 2\right) - 1\right| = 2\left|x^2 + 1\right| = 2x^2 + 2$ (since $x^2 + 1 > 0$).

29. Since $-2 < 0$, we have $f(-2) = (-2)^2 = 4$. Since $-1 < 0$, we have $f(-1) = (-1)^2 = 1$. Since $0 \geq 0$, we have $f(0) = 0 + 1 = 1$. Since $1 \geq 0$, we have $f(1) = 1 + 1 = 2$. Since $2 \geq 0$, we have $f(2) = 2 + 1 = 3$.

31. Since $-4 \leq -1$, we have $f(-4) = (-4)^2 + 2(-4) = 16 - 8 = 8$. Since $-\frac{3}{2} \leq -1$, we have $f\left(-\frac{3}{2}\right) = \left(-\frac{3}{2}\right)^2 + 2\left(-\frac{3}{2}\right) = \frac{9}{4} - 3 = -\frac{3}{4}$. Since $-1 \leq -1$, we have $f(-1) = (-1)^2 + 2(-1) = 1 - 2 = -1$. Since $-1 < 0 \leq 1$, we have $f(0) = 0$. Since $25 > 1$, we have $f(25) = -1$.

33. $f(x+2) = (x+2)^2 + 1 = x^2 + 4x + 4 + 1 = x^2 + 4x + 5$; $f(x) + f(2) = x^2 + 1 + (2)^2 + 1 = x^2 + 1 + 4 + 1 = x^2 + 6$.

35. $f\left(x^2\right) = x^2 + 4$; $[f(x)]^2 = [x + 4]^2 = x^2 + 8x + 16$.

37. $f(x) = 3x - 2$, so $f(1) = 3(1) - 2 = 1$ and $f(5) = 3(5) - 2 = 13$. Thus, the net change is $f(5) - f(1) = 13 - 1 = 12$.

39. $g(t) = 1 - t^2$, so $g(-2) = 1 - (-2)^2 = 1 - 4 = -3$ and $g(5) = 1 - 5^2 = -24$. Thus, the net change is $g(5) - g(-2) = -24 - (-3) = -21$.

41. $f(a) = 3(a) + 2 = 3a + 2$; $f(a+h) = 3(a+h) + 2 = 3a + 3h + 2$;
$$\frac{f(a+h) - f(a)}{h} = \frac{(3a + 3h + 2) - (3a + 2)}{h} = \frac{3a + 3h + 2 - 3a - 2}{h} = \frac{3h}{h} = 3.$$

43. $f(a) = 5$; $f(a+h) = 5$; $\dfrac{f(a+h) - f(a)}{h} = \dfrac{5 - 5}{h} = 0$.

45. $f(a) = \dfrac{a}{a+1}$; $f(a+h) = \dfrac{a+h}{a+h+1}$;
$$\frac{f(a+h) - f(a)}{h} = \frac{\dfrac{a+h}{a+h+1} - \dfrac{a}{a+1}}{h} = \frac{\dfrac{(a+h)(a+1)}{(a+h+1)(a+1)} - \dfrac{a(a+h+1)}{(a+h+1)(a+1)}}{h}$$
$$= \frac{\dfrac{(a+h)(a+1) - a(a+h+1)}{(a+h+1)(a+1)}}{h} = \frac{a^2 + a + ah + h - \left(a^2 + ah + a\right)}{h(a+h+1)(a+1)}$$
$$= \frac{1}{(a+h+1)(a+1)}$$

47. $f(a) = 3 - 5a + 4a^2$;
$$f(a+h) = 3 - 5(a+h) + 4(a+h)^2 = 3 - 5a - 5h + 4\left(a^2 + 2ah + h^2\right)$$
$$= 3 - 5a - 5h + 4a^2 + 8ah + 4h^2;$$
$$\frac{f(a+h) - f(a)}{h} = \frac{\left(3 - 5a - 5h + 4a^2 + 8ah + 4h^2\right) - \left(3 - 5a + 4a^2\right)}{h}$$
$$= \frac{3 - 5a - 5h + 4a^2 + 8ah + 4h^2 - 3 + 5a - 4a^2}{h} = \frac{-5h + 8ah + 4h^2}{h}$$
$$= \frac{h(-5 + 8a + 4h)}{h} = -5 + 8a + 4h.$$

49. $f(x) = 2x$. Since there is no restrictions, the domain is the set of real numbers, $(-\infty, \infty)$.

51. $f(x) = 2x$. The domain is restricted by the exercise to $[-1, 5]$.

53. $f(x) = \dfrac{1}{x-3}$. Since the denominator cannot equal 0 we have $x - 3 \neq 0 \Leftrightarrow x \neq 3$. Thus the domain is $\{x \mid x \neq 3\}$. In interval notation, the domain is $(-\infty, 3) \cup (3, \infty)$.

55. $f(x) = \dfrac{x+2}{x^2-1}$. Since the denominator cannot equal 0 we have $x^2 - 1 \neq 0 \Leftrightarrow x^2 \neq 1 \Rightarrow x \neq \pm 1$. Thus the domain is $\{x \mid x \neq \pm 1\}$. In interval notation, the domain is $(-\infty, -1) \cup (-1, 1) \cup (1, \infty)$.

57. $f(x) = \sqrt{x-5}$. We require $x - 5 \geq 0 \Leftrightarrow x \geq 5$. Thus the domain is $\{x \mid x \geq 5\}$. The domain can also be expressed in interval notation as $[5, \infty)$.

59. $f(t) = \sqrt[3]{t-1}$. Since the odd root is defined for all real numbers, the domain is the set of real numbers, $(-\infty, \infty)$.

61. $h(x) = \sqrt{2x-5}$. Since the square root is defined as a real number only for nonnegative numbers, we require that $2x - 5 \geq 0 \Leftrightarrow 2x \geq 5 \Leftrightarrow x \geq \frac{5}{2}$. So the domain is $\{x \mid x \geq \frac{5}{2}\}$. In interval notation, the domain is $\left[\frac{5}{2}, \infty\right)$.

63. $g(x) = \dfrac{\sqrt{2+x}}{3-x}$. We require $2 + x \geq 0$, and the denominator cannot equal 0. Now $2 + x \geq 0 \Leftrightarrow x \geq -2$, and $3 - x \neq 0$ $\Leftrightarrow x \neq 3$. Thus the domain is $\{x \mid x \geq -2 \text{ and } x \neq 3\}$, which can be expressed in interval notation as $[-2, 3) \cup (3, \infty)$.

65. $g(x) = \sqrt[4]{x^2 - 6x}$. Since the input to an even root must be nonnegative, we have $x^2 - 6x \geq 0 \Leftrightarrow x(x - 6) \geq 0$. We make a table:

	$(-\infty, 0)$	$(0, 6)$	$(6, \infty)$
Sign of x	$-$	$+$	$+$
Sign of $x - 6$	$-$	$-$	$+$
Sign of $x(x-6)$	$+$	$-$	$+$

Thus the domain is $(-\infty, 0] \cup [6, \infty)$.

67. $f(x) = \dfrac{3}{\sqrt{x-4}}$. Since the input to an even root must be nonnegative and the denominator cannot equal 0, we have $x - 4 > 0 \Leftrightarrow x > 4$. Thus the domain is $(4, \infty)$.

69. $f(x) = \dfrac{(x+1)^2}{\sqrt{2x-1}}$. Since the input to an even root must be nonnegative and the denominator cannot equal 0, we have $2x - 1 > 0 \Leftrightarrow x > \frac{1}{2}$. Thus the domain is $\left(\frac{1}{2}, \infty\right)$.

71. To evaluate $f(x)$, divide the input by 3 and add $\frac{2}{3}$ to the result.

(a) $f(x) = \dfrac{x}{3} + \dfrac{2}{3}$

(b)

x	$f(x)$
2	$\frac{4}{3}$
4	2
6	$\frac{8}{3}$
8	$\frac{10}{3}$

(c)

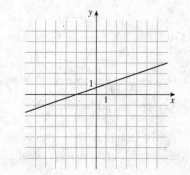

73. Let $T(x)$ be the amount of sales tax charged in Lemon County on a purchase of x dollars. To find the tax, take 8% of the purchase price.

(a) $T(x) = 0.08x$

(b)

x	$T(x)$
2	0.16
4	0.32
6	0.48
8	0.64

(c)

75. (a) $C(10) = 1500 + 3(10) + 0.02(10)^2 + 0.0001(10)^3 = 1500 + 30 + 2 + 0.1 = 1532.1$

$C(100) = 1500 + 3(100) + 0.02(100)^2 + 0.0001(100)^3 = 1500 + 300 + 200 + 100 = 2100$

(b) $C(10)$ represents the cost of producing 10 yards of fabric and $C(100)$ represents the cost of producing 100 yards of fabric.

(c) $C(0) = 1500 + 3(0) + 0.02(0)^2 + 0.0001(0)^3 = 1500$

77. (a) $V(0) = 50\left(1 - \frac{0}{20}\right)^2 = 50$ and $V(20) = 50\left(1 - \frac{20}{20}\right)^2 = 0$.

(b) $V(0) = 50$ represents the volume of the full tank at time $t = 0$, and $V(20) = 0$ represents the volume of the empty tank twenty minutes later.

(d) The net change in V as t changes from 0 minutes to 20 minutes is $V(20) - V(0) = 0 - 50 = -50$ gallons.

(c)

x	$V(x)$
0	50
5	28.125
10	12.5
15	3.125
20	0

79. (a) $v(0.1) = 18500\left(0.25 - 0.1^2\right) = 4440$,

$v(0.4) = 18500\left(0.25 - 0.4^2\right) = 1665$.

(b) They tell us that the blood flows much faster (about 2.75 times faster) 0.1 cm from the center than 0.1 cm from the edge.

(d) The net change in V as r changes from 0.1 cm to 0.5 cm is $V(0.5) - V(0.1) = 0 - 4440 = -4440$ cm/s.

(c)

r	$v(r)$
0	4625
0.1	4440
0.2	3885
0.3	2960
0.4	1665
0.5	0

81. (a) $L(0.5c) = 10\sqrt{1 - \frac{(0.5c)^2}{c^2}} \approx 8.66$ m, $L(0.75c) = 10\sqrt{1 - \frac{(0.75c)^2}{c^2}} \approx 6.61$ m, and

$L(0.9c) = 10\sqrt{1 - \frac{(0.9c)^2}{c^2}} \approx 4.36$ m.

(b) It will appear to get shorter.

83. (a) $C(75) = 75 + 15 = \$90$; $C(90) = 90 + 15 = \$105$; $C(100) = \$100$; and $C(105) = \$105$.

(b) The total price of the books purchased, including shipping.

85. **(a)** $F(x) = \begin{cases} 15\,(40-x) & \text{if } 0 < x < 40 \\ 0 & \text{if } 40 \le x \le 65 \\ 15\,(x-65) & \text{if } x > 65 \end{cases}$

(b) $F(30) = 15\,(40-10) = 15 \cdot 10 = \$150$; $F(50) = \$0$; and $F(75) = 15\,(75-65)\,15 \cdot 10 = \$150$.

(c) The fines for violating the speed limits on the freeway.

87.

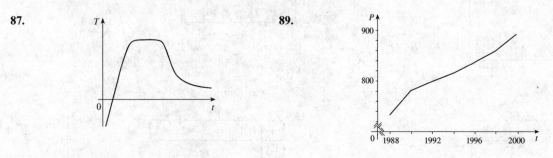

89.

91. Answers will vary.

2.2   GRAPHS OF FUNCTIONS

1. To graph the function f we plot the points $(x, f(x))$ in a coordinate plane. To graph $f(x) = x^2 - 2$ we plot the points $\left(x, x^2 - 2\right)$. So, the point $\left(3, 3^2 - 2\right) = (3, 7)$ is on the graph of f. The height of the graph of f above the x-axis when $x = 3$ is 7.

x	$f(x) = x^2 - 2$
-2	2
-1	-1
0	-2
1	-1
2	2

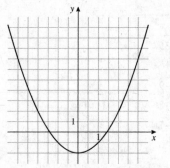

3. If the point $(2, 3)$ is on the graph of f, then $f(2) = 3$.

5.

x	$f(x) = 3$
-9	3
-6	3
-3	3
0	3
3	3
6	3

7.

x	$f(x) = x + 2$
-6	-4
-4	-2
-2	0
0	2
2	4
4	6
6	8

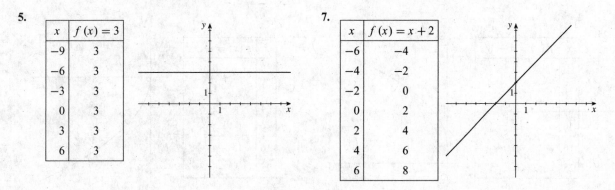

9.

x	$f(x) = -x + 3,$ $-3 \le x \le 3$
-3	6
-2	5
0	3
1	2
2	1
3	0

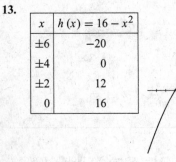

11.

x	$f(x) = -x^2$
± 4	-16
± 3	-9
± 2	-4
± 1	-1
0	0

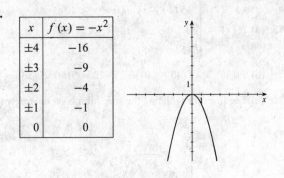

13.

x	$h(x) = 16 - x^2$
± 6	-20
± 4	0
± 2	12
0	16

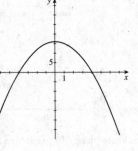

15.

x	$W(x) = x^2 + 2x + 1$
-3	4
-2	1
-1	0
0	1
1	4
2	9

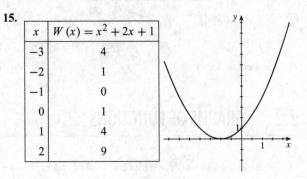

17.

x	$g(x) = x^3 - 8$
-3	-35
-2	-16
-1	-9
0	-8
1	-7
2	0
3	19

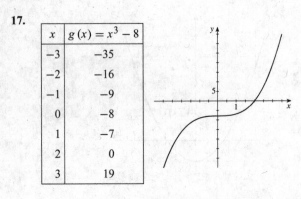

19.

x	$g(x) = x^2 - 2x$
-2	8
-1	3
0	0
1	-1
2	0
3	3
4	8

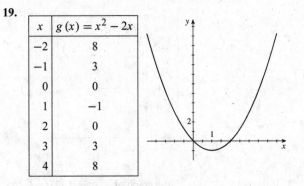

21.

x	$f(x) = 1 + \sqrt{x}$
0	1
1	2
4	3
9	4
16	5
25	6

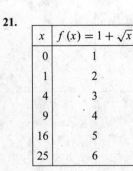

23.

x	$g(x) = -\sqrt{x}$
0	0
1	-1
4	-2
9	-3
16	-4
25	-5

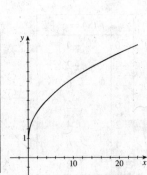

25.

x	$H(x) = |2x|$
±5	10
±4	8
±3	6
±2	4
±1	2
0	0

27.

x	$G(x) = |x| + x$
−5	0
−2	0
0	0
1	2
2	4
5	10

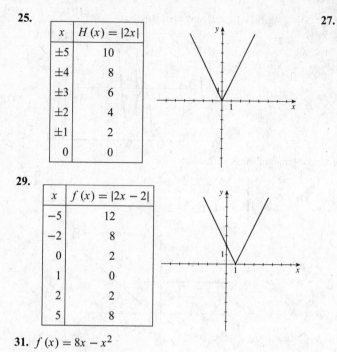

29.

x	$f(x) = |2x - 2|$
−5	12
−2	8
0	2
1	0
2	2
5	8

31. $f(x) = 8x - x^2$

(a) $[-5, 5]$ by $[-5, 5]$

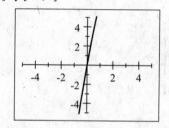

(b) $[-10, 10]$ by $[-10, 10]$

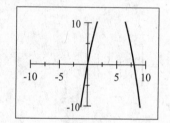

(c) $[-2, 10]$ by $[-5, 20]$

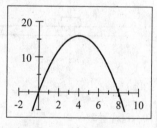

(d) $[-10, 10]$ by $[-100, 100]$

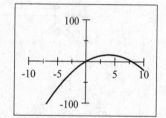

The viewing rectangle in part (c) produces the most appropriate graph of the equation.

33. $h(x) = x^3 - 5x - 4$

(a) $[-2, 2]$ by $[-2, 2]$

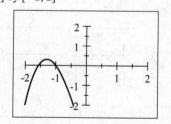

(b) $[-3, 3]$ by $[-10, 10]$

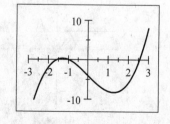

(c) $[-3, 3]$ by $[-10, 5]$

(d) $[-10, 10]$ by $[-10, 10]$

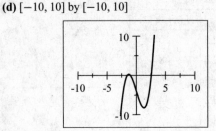

The viewing rectangle in part (c) produces the most appropriate graph of the equation.

35. $f(x) = \begin{cases} 0 & \text{if } x < 2 \\ 1 & \text{if } x \geq 2 \end{cases}$

37. $f(x) = \begin{cases} 3 & \text{if } x < 2 \\ x - 1 & \text{if } x \geq 2 \end{cases}$

39. $f(x) = \begin{cases} x & \text{if } x \leq 0 \\ x + 1 & \text{if } x > 0 \end{cases}$

41. $f(x) = \begin{cases} -1 & \text{if } x < -1 \\ 1 & \text{if } -1 \leq x \leq 1 \\ -1 & \text{if } x > 1 \end{cases}$

43. $f(x) = \begin{cases} 2 & \text{if } x \leq -1 \\ x^2 & \text{if } x > -1 \end{cases}$

45. $f(x) = \begin{cases} 0 & \text{if } |x| \leq 2 \\ 3 & \text{if } |x| > 2 \end{cases}$

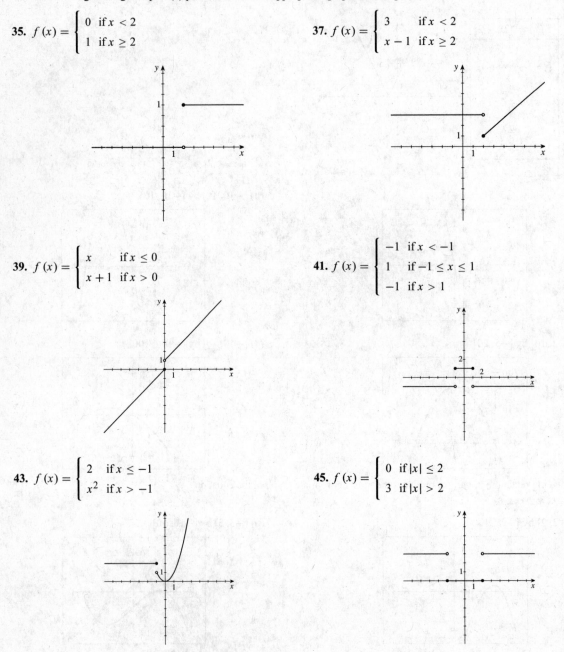

47. $f(x) = \begin{cases} 4 & \text{if } x < -2 \\ x^2 & \text{if } -2 \le x \le 2 \\ -x+6 & \text{if } x > 2 \end{cases}$

49. $f(x) = \begin{cases} x+2 & \text{if } x \le -1 \\ x^2 & \text{if } x > -1 \end{cases}$

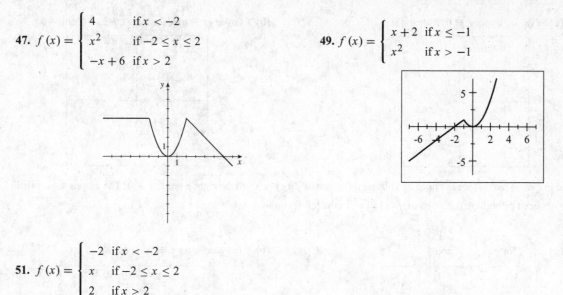

51. $f(x) = \begin{cases} -2 & \text{if } x < -2 \\ x & \text{if } -2 \le x \le 2 \\ 2 & \text{if } x > 2 \end{cases}$

53. The curves in parts (a) and (c) are graphs of a function of x, by the Vertical Line Test.

55. The given curve is the graph of a function of x, by the Vertical Line Test. Domain: $[-3, 2]$. Range: $[-2, 2]$.

57. No, the given curve is not the graph of a function of x, by the Vertical Line Test.

59. Solving for y in terms of x gives $x^2 + 2y = 4 \Leftrightarrow 2y = 4 - x^2 \Leftrightarrow y = 2 - \frac{1}{2}x^2$. This defines y as a function of x.

61. Solving for y in terms of x gives $x = y^2 \Leftrightarrow y = \pm\sqrt{x}$. The last equation gives two values of y for a given value of x. Thus, this equation does not define y as a function of x.

63. Solving for y in terms of x gives $x + y^2 = 9 \Leftrightarrow y^2 = 9 - x \Leftrightarrow y = \pm\sqrt{9-x}$. The last equation gives two values of y for a given value of x. Thus, this equation does not define y as a function of x.

65. Solving for y in terms of x gives $x^2 y + y = 1 \Leftrightarrow y\left(x^2 + 1\right) = 1 \Leftrightarrow y = \dfrac{1}{x^2 + 1}$. This defines y as a function of x.

67. Solving for y in terms of x gives $2|x| + y = 0 \Leftrightarrow y = -2|x|$. This defines y as a function of x.

69. Solving for y in terms of x gives $x = y^3 \Leftrightarrow y = \sqrt[3]{x}$. This defines y as a function of x.

71. (a) $f(x) = x^2 + c$, for $c = 0, 2, 4,$ and 6.

(b) $f(x) = x^2 + c$, for $c = 0, -2, -4,$ and -6.

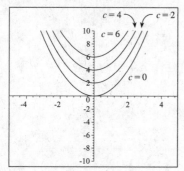

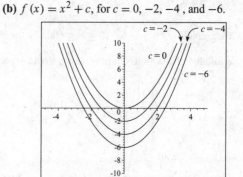

(c) The graphs in part (a) are obtained by shifting the graph of $f(x) = x^2$ upward c units, $c > 0$. The graphs in part (b) are obtained by shifting the graph of $f(x) = x^2$ downward c units.

73. (a) $f(x) = (x-c)^3$, for $c = 0, 2, 4,$ and 6.

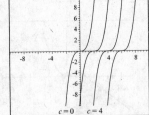

(b) $f(x) = (x-c)^3$, for $c = 0, -2, -4,$ and -6.

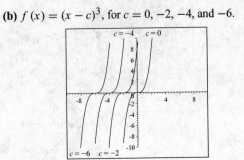

(c) The graphs in part (a) are obtained by shifting the graph of $f(x) = x^3$ to the right c units, $c > 0$. The graphs in part (b) are obtained by shifting the graph of $f(x) = x^3$ to the left $|c|$ units, $c < 0$.

75. (a) $f(x) = x^c$, for $c = \frac{1}{2}, \frac{1}{4},$ and $\frac{1}{6}$.

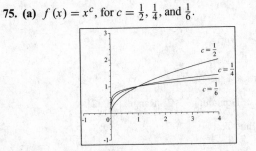

(b) $f(x) = x^c$, for $c = 1, \frac{1}{3},$ and $\frac{1}{5}$.

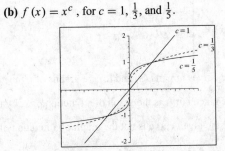

(c) Graphs of even roots are similar to $y = \sqrt{x}$, graphs of odd roots are similar to $y = \sqrt[3]{x}$. As c increases, the graph of $y = \sqrt[c]{x}$ becomes steeper near $x = 0$ and flatter when $x > 1$.

77. The slope of the line segment joining the points $(-2, 1)$ and $(4, -6)$ is $m = \dfrac{-6-1}{4-(-2)} = -\dfrac{7}{6}$. Using the point-slope form, we have $y - 1 = -\frac{7}{6}(x+2) \Leftrightarrow y = -\frac{7}{6}x - \frac{7}{3} + 1 \Leftrightarrow y = -\frac{7}{6}x - \frac{4}{3}$. Thus the function is $f(x) = -\frac{7}{6}x - \frac{4}{3}$ for $-2 \le x \le 4$.

79. First solve the circle for y: $x^2 + y^2 = 9 \Leftrightarrow y^2 = 9 - x^2 \Rightarrow y = \pm\sqrt{9-x^2}$. Since we seek the top half of the circle, we choose $y = \sqrt{9-x^2}$. So the function is $f(x) = \sqrt{9-x^2}$, $-3 \le x \le 3$.

81. We graph $T(r) = \dfrac{0.5}{r^2}$ for $10 \le r \le 100$. As the balloon is inflated, the skin gets thinner, as we would expect.

83. (a) $E(x) = \begin{cases} 6.00 + 0.10x & \text{if } 0 \leq x \leq 300 \\ 36.00 + 0.06(x - 300) & \text{if } 300 < x \end{cases}$

(b)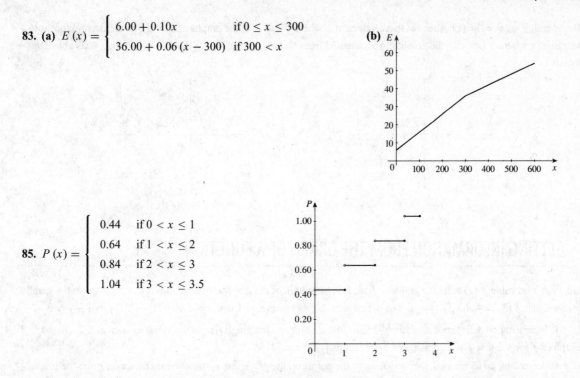

85. $P(x) = \begin{cases} 0.44 & \text{if } 0 < x \leq 1 \\ 0.64 & \text{if } 1 < x \leq 2 \\ 0.84 & \text{if } 2 < x \leq 3 \\ 1.04 & \text{if } 3 < x \leq 3.5 \end{cases}$

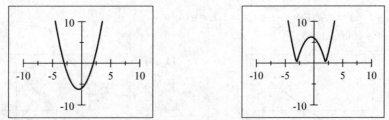

87. Answers will vary. Some examples are almost anything we purchase based on weight, volume, length, or time, for example gasoline. Although the amount delivered by the pump is continuous, the amount we pay is rounded to the penny. An example involving time would be the cost of a telephone call.

89. (a) The graphs of $f(x) = x^2 + x - 6$ and $g(x) = \left| x^2 + x - 6 \right|$ are shown in the viewing rectangle $[-10, 10]$ by $[-10, 10]$.

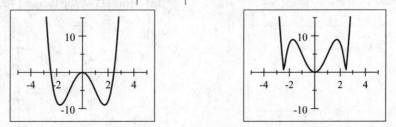

For those values of x where $f(x) \geq 0$, the graphs of f and g coincide, and for those values of x where $f(x) < 0$, the graph of g is obtained from that of f by reflecting the part below the x-axis about the x-axis.

(b) The graphs of $f(x) = x^4 - 6x^2$ and $g(x) = \left| x^4 - 6x^2 \right|$ are shown in the viewing rectangle $[-5, 5]$ by $[-10, 15]$.

For those values of x where $f(x) \geq 0$, the graphs of f and g coincide, and for those values of x where $f(x) < 0$, the graph of g is obtained from that of f by reflecting the part below the x-axis above the x-axis.

(c) In general, if $g(x) = |f(x)|$, then for those values of x where $f(x) \geq 0$, the graphs of f and g coincide, and for those values of x where $f(x) < 0$, the graph of g is obtained from that of f by reflecting the part below the x-axis above the x-axis.

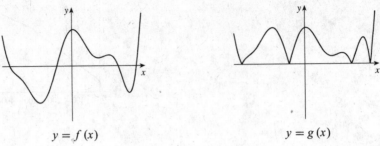

$y = f(x)$ $y = g(x)$

2.3   GETTING INFORMATION FROM THE GRAPH OF A FUNCTION

1. To find a function value $f(a)$ from the graph of f we find the height of the graph above the x-axis at $x = a$. From the graph of f we see that $f(3) = 4$ and $f(1) = 1$. The net change in f between $x = 1$ and $x = 3$ is $f(3) - f(1) = 4 - 1 = 3$.

3. (a) If f is increasing on an interval, then the y-values of the points on the graph *rise* as the x-values increase. From the graph of f we see that f is increasing on the intervals $[1, 2]$ and $[4, 5]$.

(b) If f is decreasing on an interval, then y-values of the points on the graph *fall* as the x-values increase. From the graph of f we see that f is decreasing on the intervals $[2, 4]$ and $[5, 6]$.

5. (a) $h(-2) = 1, h(0) = -1, h(2) = 3$, and $h(3) = 4$.

(b) Domain: $[-3, 4]$. Range: $[-1, 4]$.

(c) $h(-3) = 3, h(2) = 3$, and $h(4) = 3$, so $h(x) = 3$ when $x = -3, x = 2$, or $x = 4$.

(d) The graph of h lies below or on the horizontal line $y = 3$ when $-3 \leq x \leq 2$ or $x = 4$, so $h(x) \leq 3$ for those values of x.

(e) The net change in h between $x = -3$ and $x = 3$ is $h(3) - h(-3) = 4 - 3 = 1$.

7. (a) $g(-4) = 3, g(-2) = 2, g(0) = -2, g(2) = 1$, and $g(4) = 0$.

(b) Domain: $[-4, 4]$. Range: $[-2, 3]$.

9. (a)

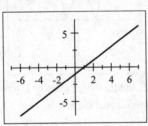

(b) Domain: $(-\infty, \infty)$; Range: $(-\infty, \infty)$

11. (a)

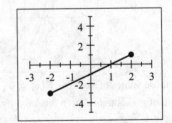

(b) Domain: $[-2, 2]$; Range: $[-3, 1]$

13. (a)

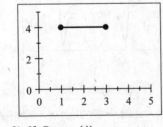

(b) Domain: $[1, 3]$; Range: $\{4\}$

15. (a)

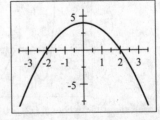

(b) Domain: $(-\infty, \infty)$; Range: $(-\infty, 4]$

17. (a)

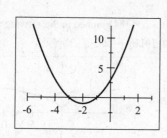

(b) Domain: $(-\infty, \infty)$; Range: $[-1, \infty)$

19. (a)

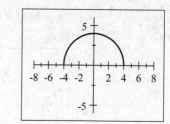

(b) Domain: $[-4, 4]$; Range: $[0, 4]$

21. (a)

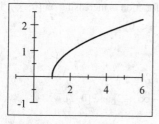

(b) Domain: $[1, \infty)$; Range: $[0, \infty)$

23. (a) The domain is $[-1, 4]$ and the range is $[-1, 3]$.

(b) The function is increasing on $[-1, 1]$ and $[2, 4]$ and decreasing on $[1, 2]$.

25. (a) The domain is $[-3, 3]$ and the range is $[-2, 2]$.

(b) The function is increasing on $[-2, -1]$ and $[1, 2]$ and decreasing on $[-3, -2]$, $[-1, 1]$, and $[2, 3]$.

27. (a) $f(x) = x^2 - 5x$ is graphed in the viewing rectangle $[-2, 7]$ by $[-10, 10]$.

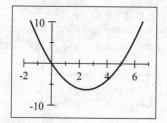

(b) The domain is $(-\infty, \infty)$ and the range is $[-6.25, \infty)$.

(c) The function is increasing on $[2.5, \infty)$. It is decreasing on $(-\infty, 2.5]$.

29. (a) $f(x) = 2x^3 - 3x^2 - 12x$ is graphed in the viewing rectangle $[-3, 5]$ by $[-25, 20]$.

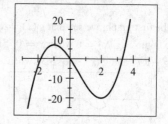

(b) The domain and range are $(-\infty, \infty)$.

(c) The function is increasing on $(-\infty, -1]$ and $[2, \infty)$. It is decreasing on $[-1, 2]$.

31. (a) $f(x) = x^3 + 2x^2 - x - 2$ is graphed in the viewing rectangle $[-5, 5]$ by $[-3, 3]$.

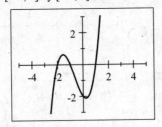

(b) The domain and range are $(-\infty, \infty)$.

(c) The function is increasing on $(-\infty, -1.55]$ and $[0.22, \infty)$. It is decreasing on $[-1.55, 0.22]$.

33. (a) $f(x) = x^{2/5}$ is graphed in the viewing rectangle $[-10, 10]$ by $[-5, 5]$.

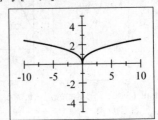

(b) The domain is $(-\infty, \infty)$ and the range is $[0, \infty)$.

(c) The function is increasing on $[0, \infty)$. It is decreasing on $(-\infty, 0]$.

35. (a) Local maximum: 2 at $x = 0$. Local minimum: -1 at $x = -2$ and 0 at $x = 2$.

(b) The function is increasing on $[-2, 0]$ and $[2, \infty)$ and decreasing on $(-\infty, -2]$ and $[0, 2]$.

37. (a) Local maximum: 0 at $x = 0$ and 1 at $x = 3$. Local minimum: -2 at $x = -2$ and -1 at $x = 1$.

(b) The function is increasing on $[-2, 0]$ and $[1, 3]$ and decreasing on $(-\infty, -2]$, $[0, 1]$, and $[3, \infty)$.

39. (a) In the first graph, we see that $f(x) = x^3 - x$ has a local minimum and a local maximum. Smaller x- and y-ranges show that $f(x)$ has a local maximum of about 0.38 when $x \approx -0.58$ and a local minimum of about -0.38 when $x \approx 0.58$.

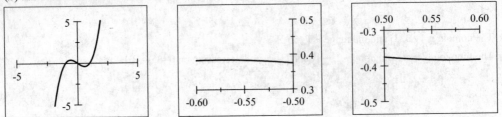

(b) The function is increasing on $(-\infty, -0.58]$ and $[0.58, \infty)$ and decreasing on $[-0.58, 0.58]$.

41. (a) In the first graph, we see that $g(x) = x^4 - 2x^3 - 11x^2$ has two local minimums and a local maximum. The local maximum is $g(x) = 0$ when $x = 0$. Smaller x- and y-ranges show that local minima are $g(x) \approx -13.61$ when $x \approx -1.71$ and $g(x) \approx -73.32$ when $x \approx 3.21$.

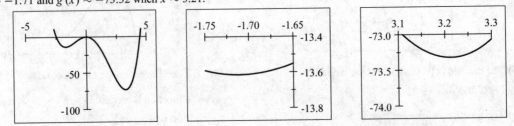

(b) The function is increasing on $[-1.71, 0]$ and $[3.21, \infty)$ and decreasing on $(-\infty, -1.71]$ and $[0, 3.21]$.

43. (a) In the first graph, we see that $U(x) = x\sqrt{6-x}$ has only a local maximum. Smaller x- and y-ranges show that $U(x)$ has a local maximum of about 5.66 when $x \approx 4.00$.

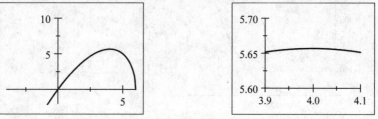

(b) The function is increasing on $(-\infty, 4.00]$ and decreasing on $[4.00, 6]$.

45. (a) In the first graph, we see that $V(x) = \dfrac{1-x^2}{x^3}$ has a local minimum and a local maximum. Smaller x- and y-ranges show that $V(x)$ has a local maximum of about 0.38 when $x \approx -1.73$ and a local minimum of about -0.38 when $x \approx 1.73$.

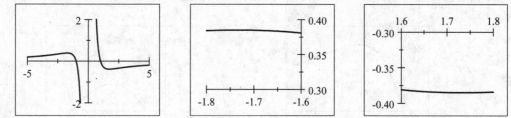

(b) The function is increasing on $(-\infty, -1.73]$ and $[1.73, \infty)$ and decreasing on $[-1.73, 0)$ and $(0, 1.73]$.

47. (a) At 6 A.M. the graph shows that the power consumption is about 500 megawatts. Since $t = 18$ represents 6 P.M., the graph shows that the power consumption at 6 P.M. is about 725 megawatts.

(b) The power consumption is lowest between 3 A.M. and 4 A.M..

(c) The power consumption is highest just before 12 noon.

(d) The net change in power consumption from 9 A.M. to 7 P.M. is $P(19) - P(9) \approx 690 - 790 \approx -100$ megawatts.

49. (a) This person appears to be gaining weight steadily until the age of 21 when this person's weight gain slows down. The person continues to gain weight until the age of 30, at which point this person experiences a sudden weight loss. Weight gain resumes around the age of 32, and the person dies at about age 68. Thus, the person's weight W is increasing on $[0, 30]$ and $[32, 68]$ and decreasing on $[30, 32]$

(b) The sudden weight loss could be due to a number of reasons, among them major illness, a weight loss program, etc.

(c) The net change in the person's weight from age 10 to age 20 is $W(20) - W(10) = 150 - 50 = 100$ lb.

51. (a) The function W is increasing on $[0, 150]$ and $[300, \infty)$ and decreasing on $[150, 300]$.

(b) W has a local maximum at $x = 150$ and a local minimum at $x = 300$.

(c) The net change in the depth W from 100 days to 300 days is $W(300) - W(100) = 25 - 75 = -50$ ft.

53. Runner A won the race. All runners finished the race. Runner B fell, but got up and finished the race.

55. (a)

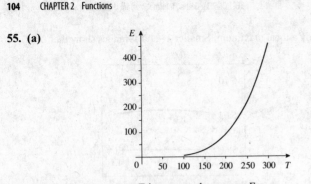

(b) As the temperature T increases, the energy E
increases. The rate of increase gets larger as the
temperature increases.

57. In the first graph, we see the general location of the maximum of $N(s) = \dfrac{88s}{17 + 17\left(\dfrac{s}{20}\right)^2}$. In the second graph we isolate

the maximum, and from this graph we see that at the speed of 20 mi/h the largest number of cars that can use the highway
safely is 52.

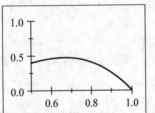
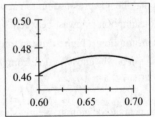

59. In the first graph, we see the general location of the maximum of $v(r) = 3.2(1-r)r^2$ is around $r = 0.7$ cm. In the second
graph, we isolate the maximum, and from this graph we see that at the maximum velocity is approximately 0.47 when
$r \approx 0.67$ cm.

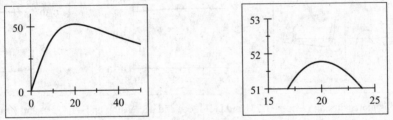

61. Numerous answers are possible.

2.4   AVERAGE RATE OF CHANGE OF A FUNCTION

1. If you travel 100 miles in two hours then your average speed for the trip is average speed $= \dfrac{100 \text{ miles}}{2 \text{ hours}} = 50$ mi/h.

3. The average rate of change of the function $f(x) = x^2$ between $x = 1$ and $x = 5$ is

average rate of change $= \dfrac{f(5) - f(1)}{5 - 1} = \dfrac{5^2 - 1^2}{4} = \dfrac{25 - 1}{4} = \dfrac{24}{4} = 6$.

5. (a) The net change is $f(4) - f(1) = 5 - 3 = 2$.

(b) We use the points $(1, 3)$ and $(4, 5)$, so the average rate of change is $\dfrac{5 - 3}{4 - 1} = \dfrac{2}{3}$.

7. (a) The net change is $f(5) - f(0) = 2 - 6 = -4$.

  (b) We use the points $(0, 6)$ and $(5, 2)$, so the average rate of change is $\dfrac{2 - 6}{5 - 0} = \dfrac{-4}{5}$.

9. (a) The net change is $f(3) - f(2) = [3(3) - 2] - [3(2) - 2] = 7 - 4 = 3$.

  (b) The average rate of change is $\dfrac{f(3) - f(2)}{3 - 2} = \dfrac{3}{1} = 3$.

11. (a) The net change is $h(1) - h(-4) = \left[-1 + \frac{3}{2}\right] - \left[-(-4) + \frac{3}{2}\right] = \frac{1}{2} - \frac{11}{2} = -5$.

  (b) The average rate of change is $\dfrac{h(1) - h(-4)}{1 - (-4)} = \dfrac{-5}{5} = -1$.

13. (a) The net change is $h(4) - h(-1) = \left[4^2 + 2(4)\right] - \left[(-1)^2 + 2(-1)\right] = 24 - (-1) = 25$.

  (b) The average rate of change is $\dfrac{h(4) - h(-1)}{4 - (-1)} = \dfrac{25}{5} = 5$.

15. (a) The net change is $f(10) - f(0) = \left[10^3 - 4\left(10^2\right)\right] - \left[0^3 - 4\left(0^2\right)\right] = 600 - 0 = 600$.

  (b) The average rate of change is $\dfrac{f(10) - f(0)}{10 - 0} = \dfrac{600}{10} = 60$.

17. (a) The net change is $f(2 + h) - f(2) = \left[3(2 + h)^2\right] - \left[3\left(2^2\right)\right] = 12 + 12h + 3h^2 - 12 = 12h + 3h^2$.

  (b) The average rate of change is $\dfrac{f(2 + h) - f(2)}{(2 + h) - 2} = \dfrac{12h + 3h^2}{h} = \dfrac{h(12 + 3h)}{h} = 12 + 3h$.

19. (a) The net change is $g(a) - g(1) = \dfrac{1}{a} - \dfrac{1}{1} = \dfrac{1 - a}{a}$.

  (b) The average rate of change is $\dfrac{g(a) - g(1)}{a - 1} = \dfrac{\frac{1 - a}{a}}{a - 1} = \dfrac{1 - a}{a(a - 1)} = -\dfrac{1}{a}$.

21. (a) The net change is $f(a + h) - f(a) = \dfrac{2}{a + h} - \dfrac{2}{a} = -\dfrac{2h}{a(a + h)}$.

  (b) The average rate of change is

$$\dfrac{f(a + h) - f(a)}{(a + h) - a} = \dfrac{-\frac{2h}{a(a + h)}}{h} = -\dfrac{2h}{ah(a + h)} = -\dfrac{2}{a(a + h)}.$$

23. (a) The average rate of change is

$$\dfrac{f(a + h) - f(a)}{(a + h) - a} = \dfrac{\left[\frac{1}{2}(a + h) + 3\right] - \left[\frac{1}{2}a + 3\right]}{h} = \dfrac{\frac{1}{2}a + \frac{1}{2}h + 3 - \frac{1}{2}a - 3}{h} = \dfrac{\frac{1}{2}h}{h} = \dfrac{1}{2}.$$

  (b) The slope of the line $f(x) = \frac{1}{2}x + 3$ is $\frac{1}{2}$, which is also the average rate of change.

25. The average rate of change is $\dfrac{W(200) - W(100)}{200 - 100} = \dfrac{50 - 75}{200 - 100} = \dfrac{-25}{100} = -\dfrac{1}{4}$ ft/day.

27. (a) The average rate of change of population is $\dfrac{1{,}591 - 856}{2001 - 1998} = \dfrac{735}{3} = 245$ persons/yr.

  (b) The average rate of change of population is $\dfrac{826 - 1{,}483}{2004 - 2002} = \dfrac{-657}{2} = -328.5$ persons/yr.

  (c) The population was increasing from 1997 to 2001.

  (d) The population was decreasing from 2001 to 2006.

29. (a) The average rate of change of sales is $\dfrac{584 - 512}{2003 - 1993} = \dfrac{72}{10} = 7.2$ units/yr.

  (b) The average rate of change of sales is $\dfrac{520 - 512}{1994 - 1993} = \dfrac{8}{1} = 8$ units/yr.

(c) The average rate of change of sales is $\dfrac{410 - 520}{1996 - 1994} = \dfrac{-110}{2} = -55$ units/yr.

(d)

Year	CD players sold	Change in sales from previous year
1993	512	—
1994	520	8
1995	413	−107
1996	410	−3
1997	468	58
1998	510	42
1999	590	80
2000	607	17
2001	732	125
2002	612	−120
2003	584	−28

Sales increased most quickly between 2000 and 2001. Sales decreased most quickly between 2001 and 2002.

31. The average rate of change of the temperature of the soup over the first 20 minutes is
$\dfrac{T(20) - T(0)}{20 - 0} = \dfrac{119 - 200}{20 - 0} = \dfrac{-81}{20} = -4.05°$ F/min. Over the next 20 minutes, it is
$\dfrac{T(40) - T(20)}{40 - 20} = \dfrac{89 - 119}{40 - 20} = -\dfrac{30}{20} = -1.5°$ F/min. The first 20 minutes had a higher average rate of change of temperature (in absolute value).

33. (a) For all three runners, the average rate of change is $\dfrac{d(10) - d(0)}{10 - 0} = \dfrac{100}{10} = 10$.

(b) Runner A gets a great jump out of the blocks but tires at the end of the race. Runner B runs a steady race. Runner C is slow at the beginning but accelerates down the track.

35. We first multiply both sides of the given expression by $x - a$: $c = \dfrac{f(x) - f(a)}{x - a} \Leftrightarrow (x - a)c = (x - a)\dfrac{f(x) - f(a)}{x - a} \Leftrightarrow$
$cx - ca = f(x) - f(a) \Leftrightarrow f(x) = cx + (f(a) - ca)$. The expression in parentheses is a constant, so this last equation is an equation of a line in slope-intercept form. Thus, f is a linear function.

2.5   TRANSFORMATIONS OF FUNCTIONS

1. (a) The graph of $y = f(x) + 3$ is obtained from the graph of $y = f(x)$ by shifting *upward* 3 units.

(b) The graph of $y = f(x + 3)$ is obtained from the graph of $y = f(x)$ by shifting *left* 3 units.

3. (a) The graph of $y = -f(x)$ is obtained from the graph of $y = f(x)$ by reflecting in the *x-axis*.

(b) The graph of $y = f(-x)$ is obtained from the graph of $y = f(x)$ by reflecting in the *y-axis*.

5. (a) The graph of $y = f(x) + 3$ can be obtained by shifting the graph of $y = f(x)$ upward 3 units.

(b) The graph of $y = f(x + 3)$ can be obtained by shifting the graph of $y = f(x)$ to the left 3 units.

7. (a) The graph of $y = -f(x)$ can be obtained by reflecting the graph of $y = f(x)$ in the x-axis.

(b) The graph of $y = f(-x)$ can be obtained by reflecting the graph of $y = f(x)$ in the y-axis.

9. (a) The graph of $y = f(x - 5) + 2$ can be obtained by shifting the graph of $y = f(x)$ to the right 5 units and upward 2 units.

(b) The graph of $y = f(x + 1) - 1$ can be obtained by shifting the graph of $y = f(x)$ to the left 1 unit and downward 1 unit.

11. (a) The graph of $y = -f(x) + 5$ can be obtained by reflecting the graph of $y = f(x)$ in the x-axis, then shifting the resulting graph upward 5 units.

(b) The graph of $y = 3f(x) - 5$ can be obtained by stretching the graph of $y = f(x)$ vertically by a factor of 3, then shifting the resulting graph downward 5 units.

13. (a) The graph of $y = 2f(x + 1) - 3$ can be obtained by shifting the graph of $y = f(x)$ to the left 1 unit, stretching it vertically by a factor of 2, and shifting it downward 3 units.

(b) The graph of $y = 2f(x - 1) + 3$ can be obtained by shifting the graph of $y = f(x)$ to the right 1 unit, stretching it vertically by a factor of 2, and shifting it upward 3 units.

15. (a) The graph of $y = f(4x)$ can be obtained by shrinking the graph of $y = f(x)$ horizontally by a factor of $\frac{1}{4}$.

(b) The graph of $y = f\left(\frac{1}{4}x\right)$ can be obtained by stretching the graph of $y = f(x)$ horizontally by a factor of 4.

17. (a) The graph of $g(x) = (x + 2)^2$ is obtained by shifting the graph of $f(x)$ to the left 2 units.

(b) The graph of $g(x) = x^2 + 2$ is obtained by shifting the graph of $f(x)$ upward 2 units.

19. (a) The graph of $g(x) = |x + 2| - 2$ is obtained by shifting the graph of $f(x)$ to the left 2 units and downward 2 units.

(b) The graph of $g(x) = g(x) = |x - 2| + 2$ is obtained from by shifting the graph of $f(x)$ to the right 2 units and upward 2 units.

21. (a)

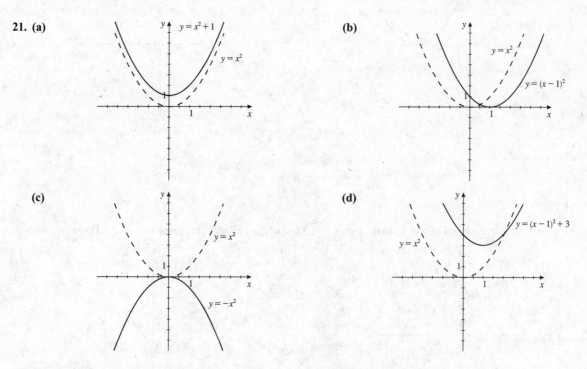

23. The graph of $y = |x + 1|$ is obtained from that of $y = |x|$ by shifting to the left 1 unit, so it has graph II.

25. The graph of $y = |x| - 1$ is obtained from that of $y = |x|$ by shifting downward 1 unit, so it has graph I.

27. $f(x) = x^2 - 1$. Shift the graph of $y = x^2$ downward 1 unit.

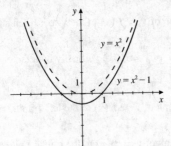

29. $f(x) = \sqrt{x} + 1$. Shift the graph of $y = \sqrt{x}$ upward 1 unit.

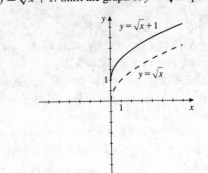

31. $f(x) = (x-5)^2$. Shift the graph of $y = x^2$ to the right 5 units.

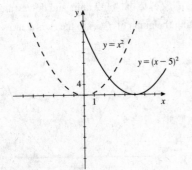

33. $f(x) = \sqrt{x+4}$. Shift the graph of $y = \sqrt{x}$ to the left 4 units.

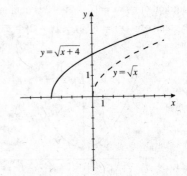

35. $f(x) = -x^3$. Reflect the graph of $y = x^3$ in the x-axis.

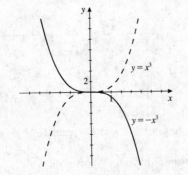

37. $y = \sqrt[4]{-x}$. Reflect the graph of $y = \sqrt[4]{x}$ in the y-axis.

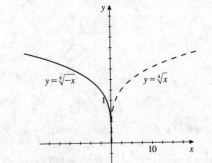

39. $y = \frac{1}{4}x^2$. Shrink the graph of $y = x^2$ vertically by a factor of $\frac{1}{4}$.

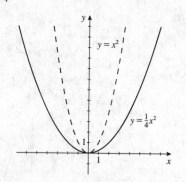

41. $y = 3\,|x|$. Stretch the graph of $y = |x|$ vertically by a factor of 3.

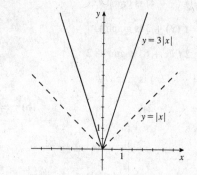

43. $y = (x-3)^2 + 5$. Shift the graph of $y = x^2$ to the right 3 units and upward 5 units.

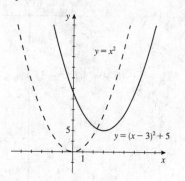

45. $y = 3 - \frac{1}{2}(x-1)^2$. Shift the graph of $y = x^2$ to the right one unit, shrink vertically by a factor of $\frac{1}{2}$, reflect in the x-axis, then shift upward 3 units.

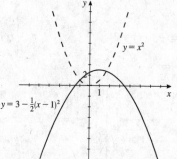

47. $y = |x+2| + 2$. Shift the graph of $y = |x|$ to the left 2 units and upward 2 units.

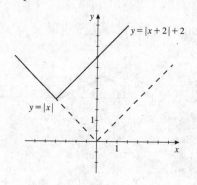

49. $y = \frac{1}{2}\sqrt{x+4} - 3$. Shrink the graph of $y = \sqrt{x}$ vertically by a factor of $\frac{1}{2}$, then shift the result to the left 4 units and downward 3 units.

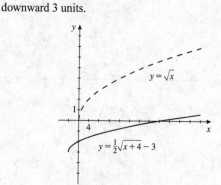

51. $y = f(x) + 3$. When $f(x) = x^2$, $y = x^2 + 3$.

53. $y = f(x+2)$. When $f(x) = \sqrt{x}$, $y = \sqrt{x+2}$.

55. $y = f(x-3) + 1$. When $f(x) = |x|$, $y = |x-3| + 1$.

57. $y = f(-x) + 1$. When $f(x) = \sqrt[4]{x}$, $y = \sqrt[4]{-x} + 1$.

59. $y = 2f(x-3) - 2$. When $f(x) = x^2$,
 $y = 2(x-3)^2 - 2$.

61. $g(x) = f(x-2) = (x-2)^2 = x^2 - 4x + 4$

63. $g(x) = f(x + 1) + 2 = |x + 1| + 2$

65. $g(x) = -f(x + 2) = -\sqrt{x + 2}$

67. **(a)** $y = f(x - 4)$ is graph #3.

   (b) $y = f(x) + 3$ is graph #1.

   (c) $y = 2f(x + 6)$ is graph #2.

   (d) $y = -f(2x)$ is graph #4.

69. **(a)** $y = f(x - 2)$

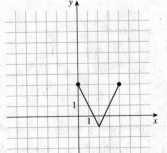

(b) $y = f(x) - 2$

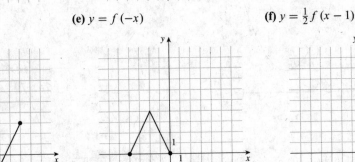

(c) $y = 2f(x)$

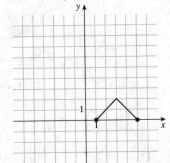

   (d) $y = -f(x) + 3$

   (e) $y = f(-x)$

   (f) $y = \frac{1}{2} f(x - 1)$

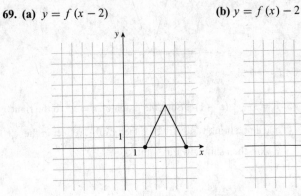

71. **(a)** $y = g(2x)$

(b) $y = g\left(\frac{1}{2}x\right)$

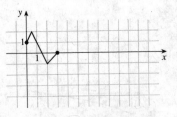

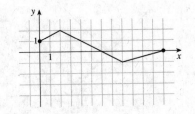

73. $y = [\![2x]\!]$

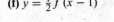

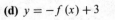

75.

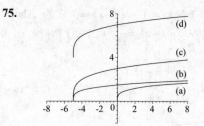

For part (b), shift the graph in (a) to the left 5 units; for part (c), shift the graph in (a) to the left 5 units, and stretch it vertically by a factor of 2; for part (d), shift the graph in (a) to the left 5 units, stretch it vertically by a factor of 2, and then shift it upward 4 units.

77.

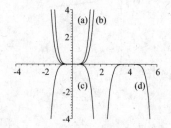

For part (b), shrink the graph in (a) vertically by a factor of $\frac{1}{3}$; for part (c), shrink the graph in (a) vertically by a factor of $\frac{1}{3}$, and reflect it in the x-axis; for part (d), shift the graph in (a) to the right 4 units, shrink vertically by a factor of $\frac{1}{3}$, and then reflect it in the x-axis.

79. (a) $y = f(x) = \sqrt{2x - x^2}$ **(b)** $y = f(2x) = \sqrt{2(2x) - (2x)^2}$ **(c)** $y = f\left(\frac{1}{2}x\right) = \sqrt{2\left(\frac{1}{2}x\right) - \left(\frac{1}{2}x\right)^2}$

$\qquad\qquad\qquad\qquad\qquad\quad = \sqrt{4x - 4x^2}$ $\qquad\qquad\quad = \sqrt{x - \frac{1}{4}x^2}$

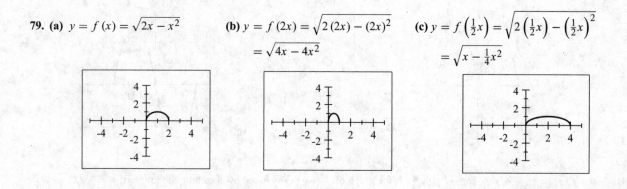

The graph in part (b) is obtained by horizontally shrinking the graph in part (a) by a factor of $\frac{1}{2}$ (so the graph is half as wide). The graph in part (c) is obtained by horizontally stretching the graph in part (a) by a factor of 2 (so the graph is twice as wide).

81. $f(x) = x^4.$ $f(-x) = (-x)^4 = x^4 = f(x).$ Thus $f(x)$ is even.

83. $f(x) = x^2 + x.$ $f(-x) = (-x)^2 + (-x) = x^2 - x.$ Thus $f(-x) \neq f(x).$ Also, $f(-x) \neq -f(x),$ so $f(x)$ is neither odd nor even.

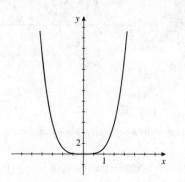

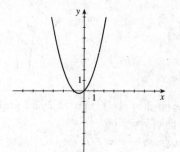

85. $f(x) = x^3 - x$.

$$f(-x) = (-x)^3 - (-x) = -x^3 + x$$
$$= -\left(x^3 - x\right) = -f(x).$$

Thus $f(x)$ is odd.

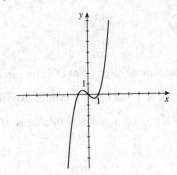

87. $f(x) = 1 - \sqrt[3]{x}$. $f(-x) = 1 - \sqrt[3]{(-x)} = 1 + \sqrt[3]{x}$. Thus $f(-x) \neq f(x)$. Also $f(-x) \neq -f(x)$, so $f(x)$ is neither odd nor even.

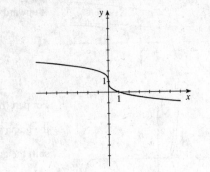

89. (a) Even

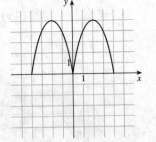

(b) Odd

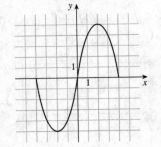

91. Since $f(x) = x^2 - 4 < 0$, for $-2 < x < 2$, the graph of $y = g(x)$ is found by sketching the graph of $y = f(x)$ for $x \leq -2$ and $x \geq 2$, then reflecting in the x-axis the part of the graph of $y = f(x)$ for $-2 < x < 2$.

93. (a) $f(x) = 4x - x^2$

(b) $f(x) = \left|4x - x^2\right|$

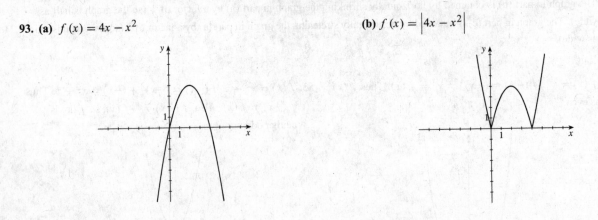

95. (a) The graph of $y = t^2$ must be shrunk vertically by a factor of 0.01 and shifted vertically 4 units upward to obtain the graph of $y = f(t)$.

(b) The graph of $y = f(t)$ must be shifted horizontally 10 units to the left to obtain the graph of $y = g(t)$. So $g(t) = f(t + 10) = 4 + 0.01(t + 10)^2 = 5 + 0.2t + 0.01t^2$.

97. f even implies $f(-x) = f(x)$; g even implies $g(-x) = g(x)$; f odd implies $f(-x) = -f(x)$; and g odd implies $g(-x) = -g(x)$

If f and g are both even, then $(f+g)(-x) = f(-x) + g(-x) = f(x) + g(x) = (f+g)(x)$ and $f+g$ is even.

If f and g are both odd, then $(f+g)(-x) = f(-x) + g(-x) = -f(x) - g(x) = -(f+g)(x)$ and $f+g$ is odd.

If f odd and g even, then $(f+g)(-x) = f(-x) + g(-x) = -f(x) + g(x)$, which is neither odd nor even.

99. $f(x) = x^n$ is even when n is an even integer and $f(x) = x^n$ is odd when n is an odd integer.

These names were chosen because polynomials with only terms with odd powers are odd functions, and polynomials with only terms with even powers are even functions.

2.6   COMBINING FUNCTIONS

1. From the graphs of f and g in the figure, we find $(f+g)(2) = f(2) + g(2) = 3 + 5 = 8$, $(f-g)(2) = f(2) - g(2) = 3 - 5 = -2$, $(fg)(2) = f(2)g(2) = 3 \cdot 5 = 15$, and $\left(\dfrac{f}{g}\right)(2) = \dfrac{f(2)}{g(2)} = \dfrac{3}{5}$.

3. If the rule of the function f is "add one" and the rule of the function g is "multiply by 2" then the rule of $f \circ g$ is "*multiply by 2, then add one*" and the rule of $g \circ f$ is "*add one, then multiply by* 2."

5. $f(x) = x$ has domain $(-\infty, \infty)$. $g(x) = 2x$ has domain $(-\infty, \infty)$. The intersection of the domains of f and g is $(-\infty, \infty)$.

$(f+g)(x) = x + 2x = 3x$, and the domain is $(-\infty, \infty)$. $(f-g)(x) = x - 2x = -x$, and the domain is $(-\infty, \infty)$.

$(fg)(x) = x(2x) = 2x^2$, and the domain is $(-\infty, \infty)$. $\left(\dfrac{f}{g}\right)(x) = \dfrac{x}{2x} = \dfrac{1}{2}$, and the domain is $(-\infty, 0) \cup (0, \infty)$.

7. $f(x) = x$ has domain $(-\infty, \infty)$. $g(x) = x^2$ has domain $(-\infty, \infty)$. The intersection of the domains of f and g is $(-\infty, \infty)$.

$(f+g)(x) = x + x^2$, and the domain is $(-\infty, \infty)$. $(f-g)(x) = x - x^2$, and the domain is $(-\infty, \infty)$.

$(fg)(x) = x\left(x^2\right) = x^3$, and the domain is $(-\infty, \infty)$. $\left(\dfrac{f}{g}\right)(x) = \dfrac{x}{x^2} = \dfrac{1}{x}$, and the domain is $(-\infty, 0) \cup (0, \infty)$.

9. $f(x) = x - 3$ has domain $(-\infty, \infty)$. $g(x) = x^2$ has domain $(-\infty, \infty)$. The intersection of the domains of f and g is $(-\infty, \infty)$.

$(f+g)(x) = (x-3) + \left(x^2\right) = x^2 + x - 3$, and the domain is $(-\infty, \infty)$.

$(f-g)(x) = (x-3) - \left(x^2\right) = -x^2 + x - 3$, and the domain is $(-\infty, \infty)$.

$(fg)(x) = (x-3)\left(x^2\right) = x^3 - 3x^2$, and the domain is $(-\infty, \infty)$.

$\left(\dfrac{f}{g}\right)(x) = \dfrac{x-3}{x^2}$, and the domain is $\{x \mid x \neq 0\}$.

11. $f(x) = \sqrt{4 - x^2}$, has domain $[-2, 2]$. $g(x) = \sqrt{1+x}$, has domain $[-1, \infty)$. The intersection of the domains of f and g is $[-1, 2]$.

$(f+g)(x) = \sqrt{4 - x^2} + \sqrt{1+x}$, and the domain is $[-1, 2]$.

$(f-g)(x) = \sqrt{4 - x^2} - \sqrt{1+x}$, and the domain is $[-1, 2]$.

$(fg)(x) = \sqrt{4 - x^2}\sqrt{1+x} = \sqrt{-x^3 - x^2 + 4x + 4}$, and the domain is $[-1, 2]$.

$\left(\dfrac{f}{g}\right)(x) = \dfrac{\sqrt{4 - x^2}}{\sqrt{1+x}} = \sqrt{\dfrac{4 - x^2}{1+x}}$, and the domain is $(-1, 2]$.

13. $f(x) = \dfrac{2}{x}$ has domain $x \neq 0$. $g(x) = \dfrac{4}{x+4}$, has domain $x \neq -4$. The intersection of the domains of f and g is

$\{x \mid x \neq 0, -4\}$; in interval notation, this is $(-\infty, -4) \cup (-4, 0) \cup (0, \infty)$.

$(f+g)(x) = \dfrac{2}{x} + \dfrac{4}{x+4} = \dfrac{2}{x} + \dfrac{4}{x+4} = \dfrac{2(3x+4)}{x(x+4)}$, and the domain is $(-\infty, -4) \cup (-4, 0) \cup (0, \infty)$.

$(f-g)(x) = \dfrac{2}{x} - \dfrac{4}{x+4} = -\dfrac{2(x-4)}{x(x+4)}$, and the domain is $(-\infty, -4) \cup (-4, 0) \cup (0, \infty)$.

$(fg)(x) = \dfrac{2}{x} \cdot \dfrac{4}{x+4} = \dfrac{8}{x(x+4)}$, and the domain is $(-\infty, -4) \cup (-4, 0) \cup (0, \infty)$.

$\left(\dfrac{f}{g}\right)(x) = \dfrac{\frac{2}{x}}{\frac{4}{x+4}} = \dfrac{x+4}{2x}$, and the domain is $(-\infty, -4) \cup (-4, 0) \cup (0, \infty)$.

15. $f(x) = \sqrt{x} + \sqrt{1-x}$. The domain of $\sqrt{x}$ is $[0, \infty)$, and the domain of $\sqrt{1-x}$ is $(-\infty, 1]$. Thus the domain is

$(-\infty, 1] \cap [0, \infty) = [0, 1]$.

17. $h(x) = (x-3)^{-1/4} = \dfrac{1}{(x-3)^{1/4}}$. Since $1/4$ is an even root and the denominator can not equal 0, $x - 3 > 0 \Leftrightarrow x > 3$.

So the domain is $(3, \infty)$.

19. **21.** **23.**

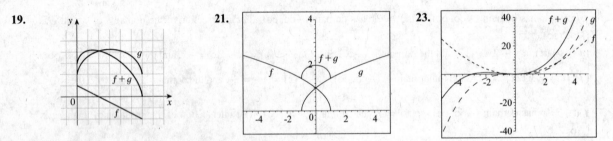

25. (a) $f(g(0)) = f\left(2 - (0)^2\right) = f(2) = 3(2) - 5 = 1$

   (b) $g(f(0)) = g(3(0) - 5) = g(-5) = 2 - (-5)^2 = -23$

27. (a) $(f \circ g)(-2) = f(g(-2)) = f\left(2 - (-2)^2\right) = f(-2) = 3(-2) - 5 = -11$

   (b) $(g \circ f)(-2) = g(f(-2)) = g(3(-2) - 5) = g(-11) = 2 - (-11)^2 = -119$

29. (a) $(f \circ g)(x) = f(g(x)) = f\left(2 - x^2\right) = 3\left(2 - x^2\right) - 5 = 6 - 3x^2 - 5 = 1 - 3x^2$

   (b) $(g \circ f)(x) = g(f(x)) = g(3x - 5) = 2 - (3x - 5)^2 = 2 - \left(9x^2 - 30x + 25\right) = -9x^2 + 30x - 23$

31. $f(g(2)) = f(5) = 4$ **33.** $(g \circ f)(4) = g(f(4)) = g(2) = 5$

35. $(g \circ g)(-2) = g(g(-2)) = g(1) = 4$

37. $f(x) = 2x + 3$, has domain $(-\infty, \infty)$; $g(x) = 4x - 1$, has domain $(-\infty, \infty)$.

   $(f \circ g)(x) = f(4x - 1) = 2(4x - 1) + 3 = 8x + 1$, and the domain is $(-\infty, \infty)$.

   $(g \circ f)(x) = g(2x + 3) = 4(2x + 3) - 1 = 8x + 11$, and the domain is $(-\infty, \infty)$.

   $(f \circ f)(x) = f(2x + 3) = 2(2x + 3) + 3 = 4x + 9$, and the domain is $(-\infty, \infty)$.

   $(g \circ g)(x) = g(4x - 1) = 4(4x - 1) - 1 = 16x - 5$, and the domain is $(-\infty, \infty)$.

39. $f(x) = x^2$, has domain $(-\infty, \infty)$; $g(x) = x + 1$, has domain $(-\infty, \infty)$.

$(f \circ g)(x) = f(x + 1) = (x + 1)^2 = x^2 + 2x + 1$, and the domain is $(-\infty, \infty)$.

$(g \circ f)(x) = g(x^2) = (x^2) + 1 = x^2 + 1$, and the domain is $(-\infty, \infty)$.

$(f \circ f)(x) = f(x^2) = (x^2)^2 = x^4$, and the domain is $(-\infty, \infty)$.

$(g \circ g)(x) = g(x + 1) = (x + 1) + 1 = x + 2$, and the domain is $(-\infty, \infty)$.

41. $f(x) = \dfrac{1}{x}$, has domain $\{x \mid x \neq 0\}$; $g(x) = 2x + 4$, has domain $(-\infty, \infty)$.

$(f \circ g)(x) = f(2x + 4) = \dfrac{1}{2x + 4}$. $(f \circ g)(x)$ is defined for $2x + 4 \neq 0 \Leftrightarrow x \neq -2$. So the domain is

$\{x \mid x \neq -2\} = (-\infty, -2) \cup (-2, \infty)$.

$(g \circ f)(x) = g\left(\dfrac{1}{x}\right) = 2\left(\dfrac{1}{x}\right) + 4 = \dfrac{2}{x} + 4$, the domain is $\{x \mid x \neq 0\} = (-\infty, 0) \cup (0, \infty)$.

$(f \circ f)(x) = f\left(\dfrac{1}{x}\right) = \dfrac{1}{\left(\dfrac{1}{x}\right)} = x$. $(f \circ f)(x)$ is defined whenever both $f(x)$ and $f(f(x))$ are defined; that is,

whenever $\{x \mid x \neq 0\} = (-\infty, 0) \cup (0, \infty)$.

$(g \circ g)(x) = g(2x + 4) = 2(2x + 4) + 4 = 4x + 8 + 4 = 4x + 12$, and the domain is $(-\infty, \infty)$.

43. $f(x) = |x|$, has domain $(-\infty, \infty)$; $g(x) = 2x + 3$, has domain $(-\infty, \infty)$

$(f \circ g)(x) = f(2x + 4) = |2x + 3|$, and the domain is $(-\infty, \infty)$.

$(g \circ f)(x) = g(|x|) = 2|x| + 3$, and the domain is $(-\infty, \infty)$.

$(f \circ f)(x) = f(|x|) = ||x|| = |x|$, and the domain is $(-\infty, \infty)$.

$(g \circ g)(x) = g(2x + 3) = 2(2x + 3) + 3 = 4x + 6 + 3 = 4x + 9$. Domain is $(-\infty, \infty)$.

45. $f(x) = \dfrac{x}{x + 1}$, has domain $\{x \mid x \neq -1\}$; $g(x) = 2x - 1$, has domain $(-\infty, \infty)$

$(f \circ g)(x) = f(2x - 1) = \dfrac{2x - 1}{(2x - 1) + 1} = \dfrac{2x - 1}{2x}$, and the domain is $\{x \mid x \neq 0\} = (-\infty, 0) \cup (0, \infty)$.

$(g \circ f)(x) = g\left(\dfrac{x}{x + 1}\right) = 2\left(\dfrac{x}{x + 1}\right) - 1 = \dfrac{2x}{x + 1} - 1$, and the domain is $\{x \mid x \neq -1\} = (-\infty, -1) \cup (-1, \infty)$

$(f \circ f)(x) = f\left(\dfrac{x}{x + 1}\right) = \dfrac{\dfrac{x}{x + 1}}{\dfrac{x}{x + 1} + 1} \cdot \dfrac{x + 1}{x + 1} = \dfrac{x}{x + x + 1} = \dfrac{x}{2x + 1}$. $(f \circ f)(x)$ is defined whenever both $f(x)$ and

$f(f(x))$ are defined; that is, whenever $x \neq -1$ and $2x + 1 \neq 0 \Rightarrow x \neq -\frac{1}{2}$, which is $(-\infty, -1) \cup \left(-1, -\frac{1}{2}\right) \cup \left(-\frac{1}{2}, \infty\right)$.

$(g \circ g)(x) = g(2x - 1) = 2(2x - 1) - 1 = 4x - 2 - 1 = 4x - 3$, and the domain is $(-\infty, \infty)$.

47. $f(x) = \dfrac{x}{x+1}$, has domain $\{x \mid x \neq -1\}$; $g(x) = \dfrac{1}{x}$ has domain $\{x \mid x \neq 0\}$.

$(f \circ g)(x) = f\left(\dfrac{1}{x}\right) = \dfrac{\frac{1}{x}}{\frac{1}{x}+1} = \dfrac{1}{x\left(\frac{1}{x}+1\right)} = \dfrac{1}{x+1}$. $(f \circ g)(x)$ is defined whenever both $g(x)$ and $f(g(x))$ are

defined, so the domain is $\{x \mid x \neq -1, 0\}$.

$(g \circ f)(x) = g\left(\dfrac{x}{x+1}\right) = \dfrac{1}{\frac{x}{x+1}} = \dfrac{x+1}{x}$. $(g \circ f)(x)$ is defined whenever both $f(x)$ and $g(f(x))$ are defined, so the

domain is $\{x \mid x \neq -1, 0\}$.

$(f \circ f)(x) = f\left(\dfrac{x}{x+1}\right) = \dfrac{\frac{x}{x+1}}{\frac{x}{x+1}+1} = \dfrac{x}{(x+1)\left(\frac{x}{x+1}+1\right)} = \dfrac{x}{2x+1}$. $(f \circ f)(x)$ is defined whenever both $f(x)$ and

$f(f(x))$ are defined, so the domain is $\left\{x \mid x \neq -1, -\frac{1}{2}\right\}$.

$(g \circ g)(x) = g\left(\dfrac{1}{x}\right) = \dfrac{1}{\frac{1}{x}} = x$. $(g \circ g)(x)$ is defined whenever both $g(x)$ and $g(g(x))$ are defined, so the domain is

$\{x \mid x \neq 0\}$.

49. $(f \circ g \circ h)(x) = f(g(h(x))) = f(g(x-1)) = f\left(\sqrt{x-1}\right) = \sqrt{x-1} - 1$

51. $(f \circ g \circ h)(x) = f(g(h(x))) = f\left(g\left(\sqrt{x}\right)\right) = f\left(\sqrt{x}-5\right) = \left(\sqrt{x}-5\right)^4 + 1$

For Exercises 49–58, many answers are possible.

53. $F(x) = (x-9)^5$. Let $f(x) = x^5$ and $g(x) = x-9$, then $F(x) = (f \circ g)(x)$.

55. $G(x) = \dfrac{x^2}{x^2+4}$. Let $f(x) = \dfrac{x}{x+4}$ and $g(x) = x^2$, then $G(x) = (f \circ g)(x)$.

57. $H(x) = \left|1-x^3\right|$. Let $f(x) = |x|$ and $g(x) = 1-x^3$, then $H(x) = (f \circ g)(x)$.

59. $F(x) = \dfrac{1}{x^2+1}$. Let $f(x) = \dfrac{1}{x}$, $g(x) = x+1$, and $h(x) = x^2$, then $F(x) = (f \circ g \circ h)(x)$.

61. $G(x) = \left(4 + \sqrt[3]{x}\right)^9$. Let $f(x) = x^9$, $g(x) = 4+x$, and $h(x) = \sqrt[3]{x}$, then $G(x) = (f \circ g \circ h)(x)$.

63. The price per sticker is $0.15 - 0.000002x$ and the number sold is x, so the revenue is

$R(x) = (0.15 - 0.000002x)\,x = 0.15x - 0.000002x^2$.

65. (a) Because the ripple travels at a speed of 60 cm/s, the distance traveled in t seconds is the radius, so $g(t) = 60t$.

(b) The area of a circle is πr^2, so $f(r) = \pi r^2$.

(c) $f \circ g = \pi\,(g(t))^2 = \pi\,(60t)^2 = 3600\pi t^2$ cm^2. This function represents the area of the ripple as a function of time.

67. Let r be the radius of the spherical balloon in centimeters. Since the radius is increasing at a rate of 2 cm/s, the radius is $r = 2t$

after t seconds. Therefore, the surface area of the balloon can be written as $S = 4\pi r^2 = 4\pi\,(2t)^2 = 4\pi\left(4t^2\right) = 16\pi t^2$.

69. (a) $f(x) = 0.90x$

(b) $g(x) = x - 100$

(c) $f \circ g = f(x - 100) = 0.90(x - 100) = 0.90x - 90$. $f \circ g$ represents applying the \$100 coupon, then the
10% discount. $g \circ f = g(0.90x) = 0.90x - 100$. $g \circ f$ represents applying the 10% discount, then the \$100 coupon.
So applying the 10% discount, then the \$100 coupon gives the lower price.

71. $A(x) = 1.05x.$ $(A \circ A)(x) = A(A(x)) = A(1.05x) = 1.05(1.05x) = (1.05)^2 x.$

$(A \circ A \circ A)(x) = A(A \circ A(x)) = A\left((1.05)^2 x\right) = 1.05\left[(1.05)^2 x\right] = (1.05)^3 x.$

$(A \circ A \circ A \circ A)(x) = A(A \circ A \circ A(x)) = A\left((1.05)^3 x\right) = 1.05\left[(1.05)^3 x\right] = (1.05)^4 x.$ A represents the amount in the account after 1 year; $A \circ A$ represents the amount in the account after 2 years; $A \circ A \circ A$ represents the amount in the account after 3 years; and $A \circ A \circ A \circ A$ represents the amount in the account after 4 years. We can see that if we compose n copies of A, we get $(1.05)^n x.$

73. $g(x) = 2x + 1$ and $h(x) = 4x^2 + 4x + 7.$

Method 1: Notice that $(2x + 1)^2 = 4x^2 + 4x + 1.$ We see that adding 6 to this quantity gives $(2x + 1)^2 + 6 = 4x^2 + 4x + 1 + 6 = 4x^2 + 4x + 7,$ which is $h(x).$ So let $f(x) = x^2 + 6,$ and we have $(f \circ g)(x) = (2x + 1)^2 + 6 = h(x).$

Method 2: Since $g(x)$ is linear and $h(x)$ is a second degree polynomial, $f(x)$ must be a second degree polynomial, that is, $f(x) = ax^2 + bx + c$ for some $a, b,$ and $c.$ Thus $f(g(x)) = f(2x + 1) = a(2x + 1)^2 + b(2x + 1) + c \Leftrightarrow$ $4ax^2 + 4ax + a + 2bx + b + c = 4ax^2 + (4a + 2b)x + (a + b + c) = 4x^2 + 4x + 7.$ Comparing this with $f(g(x)),$ we have $4a = 4$ (the x^2 coefficients), $4a + 2b = 4$ (the x coefficients), and $a + b + c = 7$ (the constant terms) $\Leftrightarrow a = 1$ and $2a + b = 2$ and $a + b + c = 7 \Leftrightarrow a = 1, b = 0, c = 6.$ Thus $f(x) = x^2 + 6.$

$f(x) = 3x + 5$ and $h(x) = 3x^2 + 3x + 2.$

Note since $f(x)$ is linear and $h(x)$ is quadratic, $g(x)$ must also be quadratic. We can then use trial and error to find $g(x).$ Another method is the following: We wish to find g so that $(f \circ g)(x) = h(x).$ Thus $f(g(x)) = 3x^2 + 3x + 2 \Leftrightarrow$ $3(g(x)) + 5 = 3x^2 + 3x + 2 \Leftrightarrow 3(g(x)) = 3x^2 + 3x - 3 \Leftrightarrow g(x) = x^2 + x - 1.$

2.7 ONE-TO-ONE FUNCTIONS AND THEIR INVERSES

1. A function f is one-to-one if different inputs produce *different* outputs. You can tell from the graph that a function is one-to-one by using the *Horizontal Line* Test.

3. (a) Proceeding backward through the description of $f,$ we can describe f^{-1} as follows: "Take the third root, subtract 5, then divide by 3."

   (b) $f(x) = (3x + 5)^3$ and $f^{-1}(x) = \dfrac{\sqrt[3]{x} - 5}{3}.$

5. If the point $(3, 4)$ is on the graph of $f,$ then the point $(4, 3)$ is on the graph of $f^{-1}.$ [This is another way of saying that $f(3) = 4 \Leftrightarrow f^{-1}(4) = 3.$]

7. By the Horizontal Line Test, f is not one-to-one.   **9.** By the Horizontal Line Test, f is one-to-one.

11. By the Horizontal Line Test, f is not one-to-one.

13. $f(x) = -2x + 4.$ If $x_1 \neq x_2,$ then $-2x_1 \neq -2x_2$ and $-2x_1 + 4 \neq -2x_2 + 4.$ So f is a one-to-one function.

15. $g(x) = \sqrt{x}.$ If $x_1 \neq x_2,$ then $\sqrt{x_1} \neq \sqrt{x_2}$ because two different numbers cannot have the same square root. Therefore, g is a one-to-one function.

17. $h(x) = x^2 - 2x.$ Because $h(0) = 0$ and $h(2) = (2) - 2(2) = 0$ we have $h(0) = h(2).$ So f is not a one-to-one function.

19. $f(x) = x^4 + 5.$ Every nonzero number and its negative have the same fourth power. For example, $(-1)^4 = 1 = (1)^4,$ so $f(-1) = f(1).$ Thus f is not a one-to-one function.

21. $f(x) = \dfrac{1}{x^2}.$ Every nonzero number and its negative have the same square. For example, $\dfrac{1}{(-1)^2} = 1 = \dfrac{1}{(1)^2},$ so $f(-1) = f(1).$ Thus f is not a one-to-one function.

23. (a) $f(2) = 7$. Since f is one-to-one, $f^{-1}(7) = 2$.

   (b) $f^{-1}(3) = -1$. Since f is one-to-one, $f(-1) = 3$.

25. $f(x) = 5 - 2x$. Since f is one-to-one and $f(1) = 5 - 2(1) = 3$, then $f^{-1}(3) = 1$. (Find 1 by solving the equation $5 - 2x = 3$.)

27. (a) Because $f(6) = 2$, $f^{-1}(2) = 6$.   **(b)** Because $f(2) = 5$, $f^{-1}(5) = 2$.   **(c)** Because $f(0) = 6$, $f^{-1}(6) = 0$.

29. $f(g(x)) = f(x + 6) = (x + 6) - 6 = x$ for all x.

   $g(f(x)) = g(x - 6) = (x - 6) + 6 = x$ for all x. Thus f and g are inverses of each other.

31. $f(g(x)) = f\left(\dfrac{x + 5}{2}\right) = 2\left(\dfrac{x + 5}{2}\right) - 5 = x + 5 - 5 = x$ for all x.

   $g(f(x)) = g(2x - 5) = \dfrac{(2x - 5) + 5}{2} = x$ for all x. Thus f and g are inverses of each other.

33. $f(g(x)) = f\left(\dfrac{1}{x}\right) = \dfrac{1}{1/x} = x$ for all $x \neq 0$. Since $f(x) = g(x)$, we also have $g(f(x)) = x$ for all $x \neq 0$. Thus f and

   g are inverses of each other.

35. $f(g(x)) = f\left(\sqrt{x + 4}\right) = \left(\sqrt{x + 4}\right)^2 - 4 = x + 4 - 4 = x$ for all $x \geq -4$.

   $g(f(x)) = g\left(x^2 - 4\right) = \sqrt{(x^2 - 4) + 4} = \sqrt{x^2} = x$ for all $x \geq 0$. Thus f and g are inverses of each other.

37. $f(g(x)) = f\left(\dfrac{1}{x} + 1\right) = \dfrac{1}{\left(\dfrac{1}{x} + 1\right) - 1} = x$ for all $x \neq 0$.

   $g(f(x)) = g\left(\dfrac{1}{x - 1}\right) = \dfrac{1}{\left(\dfrac{1}{x - 1}\right)} + 1 = (x - 1) + 1 = x$ for all $x \neq 1$. Thus f and g are inverses of each other.

39. $f(g(x)) = f\left(\dfrac{2x + 2}{x - 1}\right) = \dfrac{\frac{2x+2}{x-1} + 2}{\frac{2x+2}{x-1} - 2} = \dfrac{2x + 2 + 2(x - 1)}{2x + 2 - 2(x - 1)} = \dfrac{4x}{4} = x$ for all $x \neq 1$.

   $g(f(x)) = g\left(\dfrac{x + 2}{x - 2}\right) = \dfrac{2\left(\frac{x+2}{x-2}\right) + 2}{\frac{x+2}{x-2} - 1} = \dfrac{2(x + 2) + 2(x - 2)}{x + 2 - 1(x - 2)} = \dfrac{4x}{4} = x$ for all $x \neq 2$. Thus f and g are inverses of

   each other.

41. $f(x) = 2x + 1$. $y = 2x + 1 \Leftrightarrow 2x = y - 1 \Leftrightarrow x = \frac{1}{2}(y - 1)$. So $f^{-1}(x) = \frac{1}{2}(x - 1)$.

43. $f(x) = 4x + 7$. $y = 4x + 7 \Leftrightarrow 4x = y - 7 \Leftrightarrow x = \frac{1}{4}(y - 7)$. So $f^{-1}(x) = \frac{1}{4}(x - 7)$.

45. $f(x) = 5 - 4x^3$. $y = 5 - 4x^3 \Leftrightarrow 4x^3 = 5 - y \Leftrightarrow x^3 = \frac{1}{4}(5 - y) \Leftrightarrow x = \sqrt[3]{\frac{1}{4}(5 - y)}$. So $f^{-1}(x) = \sqrt[3]{\frac{1}{4}(5 - x)}$.

47. $f(x) = \dfrac{1}{x + 2}$. $y = \dfrac{1}{x + 2} \Leftrightarrow x + 2 = \dfrac{1}{y} \Leftrightarrow x = \dfrac{1}{y} - 2$. So $f^{-1}(x) = \dfrac{1}{x} - 2$.

49. $f(x) = \dfrac{x}{x + 4}$. $y = \dfrac{x}{x + 4} \Leftrightarrow y(x + 4) = x \Leftrightarrow xy + 4y = x \Leftrightarrow x - xy = 4y \Leftrightarrow x(1 - y) = 4y \Leftrightarrow x = \dfrac{4y}{1 - y}$. So

   $f^{-1}(x) = \dfrac{4x}{1 - x}$.

51. $f(x) = \dfrac{2x + 5}{x - 7}$. $y = \dfrac{2x + 5}{x - 7} \Leftrightarrow y(x - 7) = 2x + 5 \Leftrightarrow xy - 7y = 2x + 5 \Leftrightarrow xy - 2x = 7y + 5 \Leftrightarrow x(y - 2) = 7y + 5$

   $\Leftrightarrow x = \dfrac{7y + 5}{y - 2}$. So $f^{-1}(x) = \dfrac{7x + 5}{x - 2}$.

53. $f(x) = \dfrac{1 + 3x}{5 - 2x}$. $y = \dfrac{1 + 3x}{5 - 2x} \Leftrightarrow y(5 - 2x) = 1 + 3x \Leftrightarrow 5y - 2xy = 1 + 3x \Leftrightarrow 3x + 2xy = 5y - 1 \Leftrightarrow x(3 + 2y) = 5y - 1$

   $\Leftrightarrow x = \dfrac{5y - 1}{2y + 3}$. So $f^{-1}(x) = \dfrac{5x - 1}{2x + 3}$.

55. $f(x) = \sqrt{2+5x}$, $x \geq -\frac{2}{5}$. $y = \sqrt{2+5x}$, $y \geq 0 \Leftrightarrow y^2 = 2+5x \Leftrightarrow 5x = y^2 - 2 \Leftrightarrow x = \frac{1}{5}\left(y^2 - 2\right)$ and $y \geq 0$. So $f^{-1}(x) = \frac{1}{5}\left(x^2 - 2\right)$, $x \geq 0$.

57. $f(x) = 4 - x^2$, $x \geq 0$. $y = 4 - x^2 \Leftrightarrow x^2 = 4 - y \Leftrightarrow x = \sqrt{4-y}$. So $f^{-1}(x) = \sqrt{4-x}$, $x \leq 4$. (Note that $x \geq 0 \Rightarrow$ $f(x) \leq 4$.)

59. $f(x) = 4 + \sqrt[3]{x}$. $y = 4 + \sqrt[3]{x} \Leftrightarrow \sqrt[3]{x} = y - 4 \Leftrightarrow x = (y-4)^3$. So $f^{-1}(x) = (x-4)^3$.

61. $f(x) = 1 + \sqrt{1+x}$. $y = 1 + \sqrt{1+x}$, $y \geq 1 \Leftrightarrow \sqrt{1+x} = y - 1 \Leftrightarrow 1 + x = (y-1)^2 \Leftrightarrow x = (y-1)^2 - 1 = y^2 - 2y$. So $f^{-1}(x) = x^2 - 2x$, $x \geq 1$.

63. $f(x) = x^4$, $x \geq 0$. $y = x^4$, $y \geq 0 \Leftrightarrow x = \sqrt[4]{y}$. So $f^{-1}(x) = \sqrt[4]{x}$, $x \geq 0$.

65. (a), (b) $f(x) = 3x - 6$

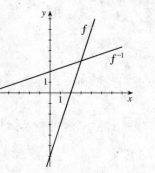

(c) $f(x) = 3x - 6$. $y = 3x - 6 \Leftrightarrow 3x = y + 6 \Leftrightarrow$ $x = \frac{1}{3}(y + 6)$. So $f^{-1}(x) = \frac{1}{3}(x + 6)$.

67. (a), (b) $f(x) = \sqrt{x+1}$

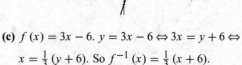

(c) $f(x) = \sqrt{x+1}$, $x \geq -1$. $y = \sqrt{x+1}$, $y \geq 0$ $\Leftrightarrow y^2 = x + 1 \Leftrightarrow x = y^2 - 1$ and $y \geq 0$. So $f^{-1}(x) = x^2 - 1$, $x \geq 0$.

69. $f(x) = x^3 - x$. Using a graphing device and the Horizontal Line Test, we see that f is not a one-to-one function. For example, $f(0) = 0 = f(-1)$.

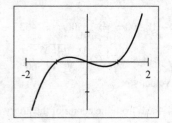

71. $f(x) = \dfrac{x+12}{x-6}$. Using a graphing device and the Horizontal Line Test, we see that f is a one-to-one function.

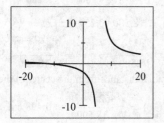

73. $f(x) = |x| - |x - 6|$. Using a graphing device and the Horizontal Line Test, we see that f is not a one-to-one function. For example $f(0) = -6 = f(-2)$.

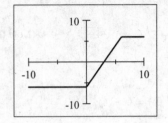

75. (a) $y = f(x) = 2 + x \Leftrightarrow x = y - 2$. So $f^{-1}(x) = x - 2$.

(b)

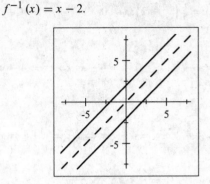

77. (a) $y = g(x) = \sqrt{x + 3}$, $y \geq 0 \Leftrightarrow x + 3 = y^2$, $y \geq 0$ $\Leftrightarrow x = y^2 - 3$, $y \geq 0$. So $g^{-1}(x) = x^2 - 3$, $x \geq 0$.

(b)

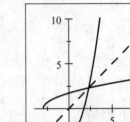

79. If we restrict the domain of $f(x)$ to $[0, \infty)$, then $y = 4 - x^2 \Leftrightarrow x^2 = 4 - y \Rightarrow x = \sqrt{4 - y}$ (since $x \geq 0$, we take the positive square root). So $f^{-1}(x) = \sqrt{4 - x}$.

If we restrict the domain of $f(x)$ to $(-\infty, 0]$, then $y = 4 - x^2 \Leftrightarrow x^2 = 4 - y \Rightarrow x = -\sqrt{4 - y}$ (since $x \leq 0$, we take the negative square root). So $f^{-1}(x) = -\sqrt{4 - x}$.

81. If we restrict the domain of $h(x)$ to $[-2, \infty)$, then $y = (x + 2)^2 \Rightarrow x + 2 = \sqrt{y}$ (since $x \geq -2$, we take the positive square root) $\Leftrightarrow x = -2 + \sqrt{y}$. So $h^{-1}(x) = -2 + \sqrt{x}$.

If we restrict the domain of $h(x)$ to $(-\infty, -2]$, then $y = (x + 2)^2 \Rightarrow x + 2 = -\sqrt{y}$ (since $x \leq -2$, we take the negative square root) $\Leftrightarrow x = -2 - \sqrt{y}$. So $h^{-1}(x) = -2 - \sqrt{x}$.

83.

85. (a) $f(x) = 500 + 80x$.

(b) $f(x) = 500 + 80x$. $y = 500 + 80x \Leftrightarrow 80x = y - 500 \Leftrightarrow x = \dfrac{y - 500}{80}$. So $f^{-1}(x) = \dfrac{x - 500}{80}$. f^{-1} represents the number of hours of investigation the investigate spends on a case for x dollars.

(c) $f^{-1}(1220) = \dfrac{1220 - 500}{80} = \dfrac{720}{80} = 9$. The investigator spent 9 hours investigating this case.

87. (a) $v(r) = 18{,}500 \left(0.25 - r^2\right)$. $t = 18{,}500 \left(0.25 - r^2\right) \Leftrightarrow t = 4625 - 18{,}500r^2 \Leftrightarrow 18500r^2 = 4625 - t \Leftrightarrow$ $r^2 = \dfrac{4625 - t}{18{,}500} \Rightarrow r = \pm\sqrt{\dfrac{4625 - t}{18{,}500}}$. Since r represents a distance, $r \geq 0$, so $v^{-1}(t) = \sqrt{\dfrac{4625 - t}{18{,}500}}$. v^{-1} represents the radial distance from the center of the vein at which the blood has velocity v.

(b) $v^{-1}(30) = \sqrt{\dfrac{4625 - 30}{18{,}500}} \approx 0.498$ cm. The velocity is 30 cm/s at 0.498 cm from the center of the artery or vein.

89. (a) $F(x) = \frac{9}{5}x + 32$. $y = \frac{9}{5}x + 32 \Leftrightarrow \frac{9}{5}x = y - 32 \Leftrightarrow x = \frac{5}{9}(y - 32)$. So $F^{-1}(x) = \frac{5}{9}(x - 32)$. F^{-1} represents the Celsius temperature that corresponds to the Fahrenheit temperature of F.

(b) $F^{-1}(86) = \frac{5}{9}(86 - 32) = \frac{5}{9}(54) = 30$. So $86°$ Fahrenheit is the same as $30°$ Celsius.

91. (a) $f(x) = \begin{cases} 0.1x, & \text{if } 0 \leq x \leq 20{,}000 \\ 2000 + 0.2(x - 20{,}000) & \text{if } x > 20{,}000 \end{cases}$

(b) We will find the inverse of each piece of the function f.

$f_1(x) = 0.1x$. $y = 0.1x \Leftrightarrow x = 10y$. So $f_1^{-1}(x) = 10x$.

$f_2(x) = 2000 + 0.2(x - 20{,}000) = 0.2x - 2000$. $y = 0.2x - 2000 \Leftrightarrow 0.2x = y + 2000 \Leftrightarrow x = 5y + 10{,}000$. So $f_2^{-1}(x) = 5x + 10{,}000$.

Since $f(0) = 0$ and $f(20{,}000) = 2000$ we have $f^{-1}(x) = \begin{cases} 10x, & \text{if } 0 \leq x \leq 2000 \\ 5x + 10{,}000 & \text{if } x > 2000 \end{cases}$ It represents the taxpayer's income.

(c) $f^{-1}(10{,}000) = 5(10{,}000) + 10{,}000 = 60{,}000$. The required income is €60,000.

93. $f(x) = 7 + 2x$. $y = 7 + 2x \Leftrightarrow 2x = y - 7 \Leftrightarrow x = \dfrac{y - 7}{2}$. So $f^{-1}(x) = \dfrac{x - 7}{2}$. f^{-1} is the number of toppings on a pizza that costs x dollars.

95. (a) $f(x) = \dfrac{2x + 1}{5}$ is "multiply by 2, add 1, and then divide by 5". So the reverse is "multiply by 5, subtract 1, and then divide by 2" or $f^{-1}(x) = \dfrac{5x - 1}{2}$. Check: $f \circ f^{-1}(x) = f\left(\dfrac{5x - 1}{2}\right) = \dfrac{2\left(\dfrac{5x - 1}{2}\right) + 1}{5} = \dfrac{5x - 1 + 1}{5} = \dfrac{5x}{5} = x$

and $f^{-1} \circ f(x) = f^{-1}\left(\dfrac{2x + 1}{5}\right) = \dfrac{5\left(\dfrac{2x + 1}{5}\right) - 1}{2} = \dfrac{2x + 1 - 1}{2} = \dfrac{2x}{2} = x$.

(b) $f(x) = 3 - \dfrac{1}{x} = \dfrac{-1}{x} + 3$ is "take the negative reciprocal and add 3". Since the reverse of "take the negative reciprocal" is "take the negative reciprocal", $f^{-1}(x)$ is "subtract 3 and take the negative reciprocal", that is, $f^{-1}(x) = \dfrac{-1}{x - 3}$. Check: $f \circ f^{-1}(x) = f\left(\dfrac{-1}{x - 3}\right) = 3 - \dfrac{1}{\dfrac{-1}{x - 3}} = 3 - \left(1 \cdot \dfrac{x - 3}{-1}\right) = 3 + x - 3 = x$ and

$f^{-1} \circ f(x) = f^{-1}\left(3 - \dfrac{1}{x}\right) = \dfrac{-1}{\left(3 - \dfrac{1}{x}\right) - 3} = \dfrac{-1}{-\dfrac{1}{x}} = -1 \cdot \dfrac{x}{-1} = x$.

(c) $f(x) = \sqrt{x^3 + 2}$ is "cube, add 2, and then take the square root". So the reverse is "square, subtract 2, then take the cube root" or $f^{-1}(x) = \sqrt[3]{x^2 - 2}$. Domain for $f(x)$ is $\left[-\sqrt[3]{2}, \infty\right)$; domain for $f^{-1}(x)$ is $[0, \infty)$. Check:

$f \circ f^{-1}(x) = f\left(\sqrt[3]{x^2 - 2}\right) = \sqrt{\left(\sqrt[3]{x^2 - 2}\right)^3 + 2} = \sqrt{x^2 - 2 + 2} = \sqrt{x^2} = x$ (on the appropriate domain) and

$f^{-1} \circ f(x) = f^{-1}\left(\sqrt{x^3 + 2}\right) = \sqrt[3]{\left(\sqrt{x^3 + 2}\right)^2 - 2} = \sqrt[3]{x^3 + 2 - 2} = \sqrt[3]{x^3} = x$ (on the appropriate domain).

(d) $f(x) = (2x - 5)^3$ is "double, subtract 5, and then cube". So the reverse is "take the cube root, add 5, and divide by 2" or $f^{-1}(x) = \dfrac{\sqrt[3]{x} + 5}{2}$ Domain for both $f(x)$ and $f^{-1}(x)$ is $(-\infty, \infty)$. Check:

$$f \circ f^{-1}(x) = f\left(\frac{\sqrt[3]{x} + 5}{2}\right) = \left[2\left(\frac{\sqrt[3]{x} + 5}{2}\right) - 5\right]^3 = (\sqrt[3]{x} + 5 - 5)^3 = (\sqrt[3]{x})^3 = \sqrt[3]{x^3} = x \text{ and}$$

$$f^{-1} \circ f(x) = f^{-1}\left((2x - 5)^3\right) = \frac{\sqrt[x]{(2x - 5)^3} + 5}{2} = \frac{(2x - 5) + 5}{2} = \frac{2x}{2} = x.$$

In a function like $f(x) = 3x - 2$, the variable occurs only once and it easy to see how to reverse the operations step by step. But in $f(x) = x^3 + 2x + 6$, you apply two different operations to the variable x (cubing and multiplying by 2) and then add 6, so it is not possible to reverse the operations step by step.

97. (a) We find $g^{-1}(x)$: $y = 2x + 1 \Leftrightarrow 2x = y - 1 \Leftrightarrow x = \frac{1}{2}(y - 1)$. So $g^{-1}(x) = \frac{1}{2}(x - 1)$. Thus

$$f(x) = h \circ g^{-1}(x) = h\left(\frac{1}{2}(x - 1)\right) = 4\left[\frac{1}{2}(x - 1)\right]^2 + 4\left[\frac{1}{2}(x - 1)\right] + 7 = x^2 - 2x + 1 + 2x - 2 + 7 = x^2 + 6.$$

(b) $f \circ g = h \Leftrightarrow f^{-1} \circ f \circ g = f^{-1} \circ h \Leftrightarrow I \circ g = f^{-1} \circ h \Leftrightarrow g = f^{-1} \circ h$. Note that we compose with f^{-1} on the left on each side of the equation. We find f^{-1}: $y = 3x + 5 \Leftrightarrow 3x = y - 5 \Leftrightarrow x = \frac{1}{3}(y - 5)$. So $f^{-1}(x) = \frac{1}{3}(x - 5)$.

Thus $g(x) = f^{-1} \circ h(x) = f^{-1}\left(3x^2 + 3x + 2\right) = \frac{1}{3}\left[\left(3x^2 + 3x + 2\right) - 5\right] = \frac{1}{3}\left[3x^2 + 3x - 3\right] = x^2 + x - 1.$

CHAPTER 2 REVIEW

1. "Square, then subtract 5" can be represented by the function $f(x) = x^2 - 5$.

3. $f(x) = 3(x + 10)$: "Add 10, then multiply by 3."

5. $g(x) = x^2 - 4x$

x	$g(x)$
-1	5
0	0
1	-3
2	-4
3	-3

7. $C(x) = 5000 + 30x - 0.001x^2$

(a) $C(1000) = 5000 + 30(1000) - 0.001(1000)^2 = \$34{,}000$ and

$C(10{,}000) = 5000 + 30(10{,}000) - 0.001(10{,}000)^2 = \$205{,}000.$

(b) From part (a), we see that the total cost of printing 1000 copies of the book is \$34,000 and the total cost of printing 10,000 copies is \$205,000.

(c) $C(0) = 5000 + 30(0) - 0.001(0)^2 = \$5000$. This represents the fixed costs associated with getting the print run ready.

(d) The net change in C as x changes from 1000 to 10,000 is $C(10{,}000) - C(1000) = 205{,}000 - 34{,}000 = \$171{,}000$.

9. $f(x) = x^2 - 4x + 6$; $f(0) = (0)^2 - 4(0) + 6 = 6$; $f(2) = (2)^2 - 4(2) + 6 = 2$; $f(-2) = (-2)^2 - 4(-2) + 6 = 18$; $f(a) = (a)^2 - 4(a) + 6 = a^2 - 4a + 6$; $f(-a) = (-a)^2 - 4(-a) + 6 = a^2 + 4a + 6$; $f(x + 1) = (x + 1)^2 - 4(x + 1) + 6 = x^2 + 2x + 1 - 4x - 4 + 6 = x^2 - 2x + 3$; $f(2x) = (2x)^2 - 4(2x) + 6 = 4x^2 - 8x + 6$.

11. The net change in $f(x) = x^4 - 3x^2$ from -2 to 1 is $f(1) - f(-2) = \left[1^4 - 3(1)^2\right] - \left[(-2)^4 - 3(-2)^2\right] = -2 - 4 = -6$.

13. By the Vertical Line Test, figures (b) and (c) are graphs of functions. By the Horizontal Line Test, figure (c) is the graph of a one-to-one function.

15. Domain: We must have $x + 3 \geq 0 \Leftrightarrow x \geq -3$. In interval notation, the domain is $[-3, \infty)$.

Range: For x in the domain of f, we have $x \geq -3 \Leftrightarrow x + 3 \geq 0 \Leftrightarrow \sqrt{x+3} \geq 0 \Leftrightarrow f(x) \geq 0$. So the range is $[0, \infty)$.

17. $f(x) = 7x + 15$. The domain is all real numbers, $(-\infty, \infty)$.

19. $f(x) = \sqrt{x+4}$. We require $x + 4 \geq 0 \Leftrightarrow x \geq -4$. Thus the domain is $[-4, \infty)$.

21. $f(x) = \dfrac{1}{x} + \dfrac{1}{x+1} + \dfrac{1}{x+2}$. The denominators cannot equal 0, therefore the domain is $\{x \mid x \neq 0, -1, -2\}$.

23. $h(x) = \sqrt{4-x} + \sqrt{x^2 - 1}$. We require the expression inside the radicals be nonnegative. So $4 - x \geq 0 \Leftrightarrow 4 \geq x$; also $x^2 - 1 \geq 0 \Leftrightarrow (x-1)(x+1) \geq 0$. We make a table:

Interval	$(-\infty, -1)$	$(-1, 1)$	$(1, \infty)$
Sign of $x - 1$	$-$	$-$	$+$
Sign of $x + 1$	$-$	$+$	$+$
Sign of $(x-1)(x+1)$	$+$	$-$	$+$

Thus the domain is $(-\infty, 4] \cap \{(-\infty, -1] \cup [1, \infty)\} = (-\infty, -1] \cup [1, 4]$.

25. $f(x) = 1 - 2x$

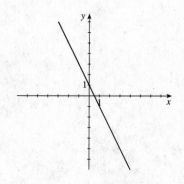

27. $f(t) = 1 - \frac{1}{2}t^2$

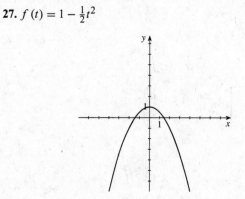

29. $f(x) = x^2 - 6x + 6$

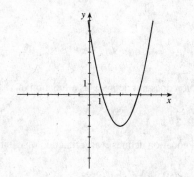

31. $g(x) = 1 - \sqrt{x}$

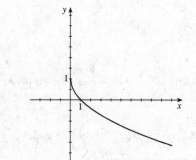

33. $h(x) = \frac{1}{2}x^3$

35. $h(x) = \sqrt[3]{x}$

37. $g(x) = \dfrac{1}{x^2}$

39. $f(x) = \begin{cases} 1-x & \text{if } x < 0 \\ 1 & \text{if } x \geq 0 \end{cases}$

41. $f(x) = \begin{cases} x+6 & \text{if } x < -2 \\ x^2 & \text{if } x \geq -2 \end{cases}$

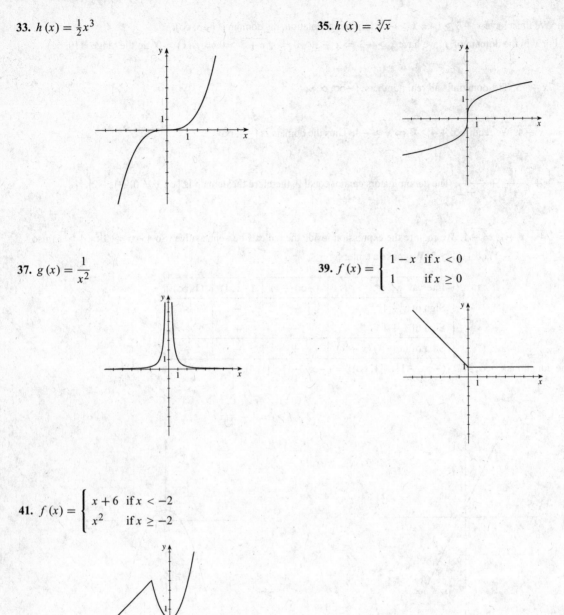

43. $x + y^2 = 14 \Rightarrow y^2 = 14 - x \Rightarrow y = \pm\sqrt{14-x}$, so the original equation does not define y as a function of x.

45. $x^3 - y^3 = 27 \Leftrightarrow y^3 = x^3 - 27 \Leftrightarrow y = \left(x^3 - 27\right)^{1/3}$, so the original equation defines y as a function of x (since the cube root function is one-to-one).

47. $f(x) = 6x^3 - 15x^2 + 4x - 1$

(i) $[-2, 2]$ by $[-2, 2]$

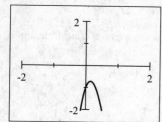

(ii) $[-8, 8]$ by $[-8, 8]$

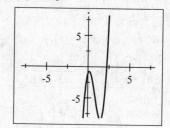

(iii) $[-4, 4]$ by $[-12, 12]$

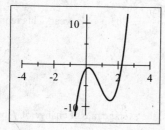

(iv) $[-100, 100]$ by $[-100, 100]$

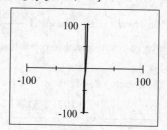

From the graphs, we see that the viewing rectangle in (iii) produces the most appropriate graph.

49. (a) We graph $f(x) = \sqrt{9 - x^2}$ in the viewing rectangle $[-4, 4]$ by $[-1, 4]$.

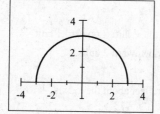

(b) From the graph, the domain of f is $[-3, 3]$ and the range of f is $[0, 3]$.

51. (a) We graph $f(x) = -2$, $-2 \le x \le 3$ in the viewing rectangle $[-4, 5]$ by $[-3, 1]$.

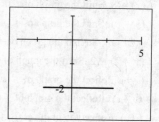

(b) From the graph, the domain of f is $[-2, 3]$ and the range of f is $\{-2\}$.

53. (a) We graph $f(x) = \sqrt{x^3 - 4x + 1}$ in the viewing rectangle $[-5, 5]$ by $[-1, 5]$.

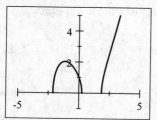

(b) From the graph, the domain of f is approximately $[-2.11, 0.25] \cup [1.86, \infty)$ and the range of f is $[0, \infty)$.

55. $f(x) = x^3 - 4x^2$ is graphed in the viewing rectangle $[-5, 5]$ by $[-20, 10]$. $f(x)$ is increasing on $(-\infty, 0]$ and $[2.67, \infty)$. It is decreasing on $[0, 2.67]$.

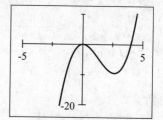

57. The net change is $f(2) - f(0) = \left[(2)^2 + 3(2)\right] - \left[0^2 + 3(0)\right] = 4 + 6 = 10$ and the average rate of change is
$$\frac{f(2) - f(0)}{2 - 0} = \frac{10}{2} = 5.$$

59. The net change is $f(3 + h) - f(3) = \frac{1}{3 + h} - \frac{1}{3} = -\frac{h}{3(3 + h)}$ and the average rate of change is
$$\frac{f(3 + h) - f(3)}{(3 + h) - 3} = \frac{-\dfrac{h}{3(3 + h)}}{h} = -\frac{1}{3(3 + h)}.$$

61. $P(t) = 3000 + 200t + 0.1t^2$

(a) $P(10) = 3000 + 200(10) + 0.1(10)^2 = 5010$ represents the population in its 10th year (that is, in 1995), and
$P(20) = 3000 + 200(20) + 0.1(20)^2 = 7040$ represents its population in its 20th year (in 2005).

(b) The average rate of change is $\dfrac{P(20) - P(10)}{20 - 10} = \dfrac{7040 - 5010}{10} = \dfrac{2030}{10} = 203$ people/year. This represents the average yearly change in population between 1995 and 2005.

63. $f(x) = \frac{1}{2}x - 6$

(a) The average rate of change of f between $x = 0$ and $x = 2$ is
$$\frac{f(2) - f(0)}{2 - 0} = \frac{\left[\frac{1}{2}(2) - 6\right] - \left[\frac{1}{2}(0) - 6\right]}{2} = \frac{-5 - (-6)}{2} = \frac{1}{2}, \text{ and the average rate of change of } f \text{ between } x = 15$$
and $x = 50$ is
$$\frac{f(50) - f(15)}{50 - 15} = \frac{\left[\frac{1}{2}(50) - 6\right] - \left[\frac{1}{2}(15) - 6\right]}{35} = \frac{19 - \frac{3}{2}}{35} = \frac{1}{2}.$$

(b) The rates of change are the same because f is a linear function.

65. (a) $y = f(x) + 8$. Shift the graph of $f(x)$ upward 8 units.

(b) $y = f(x + 8)$. Shift the graph of $f(x)$ to the left 8 units.

(c) $y = 1 + 2f(x)$. Stretch the graph of $f(x)$ vertically by a factor of 2, then shift it upward 1 unit.

(d) $y = f(x - 2) - 2$. Shift the graph of $f(x)$ to the right 2 units, then downward 2 units.

(e) $y = f(-x)$. Reflect the graph of $f(x)$ about the y-axis.

(f) $y = -f(-x)$. Reflect the graph of $f(x)$ first about the y-axis, then reflect about the x-axis.

(g) $y = -f(x)$. Reflect the graph of $f(x)$ about the x-axis.

(h) $y = f^{-1}(x)$. Reflect the graph of $f(x)$ about the line $y = x$.

67. (a) $f(x) = 2x^5 - 3x^2 + 2$. $f(-x) = 2(-x)^5 - 3(-x)^2 + 2 = -2x^5 - 3x^2 + 2$. Since $f(x) \neq f(-x)$, f is not even.
$-f(x) = -2x^5 + 3x^2 - 2$. Since $-f(x) \neq f(-x)$, f is not odd.

(b) $f(x) = x^3 - x^7$. $f(-x) = (-x)^3 - (-x)^7 = -\left(x^3 - x^7\right) = -f(x)$, hence f is odd.

(c) $f(x) = \dfrac{1 - x^2}{1 + x^2}$. $f(-x) = \dfrac{1 - (-x)^2}{1 + (-x)^2} = \dfrac{1 - x^2}{1 + x^2} = f(x)$. Since $f(x) = f(-x)$, f is even.

(d) $f(x) = \dfrac{1}{x + 2}$. $f(-x) = \dfrac{1}{(-x) + 2} = \dfrac{1}{2 - x}$. $-f(x) = -\dfrac{1}{x + 2}$. Since $f(x) \neq f(-x)$, f is not even, and since
$f(-x) \neq -f(x)$, f is not odd.

69. $g(x) = 2x^2 + 4x - 5 = 2\left(x^2 + 2x\right) - 5 = 2\left(x^2 + 2x + 1\right) - 5 - 2 = 2(x + 1)^2 - 7$. So the minimum value is
$g(-1) = -7$.

71. $h(t) = -16t^2 + 48t + 32 = -16\left(t^2 - 3t\right) + 32 = -16\left(t^2 - 3t + \frac{9}{4}\right) + 32 + 36$
$$= -16\left(t^2 - 3t + \frac{9}{4}\right) + 68 = -16\left(t - \frac{3}{2}\right)^2 + 68$$
The stone reaches a maximum height of 68 feet.

73. $f(x) = 3.3 + 1.6x - 2.5x^3$. In the first viewing rectangle, $[-2, 2]$ by $[-4, 8]$, we see that $f(x)$ has a local maximum and a local minimum. In the next viewing rectangle, $[0.4, 0.5]$ by $[3.78, 3.80]$, we isolate the local maximum value as approximately 3.79 when $x \approx 0.46$. In the last viewing rectangle, $[-0.5, -0.4]$ by $[2.80, 2.82]$, we isolate the local minimum value as 2.81 when $x \approx -0.46$.

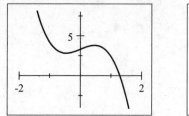 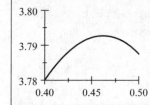 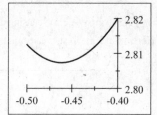

75. $f(x) = x + 2$, $g(x) = x^2$

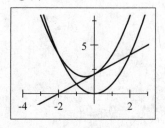

77. $f(x) = x^2 - 3x + 2$ and $g(x) = 4 - 3x$.

(a) $(f + g)(x) = \left(x^2 - 3x + 2\right) + (4 - 3x) = x^2 - 6x + 6$

(b) $(f - g)(x) = \left(x^2 - 3x + 2\right) - (4 - 3x) = x^2 - 2$

(c) $(fg)(x) = \left(x^2 - 3x + 2\right)(4 - 3x) = 4x^2 - 12x + 8 - 3x^3 + 9x^2 - 6x = -3x^3 + 13x^2 - 18x + 8$

(d) $\left(\dfrac{f}{g}\right)(x) = \dfrac{x^2 - 3x + 2}{4 - 3x}$, $x \neq \dfrac{4}{3}$

(e) $(f \circ g)(x) = f(4 - 3x) = (4 - 3x)^2 - 3(4 - 3x) + 2 = 16 - 24x + 9x^2 - 12 + 9x + 2 = 9x^2 - 15x + 6$

(f) $(g \circ f)(x) = g\left(x^2 - 3x + 2\right) = 4 - 3\left(x^2 - 3x + 2\right) = -3x^2 + 9x - 2$

79. $f(x) = 3x - 1$ and $g(x) = 2x - x^2$.

$(f \circ g)(x) = f\left(2x - x^2\right) = 3\left(2x - x^2\right) - 1 = -3x^2 + 6x - 1$, and the domain is $(-\infty, \infty)$.

$(g \circ f)(x) = g(3x - 1) = 2(3x - 1) - (3x - 1)^2 = 6x - 2 - 9x^2 + 6x - 1 = -9x^2 + 12x - 3$, and the domain is $(-\infty, \infty)$

$(f \circ f)(x) = f(3x - 1) = 3(3x - 1) - 1 = 9x - 4$, and the domain is $(-\infty, \infty)$.

$(g \circ g)(x) = g\left(2x - x^2\right) = 2\left(2x - x^2\right) - \left(2x - x^2\right)^2 = 4x - 2x^2 - 4x^2 + 4x^3 - x^4 = -x^4 + 4x^3 - 6x^2 + 4x$, and domain is $(-\infty, \infty)$.

81. $f(x) = \sqrt{1 - x}$, $g(x) = 1 - x^2$ and $h(x) = 1 + \sqrt{x}$.

$(f \circ g \circ h)(x) = f(g(h(x))) = f\left(g\left(1 + \sqrt{x}\right)\right) = f\left(1 - \left(1 + \sqrt{x}\right)^2\right) = f\left(1 - \left(1 + 2\sqrt{x} + x\right)\right)$

$\qquad = f\left(-x - 2\sqrt{x}\right) = \sqrt{1 - \left(-x - 2\sqrt{x}\right)} = \sqrt{1 + 2\sqrt{x} + x} = \sqrt{\left(1 + \sqrt{x}\right)^2} = 1 + \sqrt{x}$

83. $f(x) = 3 + x^3$. If $x_1 \neq x_2$, then $x_1^3 \neq x_2^3$ (unequal numbers have unequal cubes), and therefore $3 + x_1^3 \neq 3 + x_2^3$. Thus f is a one-to-one function.

85. $h(x) = \dfrac{1}{x^4}$. Since the fourth powers of a number and its negative are equal, h is not one-to-one. For example,

$h(-1) = \dfrac{1}{(-1)^4} = 1$ and $h(1) = \dfrac{1}{(1)^4} = 1$, so $h(-1) = h(1)$.

87. $p(x) = 3.3 + 1.6x - 2.5x^3$. Using a graphing device and the Horizontal Line Test, we see that p is not a one-to-one function.

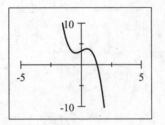

89. $f(x) = 3x - 2 \Leftrightarrow y = 3x - 2 \Leftrightarrow 3x = y + 2 \Leftrightarrow x = \frac{1}{3}(y + 2)$. So $f^{-1}(x) = \frac{1}{3}(x + 2)$.

91. $f(x) = (x+1)^3 \Leftrightarrow y = (x+1)^3 \Leftrightarrow x + 1 = \sqrt[3]{y} \Leftrightarrow x = \sqrt[3]{y} - 1$. So $f^{-1}(x) = \sqrt[3]{x} - 1$.

93. The graph passes the Horizontal Line Test, so f has an inverse. Because $f(1) = 0$, $f^{-1}(0) = 1$, and because $f(3) = 4$, $f^{-1}(4) = 3$.

95. (a), (b) $f(x) = x^2 - 4$, $x \geq 0$

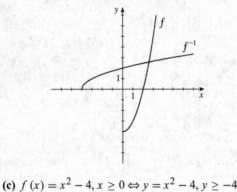

(c) $f(x) = x^2 - 4$, $x \geq 0 \Leftrightarrow y = x^2 - 4$, $y \geq -4$

$\Leftrightarrow x^2 = y + 4 \Leftrightarrow x = \sqrt{y+4}$. So

$f^{-1}(x) = \sqrt{x+4}$, $x \geq -4$.

CHAPTER 2 TEST

1. By the Vertical Line Test, figures (a) and (b) are graphs of functions. By the Horizontal Line Test, only figure (a) is the graph of a one-to-one function.

3. (a) "Subtract 2, then cube the result" can be expressed algebraically as $f(x) = (x-2)^3$.

(c)

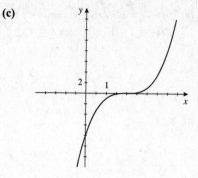

(b)

x	$f(x)$
-1	-27
0	-8
1	-1
2	0
3	1
4	8

(d) We know that f has an inverse because it passes the Horizontal Line Test. A verbal description for f^{-1} is, "Take the cube root, then add 2."

(e) $y = (x-2)^3 \Leftrightarrow \sqrt[3]{y} = x-2 \Leftrightarrow x = \sqrt[3]{y} + 2$. Thus, a formula for f^{-1} is $f^{-1}(x) = \sqrt[3]{x} + 2$.

5. $R(x) = -500x^2 + 3000x$

(a) $R(2) = -500(2)^2 + 3000(2) = \$4000$ represents their total sales revenue when their price is $2 per bar and

$R(4) = -500(4)^2 + 3000(4) = \$4000$ represents their total sales revenue when their price is $4 per bar

(c) The maximum revenue is $4500, and it is achieved at a price of $x = \$3$.

(b)

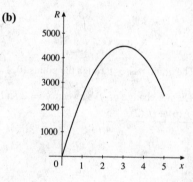

7. (a) $f(x) = x^3$

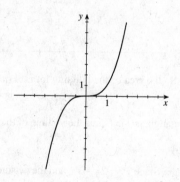

(b) $g(x) = (x-1)^3 - 2$. To obtain the graph of g, shift the graph of f to the right 1 unit and downward 2 units.

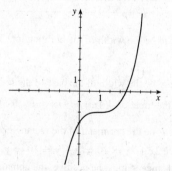

9. (a) $f(-2) = 1 - (-2) = 1 + 2 = 3$ (since $-2 \leq 1$).

$f(1) = 1 - 1 = 0$ (since $1 \leq 1$).

(b)

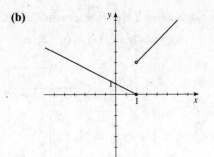

11. (a) $f(x) = \sqrt{3-x}, x \leq 3 \Leftrightarrow y = \sqrt{3-x} \Leftrightarrow$

$y^2 = 3 - x \Leftrightarrow x = 3 - y^2$. Thus

$f^{-1}(x) = 3 - x^2, x \geq 0$.

(b) $f(x) = \sqrt{3-x}, x \leq 3$ and $f^{-1}(x) = 3 - x^2$,

$x \geq 0$

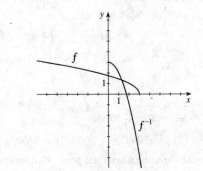

13. The graph passes through the points $(0, 1)$ and $(4, 3)$, so $f(0) = 1$ and $f(4) = 3$.

15. The net change of f between $x = 2$ and $x = 6$ is $f(6) - f(2) = 7 - 2 = 5$ and the average rate of change is

$$\frac{f(6) - f(2)}{6 - 2} = \frac{5}{4}.$$

17.

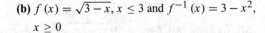

FOCUS ON MODELING   Modeling with Functions

1. Let w be the width of the building lot. Then the length of the lot is $3w$. So the area of the building lot is $A(w) = 3w^2$, $w > 0$.

3. Let w be the width of the base of the rectangle. Then the height of the rectangle is $\frac{1}{2}w$. Thus the volume of the box is given by the function $V(w) = \frac{1}{2}w^3$, $w > 0$.

5. Let P be the perimeter of the rectangle and y be the length of the other side. Since $P = 2x + 2y$ and the perimeter is 20, we have $2x + 2y = 20 \Leftrightarrow x + y = 10 \Leftrightarrow y = 10 - x$. Since area is $A = xy$, substituting gives $A(x) = x(10 - x) = 10x - x^2$, and since A must be positive, the domain is $0 < x < 10$.

7.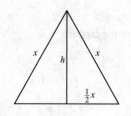

Let h be the height of an altitude of the equilateral triangle whose side has length x, as shown in the diagram. Thus the area is given by $A = \frac{1}{2}xh$. By the Pythagorean Theorem, $h^2 + \left(\frac{1}{2}x\right)^2 = x^2 \Leftrightarrow h^2 + \frac{1}{4}x^2 = x^2 \Leftrightarrow h^2 = \frac{3}{4}x^2 \Leftrightarrow h = \frac{\sqrt{3}}{2}x$. Substituting into the area of a triangle, we get

$$A(x) = \frac{1}{2}xh = \frac{1}{2}x\left(\frac{\sqrt{3}}{2}x\right) = \frac{\sqrt{3}}{4}x^2, x > 0.$$

9. We solve for r in the formula for the area of a circle. This gives $A = \pi r^2 \Leftrightarrow r^2 = \frac{A}{\pi} \Rightarrow r = \sqrt{\frac{A}{\pi}}$, so the model is

$$r(A) = \sqrt{\frac{A}{\pi}}, A > 0.$$

11. Let h be the height of the box in feet. The volume of the box is $V = 60$. Then $x^2 h = 60 \Leftrightarrow h = \frac{60}{x^2}$. The surface area, S, of the box is the sum of the area of the 4 sides and the area of the base and top. Thus $S = 4xh + 2x^2 = 4x\left(\frac{60}{x^2}\right) + 2x^2 = \frac{240}{x} + 2x^2$, so the model is $S(x) = \frac{240}{x} + 2x^2, x > 0$.

13.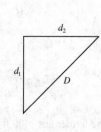

Let d_1 be the distance traveled south by the first ship and d_2 be the distance traveled east by the second ship. The first ship travels south for t hours at 5 mi/h, so $d_1 = 15t$ and, similarly, $d_2 = 20t$. Since the ships are traveling at right angles to each other, we can apply the Pythagorean Theorem to get

$$D(t) = \sqrt{d_1^2 + d_2^2} = \sqrt{(15t)^2 + (20t)^2} = \sqrt{225t^2 + 400t^2} = 25t.$$

15.

Let b be the length of the base, l be the length of the equal sides, and h be the height in centimeters. Since the perimeter is 8, $2l + b = 8 \Leftrightarrow 2l = 8 - b \Leftrightarrow$ $l = \frac{1}{2}(8 - b)$. By the Pythagorean Theorem, $h^2 + \left(\frac{1}{2}b\right)^2 = l^2 \Leftrightarrow$ $h = \sqrt{l^2 - \frac{1}{4}b^2}$. Therefore the area of the triangle is

$$A = \frac{1}{2} \cdot b \cdot h = \frac{1}{2} \cdot b\sqrt{l^2 - \frac{1}{4}b^2} = \frac{b}{2}\sqrt{\frac{1}{4}(8-b)^2 - \frac{1}{4}b^2}$$
$$= \frac{b}{4}\sqrt{64 - 16b + b^2 - b^2} = \frac{b}{4}\sqrt{64 - 16b} = \frac{b}{4} \cdot 4\sqrt{4 - b} = b\sqrt{4 - b}$$

so the model is $A(b) = b\sqrt{4 - b}, 0 < b < 4$.

17. Let w be the length of the rectangle. By the Pythagorean Theorem, $\left(\frac{1}{2}w\right)^2 + h^2 = 10^2 \Leftrightarrow \frac{w^2}{4} + h^2 = 10^2 \Leftrightarrow$ $w^2 = 4\left(100 - h^2\right) \Leftrightarrow w = 2\sqrt{100 - h^2}$ (since $w > 0$). Therefore the area of the rectangle is $A = wh = 2h\sqrt{100 - h^2}$, so the model is $A(h) = 2h\sqrt{100 - h^2}, 0 < h < 10$.

19. (a) We complete the table.

First number	Second number	Product
1	18	18
2	17	34
3	16	48
4	15	60
5	14	70
6	13	78
7	12	84
8	11	88
9	10	90
10	9	90
11	8	88

From the table we conclude that the numbers is still increasing, the numbers whose product is a maximum should both be 9.5.

(b) Let x be one number: then $19 - x$ is the other number, and so the product, p, is

$$p(x) = x(19 - x) = 19x - x^2.$$

(c) $p(x) = 19x - x^2 = -\left(x^2 - 19x\right)$

$$= -\left[x^2 - 19x + \left(\tfrac{19}{2}\right)^2\right] + \left(\tfrac{19}{2}\right)^2$$

$$= -(x - 9.5)^2 + 90.25$$

So the product is maximized when the numbers are both 9.5.

21. (a) Let x be the width of the field (in feet) and l be the length of the field (in feet). Since the farmer has 2400 ft of fencing we must have $2x + l = 2400$.

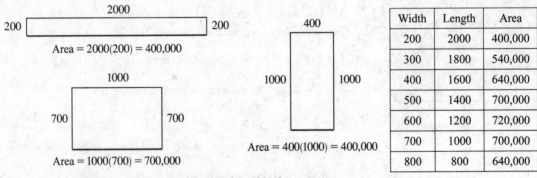

Width	Length	Area
200	2000	400,000
300	1800	540,000
400	1600	640,000
500	1400	700,000
600	1200	720,000
700	1000	700,000
800	800	640,000

It appears that the field of largest area is about 600 ft × 1200 ft.

(b) Let x be the width of the field (in feet) and l be the length of the field (in feet). Since the farmer has 2400 ft of fencing we must have $2x + l = 2400 \Leftrightarrow l = 2400 - 2x$. The area of the fenced-in field is given by

$$A(x) = l \cdot x = (2400 - 2x)x = -2x^2 + 2400x = -2\left(x^2 - 1200x\right).$$

(c) The area is $A(x) = -2\left(x^2 - 1200x + 600^2\right) + 2\left(600^2\right) = -2(x - 600)^2 + 720,000$. So the maximum area occurs when $x = 600$ feet and $l = 2400 - 2(600) = 1200$ feet.

23. (a) Let x be the length of the fence along the road. If the area is 1200, we have $1200 = x \cdot$ width, so the width of the garden is $\dfrac{1200}{x}$. Then the cost of the fence is given by the function $C(x) = 5(x) + 3\left[x + 2 \cdot \dfrac{1200}{x}\right] = 8x + \dfrac{7200}{x}$.

(b) We graph the function $y = C(x)$ in the viewing rectangle $[0, 75] \times [0, 800]$. From this we get the cost is minimized when $x = 30$ ft. Then the width is $\frac{1200}{30} = 40$ ft. So the length is 30 ft and the width is 40 ft.

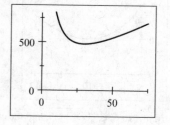

(c) We graph the function $y = C(x)$ and $y = 600$ in the viewing rectangle $[10, 65] \times [450, 650]$. From this we get that the cost is at most \$600 when $15 \le x \le 60$. So the range of lengths he can fence along the road is 15 feet to 60 feet.

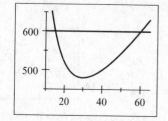

25. (a) Let h be the height in feet of the straight portion of the window. The circumference of the semicircle is $C = \frac{1}{2}\pi x$. Since the perimeter of the window is 30 feet, we have $x + 2h + \frac{1}{2}\pi x = 30$. Solving for h, we get $2h = 30 - x - \frac{1}{2}\pi x \Leftrightarrow h = 15 - \frac{1}{2}x - \frac{1}{4}\pi x$. The area of the window is
$$A(x) = xh + \frac{1}{2}\pi\left(\frac{1}{2}x\right)^2 = x\left(15 - \frac{1}{2}x - \frac{1}{4}\pi x\right) + \frac{1}{8}\pi x^2 = 15x - \frac{1}{2}x^2 - \frac{1}{8}\pi x^2.$$

(b) $A(x) = 15x - \frac{1}{8}(\pi + 4)x^2 = -\frac{1}{8}(\pi + 4)\left[x^2 - \dfrac{120}{\pi + 4}x\right]$
$$= -\frac{1}{8}(\pi + 4)\left[x^2 - \dfrac{120}{\pi + 4}x + \left(\dfrac{60}{\pi + 4}\right)^2\right] + \dfrac{450}{\pi + 4} = -\frac{1}{8}(\pi + 4)\left(x - \dfrac{60}{\pi + 4}\right)^2 + \dfrac{450}{\pi + 4}$$

The area is maximized when $x = \dfrac{60}{\pi + 4} \approx 8.40$, and hence $h \approx 15 - \frac{1}{2}(8.40) - \frac{1}{4}\pi(8.40) \approx 4.20$.

27. (a) Let x be the length of one side of the base and let h be the height of the box in feet. Since the volume of the box is $V = x^2 h = 12$, we have $x^2 h = 12 \Leftrightarrow h = \dfrac{12}{x^2}$. The surface area, A, of the box is sum of the area of the four sides and the area of the base. Thus the surface area of the box is given by the formula
$$A(x) = 4xh + x^2 = 4x\left(\dfrac{12}{x^2}\right) + x^2 = \dfrac{48}{x} + x^2, \, x > 0.$$

(b) The function $y = A(x)$ is shown in the first viewing rectangle below. In the second viewing rectangle, we isolate the minimum, and we see that the amount of material is minimized when x (the length and width) is 2.88 ft. Then the height is $h = \dfrac{12}{x^2} \approx 1.44$ ft.

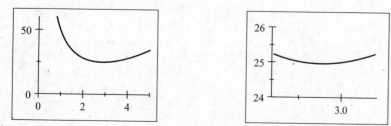

29. (a) Let w be the width of the pen and l be the length in meters. We use the area to establish a relationship between w and l. Since the area is 100 m^2, we have $l \cdot w = 100 \Leftrightarrow l = \dfrac{100}{w}$. So the amount of fencing used is

$$F = 2l + 2w = 2\left(\frac{100}{w}\right) + 2w = \frac{200 + 2w^2}{w}.$$

(b) Using a graphing device, we first graph F in the viewing rectangle $[0, 40]$ by $[0, 100]$, and locate the approximate location of the minimum value. In the second viewing rectangle, $[8, 12]$ by $[39, 41]$, we see that the minimum value of F occurs when $w = 10$. Therefore the pen should be a square with side 10 m.

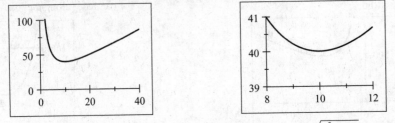

31. (a) Let x be the distance from point B to C, in miles. Then the distance from A to C is $\sqrt{x^2 + 25}$, and the energy used in flying from A to C then C to D is $f(x) = 14\sqrt{x^2 + 25} + 10(12 - x)$.

(b) By using a graphing device, the energy expenditure is minimized when the distance from B to C is about 5.1 miles.

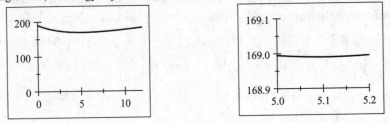

3 POLYNOMIAL AND RATIONAL FUNCTIONS

3.1 QUADRATIC FUNCTIONS AND MODELS

1. To put the quadratic function $f(x) = ax^2 + bx + c$ in standard form we complete the *square*.

3. The graph of $f(x) = 2(x-3)^2 + 5$ is a parabola that opens *upward*, with its vertex at $(3, 5)$, and $f(3) = 5$ is the *minimum* value of f.

5. (a) Vertex: $(3, 4)$

    (b) Maximum value of f: 4

    (c) Domain $(-\infty, \infty)$, range: $(-\infty, 4]$

7. (a) Vertex: $(1, -3)$

    (b) Minimum value of f: -3

    (c) Domain: $(-\infty, \infty)$, range: $[-3, \infty)$

9. (a) $f(x) = x^2 - 2x + 3 = (x-1)^2 - 1 + 3 = (x-1)^2 + 2$

    (b) The vertex is at $(1, 2)$.

      x-intercepts: $y = 0 \Rightarrow 0 = (x-1)^2 + 2 \Rightarrow (x-1)^2 = -2$. This has no real solution, so there is no x-intercept.

      y-intercept: $x = 0 \Rightarrow y = (0-1)^2 + 2 = 3$. The y-intercept is 3.

(c)

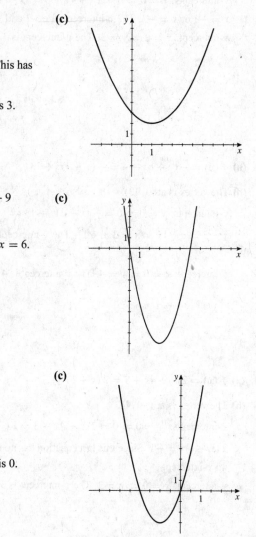

11. (a) $f(x) = x^2 - 6x = x^2 - 6x = x^2 - 6x + 9 - 9 = (x-3)^2 - 9$

    (b) The vertex is at $(3, -9)$.

      x-intercepts: $y = 0 \Rightarrow 0 = x^2 - 6x = x(x-6)$. So $x = 0$ or $x = 6$. The x-intercepts are 0 and 6.

      y-intercept: $x = 0 \Rightarrow y = 0$. The y-intercept is 0.

(c)

13. (a) $f(x) = 3x^2 + 6x = 3\left(x^2 + 2x\right) = 3(x+1)^2 - 3$

    (b) The vertex is at $(-1, -3)$.

      x-intercepts: $y = 0 \Rightarrow 0 = 3(x+1)^2 - 3 \Rightarrow (x+1)^2 = 1 \Rightarrow x = -2$ or 0. The x-intercepts are -2 and 0.

      y-intercept: $x = 0 \Rightarrow y = 3(0)^2 + 6(0) = 0$. The y-intercept is 0.

(c)

15. (a) $f(x) = -x^2 + 6x = -\left(x^2 - 6x\right) = -(x-3)^2 + 9$

(c)

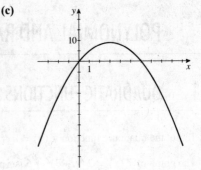

(b) The vertex is at $(3, 9)$.

x-intercepts: $y = 0 \Rightarrow 0 = -(x-3)^2 + 9 \Rightarrow (x-3)^2 = 9 \Rightarrow$

$x = 0$ or 6. The x-intercepts are 0 and 6.

y-intercept: $x = 0 \Rightarrow y = -0^2 + 6(0) = 0$. The y-intercept is 0.

17. (a) $f(x) = x^2 + 4x + 3 = (x+2)^2 - 1$

(c)

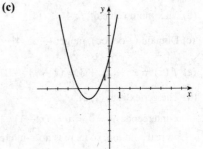

(b) The vertex is at $(-2, -1)$.

x-intercepts: $y = 0 \Rightarrow 0 = x^2 + 4x + 3 = (x+1)(x+3)$. So

$x = -1$ or $x = -3$. The x-intercepts are -1 and -3.

y-intercept: $x = 0 \Rightarrow y = 3$. The y-intercept is 3.

19. (a) $f(x) = -x^2 + 6x + 4 = -(x-3)^2 + 13$

(c)

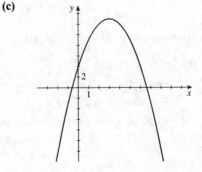

(b) The vertex is at $(3, 13)$.

x-intercepts: $y = 0 \Rightarrow 0 = -(x-3)^2 + 13 \Leftrightarrow (x-3)^2 = 13 \Rightarrow$

$x - 3 = \pm\sqrt{13} \Leftrightarrow x = 3 \pm \sqrt{13}$. The x-intercepts are $3 - \sqrt{13}$ and

$3 + \sqrt{13}$.

y-intercept: $x = 0 \Rightarrow y = 4$. The y-intercept is 4.

21. (a) $f(x) = 2x^2 + 4x + 3 = 2(x+1)^2 + 1$

(c)

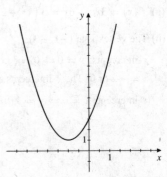

(b) The vertex is at $(-1, 1)$.

x-intercepts: $y = 0 \Rightarrow 0 = 2x^2 + 4x + 3 = 2(x+1)^2 + 1 \Leftrightarrow$

$2(x+1)^2 = -1$. Since this last equation has no real solution, there is

no x-intercept.

y-intercept: $x = 0 \Rightarrow y = 3$. The y-intercept is 3.

23. (a) $f(x) = 2x^2 - 20x + 57 = 2(x-5)^2 + 7$

(b) The vertex is at $(5, 7)$.

x-intercepts: $y = 0 \Rightarrow 0 = 2x^2 - 20x + 57 = 2(x-5)^2 + 7 \Leftrightarrow$

$2(x-5)^2 = -7$. Since this last equation has no real solution, there is

no x-intercept.

y-intercept: $x = 0 \Rightarrow y = 57$. The y-intercept is 57.

(c)

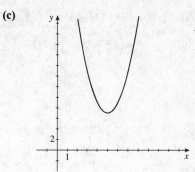

25. (a) $f(x) = -4x^2 - 12x + 1 = -4\left(x^2 + 3\right) + 1$

$$= -4\left(x + \tfrac{3}{2}\right)^2 + 9 + 1 = -4\left(x + \tfrac{3}{2}\right)^2 + 10$$

(b) The vertex is at $\left(-\tfrac{3}{2}, 10\right)$.

x-intercepts: $y = 0 \Rightarrow 0 = -4\left(x + \tfrac{3}{2}\right)^2 + 10 \Rightarrow \left(x + \tfrac{3}{2}\right)^2 = \tfrac{5}{2} \Rightarrow$

$x + \tfrac{3}{2} = \pm\sqrt{\tfrac{5}{2}} \Rightarrow x = -\tfrac{3}{2} \pm \sqrt{\tfrac{5}{2}} = -\tfrac{3}{2} \pm \tfrac{\sqrt{10}}{2}$. The x-intercepts are

$-\tfrac{3}{2} - \tfrac{\sqrt{10}}{2}$ and $-\tfrac{3}{2} + \tfrac{\sqrt{10}}{2}$.

y-intercept: $x = 0 \Rightarrow y = -4(0)^2 - 12(0) + 1 = 1$. The y-intercept

is 1.

(c)

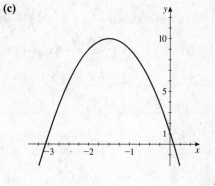

27. (a) $f(x) = x^2 + 2x - 1 = \left(x^2 + 2x\right) - 1$

$$= \left(x^2 + 2x + 1\right) - 1 - 1 = (x+1)^2 - 2$$

(b)

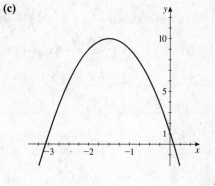

(c) The minimum value is $f(-1) = -2$.

29. (a) $f(x) = 3x^2 - 6x + 1 = 3\left(x^2 - 2x\right) + 1$

$$= 3\left(x^2 - 2x + 1\right) + 1 - 3$$

$$= 3(x-1)^2 - 2$$

(b)

(c) The minimum value is $f(1) = -2$.

31. (a) $f(x) = -x^2 - 3x + 3 = -\left(x^2 + 3x\right) + 3$

$\qquad = -\left(x^2 + 3x + \frac{9}{4}\right) + 3 + \frac{9}{4}$

$\qquad = -\left(x + \frac{3}{2}\right)^2 + \frac{21}{4}$

(b)

(c) The maximum value is $f\left(-\frac{3}{2}\right) = \frac{21}{4}$.

33. (a) $g(x) = 3x^2 - 12x + 13 = 3\left(x^2 - 4x\right) + 13$

$\qquad = 3\left(x^2 - 4x + 4\right) + 13 - 12$

$\qquad = 3(x - 2)^2 + 1$

(b)

(c) The minimum value is $g(2) = 1$.

35. (a) $h(x) = 1 - x - x^2 = -\left(x^2 + x\right) + 1$

$\qquad = -\left(x^2 + x + \frac{1}{4}\right) + 1 + \frac{1}{4}$

$\qquad = -\left(x + \frac{1}{2}\right)^2 + \frac{5}{4}$

(b)

(c) The maximum value is $h\left(-\frac{1}{2}\right) = \frac{5}{4}$.

37. $f(x) = x^2 + x + 1 = \left(x^2 + x\right) + 1$

$\qquad = \left(x^2 + x + \frac{1}{4}\right) + 1 + \frac{1}{4} = \left(x + \frac{1}{2}\right)^2 + \frac{3}{4}$

Therefore, the minimum value is $f\left(-\frac{1}{2}\right) = \frac{3}{4}$.

39. $f(t) = 100 - 49t - 7t^2 = -7\left(t^2 + 7t\right) + 100 = -7\left(t^2 + 7t + \frac{49}{4}\right) + 100 + \frac{343}{4} = -7\left(t + \frac{7}{2}\right)^2 + \frac{743}{4}$.

Therefore, the maximum value is $f\left(-\frac{7}{2}\right) = \frac{743}{4} = 185.75$.

41. $f(s) = s^2 - 1.2s + 16 = \left(s^2 - 1.2s\right) + 16 = \left(s^2 - 1.2s + 0.36\right) + 16 - 0.36 = (s - 0.6)^2 + 15.64$.

Therefore, the minimum value is $f(0.6) = 15.64$.

43. $h(x) = \frac{1}{2}x^2 + 2x - 6 = \frac{1}{2}\left(x^2 + 4x\right) - 6 = \frac{1}{2}\left(x^2 + 4x + 4\right) - 6 - 2 = \frac{1}{2}(x + 2)^2 - 8$.

Therefore, the minimum value is $h(-2) = -8$.

45. $f(x) = 3 - x - \frac{1}{2}x^2 = -\frac{1}{2}\left(x^2 + 2x\right) + 3 = -\frac{1}{2}\left(x^2 + 2x + 1\right) + 3 + \frac{1}{2} = -\frac{1}{2}(x + 1) + \frac{7}{2}$. Therefore, the maximum

value is $f(-1) = \frac{7}{2}$.

47. Since the vertex is at $(1, -2)$, the function is of the form $f(x) = a(x - 1)^2 - 2$. Substituting the point $(4, 16)$, we get

$16 = a(4 - 1)^2 - 2 \Leftrightarrow 16 = 9a - 2 \Leftrightarrow 9a = 18 \Leftrightarrow a = 2$. So the function is $f(x) = 2(x - 1)^2 - 2 = 2x^2 - 4x$.

49. $f(x) = -x^2 + 4x - 3 = -\left(x^2 - 4x\right) - 3 = -\left(x^2 - 4x + 4\right) - 3 + 4 = -(x-2)^2 + 1$. So the domain of $f(x)$ is $(-\infty, \infty)$. Since $f(x)$ has a maximum value of 1, the range is $(-\infty, 1]$.

51. $f(x) = 2x^2 + 6x - 7 = 2\left(x + \frac{3}{2}\right)^2 - 7 - \frac{9}{2} = 2\left(x + \frac{3}{2}\right)^2 - \frac{23}{2}$. The domain of the function is all real numbers, and since the minimum value of the function is $f\left(-\frac{3}{2}\right) = -\frac{23}{2}$, the range of the function is $\left[-\frac{23}{2}, \infty\right)$.

53. (a) The graph of $f(x) = x^2 + 1.79x - 3.21$ is shown. The minimum value is $f(x) \approx -4.01$.

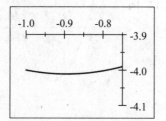

(b) $f(x) = x^2 + 1.79x - 3.21$
$$= \left[x^2 + 1.79x + \left(\frac{1.79}{2}\right)^2\right] - 3.21 - \left(\frac{1.79}{2}\right)^2$$
$$= (x + 0.895)^2 - 4.011025$$

Therefore, the exact minimum of $f(x)$ is -4.011025.

55. $y = f(t) = 40t - 16t^2 = -16\left(t^2 - \frac{5}{2}\right) = -16\left[t^2 - \frac{5}{2}t + \left(\frac{5}{4}\right)^2\right] + 16\left(\frac{5}{4}\right)^2 = -16\left(t - \frac{5}{4}\right)^2 + 25$. Thus the maximum height attained by the ball is $f\left(\frac{5}{4}\right) = 25$ feet.

57. $R(x) = 80x - 0.4x^2 = -0.4\left(x^2 - 200x\right) = -0.4\left(x^2 - 200x + 10{,}000\right) + 4{,}000 = -0.4(x - 100)^2 + 4{,}000$. So revenue is maximized at \$4,000 when 100 units are sold.

59. $E(n) = \frac{2}{3}n - \frac{1}{90}n^2 = -\frac{1}{90}\left(n^2 - 60n\right) = -\frac{1}{90}\left(n^2 - 60n + 900\right) + 10 = -\frac{1}{90}(n - 30)^2 + 10$. Since the maximum of the function occurs when $n = 30$, the viewer should watch the commercial 30 times for maximum effectiveness.

61. $A(n) = n(900 - 9n) = -9n^2 + 900n$ is a quadratic function with $a = -9$ and $b = 900$, so by the formula, the maximum or minimum value occurs at $n = -\dfrac{b}{2a} = -\dfrac{900}{2(-9)} = 50$ trees, and because $a < 0$, this gives a maximum value.

63. The area of the fenced-in field is given by $A(x) = (2400 - 2x)x = -2x^2 + 2400x$. Thus, by the formula in this section, the maximum or minimum value occurs at $x = -\dfrac{b}{2a} = -\dfrac{2400}{2(-2)} = 600$. The maximum area occurs when $x = 600$ feet and $l = 2400 - 2(600) = 1200$ feet.

65. $A(x) = 15x - \frac{1}{8}(\pi + 4)x^2$, so by the formula, the maximum area occurs when $x = -\dfrac{b}{2a} = -\dfrac{-15}{2\left[\frac{1}{8}(\pi + 4)\right]} \approx 8.4$ ft and $h \approx 15 - \frac{1}{2}(8.40) - \frac{1}{4}\pi(8.40) \approx 4.2$ ft.

67. (a) The area of the corral is $A(x) = x(1200 - x) = 1200x - x^2 = -x^2 + 1200x$.

(b) A is a quadratic function with $a = -1$ and $b = 1200$, so by the formula, it has a maximum or minimum at $x = -\dfrac{b}{2a} = -\dfrac{1200}{2(-1)} = 600$, and because $a < 0$, this gives a maximum value. The desired dimensions are 600 ft by 600 ft.

69. (a) To model the revenue, we need to find the total attendance. Let x be the ticket price. Then the amount by which the ticket price is lowered is $10 - x$, and we are given that for every dollar it is lowered, the attendance increases by 3000; that is, the increase in attendance is $3000(10 - x)$. Thus, the attendance is $27{,}000 + 3000(10 - x)$, and since each spectator pays \$$x$, the revenue is $R(x) = x[27{,}000 + 3000(10 - x)] = -3000x^2 + 57{,}000x$.

(b) Since R is a quadratic function with $a = -3000$ and $b = 57,000$, the maximum occurs at

$$x = -\frac{b}{2a} = -\frac{57,000}{2\,(-3000)} = 9.5;$$ that is, when admission is \$9.50.

(c) We solve $R(x) = 0$ for x: $-3000x^2 + 57,000x = 0 \Leftrightarrow -3000x\,(x - 19) = 0 \Leftrightarrow x = 0$ or $x = 19$. Thus, if admission is \$19, nobody will attend and no revenue will be generated.

71. Because $f(x) = (x - m)\,(x - n) = 0$ when $x = m$ or $x = n$, those are its x-intercepts. By symmetry, we expect that the vertex is halfway between these values; that is, at $x = \dfrac{m + n}{2}$. We obtain the graph shown at right.

Expanding, we see that $f(x) = x^2 - (m + n)\,x + mn$, a quadratic function with $a = 1$ and $b = -(m + n)$. Because $a > 0$, the minimum value occurs at $x = -\dfrac{b}{2a} = \dfrac{m + n}{2}$, the x-value of the vertex, as expected.

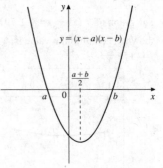

3.2   POLYNOMIAL FUNCTIONS AND THEIR GRAPHS

1. Graph I cannot be that of a polynomial because it is not smooth (it has a cusp.) Graph II could be that of a polynomial function, because it is smooth and continuous. Graph III could not be that of a polynomial function because it has a break. Graph IV could not be that of a polynomial function because it is not smooth.

3. (a) If c is a zero of the polynomial P, then $P(c) = 0$.

(b) If c is a zero of the polynomial P, then $x - c$ is a *factor* of $P(x)$.

(c) If c is a zero of the polynomial P, then c is an x-intercept of the graph of P.

5. (a) $P(x) = x^2 - 4$

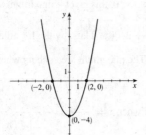

(b) $Q(x) = (x - 4)^2$

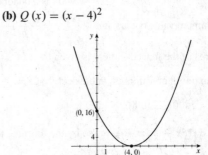

(c) $R(x) = 2x^2 - 2$

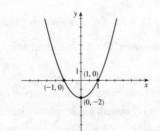

(d) $S(x) = 2\,(x - 2)^2$

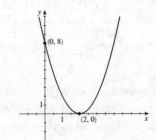

7. (a) $P(x) = x^3 - 8$

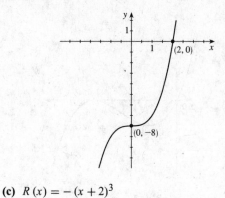

(b) $Q(x) = -x^3 + 27$

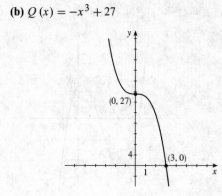

(c) $R(x) = -(x+2)^3$

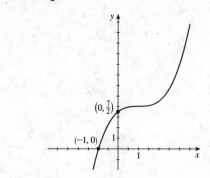

(d) $S(x) = \frac{1}{2}(x-1)^3 + 4$

9. (a) $P(x) = x\left(x^2 - 4\right) = x^3 - 4x$ has odd degree and a positive leading coefficient, so $y \to \infty$ as $x \to \infty$ and $y \to -\infty$ as $x \to -\infty$.

(b) This corresponds to graph III.

11. (a) $R(x) = -x^5 + 5x^3 - 4x$ has odd degree and a negative leading coefficient, so $y \to -\infty$ as $x \to \infty$ and $y \to \infty$ as $x \to -\infty$.

(b) This corresponds to graph V.

13. (a) $T(x) = x^4 + 2x^3$ has even degree and a positive leading coefficient, so $y \to \infty$ as $x \to \infty$ and $y \to \infty$ as $x \to -\infty$.

(b) This corresponds to graph VI.

15. $P(x) = (x-1)(x+2)$

17. $P(x) = x(x-3)(x+2)$

19. $P(x) = -(2x - 1)(x + 1)(x + 3)$

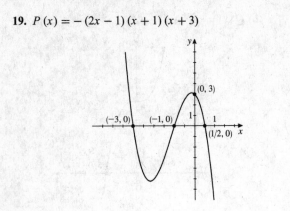

21. $P(x) = -2x(x - 2)^2$

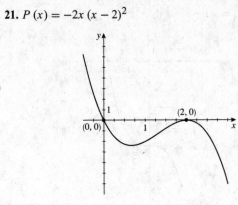

23. $P(x) = -(x + 4)(x + 3)(x - 5)^2$

25. $P(x) = \frac{1}{12}(x + 2)^2(x - 3)^2$

27. $P(x) = x^3(x + 2)(x - 3)^2$

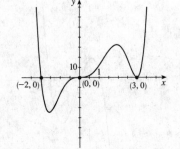

29. $P(x) = x^3 - x^2 - 6x = x(x + 2)(x - 3)$

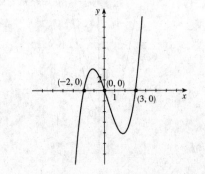

31. $P(x) = -x^3 + x^2 + 12x = -x(x+3)(x-4)$

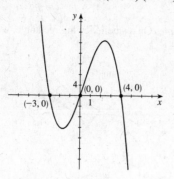

33. $P(x) = x^4 - 3x^3 + 2x^2 = x^2(x-1)(x-2)$

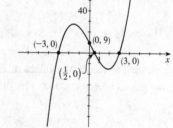

35. $P(x) = x^3 + x^2 - x - 1 = (x-1)(x+1)^2$

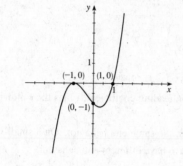

37. $P(x) = 2x^3 - x^2 - 18x + 9$

$= (x-3)(2x-1)(x+3)$

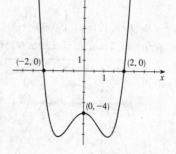

39. $P(x) = x^4 - 2x^3 - 8x + 16$

$= (x-2)^2(x^2 + 2x + 4)$

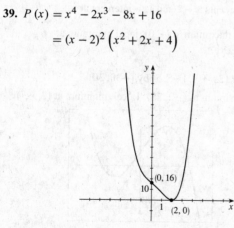

41. $P(x) = x^4 - 3x^2 - 4 = (x-2)(x+2)(x^2+1)$

43. $P(x) = 3x^3 - x^2 + 5x + 1$; $Q(x) = 3x^3$. Since P has odd degree and positive leading coefficient, it has the following end behavior: $y \to \infty$ as $x \to \infty$ and $y \to -\infty$ as $x \to -\infty$.

On a large viewing rectangle, the graphs of P and Q look almost the same. On a small viewing rectangle, we see that the graphs of P and Q have different intercepts.

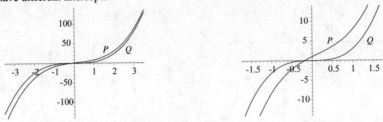

45. $P(x) = x^4 - 7x^2 + 5x + 5$; $Q(x) = x^4$. Since P has even degree and positive leading coefficient, it has the following end behavior: $y \to \infty$ as $x \to \infty$ and $y \to \infty$ as $x \to -\infty$.

On a large viewing rectangle, the graphs of P and Q look almost the same. On a small viewing rectangle, the graphs of P and Q look very different and we see that they have different intercepts.

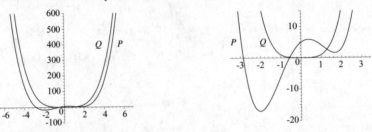

47. $P(x) = x^{11} - 9x^9$; $Q(x) = x^{11}$. Since P has odd degree and positive leading coefficient, it has the following end behavior: $y \to \infty$ as $x \to \infty$ and $y \to -\infty$ as $x \to -\infty$.

On a large viewing rectangle, the graphs of P and Q look like they have the same end behavior. On a small viewing rectangle, the graphs of P and Q look very different and seem (wrongly) to have different end behavior.

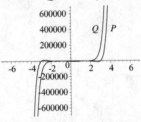
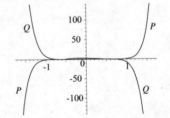

49. (a) x-intercepts at 0 and 4, y-intercept at 0.

   (b) Local maximum at $(2, 4)$, no local minimum.

51. (a) x-intercepts at -2 and 1, y-intercept at -1.

   (b) Local maximum at $(1, 0)$, local minimum at $(-1, -2)$.

53. $y = -x^2 + 8x$, $[-4, 12]$ by $[-50, 30]$

   No local minimum. Local maximum at $(4, 16)$.

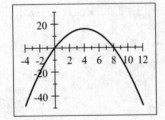

55. $y = x^3 - 12x + 9$, $[-5, 5]$ by $[-30, 30]$

   Local maximum at $(-2, 25)$. Local minimum at $(2, -7)$.

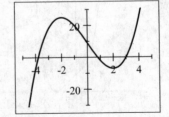

57. $y = x^4 + 4x^3$, $[-5, 5]$ by $[-30, 30]$

Local minimum at $(-3, -27)$. No local maximum.

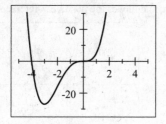

59. $y = 3x^5 - 5x^3 + 3$, $[-3, 3]$ by $[-5, 10]$

Local maximum at $(-1, 5)$. Local minimum at $(1, 1)$.

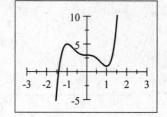

61. $y = -2x^2 + 3x + 5$ has one local maximum at $(0.75, 6.13)$.

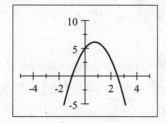

63. $y = x^3 - x^2 - x$ has one local maximum at $(-0.33, 0.19)$ and one local minimum at $(1.00, -1.00)$.

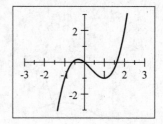

65. $y = x^4 - 5x^2 + 4$ has one local maximum at $(0, 4)$ and two local minima at $(-1.58, -2.25)$ and $(1.58, -2.25)$.

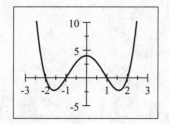

67. $y = (x - 2)^5 + 32$ has no maximum or minimum.

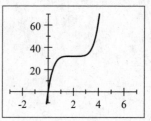

69. $y = x^8 - 3x^4 + x$ has one local maximum at $(0.44, 0.33)$ and two local minima at $(1.09, -1.15)$ and $(-1.12, -3.36)$.

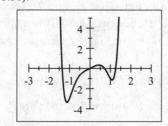

71. $y = cx^3$; $c = 1, 2, 5, \frac{1}{2}$. Increasing the value of c stretches the graph vertically.

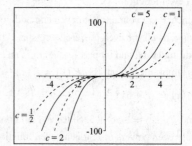

73. $P(x) = x^4 + c$; $c = -1, 0, 1$, and 2. Increasing the value of c moves the graph up.

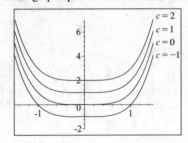

75. $P(x) = x^4 - cx$; $c = 0, 1, 8$, and 27. Increasing the value of c causes a deeper dip in the graph, in the fourth quadrant, and moves the positive x-intercept to the right.

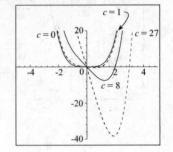

77. (a)

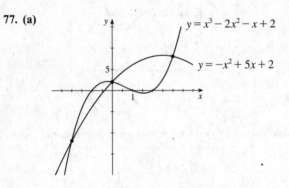

(b) The two graphs appear to intersect at 3 points.

(c) $x^3 - 2x^2 - x + 2 = -x^2 + 5x + 2 \Leftrightarrow x^3 - x^2 - 6x = 0 \Leftrightarrow$ $x\left(x^2 - x - 6\right) = 0 \Leftrightarrow x(x - 3)(x + 2) = 0$. Then either $x = 0$, $x = 3$, or $x = -2$. If $x = 0$, then $y = 2$; if $x = 3$ then $y = 8$; if $x = -2$, then $y = -12$. Hence the points where the two graphs intersect are $(0, 2)$, $(3, 8)$, and $(-2, -12)$.

79. (a) Let $P(x)$ be a polynomial containing only odd powers of x. Then each term of $P(x)$ can be written as Cx^{2n+1}, for some constant C and integer n. Since $C(-x)^{2n+1} = -Cx^{2n+1}$, each term of $P(x)$ is an odd function. Thus by part (a), $P(x)$ is an odd function.

(b) Let $P(x)$ be a polynomial containing only even powers of x. Then each term of $P(x)$ can be written as Cx^{2n}, for some constant C and integer n. Since $C(-x)^{2n} = Cx^{2n}$, each term of $P(x)$ is an even function. Thus by part (b), $P(x)$ is an even function.

(c) Since $P(x)$ contains both even and odd powers of x, we can write it in the form $P(x) = R(x) + Q(x)$, where $R(x)$ contains all the even-powered terms in $P(x)$ and $Q(x)$ contains all the odd-powered terms. By part (d), $Q(x)$ is an odd function, and by part (e), $R(x)$ is an even function. Thus, since neither $Q(x)$ nor $R(x)$ are constantly 0 (by assumption), by part (c), $P(x) = R(x) + Q(x)$ is neither even nor odd.

(d) $P(x) = x^5 + 6x^3 - x^2 - 2x + 5 = \left(x^5 + 6x^3 - 2x\right) + \left(-x^2 + 5\right) = P_O(x) + P_E(x)$ where $P_O(x) = x^5 + 6x^3 - 2x$ and $P_E(x) = -x^2 + 5$. Since $P_O(x)$ contains only odd powers of x, it is an odd function, and since $P_E(x)$ contains only even powers of x, it is an even function.

81. (a) $P(x) = (x - 2)(x - 4)(x - 5)$ has one local maximum and one local minimum.

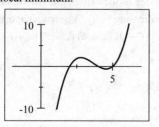

(b) Since $P(a) = P(b) = 0$, and $P(x) > 0$ for $a < x < b$ (see the table below), the graph of P must first rise and then fall on the interval (a, b), and so P must have at least one local maximum between a and b. Using similar reasoning, the fact that $P(b) = P(c) = 0$ and $P(x) < 0$ for $b < x < c$ shows that P must have at least one local minimum between b and c. Thus P has at least two local extrema.

Interval	$(-\infty, a)$	(a, b)	(b, c)	(c, ∞)
Sign of $x - a$	$-$	$+$	$+$	$+$
Sign of $x - b$	$-$	$-$	$+$	$+$
Sign of $x - c$	$-$	$-$	$-$	$+$
Sign of $(x - a)(x - b)(x - c)$	$-$	$+$	$-$	$+$

83. $P(x) = 8x + 0.3x^2 - 0.0013x^3 - 372$

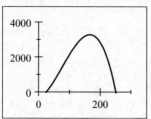

(a) For the firm to break even, $P(x) = 0$. From the graph, we see that $P(x) = 0$ when $x \approx 25.2$. Of course, the firm cannot produce fractions of a blender, so the manufacturer must produce at least 26 blenders a year.

(b) No, the profit does not increase indefinitely. The largest profit is approximately $ 3276.22, which occurs when the firm produces 166 blenders per year.

85. (a) The length of the bottom is $40 - 2x$, the width of the bottom is $20 - 2x$, and the height is x, so the volume of the box is
$$V = x(20 - 2x)(40 - 2x) = 4x^3 - 120x^2 + 800x.$$

(b) Since the height and width must be positive, we must have $x > 0$ and $20 - 2x > 0$, and so the domain of V is $0 < x < 10$.

(c) Using the domain from part (b), we graph V in the viewing rectangle $[0, 10]$ by $[0, 1600]$. The maximum volume is $V \approx 1539.6$ when $x = 4.23$.

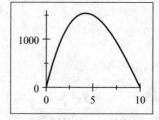

87.

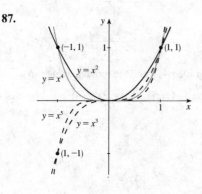

The graph of $y = x^{100}$ is close to the x-axis for $|x| < 1$, but passes through the points $(1, 1)$ and $(-1, 1)$. The graph of $y = x^{101}$ behaves similarly except that the y-values are negative for negative values of x, and it passes through $(-1, -1)$ instead of $(-1, 1)$.

89. No, it is impossible. The end behavior of a third degree polynomial is the same as that of $y = kx^3$, and for this function, the values of y go off in opposite directions as $x \to \infty$ and $x \to -\infty$. But for a function with just one extremum, the values of y would head off in the same direction (either both up or both down) on either side of the extremum. An nth-degree polynomial can have $n - 1$ extrema or $n - 3$ extrema or $n - 5$ extrema, and so on (decreasing by 2). A polynomial that has six local extrema must be of degree 7 or higher. For example, $P(x) = (x - 1)(x - 2)(x - 3)(x - 4)(x - 5)(x - 6)(x - 7)$ has six local extrema.

3.3   DIVIDING POLYNOMIALS

1. If we divide the polynomial P by the factor $x - c$, and we obtain the equation $P(x) = (x - c) Q(x) + R(x)$, then we say that $x - c$ is the *divisor*, $Q(x)$ is the *quotient*, and $R(x)$ is the *remainder*.

3.

$$
\begin{array}{r|rrr}
-3 & 1 & 4 & -8 \\
   &   & -3 & -3 \\
\hline
   & 1 & 1 & -11
\end{array}
$$

Thus the quotient is $x + 1$ and the remainder is -11, and

$$\frac{P(x)}{D(x)} = \frac{x^2 + 4x - 8}{x + 3} = (x + 1) + \frac{-11}{x + 3}.$$

5.

$$
\begin{array}{r}
2x - \frac{1}{2} \\
2x - 1 \overline{\smash{\big)}\, 4x^2 - 3x - 7} \\
\underline{4x^2 - 2x} \\
-x - 7 \\
\underline{-x + \frac{1}{2}} \\
-\frac{15}{2}
\end{array}
$$

Thus the quotient is $2x - \frac{1}{2}$ and the remainder is $-\frac{15}{2}$, and

$$\frac{P(x)}{D(x)} = \frac{4x^2 - 3x - 7}{2x - 1} = \left(2x - \frac{1}{2}\right) + \frac{-\frac{15}{2}}{2x - 1}.$$

7.

$$
\begin{array}{r}
2x^2 - x + 1 \\
x^2 + 4 \overline{\smash{\big)}\, 2x^4 - x^3 + 9x^2} \\
\underline{2x^4 \qquad + 8x^2} \\
-x^3 + x^2 \\
\underline{-x^3 \qquad - 4x} \\
x^2 + 4x \\
\underline{x^2 \qquad + 4} \\
4x - 4
\end{array}
$$

Thus the quotient is $2x^2 - x + 1$ and the remainder is $4x - 4$, and

$$\frac{P(x)}{D(x)} = \frac{2x^4 - x^3 + 9x^2}{x^2 + 4} = \left(2x^2 - x + 1\right) + \frac{4x - 4}{x^2 + 4}.$$

9.

$$
\begin{array}{r}
3x - 4 \\
x + 3 \overline{\smash{\big)}\, 3x^2 + 5x - 4} \\
\underline{3x^2 + 9x} \\
-4x - 4 \\
\underline{-4x - 12} \\
8
\end{array}
$$

Thus the quotient is $3x - 4$ and the remainder is 8, so

$$P(x) = 3x^2 + 5x - 4 = (x + 3) \cdot (3x - 4) + 8.$$

11.

$$
\begin{array}{r}
x^2 \qquad - 1 \\
2x-3 \overline{)\, 2x^3 - 3x^2 - 2x} \\
\underline{2x^3 - 3x^2} \\
-2x \\
\underline{-2x + 3} \\
-3
\end{array}
$$

Thus the quotient is $x^2 - 1$ and the remainder is -3, and

$$P(x) = 2x^3 - 3x^2 - 2x = \left(x^2 - 1\right)(2x - 3) - 3.$$

13.

$$
\begin{array}{r}
x^2 - x - 3 \\
x^2+3 \overline{)\, x^4 - x^3 + 0x^2 + 4x + 2} \\
\underline{x^4 \qquad\; + 3x^2} \\
-x^3 - 3x^2 + 4x \\
\underline{-x^2 \qquad\; - 3x} \\
-3x^2 + 7x + 2 \\
\underline{-3x^2 \qquad\quad - 9} \\
7x + 11
\end{array}
$$

Thus the quotient is $x^2 - x - 3$ and the remainder is $7x + 11$, and

$$P(x) = x^3 + 4x^2 - 6x + 1$$
$$= \left(x^2 + 3\right) \cdot \left(x^2 - x - 3\right) + (7x + 11)$$

15.

$$
\begin{array}{r}
x - 2 \\
x-4 \overline{)\, x^2 - 6x - 8} \\
\underline{x^2 - 4x} \\
-2x - 8 \\
\underline{-2x + 8} \\
-16
\end{array}
$$

Thus the quotient is $x - 2$ and the remainder is -16.

17.

$$
\begin{array}{r}
2x^2 \qquad\; - 1 \\
2x+1 \overline{)\, 4x^3 + 2x^2 - 2x - 3} \\
\underline{4x^3 + 2x^2} \\
-2x - 3 \\
\underline{-2x - 1} \\
-2
\end{array}
$$

Thus the quotient is $2x^2 - 1$ and the remainder is -2.

19.

$$
\begin{array}{r}
x + \;\; 2 \\
x^2-2x+2 \overline{)\, x^3 + 0x^2 + 6x + 3} \\
\underline{x^3 - 2x^2 + 2x} \\
2x^2 + 4x + 3 \\
\underline{2x^2 - 4x + 4} \\
8x - 1
\end{array}
$$

Thus the quotient is $x + 2$, and the remainder is $8x - 1$.

21.

$$
\begin{array}{r}
3x + \;\; 1 \\
2x^2+0x+5 \overline{)\, 6x^3 + 2x^2 + 22x + 0} \\
\underline{6x^3 \qquad\quad + 15x} \\
2x^2 + 7x + 0 \\
\underline{2x^2 \qquad\; + 5} \\
7x - 5
\end{array}
$$

Thus the quotient is $3x + 1$, and the remainder is $7x - 5$.

23.

$$
\begin{array}{r}
x^4 \qquad\qquad\quad + 1 \\
x^2+1 \overline{)\, x^6 + 0x^5 + x^4 + 0x^3 + x^2 + 0x + 1} \\
\underline{x^6 \qquad\; + x^4} \\
0 \qquad\; + x^2 \qquad + 1 \\
\underline{x^2 \qquad + 1} \\
0
\end{array}
$$

Thus the quotient is $x^4 + 1$, and the remainder is 0.

25. The synthetic division table for this problem takes the following form.

$$
\begin{array}{r|rrr}
3 & 1 & -5 & 4 \\
  &   & 3 & -6 \\
\hline
  & 1 & -2 & -2
\end{array}
$$

Thus the quotient is $x - 2$, and the remainder is -2.

27. The synthetic division table for this problem takes the following form.

$$
\begin{array}{r|rrr}
6 & 3 & 5 & 0 \\
  &   & 18 & 138 \\
\hline
  & 3 & 23 & 138
\end{array}
$$

Thus the quotient is $3x + 23$, and the remainder is 138.

29. Since $x + 2 = x - (-2)$, the synthetic division table for this problem takes the following form.

$$
\begin{array}{r|rrrr}
-2 & 1 & 2 & 2 & 1 \\
   &   & -2 & 0 & -4 \\
\hline
   & 1 & 0 & 2 & -3
\end{array}
$$

Thus the quotient is $x^2 + 2$, and the remainder is -3.

31. Since $x + 3 = x - (-3)$ and
$x^3 - 8x + 2 = x^3 + 0x^2 - 8x + 2$, the synthetic division table for this problem takes the following form.

$$
\begin{array}{r|rrrr}
-3 & 1 & 0 & -8 & 2 \\
   &   & -3 & 9 & -3 \\
\hline
   & 1 & -3 & 1 & -1
\end{array}
$$

Thus the quotient is $x^2 - 3x + 1$, and the remainder is -1.

33. Since $x^5 + 3x^3 - 6 = x^5 + 0x^4 + 3x^3 + 0x^2 + 0x - 6$, the synthetic division table for this problem takes the following form.

$$
\begin{array}{r|rrrrrr}
1 & 1 & 0 & 3 & 0 & 0 & -6 \\
  &   & 1 & 1 & 4 & 4 & 4 \\
\hline
  & 1 & 1 & 4 & 4 & 4 & -2
\end{array}
$$

Thus the quotient is $x^4 + x^3 + 4x^2 + 4x + 4$, and the remainder is -2.

35. The synthetic division table for this problem takes the following form.

$$
\begin{array}{r|rrrr}
\frac{1}{2} & 2 & 3 & -2 & 1 \\
            &   & 1 & 2 & 0 \\
\hline
            & 2 & 4 & 0 & 1
\end{array}
$$

Thus the quotient is $2x^2 + 4x$, and the remainder is 1.

37. Since $x^3 - 27 = x^3 + 0x^2 + 0x - 27$, the synthetic division table for this problem takes the following form.

$$
\begin{array}{r|rrrr}
3 & 1 & 0 & 0 & -27 \\
  &   & 3 & 9 & 27 \\
\hline
  & 1 & 3 & 9 & 0
\end{array}
$$

Thus the quotient is $x^2 + 3x + 9$, and the remainder is 0.

39. $P(x) = 4x^2 + 12x + 5, c = -1$

$$
\begin{array}{r|rrr}
-1 & 4 & 12 & 5 \\
   &   & -4 & -8 \\
\hline
   & 4 & 8 & -3
\end{array}
$$

Therefore, by the Remainder Theorem, $P(-1) = -3$.

41. $P(x) = x^3 + 3x^2 - 7x + 6, c = 2$

$$
\begin{array}{r|rrrr}
2 & 1 & 3 & -7 & 6 \\
  &   & 2 & 10 & 6 \\
\hline
  & 1 & 5 & 3 & 12
\end{array}
$$

Therefore, by the Remainder Theorem, $P(2) = 12$.

43. $P(x) = x^3 + 2x^2 - 7, c = -2$

$$
\begin{array}{r|rrrr}
-2 & 1 & 2 & 0 & -7 \\
   &   & -2 & 0 & 0 \\
\hline
   & 1 & 0 & 0 & -7
\end{array}
$$

Therefore, by the Remainder Theorem, $P(-2) = -7$.

45. $P(x) = 5x^4 + 30x^3 - 40x^2 + 36x + 14, c = -7$

$$
\begin{array}{r|rrrrr}
-7 & 5 & 30 & -40 & 36 & 14 \\
   &   & -35 & 35 & 35 & -497 \\
\hline
   & 5 & -5 & -5 & 71 & -483
\end{array}
$$

Therefore, by the Remainder Theorem, $P(-7) = -483$.

47. $P(x) = x^7 - 3x^2 - 1$

$\qquad = x^7 + 0x^6 + 0x^5 + 0x^4 + 0x^3 - 3x^2 + 0x - 1$

$c = 3$

$$
\begin{array}{c|cccccccc}
3 & 1 & 0 & 0 & 0 & 0 & -3 & 0 & -1 \\
  &   & 3 & 9 & 27 & 81 & 243 & 720 & 2160 \\
\hline
  & 1 & 3 & 9 & 27 & 81 & 240 & 720 & 2159
\end{array}
$$

Therefore by the Remainder Theorem, $P(3) = 2159$.

49. $P(x) = 3x^3 + 4x^2 - 2x + 1, c = \frac{2}{3}$

$$
\begin{array}{c|cccc}
\frac{2}{3} & 3 & 4 & -2 & 1 \\
            &   & 2 & 4 & \frac{4}{3} \\
\hline
            & 3 & 6 & 2 & \frac{7}{3}
\end{array}
$$

Therefore, by the Remainder Theorem, $P\left(\frac{2}{3}\right) = \frac{7}{3}$.

51. $P(x) = x^3 + 2x^2 - 3x - 8, c = 0.1$

$$
\begin{array}{c|cccc}
0.1 & 1 & 2 & -3 & -8 \\
    &   & 0.1 & 0.21 & -0.279 \\
\hline
    & 1 & 2.1 & -2.79 & -8.279
\end{array}
$$

Therefore, by the Remainder Theorem, $P(0.1) = -8.279$.

53. $P(x) = x^3 - 3x^2 + 3x - 1, c = 1$

$$
\begin{array}{c|cccc}
1 & 1 & -3 & 3 & -1 \\
  &   & 1 & -2 & 1 \\
\hline
  & 1 & -2 & 1 & 0
\end{array}
$$

Since the remainder is 0, $x - 1$ is a factor.

55. $P(x) = 2x^3 + 7x^2 + 6x - 5, c = \frac{1}{2}$

$$
\begin{array}{c|cccc}
\frac{1}{2} & 2 & 7 & 6 & -5 \\
            &   & 1 & 4 & 5 \\
\hline
            & 2 & 8 & 10 & 0
\end{array}
$$

Since the remainder is 0, $x - \frac{1}{2}$ is a factor.

57. $P(x) = x^3 - x^2 - 11x + 15, c = 3$

$$
\begin{array}{c|cccc}
3 & 1 & -1 & -11 & 15 \\
  &   & 3 & 6 & -15 \\
\hline
  & 1 & 2 & -5 & 0
\end{array}
$$

Since the remainder is 0, we know that 3 is a zero

$x^3 - x^2 - 11x + 15 = (x - 3)\left(x^2 + 2x - 5\right)$. Now

$x^2 + 2x - 5 = 0$ when $x = \frac{-2 \pm \sqrt{2^2 + 4(1)(5)}}{2} = -1 \pm \sqrt{6}$.

Hence, the zeros are $-1 - \sqrt{6}$, $-1 + \sqrt{6}$, and 3.

59. Since the zeros are $x = -1$, $x = 1$, and $x = 3$, the factors are $x + 1$, $x - 1$, and $x - 3$.

Thus

$P(x) = (x + 1)(x - 1)(x - 3) = x^3 - 3x^2 - x + 3$.

61. Since the zeros are $x = -1$, $x = 1$, $x = 3$, and $x = 5$, the factors are $x + 1$, $x - 1$, $x - 3$, and $x - 5$.

Thus $P(x) = (x + 1)(x - 1)(x - 3)(x - 5) = x^4 - 8x^3 + 14x^2 + 8x - 15$.

63. Since the zeros of the polynomial are 1, -2, and 3, it follows that

$P(x) = C(x - 1)(x + 2)(x - 3) = C\left(x^3 - 2x^2 - 5x + 6\right) = Cx^3 - 2Cx^2 - 5Cx + 6C$. Since the coefficient of x^2 is

to be 3, $-2C = 3$ so $C = -\frac{3}{2}$. Therefore, $P(x) = -\frac{3}{2}\left(x^3 - 2x^2 - 5x + 6\right) = -\frac{3}{2}x^3 + 3x^2 + \frac{15}{2}x - 9$ is the polynomial.

65. The y-intercept is 2 and the zeros of the polynomial are -1, 1, and 2.

It follows that $P(x) = C(x + 1)(x - 1)(x - 2) = C\left(x^3 - 2x^2 - x + 2\right)$. Since $P(0) = 2$ we have

$2 = C\left[(0)^3 - 2(0)^2 - (0) + 2\right] \Leftrightarrow 2 = 2C \Leftrightarrow C = 1$ and $P(x) = (x + 1)(x - 1)(x - 2) = x^3 - 2x^2 - x + 2$.

67. The y-intercept is 4 and the zeros of the polynomial are -2 and 1 both being degree two.

It follows that $P(x) = C(x+2)^2 (x-1)^2 = C\left(x^4 + 2x^3 - 3x^2 - 4x + 4\right)$. Since $P(0) = 4$ we have

$4 = C\left[(0)^4 + 2(0)^3 - 3(0)^2 - 4(0) + 4\right] \Leftrightarrow 4 = 4C \Leftrightarrow C = 1$.

Thus $P(x) = (x+2)^2 (x-1)^2 = x^4 + 2x^3 - 3x^2 - 4x + 4$.

69. A. By the Remainder Theorem, the remainder when $P(x) = 6x^{1000} - 17x^{562} + 12x + 26$ is divided by $x + 1$ is

$P(-1) = 6(-1)^{1000} - 17(-1)^{562} + 12(-1) + 26 = 6 - 17 - 12 + 26 = 3$.

B. If $x - 1$ is a factor of $Q(x) = x^{567} - 3x^{400} + x^9 + 2$, then $Q(1)$ must equal 0.

$Q(1) = (1)^{567} - 3(1)^{400} + (1)^9 + 2 = 1 - 3 + 1 + 2 = 1 \neq 0$, so $x - 1$ is not a factor.

3.4　REAL ZEROS OF POLYNOMIALS

1. If the polynomial function $P(x) = a_n x^n + a_{n-1} x^{n-1} + \cdots a_1 x + a_0$ has integer coefficients, then the only numbers that

could possibly be rational zeros of P are all of the form $\dfrac{p}{q}$, where p is a factor of *the constant coefficient* a_0 and q is a factor

of *the leading coefficient* a_n. The possible rational zeros of $P(x) = 6x^3 + 5x^2 - 19x - 10$ are ± 1, $\pm\frac{1}{2}$, $\pm\frac{1}{3}$, $\pm\frac{1}{6}$, ± 2, $\pm\frac{2}{3}$,

± 5, $\pm\frac{5}{2}$, $\pm\frac{5}{3}$, $\pm\frac{5}{6}$, ± 10, and $\pm\frac{10}{3}$.

3. This is true. If c is a real zero of the polynomial P, then $P(x) = (x - c) Q(x)$, and any other zero of $P(x)$ is also a zero of $Q(x) = P(x) / (x - c)$.

5. $P(x) = x^3 - 4x^2 + 3$ has possible rational zeros ± 1 and ± 3.

7. $R(x) = 2x^5 + 3x^3 + 4x^2 - 8$ has possible rational zeros ± 1, ± 2, ± 4, ± 8, $\pm\frac{1}{2}$.

9. $T(x) = 4x^4 - 2x^2 - 7$ has possible rational zeros ± 1, ± 7, $\pm\frac{1}{2}$, $\pm\frac{7}{2}$, $\pm\frac{1}{4}$, $\pm\frac{7}{4}$.

11. (a) $P(x) = 5x^3 - x^2 - 5x + 1$ has possible rational zeros ± 1, $\pm\frac{1}{5}$.

(b) From the graph, the actual zeros are -1, $\frac{1}{5}$, and 1.

13. (a) $P(x) = 2x^4 - 9x^3 + 9x^2 + x - 3$ has possible rational zeros ± 1, ± 3, $\pm\frac{1}{2}$, $\pm\frac{3}{2}$.

(b) From the graph, the actual zeros are $-\frac{1}{2}$, 1, and 3.

15. $P(x) = x^3 - 4x^2 + x + 6$. The possible rational zeros are ± 1, ± 2, ± 3, ± 6. $P(x)$ has 2 variations in sign and hence 0 or 2 positive real zeros. $P(-x) = -x^3 - 4x^2 - x + 6$ has 1 variation in sign and hence 1 negative real zero.

$$
\begin{array}{r|rrrr}
1 & 1 & -4 & 1 & 6 \\
  &   & 1 & -3 & -2 \\
\hline
  & 1 & -3 & -2 & 4 \\
\end{array}
\Rightarrow x = 1 \text{ is not a zero.}
\qquad
\begin{array}{r|rrrr}
2 & 1 & -4 & 1 & 6 \\
  &   & 2 & -4 & -6 \\
\hline
  & 1 & -2 & -3 & 0 \\
\end{array}
\Rightarrow x = 2 \text{ is a zero.}
$$

$P(x) = x^3 - 4x^2 + x + 6 = (x - 2)\left(x^2 - 2x - 3\right) = (x - 2)(x + 1)(x - 3)$. Therefore, the zeros are -1, 2, and 3.

17. $P(x) = x^3 + 3x^2 - 4$. The possible rational zeros are ± 1, ± 2, ± 4. $P(x)$ has 1 variation in sign and hence 1 positive real zero. $P(-x) = -x^3 + 3x^2 - 4$ has 2 variations in sign and hence 0 or 2 negative real zeros.

$$
\begin{array}{r|rrrr}
1 & 1 & 3 & 0 & -4 \\
  &   & 1 & 4 & 4 \\
\hline
  & 1 & 4 & 4 & 0 \\
\end{array}
\Rightarrow x = 1 \text{ is a zero.}
$$

$P(x) = x^3 + 3x^2 - 4 = (x - 1)\left(x^2 + 4x + 4\right) = (x - 1)(x + 2)^2$. Therefore, the zeros are $x = -2$ and 1.

19. $P(x) = x^3 + 4x^2 - 3x - 18$. The possible rational zeros are $\pm 1, \pm 2, \pm 3, \pm 6, \pm 9, \pm 18$. $P(x)$ has 1 variation in sign and hence 1 positive real zero. $P(-x) = -x^3 + 4x^2 + 3x - 18$ has 2 variations in sign and hence 0 or 2 negative real zeros.

$$
\begin{array}{r|rrrr}
1 & 1 & 4 & -3 & -18 \\
  &   & 1 & 5 & 2 \\
\hline
  & 1 & 5 & 2 & -16
\end{array}
\qquad
\begin{array}{r|rrrr}
2 & 1 & 4 & -3 & -18 \\
  &   & 2 & 12 & 18 \\
\hline
  & 1 & 6 & 9 & 0
\end{array}
\Rightarrow x = 2 \text{ is a zero.}
$$

$P(x) = x^3 + 4x^2 - 3x - 18 = (x-2)\left(x^2 + 6x + 9\right) = (x-2)(x+3)^2$. Therefore, the zeros are $x = -3$ and 2.

21. $P(x) = x^3 - 6x^2 + 12x - 8$. The possible rational zeros are $\pm 1, \pm 2, \pm 4, \pm 8$. $P(x)$ has 3 variations in sign and hence 1 or 3 positive real zeros. $P(-x) = -x^3 - 6x^2 - 12x - 8$ has no variations in sign and hence 0 negative real zeros.

$$
\begin{array}{r|rrrr}
1 & 1 & -6 & 12 & -8 \\
  &   & 1 & -5 & 7 \\
\hline
  & 1 & -5 & 7 & -1
\end{array}
\Rightarrow x = 1 \text{ is not a zero.}
\qquad
\begin{array}{r|rrrr}
2 & 1 & -6 & 12 & -8 \\
  &   & 2 & -8 & 8 \\
\hline
  & 1 & -4 & 4 & 0
\end{array}
\Rightarrow x = 2 \text{ is a zero.}
$$

$P(x) = x^3 - 6x^2 + 12x - 8 = (x-2)\left(x^2 - 4x + 4\right) = (x-2)^3$. Therefore, the only zero is $x = 2$.

23. $P(x) = x^3 - 4x^2 + x + 6$. The possible rational zeros are $\pm 1, \pm 2, \pm 3, \pm 6$. $P(x)$ has 2 variations in sign and hence 0 or 2 positive real zeros. $P(-x) = -x^3 - 4x^2 - x + 6$ has 1 variation in sign and hence 1 negative real zero.

$$
\begin{array}{r|rrrr}
-1 & 1 & -4 & 1 & 6 \\
   &   & -1 & 5 & -6 \\
\hline
   & 1 & -5 & 6 & 0
\end{array}
\Rightarrow x = -1 \text{ is a zero.}
$$

$P(x) = x^3 - 4x^2 + x + 6 = (x+1)\left(x^2 - 5x + 6\right) = (x+1)(x-3)(x-2)$. Therefore, the zeros are $x = -1, 2$, and 3.

25. $P(x) = x^3 + 3x^2 - x - 3$. The possible rational zeros are $\pm 1, \pm 3$. $P(x)$ has 1 variation in sign and hence 1 positive real zero. $P(-x) = -x^3 + 3x^2 + x - 4$ has 2 variations in sign and hence 0 or 2 negative real zeros.

$$
\begin{array}{r|rrrr}
-1 & 1 & 3 & -1 & -3 \\
   &   & -1 & -2 & 3 \\
\hline
   & 1 & 2 & -3 & 0
\end{array}
\Rightarrow x = -1 \text{ is a zero.}
$$

So $P(x) = x^3 + 3x^2 - x - 3 = (x+1)\left(x^2 + 2x - 3\right) = (x+1)(x+3)(x-1)$. Therefore, the zeros are $-1, -3$, and 1.

27. *Method 1:* $P(x) = x^4 - 5x^2 + 4$. The possible rational zeros are ± 1, ± 2, ± 4. $P(x)$ has 1 variation in sign and hence 1 positive real zero. $P(-x) = x^4 - 5x^2 + 4$ has 2 variations in sign and hence 0 or 2 negative real zeros.

$$
\begin{array}{r|rrrrr}
1 & 1 & 0 & -5 & 0 & 4 \\
  &   & 1 & 1 & -4 & -4 \\
\hline
  & 1 & 1 & -4 & -4 & 0
\end{array} \Rightarrow x = 1 \text{ is a zero.}
$$

Thus $P(x) = x^4 - 5x^2 + 4 = (x-1)\left(x^3 + x^2 - 4x - 4\right)$. Continuing with the quotient we have:

$$
\begin{array}{r|rrrr}
-1 & 1 & 1 & -4 & -4 \\
   &   & -1 & 0 & 4 \\
\hline
   & 1 & 0 & -4 & 0
\end{array} \Rightarrow x = -1 \text{ is a zero.}
$$

$P(x) = x^4 - 5x^2 + 4 = (x-1)(x+1)\left(x^2 - 4\right) = (x-1)(x+1)(x-2)(x+2)$. Therefore, the zeros are $x = \pm 1$, ± 2.

Method 2: Substituting $u = x^2$, the polynomial becomes $P(u) = u^2 - 5u + 4$, which factors:

$u^2 - 5u + 4 = (u-1)(u-4) = \left(x^2 - 1\right)\left(x^2 - 4\right)$, so either $x^2 = 1$ or $x^2 = 4$. If $x^2 = 1$, then $x = \pm 1$; if $x^2 = 4$, then $x = \pm 2$. Therefore, the zeros are $x = \pm 1$ and ± 2.

29. $P(x) = x^4 + 6x^3 + 7x^2 - 6x - 8$. The possible rational zeros are ± 1, ± 2, ± 4, ± 8. $P(x)$ has 1 variation in sign and hence 1 positive real zero. $P(-x) = x^4 - 6x^3 + 7x^2 + 6x - 8$ has 3 variations in sign and hence 1 or 3 negative real zeros.

$$
\begin{array}{r|rrrrr}
1 & 1 & 6 & 7 & -6 & -8 \\
  &   & 1 & 7 & 14 & 8 \\
\hline
  & 1 & 7 & 14 & 8 & 0
\end{array} \Rightarrow x = 1 \text{ is a zero}
$$

and there are no other positive zeros. Thus $P(x) = x^4 + 6x^3 + 7x^2 - 6x - 8 = (x-1)\left(x^3 + 7x^2 + 14x + 8\right)$. Continuing by factoring the quotient, we have:

$$
\begin{array}{r|rrrr}
-1 & 1 & 7 & 14 & 8 \\
   &   & -1 & -6 & -8 \\
\hline
   & 1 & 6 & 8 & 0
\end{array} \Rightarrow x = -1 \text{ is a zero.}
$$

So $P(x) = x^4 + 6x^3 + 7x^2 - 6x - 8 = (x-1)(x+1)\left(x^2 + 6x + 8\right) = (x-1)(x+1)(x+2)(x+4)$. Therefore, the zeros are $x = -4$, -2, and ± 1.

31. $P(x) = 4x^4 - 25x^2 + 36$ has possible rational zeros ± 1, ± 2, ± 3, ± 4, ± 6, ± 9, ± 12, ± 18, ± 36, $\pm \frac{1}{2}$, $\pm \frac{1}{4}$, $\pm \frac{3}{2}$, $\pm \frac{3}{4}$, $\pm \frac{9}{2}$, $\pm \frac{9}{4}$. Since $P(x)$ has 2 variations in sign, there are 0 or 2 positive real zeros. Since $P(-x) = 4x^4 - 25x^2 + 36$ has 2 variations in sign, there are 0 or 2 negative real zeros.

$$
\begin{array}{r|rrrrr}
1 & 4 & 0 & -25 & 0 & 36 \\
  &   & 4 & 4 & -21 & -21 \\
\hline
  & 4 & 4 & -21 & -21 & 15
\end{array}
\qquad
\begin{array}{r|rrrrr}
2 & 4 & 0 & -25 & 0 & 36 \\
  &   & 8 & 16 & -18 & -36 \\
\hline
  & 4 & 8 & -9 & -18 & 0
\end{array} \Rightarrow x = 2 \text{ is a zero.}
$$

$$
\begin{array}{r|rrrr}
2 & 4 & 8 & -9 & -18 \\
  &   & 8 & 32 & 46 \\
\hline
  & 4 & 16 & 23 & 28
\end{array} \Rightarrow \text{all positive, } x = 2 \text{ is an upper bound.}
\qquad
\begin{array}{r|rrrr}
\frac{1}{2} & 4 & 8 & -9 & -18 \\
  &   & 2 & 5 & -2 \\
\hline
  & 4 & 10 & -4 & -20
\end{array}
$$

$$
\begin{array}{r|rrrr}
\frac{1}{4} & 4 & 8 & -9 & -18 \\
  &   & 1 & \frac{9}{4} & -\frac{27}{16} \\
\hline
  & 4 & 9 & -\frac{27}{4} & -\frac{315}{16}
\end{array}
\qquad
\begin{array}{r|rrrr}
\frac{3}{2} & 4 & 8 & -9 & -18 \\
  &   & 6 & 21 & 18 \\
\hline
  & 4 & 14 & 12 & 0
\end{array} \Rightarrow x = \frac{3}{2} \text{ is a zero.}
$$

$P(x) = (x-2)(2x-3)\left(2x^2 + 7x + 6\right) = (x-2)(2x-3)(2x+3)(x+2)$. Therefore, the zeros are $x = \pm 2$ and $\pm \frac{3}{2}$.

Note: Since $P(x)$ has only even terms, factoring by substitution also works. Let $x^2 = u$; then $P(u) = 4u^2 - 25u + 36 = (u-4)(4u-9) = \left(x^2 - 4\right)\left(4x^2 - 9\right)$, which gives the same results.

33. $P(x) = 3x^4 - 10x^3 - 9x^2 + 40x - 12$. The possible rational zeros are ± 1, ± 2, ± 3, ± 4, ± 6, ± 12. $P(x)$ has 3 variations in sign and hence 1, 3, or 5 positive real zeros. $P(-x) = 3x^4 + 10x^3 - 9x^2 - 40x - 12$ has 1 variation in sign and hence 1 negative real zero.

$$
\begin{array}{r|rrrrr}
1 & 3 & -10 & -9 & 40 & -12 \\
  &   & 3 & -7 & -16 & 24 \\
\hline
  & 3 & -7 & -16 & 24 & 24
\end{array} \Rightarrow x = 1 \text{ is not a zero.}
\qquad
\begin{array}{r|rrrrr}
2 & 3 & -10 & -9 & 40 & -12 \\
  &   & 6 & -8 & -34 & 12 \\
\hline
  & 3 & -4 & -17 & 6 & 0
\end{array} \Rightarrow x = 2 \text{ is a zero.}
$$

Thus $P(x) = 3x^4 - 10x^3 - 9x^2 + 40x - 12 = (x-2)\left(3x^3 - 4x^2 - 17x + 6\right)$. Continuing by factoring the quotient, we have

$$
\begin{array}{r|rrrr}
3 & 3 & -4 & -17 & 6 \\
  &   & 9 & 15 & -6 \\
\hline
  & 3 & 5 & -2 & 0
\end{array} \Rightarrow x = 3 \text{ is a zero.}
$$

Thus $P(x) = (x-3)(x-2)\left(3x^2 + 5x - 2\right) = (x-3)(x-2)(3x-1)(x+2)$. Therefore, the zeros are -2, $\frac{1}{3}$, 2, and 3.

35. Factoring by grouping can be applied to this exercise. $4x^3 + 4x^2 - x - 1 = 4x^2(x+1) - (x+1) = (x+1)\left(4x^2 - 1\right) = (x+1)(2x+1)(2x-1)$. Therefore, the zeros are $x = -1$ and $\pm \frac{1}{2}$.

37. $P(x) = 4x^3 - 7x + 3$. The possible rational zeros are ± 1, ± 3, $\pm \frac{1}{2}$, $\pm \frac{3}{2}$, $\pm \frac{1}{4}$, $\pm \frac{3}{4}$. Since $P(x)$ has 2 variations in sign, there are 0 or 2 positive zeros. Since $P(-x) = -4x^3 + 7x + 3$ has 1 variation in sign, there is 1 negative zero.

$$
\begin{array}{r|rrrr}
\frac{1}{2} & 4 & 0 & -7 & 3 \\
  &   & 2 & 1 & -3 \\
\hline
  & 4 & 2 & -6 & 0
\end{array} \Rightarrow x = \frac{1}{2} \text{ is a zero.}
$$

$P(x) = \left(x - \frac{1}{2}\right)\left(4x^2 + 2x - 6\right) = (2x-1)\left(2x^2 + x - 3\right) = (2x-1)(x-1)(2x+3) = 0$. Thus, the zeros are $x = -\frac{3}{2}, \frac{1}{2}$, and 1.

39. $P(x) = 4x^3 + 8x^2 - 11x - 15$. The possible rational zeros are ± 1, ± 3, ± 5, $\pm\frac{1}{2}$, $\pm\frac{1}{4}$, $\pm\frac{3}{2}$, $\pm\frac{3}{4}$, $\pm\frac{5}{2}$, $\pm\frac{5}{4}$. $P(x)$ has 1 variation in sign and hence 1 positive real zero. $P(-x) = -4x^3 + 8x^2 + 11x - 15$ has 2 variations in sign, so P has 0 or 2 negative real zeros.

$$
\begin{array}{r|rrrr}
1 & 4 & 8 & -11 & -15 \\
  &   & 4 & 12 & 1 \\
\hline
  & 4 & 12 & 1 & -14
\end{array}
\Rightarrow x = 1 \text{ is not a zero.}
\qquad
\begin{array}{r|rrrr}
3 & 4 & 8 & -11 & -15 \\
  &   & 12 & 60 & 147 \\
\hline
  & 4 & 20 & 49 & 132
\end{array}
\Rightarrow x = 3 \text{ is not a zero.}
$$

$$
\begin{array}{r|rrrr}
5 & 4 & 8 & -11 & -15 \\
  &   & 20 & 140 & 645 \\
\hline
  & 4 & 28 & 129 & 630
\end{array}
\Rightarrow x = 5 \text{ is not a zero.}
\qquad
\begin{array}{r|rrrr}
3 & 4 & 8 & -11 & -15 \\
  &   & 12 & 60 & 147 \\
\hline
  & 4 & 20 & 49 & 132
\end{array}
\Rightarrow x = 3 \text{ is not a zero.}
$$

$$
\begin{array}{r|rrrr}
\frac{1}{2} & 4 & 8 & -11 & -15 \\
  &   & 2 & 5 & -3 \\
\hline
  & 4 & 10 & -6 & -18
\end{array}
\Rightarrow x = \tfrac{1}{2} \text{ is not a zero.}
\qquad
\begin{array}{r|rrrr}
\frac{1}{4} & 4 & 8 & -11 & -15 \\
  &   & 1 & \frac{9}{4} & -\frac{35}{16} \\
\hline
  & 4 & 9 & -\frac{35}{4} & -\frac{275}{16}
\end{array}
\Rightarrow x = \tfrac{1}{4} \text{ is not a zero.}
$$

$$
\begin{array}{r|rrrr}
\frac{3}{2} & 4 & 8 & -11 & -15 \\
  &   & 6 & 21 & 15 \\
\hline
  & 4 & 14 & 10 & 0
\end{array}
\Rightarrow x = \tfrac{3}{2} \text{ is a zero.}
$$

Thus $P(x) = 4x^3 + 8x^2 - 11x - 15 = \left(x - \frac{3}{2}\right)\left(4x^2 + 14x + 10\right)$. Continuing by factoring the quotient, whose possible rational zeros are -1, -5, $-\frac{1}{2}$, $-\frac{1}{4}$, $-\frac{5}{2}$, and $-\frac{5}{4}$, we have

$$
\begin{array}{r|rrr}
-1 & 4 & 14 & 10 \\
   &   & -4 & -10 \\
\hline
   & 4 & 10 & 0
\end{array}
\Rightarrow x = -1 \text{ is a zero.}
$$

Thus $P(x) = (2x - 3)(x + 1)(2x + 5)$ has zeros $\frac{3}{2}$, -1, and $-\frac{5}{2}$.

41. $P(x) = 20x^3 - 8x^2 - 5x + 2$. The possible rational zeros are ± 1, ± 2, $\pm\frac{1}{2}$, $\pm\frac{1}{4}$, $\pm\frac{1}{5}$, $\pm\frac{1}{10}$, $\pm\frac{1}{20}$, $\pm\frac{2}{5}$. $P(x)$ has 2 variations in sign and hence 0 or 2 positive real zeros. $P(-x) = -20x^3 - 8x^2 + 5x + 2$ has 1 variations in sign and hence 1 negative real zero.

$$
\begin{array}{r|rrrr}
1 & 20 & -8 & -5 & 2 \\
  &   & 20 & 12 & 7 \\
\hline
  & 20 & 12 & 7 & 9
\end{array}
\Rightarrow x = 1 \text{ is not a zero.}
\qquad
\begin{array}{r|rrrr}
2 & 20 & -8 & -5 & 2 \\
  &   & 40 & 64 & 118 \\
\hline
  & 20 & 32 & 59 & 120
\end{array}
\Rightarrow x = 2 \text{ is not a zero.}
$$

$$
\begin{array}{r|rrrr}
\frac{1}{2} & 20 & -8 & -5 & 2 \\
  &   & 10 & 1 & -2 \\
\hline
  & 20 & 2 & -4 & 0
\end{array}
\Rightarrow x = \tfrac{1}{2} \text{ is a zero.}
$$

Thus, $P(x) = 20x^3 - 8x^2 - 5x + 2 = \left(x - \frac{1}{2}\right)\left(20x^2 + 2x - 4\right)$. Continuing:

$$
\begin{array}{r|rrr}
\frac{1}{4} & 20 & 2 & -4 \\
   &   & 5 & \frac{7}{4} \\
\hline
   & 20 & 7 & -\frac{9}{4}
\end{array}
\Rightarrow x = \tfrac{1}{4} \text{ is not a zero.}
\qquad
\begin{array}{r|rrr}
\frac{1}{5} & 20 & 2 & -4 \\
   &   & 4 & \frac{6}{5} \\
\hline
   & 20 & 6 & -\frac{14}{5}
\end{array}
\Rightarrow x = \tfrac{1}{5} \text{ is not a zero.}
$$

$\dfrac{1}{10}$ | 20   2 -4

      2    $\frac{2}{5}$

20   4 $-\frac{18}{5}$ $\Rightarrow x = \frac{1}{10}$ is not a zero.

$\dfrac{1}{20}$ | 20   2 -4

      1    $\frac{3}{20}$

20   3 $-\frac{77}{20}$ $\Rightarrow x = \frac{1}{20}$ is not a zero.

$\dfrac{2}{5}$ | 20   2 -4

      8   4

20   10   0 $\Rightarrow x = \frac{2}{5}$ is a zero.

Thus, $P(x) = (2x - 1)(5x - 2)(2x + 1)$ has zeros $\frac{1}{2}$, $\frac{2}{5}$, and $-\frac{1}{2}$.

43. $P(x) = 2x^4 - 7x^3 + 3x^2 + 8x - 4$. The possible rational zeros are ± 1, ± 2, ± 4, $\pm\frac{1}{2}$. $P(x)$ has 3 variations in sign and hence 1 or 3 positive real zeros. $P(-x) = 2x^4 + 7x^3 + 3x^2 - 8x - 4$ has 1 variation in sign and hence 1 negative real zero.

1 | 2 -7 3   8 -4

     2   -5 -2 6

2   -5 -2 6   2 $\Rightarrow x = 1$ is not a zero.

$\frac{1}{2}$ | 2 -7 3   8 -4

     1   -3 0   4

2   -6 0   8   0 $\Rightarrow x = \frac{1}{2}$ is a zero.

Thus $P(x) = 2x^4 - 7x^3 + 3x^2 + 8x - 4 = \left(x - \frac{1}{2}\right)\left(2x^3 - 6x^2 + 8\right)$. Continuing by factoring the quotient, we have:

2 | 2 -6 0   8

     4   -4 -8

2   -2 -4 0 $\Rightarrow x = 2$ is a zero.

$P(x) = \left(x - \frac{1}{2}\right)(x - 2)\left(2x^2 - 2x - 4\right) = 2\left(x - \frac{1}{2}\right)(x - 2)\left(x^2 - x - 2\right) = 2\left(x - \frac{1}{2}\right)(x - 2)^2(x + 1)$. Thus, the zeros are $x = \frac{1}{2}$, 2, and -1.

45. $P(x) = x^5 + 3x^4 - 9x^3 - 31x^2 + 36$. The possible rational zeros are ± 1, ± 2, ± 3, ± 4, ± 6, ± 8, ± 9, ± 12, ± 18. $P(x)$ has 2 variations in sign and hence 0 or 2 positive real zeros. $P(-x) = -x^5 + 3x^4 + 9x^3 - 31x^2 + 36$ has 3 variations in sign and hence 1 or 3 negative real zeros.

1 | 1   3 -9 -31 0   36

     1   4 -5 -36 -36

1   4 -5 -36 -36 0 $\Rightarrow x = 1$ is a zero.

So $P(x) = x^5 + 3x^4 - 9x^3 - 31x^2 + 36 = (x - 1)\left(x^4 + 4x^3 - 5x^2 - 36x - 36\right)$. Continuing by factoring the quotient, we have:

1 | 1   4 -5 -36 -36

     1   5   0 -36

1   1   0 -36 -72

2 | 1   4 -5 -36 -36

     2   12   14 -44

1   6   7 -22 -80

3 | 1   4 -5 -36 -36

     3   21   48   36

1   7   16   12   0 $\Rightarrow x = 3$ is a zero.

So $P(x) = (x-1)(x-3)\left(x^3 + 7x^2 + 16x + 12\right)$. Since we have 2 positive zeros, there are no more positive zeros, so we continue by factoring the quotient with possible negative zeros.

$$
\begin{array}{r|rrrr}
-1 & 1 & 7 & 16 & 12 \\
   &   & -1 & -6 & -10 \\
\hline
   & 1 & 6 & 10 & 2
\end{array}
\qquad
\begin{array}{r|rrrr}
-2 & 1 & 7 & 16 & 12 \\
   &   & -2 & -10 & -12 \\
\hline
   & 1 & 5 & 6 & 0
\end{array}
\Rightarrow x = -2 \text{ is a zero.}
$$

Then $P(x) = (x-1)(x-3)(x+2)\left(x^2 + 5x + 6\right) = (x-1)(x-3)(x+2)^2(x+3)$. Thus, the zeros are $x = 1, 3$, -2, and -3.

47. $P(x) = 3x^5 - 14x^4 - 14x^3 + 36x^2 + 43x + 10$ has possible rational zeros $\pm 1, \pm 2, \pm 5, \pm 10, \pm\frac{1}{3}, \pm\frac{2}{3}, \pm\frac{5}{3}, \pm\frac{10}{3}$. Since $P(x)$ has 2 variations in sign, there are 0 or 2 positive real zeros. Since $P(-x) = -3x^5 - 14x^4 + 14x^3 + 36x^2 - 43x + 10$ has 3 variations in sign, there are 1 or 3 negative real zeros.

$$
\begin{array}{r|rrrrrr}
1 & 3 & -14 & -14 & 36 & 43 & 10 \\
  &   & 3 & -11 & -25 & 11 & 54 \\
\hline
  & 3 & -11 & -25 & 11 & 54 & 64
\end{array}
\qquad
\begin{array}{r|rrrrrr}
2 & 3 & -14 & -14 & 36 & 43 & 10 \\
  &   & 6 & -16 & -60 & -48 & -10 \\
\hline
  & 3 & -8 & -30 & -24 & -5 & 0
\end{array}
\Rightarrow x = 2 \text{ is a zero.}
$$

$P(x) = (x-2)\left(3x^4 - 8x^3 - 30x^2 - 24x - 5\right)$

$$
\begin{array}{r|rrrrr}
2 & 3 & -8 & -30 & -24 & -5 \\
  &   & 6 & -4 & -68 & -184 \\
\hline
  & 3 & -2 & -34 & -92 & -189
\end{array}
\qquad
\begin{array}{r|rrrrr}
5 & 3 & -8 & -30 & -24 & -5 \\
  &   & 15 & 35 & 25 & 5 \\
\hline
  & 3 & 7 & 5 & 1 & 0
\end{array}
\Rightarrow x = 5 \text{ is a zero.}
$$

$P(x) = (x-2)(x-5)\left(3x^3 + 7x^2 + 5x + 1\right)$. Since $3x^3 + 7x^2 + 5x + 1$ has no variation in sign, there are no more positive zeros.

$$
\begin{array}{r|rrrr}
-1 & 3 & 7 & 5 & 1 \\
   &   & -3 & -4 & -1 \\
\hline
   & 3 & 4 & 1 & 0
\end{array}
\Rightarrow x = -1 \text{ is a zero.}
$$

$P(x) = (x-2)(x-5)(x+1)\left(3x^2 + 4x + 1\right) = (x-2)(x-5)(x+1)(x+1)(3x+1)$. Therefore, the zeros are $x = -1, -\frac{1}{3}, 2$, and 5.

49. $P(x) = x^3 + 4x^2 + 3x - 2$. The possible rational zeros are $\pm 1, \pm 2$. $P(x)$ has 1 variation in sign and hence 1 positive real zero. $P(-x) = -x^3 + 4x^2 - 3x - 2$ has 2 variations in sign and hence 0 or 2 negative real zeros.

$$
\begin{array}{r|rrrr}
1 & 1 & 4 & 3 & -2 \\
  &   & 1 & 5 & 8 \\
\hline
  & 1 & 5 & 8 & 6
\end{array}
\Rightarrow x = 1 \text{ is an upper bound.}
\qquad
\begin{array}{r|rrrr}
-1 & 1 & 4 & 3 & -2 \\
   &   & -1 & -3 & 0 \\
\hline
   & 1 & 3 & 0 & -2
\end{array}
$$

$$
\begin{array}{r|rrrr}
-2 & 1 & 4 & 3 & -2 \\
   &   & -2 & -4 & 2 \\
\hline
   & 1 & 2 & -1 & 0
\end{array}
\Rightarrow x = -2 \text{ is a zero.}
$$

So $P(x) = (x+2)\left(x^2 + 2x - 1\right)$. Using the quadratic formula on the second factor, we have:

$x = \dfrac{-2 \pm \sqrt{2^2 - 4(1)(-1)}}{2(1)} = \dfrac{-2 \pm \sqrt{8}}{2} = \dfrac{-2 \pm 2\sqrt{2}}{2} = -1 \pm \sqrt{2}$. Therefore, the zeros are $x = -2, -1 + \sqrt{2}$, and $-1 - \sqrt{2}$.

51. $P(x) = x^4 - 6x^3 + 4x^2 + 15x + 4$. The possible rational zeros are $\pm 1, \pm 2, \pm 4$. $P(x)$ has 2 variations in sign and hence 0 or 2 positive real zeros. $P(-x) = x^4 + 6x^3 + 4x^2 - 15x + 4$ has 2 variations in sign and hence 0 or 2 negative real zeros.

$$
\begin{array}{r|rrrrr}
1 & 1 & -6 & 4 & 15 & 4 \\
  &   & 1 & -5 & -1 & 14 \\
\hline
  & 1 & -5 & -1 & 14 & 18
\end{array}
\qquad
\begin{array}{r|rrrrr}
2 & 1 & -6 & 4 & 15 & 4 \\
  &   & 2 & -8 & -8 & 14 \\
\hline
  & 1 & -4 & -4 & 7 & 18
\end{array}
$$

$$
\begin{array}{r|rrrrr}
4 & 1 & -6 & 4 & 15 & 4 \\
  &   & 4 & -8 & -16 & -4 \\
\hline
  & 1 & -2 & -4 & -1 & 0
\end{array}
\Rightarrow x = 4 \text{ is a zero.}
$$

So $P(x) = (x - 4)\left(x^3 - 2x^2 - 4x - 1\right)$. Continuing by factoring the quotient, we have:

$$
\begin{array}{r|rrrr}
4 & 1 & -2 & -4 & -1 \\
  &   & 4 & 8 & 16 \\
\hline
  & 1 & 2 & 4 & 15
\end{array}
\Rightarrow x = 4 \text{ is an upper bound.}
\qquad
\begin{array}{r|rrrr}
-1 & 1 & -2 & -4 & -1 \\
   &   & -1 & 3 & 1 \\
\hline
   & 1 & -3 & -1 & 0
\end{array}
\Rightarrow x = -1 \text{ is a zero.}
$$

So $P(x) = (x - 4)(x + 1)\left(x^2 - 3x - 1\right)$. Using the quadratic formula on the third factor, we have:

$x = \dfrac{-(-3) \pm \sqrt{(-3)^2 - 4(1)(-1)}}{2(1)} = \dfrac{3 \pm \sqrt{13}}{2}$. Therefore, the zeros are $x = 4, -1$, and $\dfrac{3 \pm \sqrt{13}}{2}$.

53. $P(x) = x^4 - 7x^3 + 14x^2 - 3x - 9$. The possible rational zeros are $\pm 1, \pm 3, \pm 9$. $P(x)$ has 3 variations in sign and hence 1 or 3 positive real zeros. $P(-x) = x^4 + 7x^3 + 14x^2 + 3x - 4$ has 1 variation in sign and hence 1 negative real zero.

$$
\begin{array}{r|rrrrr}
1 & 1 & -7 & 14 & -3 & -9 \\
  &   & 1 & -6 & 8 & 5 \\
\hline
  & 1 & -6 & 8 & 5 & 4
\end{array}
\qquad
\begin{array}{r|rrrrr}
3 & 1 & -7 & 14 & -3 & -9 \\
  &   & 3 & -12 & 6 & 9 \\
\hline
  & 1 & -4 & 2 & 3 & 0
\end{array}
\Rightarrow x = 3 \text{ is a zero.}
$$

So $P(x) = (x - 3)\left(x^3 - 4x^2 + 2x + 3\right)$. Since the constant term of the second term is 3, ± 9 are no longer possible zeros.

Continuing by factoring the quotient, we have:
$$
\begin{array}{r|rrrr}
3 & 1 & -4 & 2 & 3 \\
  &   & 3 & -3 & -3 \\
\hline
  & 1 & -1 & -1 & 0
\end{array}
\Rightarrow x = 3 \text{ is a zero again.}
$$

So $P(x) = (x - 3)^2 \left(x^2 - x - 1\right)$. Using the quadratic formula on the second factor, we have:

$x = \dfrac{-(-1) \pm \sqrt{(-1)^2 - 4(1)(-1)}}{2(1)} = \dfrac{1 \pm \sqrt{5}}{2}$. Therefore, the zeros are $x = 3$ and $\dfrac{1 \pm \sqrt{5}}{2}$.

55. $P(x) = 4x^3 - 6x^2 + 1$. The possible rational zeros are $\pm 1, \pm \frac{1}{2}, \pm \frac{1}{4}$. $P(x)$ has 2 variations in sign and hence 0 or 2 positive real zeros. $P(-x) = -4x^3 - 6x^2 + 1$ has 1 variation in sign and hence 1 negative real zero.

$$
\begin{array}{r|rrrr}
1 & 4 & -6 & 0 & 1 \\
  &   & 4 & -2 & -2 \\
\hline
  & 4 & -2 & -2 & -1
\end{array}
\qquad
\begin{array}{r|rrrr}
\frac{1}{2} & 4 & -6 & 0 & 1 \\
  &   & 2 & -2 & -1 \\
\hline
  & 4 & -4 & -2 & 0
\end{array}
\Rightarrow x = \tfrac{1}{2} \text{ is a zero.}
$$

So $P(x) = \left(x - \frac{1}{2}\right)\left(4x^2 - 4x - 2\right)$. Using the quadratic formula on the second factor, we have:

$x = \dfrac{-(-4) \pm \sqrt{(-4)^2 - 4(4)(-2)}}{2(4)} = \dfrac{4 \pm \sqrt{48}}{8} = \dfrac{4 \pm 4\sqrt{3}}{8} = \dfrac{1 \pm \sqrt{3}}{2}$. Therefore, the zeros are $x = \frac{1}{2}$ and $\dfrac{1 \pm \sqrt{3}}{2}$.

57. $P(x) = 2x^4 + 15x^3 + 17x^2 + 3x - 1$. The possible rational zeros are ± 1, $\pm \frac{1}{2}$. $P(x)$ has 1 variation in sign and hence 1 positive real zero. $P(-x) = 2x^4 - 15x^3 + 17x^2 - 3x - 1$ has 3 variations in sign and hence 1 or 3 negative real zeros.

$$
\begin{array}{r|rrrrr}
\frac{1}{2} & 2 & 15 & 17 & 3 & -1 \\
 & & 1 & 8 & \frac{25}{2} & \frac{31}{4} \\
\hline
 & 2 & 16 & 25 & \frac{31}{2} & \frac{27}{4}
\end{array}
\Rightarrow x = \frac{1}{2} \text{ is an upper bound.}
$$

$$
\begin{array}{r|rrrrr}
-\frac{1}{2} & 2 & 15 & 17 & 3 & -1 \\
 & & -1 & -7 & -5 & 1 \\
\hline
 & 2 & 14 & 10 & -2 & 0
\end{array}
\Rightarrow x = -\frac{1}{2} \text{ is a zero.}
$$

So $P(x) = \left(x + \frac{1}{2}\right)\left(2x^3 + 14x^2 + 10x - 2\right) = 2\left(x + \frac{1}{2}\right)\left(x^3 + 7x^2 + 5x - 1\right)$.

$$
\begin{array}{r|rrrr}
-1 & 1 & 7 & 5 & -1 \\
 & & -1 & -6 & 1 \\
\hline
 & 1 & 6 & -1 & 0
\end{array}
\Rightarrow x = -1 \text{ is a zero.}
$$

So $P(x) = \left(x + \frac{1}{2}\right)\left(2x^3 + 14x^2 + 10x - 2\right) = 2\left(x + \frac{1}{2}\right)(x + 1)\left(x^2 + 6x - 1\right)$ Using the quadratic formula on the

third factor, we have $x = \dfrac{-(6) \pm \sqrt{(6)^2 - 4(1)(-1)}}{2(1)} = \dfrac{-6 \pm \sqrt{40}}{2} = \dfrac{-6 \pm 2\sqrt{10}}{2} = -3 \pm \sqrt{10}$. Therefore, the zeros are $x = -1$, $-\frac{1}{2}$, and $-3 \pm \sqrt{10}$.

59. **(a)** $P(x) = x^3 - 3x^2 - 4x + 12$ has possible rational zeros ± 1, ± 2, ± 3, ± 4, ± 6, ± 12.

(b)

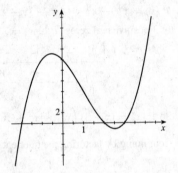

$$
\begin{array}{r|rrrr}
1 & 1 & -3 & -4 & 12 \\
 & & 1 & -2 & -6 \\
\hline
 & 1 & -2 & -6 & 6
\end{array}
$$

$$
\begin{array}{r|rrrr}
2 & 1 & -3 & -4 & 12 \\
 & & 2 & -2 & -12 \\
\hline
 & 1 & -1 & -6 & 0
\end{array}
\Rightarrow x = 2 \text{ is a zero.}
$$

So $P(x) = (x - 2)\left(x^2 - x - 6\right) = (x - 2)(x + 2)(x - 3)$. The real zeros of P are -2, 2, and 3.

61. **(a)** $P(x) = 2x^3 - 7x^2 + 4x + 4$ has possible rational zeros ± 1, ± 2, ± 4, $\pm \frac{1}{2}$.

(b)

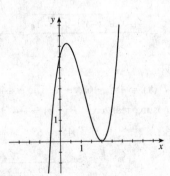

$$
\begin{array}{r|rrrr}
1 & 2 & -7 & 4 & 4 \\
 & & 2 & -5 & -1 \\
\hline
 & 2 & -5 & -1 & -3
\end{array}
\qquad
\begin{array}{r|rrrr}
2 & 2 & -7 & 4 & 4 \\
 & & 4 & -6 & -4 \\
\hline
 & 2 & -3 & -2 & 0
\end{array}
\Rightarrow x = 2 \text{ is a zero.}
$$

So $P(x) = (x - 2)\left(2x^2 - 3x - 2\right)$. Continuing:

$$
\begin{array}{r|rrr}
2 & 2 & -3 & -2 \\
 & & 4 & 2 \\
\hline
 & 2 & 1 & 0
\end{array}
\Rightarrow x = 2 \text{ is a zero again.}
$$

Thus $P(x) = (x - 2)^2 (2x + 1)$. The real zeros of P are 2 and $-\frac{1}{2}$.

63. (a) $P(x) = x^4 - 5x^3 + 6x^2 + 4x - 8$ has possible rational zeros ± 1, ± 2, ± 4, ± 8.

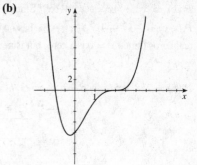

(b)

$$\begin{array}{r|rrrrr} 1 & 1 & -5 & 6 & 4 & -8 \\ & & 1 & -4 & 2 & 6 \\ \hline & 1 & -4 & 2 & 6 & -2 \end{array}$$

$$\begin{array}{r|rrrrr} 2 & 1 & -5 & 6 & 4 & -8 \\ & & 2 & -6 & 0 & 8 \\ \hline & 1 & -3 & 0 & 4 & 0 \end{array} \Rightarrow x = 2 \text{ is a zero.}$$

So $P(x) = (x - 2)\left(x^3 - 3x^2 + 4\right)$ and the possible rational zeros are restricted to -1, ± 2, ± 4.

$$\begin{array}{r|rrrr} 2 & 1 & -3 & 0 & 4 \\ & & 2 & -2 & -4 \\ \hline & 1 & -1 & -2 & 0 \end{array} \Rightarrow x = 2 \text{ is a zero again.}$$

$P(x) = (x - 2)^2 \left(x^2 - x - 2\right) = (x - 2)^2 (x - 2)(x + 1) = (x - 2)^3 (x + 1)$. So the real zeros of P are -1 and 2.

65. (a) $P(x) = x^5 - x^4 - 5x^3 + x^2 + 8x + 4$ has possible rational zeros ± 1, ± 2, ± 4.

(b)

$$\begin{array}{r|rrrrrr} 1 & 1 & -1 & -5 & 1 & 8 & 4 \\ & & 1 & 0 & -5 & -4 & 4 \\ \hline & 1 & 0 & -5 & -4 & 4 & 8 \end{array}$$

$$\begin{array}{r|rrrrrr} 2 & 1 & -1 & -5 & 1 & 8 & 4 \\ & & 2 & 2 & -6 & -10 & -4 \\ \hline & 1 & 1 & -3 & -5 & -2 & 0 \end{array} \Rightarrow x = 2 \text{ is a zero.}$$

So $P(x) = (x - 2)\left(x^4 + x^3 - 3x^2 - 5x - 2\right)$, and the possible rational zeros are restricted to -1, ± 2.

$$\begin{array}{r|rrrrr} 2 & 1 & 1 & -3 & -5 & -2 \\ & & 2 & 6 & 6 & 2 \\ \hline & 1 & 3 & 3 & 1 & 0 \end{array} \Rightarrow x = 2 \text{ is a zero again.}$$

So $P(x) = (x - 2)^2 \left(x^3 + 3x^2 + 3x + 1\right)$, and the possible rational zeros are restricted to -1.

$$\begin{array}{r|rrrr} -1 & 1 & 3 & 3 & 1 \\ & & -1 & -2 & -1 \\ \hline & 1 & 2 & 1 & 0 \end{array} \Rightarrow x = -1 \text{ is a zero.}$$

So $P(x) = (x - 2)^2 (x + 1)\left(x^2 + 2x + 1\right) = (x - 2)^2 (x + 1)^3$., and the real zeros of P are -1 and 2.

67. $P(x) = x^3 - x^2 - x - 3$. Since $P(x)$ has 1 variation in sign, P has 1 positive real zero. Since $P(-x) = -x^3 - x^2 + x - 3$ has 2 variations in sign, P has 2 or 0 negative real zeros. Thus, P has 1 or 3 real zeros.

69. $P(x) = 2x^6 + 5x^4 - x^3 - 5x - 1$. Since $P(x)$ has 1 variation in sign, P has 1 positive real zero. Since $P(-x) = 2x^6 + 5x^4 + x^3 + 5x - 1$ has 1 variation in sign, P has 1 negative real zero. Therefore, P has 2 real zeros.

71. $P(x) = x^5 + 4x^3 - x^2 + 6x$. Since $P(x)$ has 2 variations in sign, P has 2 or 0 positive real zeros. Since $P(-x) = -x^5 - 4x^3 - x^2 - 6x$ has no variation in sign, P has no negative real zero. Therefore, P has a total of 1 or 3 real zeros (since $x = 0$ is a zero, but is neither positive nor negative).

73. $P(x) = 2x^3 + 5x^2 + x - 2$; $a = -3, b = 1$

$$
\begin{array}{r|rrrr}
-3 & 2 & 5 & 1 & -2 \\
   &   & -6 & 3 & -12 \\
\hline
   & 2 & -1 & 4 & -14
\end{array}
$$ alternating signs $\Rightarrow$ lower bound.

$$
\begin{array}{r|rrrr}
1 & 2 & 5 & 1 & -2 \\
  &   & 2 & 7 & 8 \\
\hline
  & 2 & 7 & 8 & 6
\end{array}
$$ all nonnegative $\Rightarrow$ upper bound.

Therefore $a = -3$ and $b = 1$ are lower and upper bounds.

75. $P(x) = 8x^3 + 10x^2 - 39x + 9$; $a = -3, b = 2$

$$
\begin{array}{r|rrrr}
-3 & 8 & 10 & -39 & 9 \\
   &   & -24 & 42 & -9 \\
\hline
   & 8 & -14 & 3 & 0
\end{array}
$$ alternating signs $\Rightarrow$ lower bound.

$$
\begin{array}{r|rrrr}
2 & 8 & 10 & -39 & 9 \\
  &   & 16 & 52 & 26 \\
\hline
  & 8 & 26 & 13 & 35
\end{array}
$$ all nonnegative $\Rightarrow$ upper bound.

Therefore $a = -3$ and $b = 2$ are lower and upper bounds. Note that $x = -3$ is also a zero.

77. $P(x) = x^4 + 2x^3 + 3x^2 + 5x - 1$; $a = -2, b = 1$

$$
\begin{array}{r|rrrrr}
-2 & 1 & 2 & 3 & 5 & -1 \\
   &   & -2 & 0 & -6 & 2 \\
\hline
   & 1 & 0 & 3 & -1 & 1
\end{array}
$$ Alternating signs $\Rightarrow$ lower bound.

$$
\begin{array}{r|rrrrr}
1 & 1 & 2 & 3 & 5 & -1 \\
  &   & 1 & 3 & 6 & 11 \\
\hline
  & 1 & 3 & 6 & 11 & 10
\end{array}
$$ All nonnegative $\Rightarrow$ upper bound.

Therefore $a = -2$ and $b = 1$ are lower and upper bounds.

79. $P(x) = 2x^4 - 6x^3 + x^2 - 2x + 3$; $a = -1, b = 3$

$$
\begin{array}{r|rrrrr}
-1 & 2 & -6 & 1 & -2 & 3 \\
   &   & -2 & 8 & -9 & 11 \\
\hline
   & 2 & -8 & 9 & -11 & 14
\end{array}
$$ Alternating signs $\Rightarrow$ lower bound.

$$
\begin{array}{r|rrrrr}
3 & 2 & -6 & 1 & -2 & 3 \\
  &   & 6 & 0 & 3 & 3 \\
\hline
  & 2 & 0 & 1 & 1 & 6
\end{array}
$$ All nonnegative $\Rightarrow$ upper bound.

Therefore $a = -1$ and $b = 3$ are lower and upper bounds.

81. $P(x) = x^3 - 3x^2 + 4$ and use the Upper and Lower Bounds Theorem:

$$
\begin{array}{r|rrrr}
-1 & 1 & -3 & 0 & 4 \\
  &   & -1 & 4 & -4 \\
\hline
  & 1 & -4 & 4 & 0
\end{array}
\quad \text{alternating signs} \Rightarrow \text{lower bound.}
$$

$$
\begin{array}{r|rrrr}
3 & 1 & -3 & 0 & 4 \\
  &   & 3 & 0 & 0 \\
\hline
  & 1 & 0 & 0 & 4
\end{array}
\quad \text{all nonnegative} \Rightarrow \text{upper bound.}
$$

Therefore -1 is a lower bound (and a zero) and 3 is an upper bound. (There are many possible solutions.)

83. $P(x) = x^4 - 2x^3 + x^2 - 9x + 2$.

$$
\begin{array}{r|rrrrr}
1 & 1 & -2 & 1 & -9 & 2 \\
  &   & 1 & -1 & 0 & -9 \\
\hline
  & 1 & -1 & 0 & -9 & -7
\end{array}
\qquad
\begin{array}{r|rrrrr}
2 & 1 & -2 & 1 & -9 & 2 \\
  &   & 2 & 0 & 2 & -14 \\
\hline
  & 1 & 0 & 1 & -7 & -12
\end{array}
$$

$$
\begin{array}{r|rrrrr}
3 & 1 & -2 & 1 & -9 & 2 \\
  &   & 3 & 3 & 12 & 9 \\
\hline
  & 1 & 1 & 4 & 3 & 11
\end{array}
\quad \text{all positive} \Rightarrow \text{upper bound.}
$$

$$
\begin{array}{r|rrrrr}
-1 & 1 & -2 & 1 & -9 & 2 \\
  &   & -1 & 3 & -4 & 13 \\
\hline
  & 1 & -3 & 4 & -13 & 15
\end{array}
\quad \text{alternating signs} \Rightarrow \text{lower bound.}
$$

Therefore -1 is a lower bound and 3 is an upper bound. (There are many possible solutions.)

85. $P(x) = 2x^4 + 3x^3 - 4x^2 - 3x + 2$.

$$
\begin{array}{r|rrrrr}
1 & 2 & 3 & -4 & -3 & 2 \\
  &   & 2 & 5 & 1 & -2 \\
\hline
  & 2 & 5 & 1 & -2 & 0
\end{array}
\quad \Rightarrow x = 1 \text{ is a zero.}
$$

$P(x) = (x - 1)\left(2x^3 + 5x^2 + x - 2\right)$

$$
\begin{array}{r|rrrr}
-1 & 2 & 5 & 1 & -2 \\
  &   & -2 & -3 & 2 \\
\hline
  & 2 & 3 & -2 & 0
\end{array}
\quad \Rightarrow x = -1 \text{ is a zero.}
$$

$P(x) = (x - 1)(x + 1)\left(2x^2 + 3x - 2\right) = (x - 1)(x + 1)(2x - 1)(x + 2)$. Therefore, the zeros are $x = -2, \frac{1}{2}, \pm 1$.

87. *Method 1:* $P(x) = 4x^4 - 21x^2 + 5$ has 2 variations in sign, so by Descartes' rule of signs there are either 2 or 0 positive zeros. If we replace x with $(-x)$, the function does not change, so there are either 2 or 0 negative zeros. Possible rational zeros are $\pm 1, \pm \frac{1}{2}, \pm \frac{1}{4}, \pm 5, \pm \frac{5}{2}, \pm \frac{5}{4}$. By inspection, ± 1 and ± 5 are not zeros, so we must look for non-integer solutions:

$$\begin{array}{r|rrrrr} \frac{1}{2} & 4 & 0 & -21 & 0 & 5 \\ & & 2 & 1 & -10 & -5 \\ \hline & 4 & 2 & -20 & -10 & 0 \end{array} \Rightarrow x = \frac{1}{2} \text{ is a zero.}$$

$P(x) = \left(x - \frac{1}{2}\right)\left(4x^3 + 2x^2 - 20x - 10\right)$, continuing with the quotient, we have:

$$\begin{array}{r|rrrr} -\frac{1}{2} & 4 & 2 & -20 & -10 \\ & & -2 & 0 & 10 \\ \hline & 4 & 0 & -20 & 0 \end{array} \Rightarrow x = -\frac{1}{2} \text{ is a zero.}$$

$P(x) = \left(x - \frac{1}{2}\right)\left(x + \frac{1}{2}\right)\left(4x^2 - 20\right) = 0$. If $4x^2 - 20 = 0$, then $x = \pm\sqrt{5}$. Thus the zeros are $x = \pm\frac{1}{2}, \pm\sqrt{5}$.

Method 2: Substituting $u = x^2$, the equation becomes $4u^2 - 21u + 5 = 0$, which factors:

$4u^2 - 21u + 5 = (4u - 1)(u - 5) = \left(4x^2 - 1\right)\left(x^2 - 5\right)$. Then either we have $x^2 = 5$, so that $x = \pm\sqrt{5}$, or we have

$x^2 = \frac{1}{4}$, so that $x = \pm\sqrt{\frac{1}{4}} = \pm\frac{1}{2}$. Thus the zeros are $x = \pm\frac{1}{2}, \pm\sqrt{5}$.

89. $P(x) = x^5 - 7x^4 + 9x^3 + 23x^2 - 50x + 24$. The possible rational zeros are $\pm 1, \pm 2, \pm 3, \pm 4, \pm 6, \pm 8, \pm 12, \pm 24$. $P(x)$ has 4 variations in sign and hence 0, 2, or 4 positive real zeros. $P(-x) = -x^5 - 7x^4 - 9x^3 + 23x^2 + 50x + 24$ has 1 variation in sign, and hence 1 negative real zero.

$$\begin{array}{r|rrrrrr} 1 & 1 & -7 & 9 & 23 & -50 & 24 \\ & & 1 & -6 & 3 & 26 & -24 \\ \hline & 1 & -6 & 3 & 26 & -24 & 0 \end{array} \Rightarrow x = 1 \text{ is a zero.}$$

$P(x) = (x - 1)\left(x^4 - 6x^3 + 3x^2 + 26x - 24\right)$; continuing with the quotient, we try 1 again.

$$\begin{array}{r|rrrrr} 1 & 1 & -6 & 3 & 26 & -24 \\ & & 1 & -5 & -2 & 24 \\ \hline & 1 & -5 & -2 & 24 & 0 \end{array} \Rightarrow x = 1 \text{ is a zero again.}$$

$P(x) = (x - 1)^2\left(x^3 - 5x^2 - 2x + 24\right)$; continuing with the quotient, we start by trying 1 again.

$$\begin{array}{r|rrrr} 1 & 1 & -5 & -2 & 24 \\ & & 1 & -4 & -6 \\ \hline & 1 & -4 & -6 & 18 \end{array} \qquad \begin{array}{r|rrrr} 2 & 1 & -5 & -2 & 24 \\ & & 2 & -6 & -16 \\ \hline & 1 & -3 & -8 & 8 \end{array} \qquad \begin{array}{r|rrrr} 3 & 1 & -5 & -2 & 24 \\ & & 3 & -6 & -24 \\ \hline & 1 & -2 & -8 & 0 \end{array} \Rightarrow x = 3 \text{ is a zero.}$$

$P(x) = (x - 1)^2 (x - 3)\left(x^2 - 2x - 8\right) = (x - 1)^2 (x - 3)(x - 4)(x + 2)$. Therefore, the zeros are $x = -2, 1, 3, 4$.

91. $P(x) = x^3 - x - 2$. The only possible rational zeros of $P(x)$ are ± 1 and ± 2.

$$\begin{array}{r|rrrr} 1 & 1 & 0 & -1 & -2 \\ & & 1 & 1 & 0 \\ \hline & 1 & 1 & 0 & -2 \end{array} \qquad \begin{array}{r|rrrr} 2 & 1 & 0 & -1 & -2 \\ & & 2 & 4 & 6 \\ \hline & 1 & 2 & 3 & 4 \end{array} \qquad \begin{array}{r|rrrr} -1 & 1 & 0 & -1 & -2 \\ & & -1 & 1 & 0 \\ \hline & 1 & -1 & 0 & -2 \end{array}$$

Since the row that contains -1 alternates between nonnegative and nonpositive, -1 is a lower bound and there is no need to try -2. Therefore, $P(x)$ does not have any rational zeros.

93. $P(x) = 3x^3 - x^2 - 6x + 12$ has possible rational zeros $\pm 1, \pm 2, \pm 3, \pm 4, \pm 6, \pm 12, \pm\frac{1}{3}, \pm\frac{2}{3}, \pm\frac{4}{3}$.

	3	−1	−6	12
1	3	2	−4	8
2	3	5	4	20
−1	3	−4	−2	14
−2	3	−7	8	−4

all positive $\Rightarrow x = 2$ is an upper bound

alternating signs $\Rightarrow x = -2$ is a lower bound

	3	−1	−6	12
$\frac{1}{3}$	3	0	−6	10
$\frac{2}{3}$	3	1	$-\frac{16}{3}$	$\frac{76}{9}$
$\frac{4}{3}$	3	3	−2	$\frac{28}{3}$
$-\frac{1}{3}$	3	−2	$-\frac{16}{3}$	$\frac{124}{9}$
$-\frac{2}{3}$	3	−3	−4	$\frac{44}{3}$
$-\frac{4}{3}$	3	−5	$\frac{2}{3}$	$\frac{100}{9}$

Therefore, there is no rational zero.

95. $P(x) = x^3 - 3x^2 - 4x + 12$, $[-4, 4]$ by $[-15, 15]$. The possible rational zeros are $\pm 1, \pm 2, \pm 3, \pm 4, \pm 6, \pm 12$. By observing the graph of P, the rational zeros are $x = -2, 2, 3$.

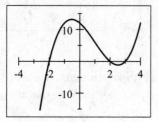

97. $P(x) = 2x^4 - 5x^3 - 14x^2 + 5x + 12$, $[-2, 5]$ by $[-40, 40]$. The possible rational zeros are $\pm 1, \pm 2, \pm 3, \pm 4, \pm 6, \pm 12, \pm\frac{1}{2}, \pm\frac{3}{2}$. By observing the graph of P, the zeros are $x = -\frac{3}{2}, -1, 1, 4$.

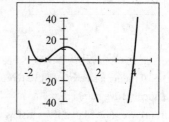

99. $x^4 - x - 4 = 0$. Possible rational solutions are $\pm 1, \pm 2, \pm 4$.

	1	0	0	−1	−4
1		1	1	1	0
	1	1	1	0	−4

	1	0	0	−1	−4
2		2	4	8	14
	1	2	4	7	10

$\Rightarrow x = 2$ is an upper bound.

	1	0	0	−1	−4
−1		−1	1	−1	2
	1	−1	1	−2	−2

	1	0	0	−1	−4
−2		−2	4	−8	18
	1	−2	4	−9	14

$\Rightarrow x = -2$ is a lower bound.

Therefore, we graph the function $P(x) = x^4 - x - 4$ in the viewing rectangle $[-2, 2]$ by $[-5, 20]$ and see there are two solutions. In the viewing rectangle $[-1.3, -1.25]$ by $[-0.1, 0.1]$, we find the solution $x \approx -1.28$. In the viewing rectangle $[1.5. 1.6]$ by $[-0.1, 0.1]$, we find the solution $x \approx 1.53$. Thus the solutions are $x \approx -1.28, 1.53$.

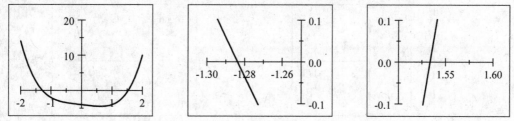

101. $4.00x^4 + 4.00x^3 - 10.96x^2 - 5.88x + 9.09 = 0.$

$$
\begin{array}{r|rrrrr}
1 & 4 & 4 & -10.96 & -5.88 & 9.09 \\
  &   & 4 & 8 & -2.96 & -8.84 \\
\hline
  & 4 & 8 & -2.96 & -8.84 & 0.25
\end{array}
\qquad
\begin{array}{r|rrrrr}
2 & 4 & 4 & -10.96 & -5.88 & 9.09 \\
  &   & 8 & 24 & 26.08 & 40.40 \\
\hline
  & 4 & 12 & 13.04 & 20.2 & 49.49 \Rightarrow x = 2 \text{ is an upper bound.}
\end{array}
$$

$$
\begin{array}{r|rrrrr}
-2 & 4 & 4 & -10.96 & -5.88 & 9.09 \\
   &   & -8 & 8 & 5.92 & -0.08 \\
\hline
   & 4 & -4 & -2.96 & 0.04 & 9.01
\end{array}
\qquad
\begin{array}{r|rrrrr}
-3 & 4 & 4 & -10.96 & -5.88 & 9.09 \\
   &   & -12 & 24 & -39.12 & 135 \\
\hline
   & 4 & -8 & 13.04 & -45 & 144.09 \Rightarrow x = -3 \text{ is a lower bound.}
\end{array}
$$

Therefore, we graph the function $P(x) = 4.00x^4 + 4.00x^3 - 10.96x^2 - 5.88x + 9.09$ in the viewing rectangle $[-3, 2]$ by $[-10, 40]$. There appear to be two solutions. In the viewing rectangle $[-1.6, -1.4]$ by $[-0.1, 0.1]$, we find the solution $x \approx -1.50$. In the viewing rectangle $[0.8, 1.2]$ by $[0, 1]$, we see that the graph comes close but does not go through the x-axis. Thus there is no solution here. Therefore, the only solution is $x \approx -1.50$.

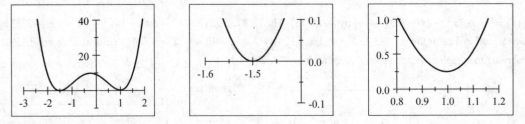

103. (a) Since $z > b$, we have $z - b > 0$. Since all the coefficients of $Q(x)$ are nonnegative, and since $z > 0$, we have $Q(z) > 0$ (being a sum of positive terms). Thus, $P(z) = (z - b) \cdot Q(z) + r > 0$, since the sum of a positive number and a nonnegative number.

(b) In part (a), we showed that if b satisfies the conditions of the first part of the Upper and Lower Bounds Theorem and $z > b$, then $P(z) > 0$. This means that no real zero of P can be larger than b, so b is an upper bound for the real zeros.

(c) Suppose $-b$ is a negative lower bound for the real zeros of $P(x)$. Then clearly b is an upper bound for $P_1(x) = P(-x)$. Thus, as in Part (a), we can write $P_1(x) = (x - b) \cdot Q(x) + r$, where $r > 0$ and the coefficients of Q are all nonnegative, and $P(x) = P_1(-x) = (-x - b) \cdot Q(-x) + r = (x + b) \cdot [-Q(-x)] + r$. Since the coefficients of $Q(x)$ are all nonnegative, the coefficients of $-Q(-x)$ will be alternately nonpositive and nonnegative, which proves the second part of the Upper and Lower Bounds Theorem.

105. Let r be the radius of the silo. The volume of the hemispherical roof is $\frac{1}{2}\left(\frac{4}{3}\pi r^3\right) = \frac{2}{3}\pi r^3$. The volume of the cylindrical section is $\pi\left(r^2\right)(30) = 30\pi r^2$. Because the total volume of the silo is $15{,}000 \text{ ft}^3$, we get the following equation: $\frac{2}{3}\pi r^3 + 30\pi r^2 = 15000 \Leftrightarrow \frac{2}{3}\pi r^3 + 30\pi r^2 - 15000 = 0 \Leftrightarrow \pi r^3 + 45\pi r^2 - 22500 = 0$. Using a graphing device, we first graph the polynomial in the viewing rectangle $[0, 15]$ by $[-10000, 10000]$. The solution, $r \approx 11.28$ ft., is shown in the viewing rectangle $[11.2, 11.4]$ by $[-1, 1]$.

107. $h(t) = 11.60t - 12.41t^2 + 6.20t^3$
$$- 1.58t^4 + 0.20t^5 - 0.01t^6$$
is shown in the viewing rectangle
$[0, 10]$ by $[0, 6]$.

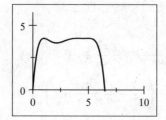

(a) It started to snow again.

(b) No, $h(t) \leq 4$.

(c) The function $h(t)$ is shown in the viewing rectangle $[6, 6.5]$ by $[0, 0.5]$. The x-intercept of the function is a little less than 6.5, which means that the snow melted just before midnight on Saturday night.

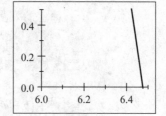

109. Let r be the radius of the cone and cylinder and let h be the height of the cone. Since the height and diameter are equal, we get $h = 2r$. So the volume of the cylinder is $V_1 = \pi r^2 \cdot \text{(cylinder height)} = 20\pi r^2$, and the volume of the cone is $V_2 = \frac{1}{3}\pi r^2 h = \frac{1}{3}\pi r^2 (2r) = \frac{2}{3}\pi r^3$. Since the total volume is $\dfrac{500\pi}{3}$, it follows that $\frac{2}{3}\pi r^3 + 20\pi r^2 = \dfrac{500\pi}{3} \Leftrightarrow r^3 + 30r^2 - 250 = 0$. By Descartes' Rule of Signs, there is 1 positive zero. Since r is between 2.76 and 2.765 (see the table), the radius should be 2.76 m (correct to two decimals).

r	$r^3 + 30r^2 - 250$
1	-219
2	-122
3	47
2.7	-11.62
2.76	-2.33
2.77	1.44
2.765	1.44
2.8	7.15

111. Let b be the width of the base, and let l be the length of the box. Then the length plus girth is $l + 4b = 108$, and the volume is $V = lb^2 = 2200$. Solving the first equation for l and substituting this value into the second equation yields $l = 108 - 4b$
$\Rightarrow V = (108 - 4b)b^2 = 2200 \Leftrightarrow 4b^3 - 108b^2 + 2200 = 0 \Leftrightarrow 4\left(b^3 - 27b^2 + 550\right) = 0$. Now $P(b) = b^3 - 27b^2 + 550$ has two variations in sign, so there are 0 or 2 positive real zeros. We also observe that since $l > 0$, $b < 27$, so $b = 27$ is an upper bound. Thus the possible positive rational real zeros are 1, 2, 3, 10, 11, 22, 25.

$$\begin{array}{r|rrrr} 1 & 1 & -27 & 0 & 550 \\ & & 1 & -26 & -26 \\ \hline & 1 & -26 & -26 & 524 \end{array} \qquad \begin{array}{r|rrrr} 2 & 1 & -27 & 0 & 550 \\ & & 2 & -50 & -100 \\ \hline & 1 & -25 & -50 & 450 \end{array}$$

$$\begin{array}{r|rrrr} 5 & 1 & -27 & 0 & 550 \\ & & 5 & -110 & -550 \\ \hline & 1 & -22 & -110 & 0 \end{array} \Rightarrow b = 5 \text{ is a zero.}$$

$P(b) = (b - 5)\left(b^2 - 22b - 110\right)$. The other zeros are $b = \dfrac{22 \pm \sqrt{484 - 4(1)(-110)}}{2} = \dfrac{22 \pm \sqrt{924}}{2} = \dfrac{22 \pm 30.397}{2}$. The positive answer from this factor is $b \approx 26.20$. Thus we have two possible solutions, $b = 5$ or $b \approx 26.20$. If $b = 5$, then $l = 108 - 4(5) = 88$; if $b \approx 26.20$, then $l = 108 - 4(26.20) = 3.20$. Thus the length of the box is either 88 in. or 3.20 in.

113. (a) Substituting $X - \frac{a}{3}$ for x we have

$$x^3 + ax^2 + bx + c = \left(X - \frac{a}{3}\right)^3 + a\left(X - \frac{a}{3}\right)^2 + b\left(X - \frac{a}{3}\right) + c$$

$$= X^3 - aX^2 + \frac{a^2}{3}X + \frac{a^3}{27} + a\left(X^2 - \frac{2a}{3}X + \frac{a^2}{9}\right) + bX - \frac{ab}{3} + c$$

$$= X^3 - aX^2 + \frac{a^2}{3}X + \frac{a^3}{27} + aX^2 - \frac{2a^2}{3}X + \frac{a^3}{9} + bX - \frac{ab}{3} + c$$

$$= X^3 + (-a + a)X^2 + \left(-\frac{a^2}{3} - \frac{2a^2}{3} + b\right)X + \left(\frac{a^3}{27} + \frac{a^3}{9} - \frac{ab}{3} + c\right)$$

$$= X^3 + \left(b - a^2\right)X + \left(\frac{4a^3}{27} - \frac{ab}{3} + c\right)$$

(b) $x^3 + 6x^2 + 9x + 4 = 0$. Setting $a = 6$, $b = 9$, and $c = 4$, we have: $X^3 + \left(9 - 6^2\right)X + (32 - 18 + 4) = X^3 - 27X + 18$.

3.5   COMPLEX NUMBERS

1. The imaginary number i has the property that $i^2 = -1$.

3. (a) The complex conjugate of $3 + 4i$ is $\overline{3 + 4i} = 3 - 4i$.

   (b) $(3 + 4i)\left(\overline{3 + 4i}\right) = 3^2 + 4^2 = 25$

5. $5 - 7i$: real part 5, imaginary part -7.

7. $\dfrac{-2 - 5i}{3} = -\dfrac{2}{3} - \dfrac{5}{3}i$: real part $-\dfrac{2}{3}$, imaginary part $-\dfrac{5}{3}$.

9. 3: real part 3, imaginary part 0.

11. $-\dfrac{2}{3}i$: real part 0, imaginary part $-\dfrac{2}{3}$.

13. $\sqrt{3} + \sqrt{-4} = \sqrt{3} + 2i$: real part $\sqrt{3}$, imaginary part 2.

15. $(3 + 2i) + 5i = 3 + (2 + 5)i = 3 + 7i$

17. $(2 - 5i) + (3 + 4i) = (2 + 3) + (-5 + 4)i = 5 - i$

19. $(-6 + 6i) + (9 - i) = (-6 + 9) + (6 - 1)i = 3 + 5i$

21. $\left(7 - \frac{1}{2}i\right) - \left(5 + \frac{3}{2}i\right) = (7 - 5) + \left(-\frac{1}{2} - \frac{3}{2}\right)i = 2 - 2i$

23. $(-12 + 8i) - (7 + 4i) = -12 + 8i - 7 - 4i = (-12 - 7) + (8 - 4)i = -19 + 4i$

25. $4(-1 + 2i) = -4 + 8i$

27. $-3i(5 - i) = -15i + 3i^2 = -3 - 15i$

29. $(7 - i)(4 + 2i) = 28 + 14i - 4i - 2i^2 = (28 + 2) + (14 - 4)i = 30 + 10i$

31. $(3 - 4i)(5 - 12i) = 15 - 36i - 20i + 48i^2 = (15 - 48) + (-36 - 20)i = -33 - 56i$

33. $(6 + 5i)(2 - 3i) = 12 - 18i + 10i - 15i^2 = (12 + 15) + (-18 + 10)i = 27 - 8i$

35. $\dfrac{1}{i} = \dfrac{1}{i} \cdot \dfrac{i}{i} = \dfrac{i}{i^2} = \dfrac{i}{-1} = -i$

37. $\dfrac{2 - 3i}{1 - 2i} = \dfrac{2 - 3i}{1 - 2i} \cdot \dfrac{1 + 2i}{1 + 2i} = \dfrac{2 + 4i - 3i - 6i^2}{1 - 4i^2} = \dfrac{(2 + 6) + (4 - 3)i}{1 + 4} = \dfrac{8 + i}{5}$ or $\dfrac{8}{5} + \dfrac{1}{5}i$

39. $\dfrac{26 + 39i}{2 - 3i} = \dfrac{26 + 39i}{2 - 3i} \cdot \dfrac{2 + 3i}{2 + 3i} = \dfrac{52 + 78i + 78i + 117i^2}{4 - 9i^2} = \dfrac{(52 - 117) + (78 + 78)i}{4 + 9} = \dfrac{-65 + 156i}{13}$

   $= \dfrac{13(-5 + 12i)}{13} = -5 + 12i$

41. $\dfrac{10i}{1 - 2i} = \dfrac{10i}{1 - 2i} \cdot \dfrac{1 + 2i}{1 + 2i} = \dfrac{10i + 20i^2}{1 - 4i^2} = \dfrac{-20 + 10i}{1 + 4} = \dfrac{5(-4 + 2i)}{5} = -4 + 2i$

43. $\dfrac{4+6i}{3i} = \dfrac{4+6i}{3i} \cdot \dfrac{3i}{3i} = \dfrac{12i+18i^2}{9i^2} = \dfrac{-18+12i}{-9} = \dfrac{-18}{-9} + \dfrac{12}{-9}i = 2 - \dfrac{4}{3}i$

45. $\dfrac{1}{1+i} - \dfrac{1}{1-i} = \dfrac{1}{1+i} \cdot \dfrac{1-i}{1-i} - \dfrac{1}{1-i} \cdot \dfrac{1+i}{1+i} = \dfrac{1-i}{1-i^2} - \dfrac{1+i}{1-i^2} = \dfrac{1-i}{2} + \dfrac{-1-i}{2} = -i$

47. $i^3 = i^2 i = -i$

49. $(3i)^5 = 3^5 \left(i^2\right)^2 i = 243\,(-1)^2\,i = 243i$

51. $i^{1000} = \left(i^4\right)^{250} = 1^{250} = 1$

53. $\sqrt{-25} = 5i$

55. $\sqrt{-3}\sqrt{-12} = i\sqrt{3} \cdot 2i\sqrt{3} = 6i^2 = -6$

57. $\left(3 - \sqrt{-5}\right)\left(1 + \sqrt{-1}\right) = \left(3 - i\sqrt{5}\right)(1+i) = 3 + 3i - i\sqrt{5} - i^2\sqrt{5} = \left(3 + \sqrt{5}\right) + \left(3 - \sqrt{5}\right)i$

59. $\dfrac{2+\sqrt{-8}}{1+\sqrt{-2}} = \dfrac{2+2i\sqrt{2}}{1+i\sqrt{2}} = \dfrac{2\left(1+i\sqrt{2}\right)}{1+i\sqrt{2}} = 2$

61. $\dfrac{\sqrt{-36}}{\sqrt{-2}\sqrt{-9}} = \dfrac{6i}{i\sqrt{2}\cdot 3i} = \dfrac{2}{i\sqrt{2}} \cdot \dfrac{i\sqrt{2}}{i\sqrt{2}} = \dfrac{2i\sqrt{2}}{2i^2} = \dfrac{i\sqrt{2}}{-1} = -i\sqrt{2}$

63. $x^2 + 49 = 0 \Leftrightarrow x^2 = -49 \Rightarrow x = \pm 7i$

65. $x^2 - 4x + 5 = 0 \Rightarrow x = \dfrac{-(-4) \pm \sqrt{(-4)^2 - 4(1)(5)}}{2(1)} = \dfrac{4 \pm \sqrt{16-20}}{2} = \dfrac{4 \pm \sqrt{-4}}{2} = \dfrac{4 \pm 2i}{2} = 2 \pm i$

67. $x^2 + 2x + 5 = 0 \Rightarrow x = \dfrac{-(2) \pm \sqrt{(2)^2 - 4(1)(5)}}{2(1)} = \dfrac{-2 \pm \sqrt{4-20}}{2} = \dfrac{-2 \pm \sqrt{-16}}{2} = \dfrac{-2 \pm 4i}{2} = -1 \pm 2i$

69. $x^2 + x + 1 = 0 \Rightarrow x = \dfrac{-(1) \pm \sqrt{(1)^2 - 4(1)(1)}}{2(1)} = \dfrac{-1 \pm \sqrt{1-4}}{2} = \dfrac{-1 \pm \sqrt{-3}}{2} = \dfrac{-1 \pm i\sqrt{3}}{2} = -\dfrac{1}{2} \pm \dfrac{\sqrt{3}}{2}i$

71. $2x^2 - 2x + 1 = 0 \Rightarrow x = \dfrac{-(-2) \pm \sqrt{(-2)^2 - 4(2)(1)}}{2(2)} = \dfrac{2 \pm \sqrt{4-8}}{4} = \dfrac{2 \pm \sqrt{-4}}{4} = \dfrac{2 \pm 2i}{4} = \dfrac{1}{2} \pm \dfrac{1}{2}i$

73. $t + 3 + \dfrac{3}{t} = 0 \Leftrightarrow t^2 + 3t + 3 = 0 \Rightarrow t = \dfrac{-(3) \pm \sqrt{(3)^2 - 4(1)(3)}}{2(1)} = \dfrac{-3 \pm \sqrt{9-12}}{2} = \dfrac{-3 \pm \sqrt{-3}}{2} = \dfrac{-3 \pm i\sqrt{3}}{2} = -\dfrac{3}{2} \pm \dfrac{\sqrt{3}}{2}i$

75. $6x^2 + 12x + 7 = 0 \Rightarrow$

$x = \dfrac{-(12) \pm \sqrt{(12)^2 - 4(6)(7)}}{2(6)} = \dfrac{-12 \pm \sqrt{144-168}}{12} = \dfrac{-12 \pm \sqrt{-24}}{12} = \dfrac{-12 \pm 2i\sqrt{6}}{12} = \dfrac{-12}{12} \pm \dfrac{2i\sqrt{6}}{12} = -1 \pm \dfrac{\sqrt{6}}{6}i$

77. $\dfrac{1}{2}x^2 - x + 5 = 0 \Rightarrow x = \dfrac{-(-1) \pm \sqrt{(-1)^2 - 4\left(\frac{1}{2}\right)(5)}}{2\left(\frac{1}{2}\right)} = \dfrac{1 \pm \sqrt{1-10}}{1} = 1 \pm \sqrt{-9} = 1 \pm 3i$

79. LHS $= \bar{z} + \bar{w} = \overline{(a+bi)} + \overline{(c+di)} = a - bi + c - di = (a+c) + (-b-d)i = (a+c) - (b+d)i$.

RHS $= \overline{z+w} = \overline{(a+bi) + (c+di)} = \overline{(a+c) + (b+d)i} = (a+c) - (b+d)i$.

Since LHS $=$ RHS, this proves the statement.

81. LHS $= (\bar{z})^2 = \left(\overline{(a+bi)}\right)^2 = (a-bi)^2 = a^2 - 2abi + b^2i^2 = \left(a^2 - b^2\right) - 2abi$.

RHS $= \overline{z^2} = \overline{(a+bi)^2} = \overline{a^2 + 2abi + b^2i^2} = \overline{\left(a^2 - b^2\right) + 2abi} = \left(a^2 - b^2\right) - 2abi$.

Since LHS $=$ RHS, this proves the statement.

83. $z + \bar{z} = (a+bi) + \overline{(a+bi)} = a + bi + a - bi = 2a$, which is a real number.

85. $z \cdot \bar{z} = (a+bi) \cdot \overline{(a+bi)} = (a+bi) \cdot (a-bi) = a^2 - b^2i^2 = a^2 + b^2$, which is a real number.

87. Using the quadratic formula, the solutions to the equation are $x = \dfrac{-b \pm \sqrt{b^2 - 4ac}}{2a}$. Since both solutions are imaginary,

we have $b^2 - 4ac < 0 \Leftrightarrow 4ac - b^2 > 0$, so the solutions are $x = \dfrac{-b}{2a} \pm \dfrac{\sqrt{4ac - b^2}}{2a}\, i$, where $\sqrt{4ac - b^2}$ is a real number.

Thus the solutions are complex conjugates of each other.

3.6   COMPLEX ZEROS AND THE FUNDAMENTAL THEOREM OF ALGEBRA

1. The polynomial $P(x) = 3(x - 5)^3 (x - 3)(x + 2)$ has degree 5. It has zeros 5, 3, and -2. The zero 5 has multiplicity 3 and the zero 3 has multiplicity 1.

3. A polynomial of degree $n \geq 1$ has exactly n zeros, if a zero of multiplicity m is counted m times.

5. (a) $x^4 + 4x^2 = 0 \Leftrightarrow x^2 \left(x^2 + 4\right) = 0$. So $x = 0$ or $x^2 + 4 = 0$. If $x^2 + 4 = 0$ then $x^2 = -4 \Leftrightarrow x = \pm 2i$. Therefore, the solutions are $x = 0$ and $\pm 2i$.

(b) To get the complete factorization, we factor the remaining quadratic factor $P(x) = x^2(x + 4) = x^2(x - 2i)(x + 2i)$.

7. (a) $x^3 - 2x^2 + 2x = 0 \Leftrightarrow x\left(x^2 - 2x + 2\right) = 0$. So $x = 0$ or $x^2 - 2x + 2 = 0$. If $x^2 - 2x + 2 = 0$ then

$x = \dfrac{-(-2) \pm \sqrt{(-2)^2 - 4(1)(2)}}{2} = \dfrac{2 \pm \sqrt{-4}}{2} = \dfrac{2 \pm 2i}{2} = 1 \pm i$. Therefore, the solutions are $x = 0, 1 \pm i$.

(b) Since $1 - i$ and $1 + i$ are zeros, $x - (1 - i) = x - 1 + i$ and $x - (1 + i) = x - 1 - i$ are the factors of $x^2 - 2x + 2$. Thus the complete factorization is $P(x) = x\left(x^2 - 2x + 2\right) = x(x - 1 + i)(x - 1 - i)$.

9. (a) $x^4 + 2x^2 + 1 = 0 \Leftrightarrow \left(x^2 + 1\right)^2 = 0 \Leftrightarrow x^2 + 1 = 0 \Leftrightarrow x^2 = -1 \Leftrightarrow x = \pm i$. Therefore the zeros of P are $x = \pm i$.

(b) Since $-i$ and i are zeros, $x + i$ and $x - i$ are the factors of $x^2 + 1$. Thus the complete factorization is

$$P(x) = \left(x^2 + 1\right)^2 = [(x + i)(x - i)]^2 = (x + i)^2 (x - i)^2.$$

11. (a) $x^4 - 16 = 0 \Leftrightarrow 0 = \left(x^2 - 4\right)\left(x^2 + 4\right) = (x - 2)(x + 2)\left(x^2 + 4\right)$. So $x = \pm 2$ or $x^2 + 4 = 0$. If $x^2 + 4 = 0$ then

$x^2 = -4 \Rightarrow x = \pm 2i$. Therefore the zeros of P are $x = \pm 2, \pm 2i$.

(b) Since $-i$ and i are zeros, $x + i$ and $x - i$ are the factors of $x^2 + 1$. Thus the complete factorization is

$$P(x) = (x - 2)(x + 2)\left(x^2 + 4\right) = (x - 2)(x + 2)(x - 2i)(x + 2i).$$

13. (a) $x^3 + 8 = 0 \Leftrightarrow (x + 2)\left(x^2 - 2x + 4\right) = 0$. So $x = -2$ or $x^2 - 2x + 4 = 0$. If $x^2 - 2x + 4 = 0$ then

$x = \dfrac{-(-2) \pm \sqrt{(-2)^2 - 4(1)(4)}}{2} = \dfrac{2 \pm \sqrt{-12}}{2} = \dfrac{2 \pm 2i\sqrt{3}}{2} = 1 \pm i\sqrt{3}$. Therefore, the zeros of P are $x = -2, 1 \pm i\sqrt{3}$.

(b) Since $1 - i\sqrt{3}$ and $1 + i\sqrt{3}$ are the zeros from the $x^2 - 2x + 4 = 0$, $x - \left(1 - i\sqrt{3}\right)$ and $x - \left(1 + i\sqrt{3}\right)$ are the factors

of $x^2 - 2x + 4$. Thus the complete factorization is

$$P(x) = (x + 2)\left(x^2 - 2x + 4\right) = (x + 2)\left[x - \left(1 - i\sqrt{3}\right)\right]\left[x - \left(1 + i\sqrt{3}\right)\right]$$
$$= (x + 2)\left(x - 1 + i\sqrt{3}\right)\left(x - 1 - i\sqrt{3}\right)$$

15. (a) $x^6 - 1 = 0 \Leftrightarrow 0 = \left(x^3 - 1\right)\left(x^3 + 1\right) = (x - 1)\left(x^2 + x + 1\right)(x + 1)\left(x^2 - x + 1\right)$. Clearly, $x = \pm 1$ are solutions.

If $x^2 + x + 1 = 0$, then $x = \dfrac{-1 \pm \sqrt{1 - 4(1)(1)}}{2} = \dfrac{-1 \pm \sqrt{-3}}{2} = -\dfrac{1}{2} \pm \dfrac{\sqrt{-3}}{2}$ so $x = -\dfrac{1}{2} \pm i\dfrac{\sqrt{3}}{2}$. And if $x^2 - x + 1 = 0$, then

$x = \dfrac{1 \pm \sqrt{1 - 4(1)(1)}}{2} = \dfrac{1 \pm \sqrt{-3}}{2} = \dfrac{1}{2} \pm \dfrac{\sqrt{-3}}{2} = \dfrac{1}{2} \pm i\dfrac{\sqrt{3}}{2}$. Therefore, the zeros of P are $x = \pm 1, -\dfrac{1}{2} \pm i\dfrac{\sqrt{3}}{2}, \dfrac{1}{2} \pm i\dfrac{\sqrt{3}}{2}$.

(b) The zeros of $x^2 + x + 1 = 0$ are $-\frac{1}{2} - i\frac{\sqrt{3}}{2}$ and $-\frac{1}{2} + i\frac{\sqrt{3}}{2}$, so $x^2 + x + 1$ factors as

$$\left[x - \left(-\frac{1}{2} - i\frac{\sqrt{3}}{2}\right)\right]\left[x - \left(-\frac{1}{2} + i\frac{\sqrt{3}}{2}\right)\right] = \left(x + \frac{1}{2} + i\frac{\sqrt{3}}{2}\right)\left(x + \frac{1}{2} - i\frac{\sqrt{3}}{2}\right).$$ Similarly, since

the zeros of $x^2 - x + 1 = 0$ are $\frac{1}{2} - i\frac{\sqrt{3}}{2}$ and $\frac{1}{2} + i\frac{\sqrt{3}}{2}$, so $x^2 - x + 1$ factors as

$$\left[x - \left(\frac{1}{2} - i\frac{\sqrt{3}}{2}\right)\right]\left[x - \left(\frac{1}{2} + i\frac{\sqrt{3}}{2}\right)\right] = \left(x - \frac{1}{2} + i\frac{\sqrt{3}}{2}\right)\left(x - \frac{1}{2} - i\frac{\sqrt{3}}{2}\right).$$ Thus the complete

factorization is

$$P(x) = (x-1)\left(x^2 + x + 1\right)(x+1)\left(x^2 - x + 1\right)$$

$$= (x-1)(x+1)\left(x + \frac{1}{2} + i\frac{\sqrt{3}}{2}\right)\left(x + \frac{1}{2} - i\frac{\sqrt{3}}{2}\right)\left(x - \frac{1}{2} + i\frac{\sqrt{3}}{2}\right)\left(x - \frac{1}{2} - i\frac{\sqrt{3}}{2}\right)$$

17. $P(x) = x^2 + 25 = (x - 5i)(x + 5i)$. The zeros of P are $5i$ and $-5i$, both multiplicity 1.

19. $Q(x) = x^2 + 2x + 2$. Using the quadratic formula $x = \frac{-(2) \pm \sqrt{(2)^2 - 4(1)(2)}}{2(1)} = \frac{-2 \pm \sqrt{-4}}{2} = \frac{-2 \pm 2i}{2} = -1 \pm i$. So $Q(x) = (x + 1 - i)(x + 1 + i)$. The zeros of Q are $-1 - i$ (multiplicity 1) and $-1 + i$ (multiplicity 1).

21. $P(x) = x^3 + 4x = x\left(x^2 + 4\right) = x(x - 2i)(x + 2i)$. The zeros of P are 0, $2i$, and $-2i$ (all multiplicity 1).

23. $Q(x) = x^4 - 1 = \left(x^2 - 1\right)\left(x^2 + 1\right) = (x - 1)(x + 1)\left(x^2 + 1\right) = (x - 1)(x + 1)(x - i)(x + i)$. The zeros of Q are $1, -1, i$, and $-i$ (all of multiplicity 1).

25. $P(x) = 16x^4 - 81 = \left(4x^2 - 9\right)\left(4x^2 + 9\right) = (2x - 3)(2x + 3)(2x - 3i)(2x + 3i)$. The zeros of P are $\frac{3}{2}, -\frac{3}{2}, \frac{3}{2}i$, and $-\frac{3}{2}i$ (all of multiplicity 1).

27. $P(x) = x^3 + x^2 + 9x + 9 = x^2(x + 1) + 9(x + 1) = (x + 1)\left(x^2 + 9\right) = (x + 1)(x - 3i)(x + 3i)$. The zeros of P are $-1, 3i$, and $-3i$ (all of multiplicity 1).

29. $Q(x) = x^4 + 2x^2 + 1 = \left(x^2 + 1\right)^2 = (x - i)^2(x + i)^2$. The zeros of Q are i and $-i$ (both of multiplicity 2).

31. $P(x) = x^4 + 3x^2 - 4 = \left(x^2 - 1\right)\left(x^2 + 4\right) = (x - 1)(x + 1)(x - 2i)(x + 2i)$. The zeros of P are $1, -1, 2i$, and $-2i$ (all of multiplicity 1).

33. $P(x) = x^5 + 6x^3 + 9x = x\left(x^4 + 6x^2 + 9\right) = x\left(x^2 + 3\right)^2 = x\left(x - i\sqrt{3}\right)^2\left(x + i\sqrt{3}\right)^2$. The zeros of P are 0 (multiplicity 1), $i\sqrt{3}$ (multiplicity 2), and $-i\sqrt{3}$ (multiplicity 2).

35. Since $1 + i$ and $1 - i$ are conjugates, the factorization of the polynomial must be $P(x) = a(x - [1 + i])(x - [1 - i]) = a\left(x^2 - 2x + 2\right)$. If we let $a = 1$, we get $P(x) = x^2 - 2x + 2$.

37. Since $2i$ and $-2i$ are conjugates, the factorization of the polynomial must be $Q(x) = b(x - 3)(x - 2i)(x + 2i] = b(x - 3)\left(x^2 + 4\right) = b\left(x^3 - 3x^2 + 4x - 12\right)$. If we let $b = 1$, we get $Q(x) = x^3 - 3x^2 + 4x - 12$.

39. Since i is a zero, by the Conjugate Roots Theorem, $-i$ is also a zero. So the factorization of the polynomial must be $P(x) = a(x - 2)(x - i)(x + i) = a\left(x^3 - 2x^2 + x - 2\right)$. If we let $a = 1$, we get $P(x) = x^3 - 2x^2 + x - 2$.

41. Since the zeros are $1 - 2i$ and 1 (with multiplicity 2), by the Conjugate Roots Theorem, the other zero is $1 + 2i$. So a factorization is

$$R(x) = c(x - [1 - 2i])(x - [1 + 2i])(x - 1)^2 = c([x - 1] + 2i)([x - 1] - 2i)(x - 1)^2$$

$$= c\left([x - 1]^2 - [2i]^2\right)\left(x^2 - 2x + 1\right) = c\left(x^2 - 2x + 1 + 4\right)\left(x^2 - 2x + 1\right) = c\left(x^2 - 2x + 5\right)\left(x^2 - 2x + 1\right)$$

$$= c\left(x^4 - 2x^3 + x^2 - 2x^3 + 4x^2 - 2x + 5x^2 - 10x + 5\right) = c\left(x^4 - 4x^3 + 10x^2 - 12x + 5\right)$$

If we let $c = 1$ we get $R(x) = x^4 - 4x^3 + 10x^2 - 12x + 5$.

43. Since the zeros are i and $1 + i$, by the Conjugate Roots Theorem, the other zeros are $-i$ and $1 - i$. So a factorization is

$$
\begin{aligned}
T(x) &= C(x - i)(x + i)(x - [1 + i])(x - [1 - i]) \\
&= C\left(x^2 - i^2\right)([x - 1] - i)([x - 1] + i) = C\left(x^2 + 1\right)\left(x^2 - 2x + 1 - i^2\right) = C\left(x^2 + 1\right)\left(x^2 - 2x + 2\right) \\
&= C\left(x^4 - 2x^3 + 2x^2 + x^2 - 2x + 2\right) = C\left(x^4 - 2x^3 + 3x^2 - 2x + 2\right) = Cx^4 - 2Cx^3 + 3Cx^2 - 2Cx + 2C
\end{aligned}
$$

Since the constant coefficient is 12, it follows that $2C = 12 \Leftrightarrow C = 6$, and so
$T(x) = 6\left(x^4 - 2x^3 + 3x^2 - 2x + 2\right) = 6x^4 - 12x^3 + 18x^2 - 12x + 12$.

45. $P(x) = x^3 + 2x^2 + 4x + 8 = x^2(x + 2) + 4(x + 2) = (x + 2)\left(x^2 + 4\right) = (x + 2)(x - 2i)(x + 2i)$. Thus the zeros are -2 and $\pm 2i$.

47. $P(x) = x^3 - 2x^2 + 2x - 1$. By inspection, $P(1) = 1 - 2 + 2 - 1 = 0$, and hence $x = 1$ is a zero.

$$
\begin{array}{r|rrrr}
1 & 1 & -2 & 2 & -1 \\
  &   & 1 & -1 & 1 \\
\hline
  & 1 & -1 & 1 & 0
\end{array}
$$

Thus $P(x) = (x - 1)\left(x^2 - x + 1\right)$. So $x = 1$ or $x^2 - x + 1 = 0$.

Using the quadratic formula, we have $x = \frac{1 \pm \sqrt{1 - 4(1)(1)}}{2} = \frac{1 \pm i\sqrt{3}}{2}$. Hence, the zeros are 1 and $\frac{1 \pm i\sqrt{3}}{2}$.

49. $P(x) = x^3 - 3x^2 + 3x - 2$.

$$
\begin{array}{r|rrrr}
2 & 1 & -3 & 3 & -2 \\
  &   & 2 & -2 & 2 \\
\hline
  & 1 & -1 & 1 & 0
\end{array}
$$

Thus $P(x) = (x - 2)\left(x^2 - x + 1\right)$. So $x = 2$ or $x^2 - x + 1 = 0$

Using the quadratic formula we have $x = \frac{1 \pm \sqrt{1 - 4(1)(1)}}{2} = \frac{1 \pm i\sqrt{3}}{2}$. Hence, the zeros are 2, and $\frac{1 \pm i\sqrt{3}}{2}$.

51. $P(x) = 2x^3 + 7x^2 + 12x + 9$ has possible rational zeros $\pm 1, \pm 3, \pm 9, \pm \frac{1}{2}, \pm \frac{3}{2}, \pm \frac{9}{2}$. Since all coefficients are positive, there are no positive real zeros.

$$
\begin{array}{r|rrrr}
-1 & 2 & 7 & 12 & 9 \\
   &   & -2 & -5 & -7 \\
\hline
   & 2 & 5 & 7 & 2
\end{array}
\qquad
\begin{array}{r|rrrr}
-2 & 2 & 7 & 12 & 9 \\
   &   & -4 & -6 & -12 \\
\hline
   & 2 & 3 & 6 & -3
\end{array}
$$

There is a zero between -1 and -2.

$$
\begin{array}{r|rrrr}
-\frac{3}{2} & 2 & 7 & 12 & 9 \\
             &   & -3 & -6 & -9 \\
\hline
             & 2 & 4 & 6 & 0
\end{array}
\Rightarrow x = -\frac{3}{2} \text{ is a zero.}
$$

$P(x) = \left(x + \frac{3}{2}\right)\left(2x^2 + 4x + 6\right) = 2\left(x + \frac{3}{2}\right)\left(x^2 + 2x + 3\right)$. Now $x^2 + 2x + 3$ has zeros
$x = \frac{-2 \pm \sqrt{4 - 4(3)(1)}}{2} = \frac{-2 \pm 2\sqrt{-2}}{2} = -1 \pm i\sqrt{2}$. Hence, the zeros are $-\frac{3}{2}$ and $-1 \pm i\sqrt{2}$.

53. $P(x) = x^4 + x^3 + 7x^2 + 9x - 18$. Since $P(x)$ has one change in sign, we are guaranteed a positive zero, and since $P(-x) = x^4 - x^3 + 7x^2 - 9x - 18$, there are 1 or 3 negative zeros.

$$
\begin{array}{r|rrrrr}
1 & 1 & 1 & 7 & 9 & -18 \\
  &   & 1 & 2 & 9 & 18 \\
\hline
  & 1 & 2 & 9 & 18 & 0
\end{array}
$$

Therefore, $P(x) = (x - 1)\left(x^3 + 2x^2 + 9x + 18\right)$. Continuing with the quotient, we try negative zeros.

$$
\begin{array}{r|rrrr}
-1 & 1 & 2 & 9 & 18 \\
   &   & -1 & -1 & -8 \\
\hline
   & 1 & 1 & 8 & 10
\end{array}
\qquad\qquad
\begin{array}{r|rrrr}
-2 & 1 & 2 & 9 & 18 \\
   &   & -2 & 0 & -18 \\
\hline
   & 1 & 0 & 9 & 0
\end{array}
$$

$P(x) = (x - 1)(x + 2)\left(x^2 + 9\right) = (x - 1)(x + 2)(x - 3i)(x + 3i)$. Therefore, the zeros are 1, -2, and $\pm 3i$.

55. We see a pattern and use it to factor by grouping. This gives

$$
P(x) = x^5 - x^4 + 7x^3 - 7x^2 + 12x - 12 = x^4(x - 1) + 7x^2(x - 1) + 12(x - 1) = (x - 1)\left(x^4 + 7x^2 + 12\right)
$$

$$
= (x - 1)\left(x^2 + 3\right)\left(x^2 + 4\right) = (x - 1)\left(x - i\sqrt{3}\right)\left(x + i\sqrt{3}\right)(x - 2i)(x + 2i)
$$

Therefore, the zeros are 1, $\pm i\sqrt{3}$, and $\pm 2i$.

57. $P(x) = x^4 - 6x^3 + 13x^2 - 24x + 36$ has possible rational zeros ± 1, ± 2, ± 3, ± 4, ± 6, ± 9, ± 12, ± 18. $P(x)$ has 4 variations in sign and $P(-x)$ has no variation in sign.

$$
\begin{array}{r|rrrrr}
1 & 1 & -6 & 13 & -24 & 36 \\
  &   & 1 & -5 & 8 & -16 \\
\hline
  & 1 & -5 & 8 & -16 & 20
\end{array}
\quad
\begin{array}{r|rrrrr}
2 & 1 & -6 & 13 & -24 & 36 \\
  &   & 2 & -8 & 10 & -28 \\
\hline
  & 1 & -4 & 5 & -14 & 8
\end{array}
\quad
\begin{array}{r|rrrrr}
3 & 1 & -6 & 13 & -24 & 36 \\
  &   & 3 & -9 & 12 & -36 \\
\hline
  & 1 & -3 & 4 & -12 & 0
\end{array} \Rightarrow x = 3 \text{ is a zero.}
$$

Continuing:

$$
\begin{array}{r|rrrr}
3 & 1 & -3 & 4 & -12 \\
  &   & 3 & 0 & 12 \\
\hline
  & 1 & 0 & 4 & 0
\end{array} \Rightarrow x = 3 \text{ is a zero.}
$$

$P(x) = (x - 3)^2\left(x^2 + 4\right) = (x - 3)^2(x - 2i)(x + 2i)$. Therefore, the zeros are 3 (multiplicity 2) and $\pm 2i$.

59. $P(x) = 4x^4 + 4x^3 + 5x^2 + 4x + 1$ has possible rational zeros ± 1, $\pm\frac{1}{2}$, $\pm\frac{1}{4}$. Since there is no variation in sign, all real zeros (if there are any) are negative.

$$
\begin{array}{r|rrrrr}
-1 & 4 & 4 & 5 & 4 & 1 \\
   &   & -4 & 0 & -5 & 1 \\
\hline
   & 4 & 0 & 5 & -1 & 2
\end{array}
\qquad
\begin{array}{r|rrrrr}
-\frac{1}{2} & 4 & 4 & 5 & 4 & 1 \\
             &   & -2 & -1 & -2 & -1 \\
\hline
             & 4 & 2 & 4 & 2 & 0
\end{array} \Rightarrow x = -\frac{1}{2} \text{ is a zero.}
$$

$P(x) = \left(x + \frac{1}{2}\right)\left(4x^3 + 2x^2 + 4x + 2\right)$. Continuing:

$$
\begin{array}{r|rrrr}
-\frac{1}{2} & 4 & 2 & 4 & 2 \\
             &   & -2 & 0 & -2 \\
\hline
             & 4 & 0 & 4 & 0
\end{array} \Rightarrow x = -\frac{1}{2} \text{ is a zero again.}
$$

$P(x) = \left(x + \frac{1}{2}\right)^2\left(4x^2 + 4\right)$. Thus, the zeros of $P(x)$ are $-\frac{1}{2}$ (multiplicity 2) and $\pm i$.

61. $P(x) = x^5 - 3x^4 + 12x^3 - 28x^2 + 27x - 9$ has possible rational zeros ± 1, ± 3, ± 9. $P(x)$ has 4 variations in sign and $P(-x)$ has 1 variation in sign.

$$
\begin{array}{r|rrrrrr}
1 & 1 & -3 & 12 & -28 & 27 & -9 \\
  &   & 1 & -2 & 10 & -18 & 9 \\
\hline
  & 1 & -2 & 10 & -18 & 9 & 0
\end{array} \Rightarrow x = 1 \text{ is a zero.}
$$

$$
\begin{array}{r|rrrrr}
1 & 1 & -2 & 10 & -18 & 9 \\
  &   & 1 & -1 & 9 & -9 \\
\hline
  & 1 & -1 & 9 & -9 & 0
\end{array} \Rightarrow x = 1 \text{ is a zero.}
\qquad
\begin{array}{r|rrrr}
1 & 1 & -1 & 9 & -9 \\
  &   & 1 & 0 & 9 \\
\hline
  & 1 & 0 & 9 & 0
\end{array} \Rightarrow x = 1 \text{ is a zero.}
$$

$P(x) = (x-1)^3 \left(x^2 + 9\right) = (x-1)^3 (x - 3i)(x + 3i)$. Therefore, the zeros are 1 (multiplicity 3) and $\pm 3i$.

63. (a) $P(x) = x^3 - 5x^2 + 4x - 20 = x^2(x-5) + 4(x-5) = (x-5)\left(x^2 + 4\right)$

(b) $P(x) = (x-5)(x-2i)(x+2i)$

65. (a) $P(x) = x^4 + 8x^2 - 9 = \left(x^2 - 1\right)\left(x^2 + 9\right) = (x-1)(x+1)\left(x^2 + 9\right)$

(b) $P(x) = (x-1)(x+1)(x-3i)(x+3i)$

67. (a) $P(x) = x^6 - 64 = \left(x^3 - 8\right)\left(x^3 + 8\right) = (x-2)\left(x^2 + 2x + 4\right)(x+2)\left(x^2 - 2x + 4\right)$

(b) $P(x) = (x-2)(x+2)\left(x + 1 - i\sqrt{3}\right)\left(x + 1 + i\sqrt{3}\right)\left(x - 1 - i\sqrt{3}\right)\left(x - 1 + i\sqrt{3}\right)$

69. (a) $x^4 - 2x^3 - 11x^2 + 12x = x\left(x^3 - 2x^2 - 11x + 12\right) = 0$. We first find the bounds for our viewing rectangle.

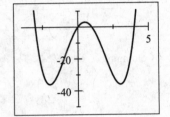

$$
\begin{array}{r|rrrr}
 & 1 & -2 & -11 & 12 \\
5 & 1 & 3 & 4 & 32 \\
-4 & 1 & -6 & 13 & -50
\end{array}
\begin{array}{l}
\Rightarrow x = 5 \text{ is an upper bound.} \\
\Rightarrow x = -4 \text{ is a lower bound.}
\end{array}
$$

We graph $P(x) = x^4 - 2x^3 - 11x^2 + 12x$ in the viewing rectangle $[-4, 5]$ by $[-50, 10]$ and see that it has 4 real solutions. Since this matches the degree of $P(x)$, $P(x)$ has no imaginary solution.

(b) $x^4 - 2x^3 - 11x^2 + 12x - 5 = 0$. We use the same bounds for our viewing rectangle, $[-4, 5]$ by $[-50, 10]$, and see that $R(x) = x^4 - 2x^3 - 11x^2 + 12x - 5$ has 2 real solutions. Since the degree of $R(x)$ is 4, $R(x)$ must have 2 imaginary solutions.

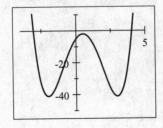

(c) $x^4 - 2x^3 - 11x^2 + 12x + 40 = 0$. We graph $T(x) = x^4 - 2x^3 - 11x^2 + 12x + 40$ in the viewing rectangle $[-4, 5]$ by $[-10, 50]$, and see that T has no real solution. Since the degree of T is 4, T must have 4 imaginary solutions.

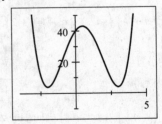

71. (a) $P(x) = x^2 - (1+i)x + (2+2i)$. So $P(2i) = (2i)^2 - (1+i)(2i) + 2 + 2i = -4 - 2i + 2 + 2 + 2i = 0$, and $P(1-i) = (1-i)^2 - (1+i)(1-i) + (2+2i) = 1 - 2i - 1 - 1 - 1 + 2 + 2i = 0$. Therefore, $2i$ and $1 - i$ are solutions of the equation $x^2 - (1+i)x + (2+2i) = 0$. However, $P(-2i) = (-2i)^2 - (1+i)(-2i) + 2 + 2i = -4 + 2i - 2 + 2 + 2i = -4 + 4i$, and $P(1+i) = (1+i)^2 - (1+i)(1+i) + 2 + 2i = 2 + 2i$. Since, $P(-2i) \neq 0$ and $P(1+i) \neq 0$, $-2i$ and $1 + i$ are not solutions.

(b) This does not violate the Conjugate Roots Theorem because the coefficients of the polynomial $P(x)$ are not all real.

73. Because P has real coefficients, the imaginary zeros come in pairs: $a \pm bi$ (by the Conjugate Roots Theorem), where $b \neq 0$. Thus there must be an even number of imaginary zeros. Since P is of odd degree, it has an odd number of zeros (counting multiplicity). It follows that P has at least one real zero.

3.7   RATIONAL FUNCTIONS

1. If the rational function $y = r(x)$ has the vertical asymptote $x = 2$, then as $x \to 2^+$, either $y \to \infty$ or $y \to -\infty$.

3. The function $r(x) = \dfrac{(x+1)(x-2)}{(x+2)(x-3)}$ has x-intercepts -1 and 2.

5. The function r has vertical asymptotes $x = -2$ and $x = 3$.

7. $r(x) = \dfrac{x}{x-2}$

(a)

x	$r(x)$
1.5	-3
1.9	-19
1.99	-199
1.999	-1999

x	$r(x)$
2.5	5
2.1	21
2.01	201
2.001	2001

x	$r(x)$
10	1.25
50	1.042
100	1.020
1000	1.002

x	$r(x)$
-10	0.833
-50	0.962
-100	0.980
-1000	0.998

(b) $r(x) \to -\infty$ as $x \to 2^-$ and $r(x) \to \infty$ as $x \to 2^+$.

(c) r has horizontal asymptote $y = 1$.

9. $r(x) = \dfrac{3x - 10}{(x-2)^2}$

(a)

x	$r(x)$
1.5	-22
1.9	-430
1.99	$-40,300$
1.999	$-4,003,000$

x	$r(x)$
2.5	-10
2.1	-370
2.01	$-39,700$
2.001	$-3,997,000$

x	$r(x)$
10	0.3125
50	0.0608
100	0.0302
1000	0.0030

x	$r(x)$
-10	-0.2778
-50	-0.0592
-100	-0.0298
-1000	-0.0030

(b) $r(x) \to -\infty$ as $x \to 2$.

(c) r has horizontal asymptote $y = 0$.

In Exercises 11–18, let $f(x) = \dfrac{1}{x}$.

11. $r(x) = \dfrac{1}{x-1} = f(x-1)$. From this form we see that the graph of r is obtained

from the graph of f by shifting 1 unit to the right. Thus r has vertical asymptote

$x = 1$ and horizontal asymptote $y = 0$. The domain of r is $(-\infty, 1) \cup (1, \infty)$ and

its range is $(-\infty, 0) \cup (0, \infty)$.

13. $s(x) = \dfrac{3}{x+1} = 3\left(\dfrac{1}{x+1}\right) = 3f(x+1)$. From this form we see that the graph

of s is obtained from the graph of f by shifting 1 unit to the left and stretching

vertically by a factor of 3. Thus s has vertical asymptote $x = -1$ and horizontal

asymptote $y = 0$. The domain of s is $(-\infty, -1) \cup (-1, \infty)$ and its range is

$(-\infty, 0) \cup (0, \infty)$.

15. $t(x) = \dfrac{2x-3}{x-2} = 2 + \dfrac{1}{x-2} = f(x-2) + 2$ (see the long

$$x-2 \,\overline{\smash{\big)}\, \begin{array}{r} 2 \\[-0.3ex] 2x-3 \\ 2x-2 \\ \hline 1 \end{array}}$$

division at right). From this form we see that the graph of t is

obtained from the graph of f by shifting 2 units to the right

and 2 units vertically. Thus t has vertical asymptote $x = 2$

and horizontal asymptote $y = 2$. The domain of t is

$(-\infty, 2) \cup (2, \infty)$ and its range is $(-\infty, 2) \cup (2, \infty)$.

17. $r(x) = \dfrac{x+2}{x+3} = 1 - \dfrac{1}{x+3} = -f(x+3) + 1$ (see the long

$$x+3 \,\overline{\smash{\big)}\, \begin{array}{r} 1 \\[-0.3ex] x+2 \\ x+3 \\ \hline -1 \end{array}}$$

division at right). From this form we see that the graph of r is

obtained from the graph of f by shifting 3 units to the left,

reflect about the x-axis, and then shifting vertically 1 unit.

Thus r has vertical asymptote $x = -3$ and horizontal

asymptote $y = 1$. The domain of r is $(-\infty, -3) \cup (-3, \infty)$

and its range is $(-\infty, 1) \cup (1, \infty)$.

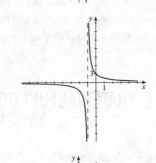

19. $r(x) = \dfrac{x-1}{x+4}$. When $x = 0$, we have $r(0) = -\frac{1}{4}$, so the y-intercept is $-\frac{1}{4}$. The numerator is 0 when $x = 1$, so the

x-intercept is 1.

21. $t(x) = \dfrac{x^2 - x - 2}{x - 6}$. When $x = 0$, we have $t(0) = \dfrac{-2}{-6} = \frac{1}{3}$, so the y-intercept is $\frac{1}{3}$. The numerator is 0 when

$x^2 - x - 2 = (x-2)(x+1) = 0$ or when $x = 2$ or $x = -1$, so the x-intercepts are 2 and -1.

23. $r(x) = \dfrac{x^2 - 9}{x^2}$. Since 0 is not in the domain of $r(x)$, there is no y-intercept. The numerator is 0 when

$x^2 - 9 = (x-3)(x+3) = 0$ or when $x = \pm 3$, so the x-intercepts are ± 3.

25. From the graph, the x-intercept is 3, the y-intercept is 3, the vertical asymptote is $x = 2$, and the horizontal asymptote is

$y = 2$.

27. From the graph, the x-intercepts are -1 and 1, the y-intercept is about $\frac{1}{4}$, the vertical asymptotes are $x = -2$ and $x = 2$, and the horizontal asymptote is $y = 1$.

29. $r(x) = \dfrac{5}{x-2}$ has a vertical asymptote where $x - 2 = 0 \Leftrightarrow x = 2$, and $y = 0$ is a horizontal asymptote because the degree of the denominator is greater than that of the numerator.

31. $r(x) = \dfrac{6x}{x^2 + 2}$ has no vertical asymptote because $x^2 + 2 > 0$ for all x. $y = 0$ is a horizontal asymptote because the degree of the denominator is greater than that of the numerator.

33. $s(x) = \dfrac{6x^2 + 1}{2x^2 + x - 1}$ has vertical asymptotes where $2x^2 + x - 1 = 0 \Leftrightarrow (x+1)(2x-1) = 0 \Leftrightarrow x = -1$ or $x = \frac{1}{2}$, and horizontal asymptote $y = \frac{6}{2} = 3$.

35. $s(x) = \dfrac{(5x-1)(x+1)}{(3x-1)(x+2)}$ has vertical asymptotes $x = \frac{1}{3}$ and $x = -2$, and horizontal asymptote $y = \frac{5}{3}$.

37. $r(x) = \dfrac{6x^3 - 2}{2x^3 + 5x^2 + 6x} = \dfrac{6x^3 - 2}{x\left(2x^2 + 5x + 6\right)}$. Because the quadratic in the denominator has no real zero, r has vertical asymptote $x = 0$ and horizontal asymptote $y = \frac{6}{2} = 3$.

39. $y = \dfrac{x^2 + 2}{x - 1}$. A vertical asymptote occurs when $x - 1 = 0 \Leftrightarrow x = 1$. There is no horizontal asymptote because the degree of the numerator is greater than the degree of the denominator.

41. $y = \dfrac{4x - 4}{x + 2}$. When $x = 0$, $y = -2$, so the y-intercept is -2. When $y = 0$,

$4x - 4 = 0 \Leftrightarrow x = 1$, so the x-intercept is 1. Since the degree of the numerator and denominator are the same, the horizontal asymptote is $y = \frac{4}{1} = 4$. A vertical asymptote occurs when $x = -2$. As $x \to -2^+$, $y = \dfrac{4x - 4}{x + 2} \to -\infty$, and as $x \to -2^-$, $y = \dfrac{4x - 4}{x + 2} \to \infty$. The domain is $\{x \mid x \neq -2\}$ and the range is $\{y \mid y \neq 4\}$.

43. $s(x) = \dfrac{4 - 3x}{x + 7}$. When $x = 0$, $y = \frac{4}{7}$, so the y-intercept is $\frac{4}{7}$. The x-intercepts occur when $y = 0 \Leftrightarrow 4 - 3x = 0 \Leftrightarrow x = \frac{4}{3}$. A vertical asymptote occurs when $x = -7$. Since the degree of the numerator and denominator are the same the horizontal asymptote is $y = \dfrac{-3}{1} = -3$. The domain is $\{x \mid x \neq -7\}$ and the range is $\{y \mid y \neq -3\}$.

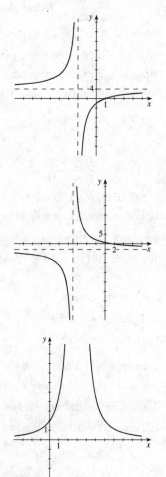

45. $r(x) = \dfrac{18}{(x - 3)^2}$. When $x = 0$, $y = \frac{18}{9} = 2$, and so the y-intercept is 2. Since the numerator can never be zero, there is no x-intercept. There is a vertical asymptote when $x - 3 = 0 \Leftrightarrow x = 3$, and because the degree of the numerator is less than the degree of the denominator, the horizontal asymptote is $y = 0$. The domain is $\{x \mid x \neq 3\}$ and the range is $\{y \mid y > 0\}$.

47. $s(x) = \dfrac{4x - 8}{(x - 4)(x + 1)}$. When $x = 0$, $y = \dfrac{-8}{(-4)(1)} = 2$, so the y-intercept is 2.

When $y = 0$, $4x - 8 = 0 \Leftrightarrow x = 2$, so the x-intercept is 2. The vertical asymptotes

are $x = -1$ and $x = 4$, and because the degree of the numerator is less than the

degree of the denominator, the horizontal asymptote is $y = 0$. The domain is

$\{x \mid x \neq -1, 4\}$ and the range is $\mathbb{R}$.

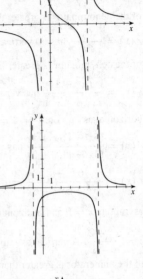

49. $s(x) = \dfrac{6}{x^2 - 5x - 6}$. When $x = 0$, $y = \dfrac{6}{-6} = -1$, so the y-intercept is -1.

Since the numerator is never zero, there is no x-intercept. The vertical asymptotes

occur when $x^2 - 5x - 6 = (x + 1)(x - 6) \Leftrightarrow x = -1$ and $x = 6$, and because the

degree of the numerator is less less than the degree of the denominator, the

horizontal asymptote is $y = 0$. The domain is $\{x \mid x \neq -1, 6\}$ and the range is

$\{y \mid y \leq -0.5 \text{ or } y > 0\}$.

51. $t(x) = \dfrac{3x + 6}{x^2 + 2x - 8}$. When $x = 0$, $y = \dfrac{6}{-8} = -\dfrac{3}{4}$, so the y-intercept is $-\dfrac{3}{4}$.

When $y = 0$, $3x + 6 = 0 \Leftrightarrow x = -2$, so the x-intercept is -2. The vertical

asymptotes occur when $x^2 + 2x - 8 = (x - 2)(x + 4) = 0 \Leftrightarrow x = 2$ and $x = -4$.

Since the degree of the numerator is less than the degree of the denominator, the

horizontal asymptote is $y = 0$. The domain is $\{x \mid x \neq -4, 2\}$ and the range is $\mathbb{R}$.

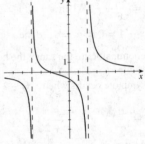

53. $r(x) = \dfrac{(x - 1)(x + 2)}{(x + 1)(x - 3)}$. When $x = 0$, $y = \dfrac{2}{3}$, so the y-intercept is $\dfrac{2}{3}$. When

$y = 0$, $(x - 1)(x + 2) = 0 \Rightarrow x = -2, 1$, so, the x-intercepts are -2 and 1. The

vertical asymptotes are $x = -1$ and $x = 3$, and because the degree of the

numerator and denominator are the same the horizontal asymptote is $y = \dfrac{1}{1} = 1$.

The domain is $\{x \mid x \neq -1, 3\}$ and the range is $\mathbb{R}$.

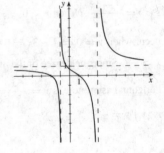

55. $r(x) = \dfrac{x^2 - 2x + 1}{x^2 + 2x + 1} = \dfrac{(x - 1)^2}{(x + 1)^2} = \left(\dfrac{x - 1}{x + 1}\right)^2$. When $x = 0$, $y = 1$, so the

y-intercept is 1. When $y = 0$, $x = 1$, so the x-intercept is 1. A vertical asymptote

occurs at $x + 1 = 0 \Leftrightarrow x = -1$. Because the degree of the numerator and

denominator are the same the horizontal asymptote is $y = \dfrac{1}{1} = 1$. The domain is

$\{x \mid x \neq -1\}$ and the range is $\{y \mid y \geq 0\}$.

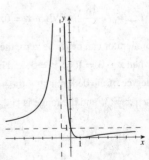

57. $r(x) = \dfrac{2x^2 + 10x - 12}{x^2 + x - 6} = \dfrac{2(x-1)(x+6)}{(x-2)(x+3)}$. When $x = 0$, $y = \dfrac{2(-1)(6)}{(-2)(3)} = 2$,

so the y-intercept is 2. When $y = 0$, $2(x-1)(x+6) = 0 \Rightarrow x = -6, 1$, so the

x-intercepts are -6 and 1. Vertical asymptotes occur when $(x-2)(x+3) = 0$

$\Leftrightarrow x = -3$ or $x = 2$. Because the degree of the numerator and denominator are the

same the horizontal asymptote is $y = \frac{2}{1} = 2$. The domain is $\{x \mid x \neq -3, 2\}$ and

the range is $\mathbb{R}$.

59. $y = \dfrac{x^2 - x - 6}{x^2 + 3x} = \dfrac{(x-3)(x+2)}{x(x+3)}$. The x-intercept occurs when $y = 0 \Leftrightarrow$

$(x-3)(x+2) = 0 \Rightarrow x = -2, 3$, so the x-intercepts are -2 and 3. There is no

y-intercept because y is undefined when $x = 0$. The vertical asymptotes are $x = 0$

and $x = -3$. Because the degree of the numerator and denominator are the same,

the horizontal asymptotes is $y = \frac{1}{1} = 1$. The domain is $\{x \mid x \neq -3, 0\}$ and the

range is $\mathbb{R}$.

61. $r(x) = \dfrac{3x^2 + 6}{x^2 - 2x - 3} = \dfrac{3(x^2 + 2)}{(x-3)(x+1)}$. When $x = 0$, $y = -2$, so the y-intercept

is -2. Since the numerator can never equal zero, there is no x-intercept. Vertical

asymptotes occur when $x = -1, 3$. Because the degree of the numerator and

denominator are the same, the horizontal asymptote is. $y = \frac{3}{1} = 3$. The domain is

$\{x \mid x \neq -1, 3\}$ and the range is $\{y \mid y \leq -1.5 \text{ or } y \geq 2.4\}$.

63. $s(x) = \dfrac{x^2 - 2x + 1}{x^3 - 3x^2} = \dfrac{(x-1)^2}{x^2(x-3)}$. Since $x = 0$ is not in the domain of $s(x)$,

there is no y-intercept. The x-intercept occurs when $y = 0 \Leftrightarrow$

$x^2 - 2x + 1 = (x-1)^2 = 0 \Rightarrow x = 1$, so the x-intercept is 1. Vertical asymptotes

occur when $x = 0, 3$. Since the degree of the numerator is less than the degree of

the denominator, the horizontal asymptote is $y = 0$. The domain is $\{x \mid x \neq 0, 3\}$

and the range is $\mathbb{R}$.

65. $r(x) = \dfrac{x^2}{x-2}$. When $x = 0$, $y = 0$, so the graph passes through the origin. There

is a vertical asymptote when $x - 2 = 0 \Leftrightarrow x = 2$, with $y \to \infty$ as $x \to 2^+$, and

$y \to -\infty$ as $x \to 2^-$. Because the degree of the numerator is greater than the

degree of the denominator, there is no horizontal asymptotes. By using long

division, we see that $y = x + 2 + \dfrac{4}{x-2}$, so $y = x + 2$ is a slant asymptote.

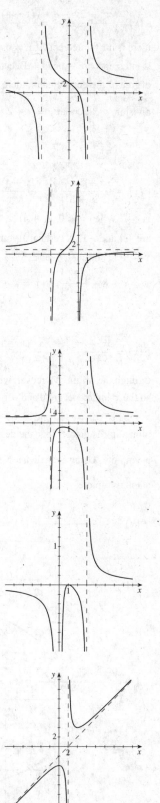

67. $r(x) = \dfrac{x^2 - 2x - 8}{x} = \dfrac{(x-4)(x+2)}{x}$. The vertical asymptote is $x = 0$, thus, there is no y-intercept. If $y = 0$, then $(x-4)(x+2) = 0 \Rightarrow x = -2, 4$, so the x-intercepts are -2 and 4. Because the degree of the numerator is greater than the degree of the denominator, there are no horizontal asymptotes. By using long division, we see that $y = x - 2 - \dfrac{8}{x}$, so $y = x - 2$ is a slant asymptote.

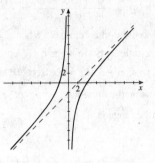

69. $r(x) = \dfrac{x^2 + 5x + 4}{x - 3} = \dfrac{(x+4)(x+1)}{x - 3}$. When $x = 0$, $y = -\dfrac{4}{3}$, so the y-intercept is $-\dfrac{4}{3}$. When $y = 0$, $(x+4)(x+1) = 0 \Leftrightarrow x = -4, -1$, so the two x-intercepts are -4 and -1. A vertical asymptote occurs when $x = 3$, with $y \to \infty$ as $x \to 3^+$, and $y \to -\infty$ as $x \to 3^-$. Using long division, we see that $y = x + 8 + \dfrac{28}{x - 3}$, so $y = x + 8$ is a slant asymptote.

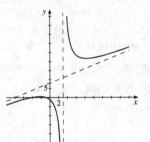

71. $r(x) = \dfrac{x^3 + x^2}{x^2 - 4} = \dfrac{x^2(x+1)}{(x-2)(x+2)}$. When $x = 0$, $y = 0$, so the graph passes through the origin. Moreover, when $y = 0$, we have $x^2(x+1) = 0 \Rightarrow x = 0, -1$, so the x-intercepts are 0 and -1. Vertical asymptotes occur when $x = \pm 2$; as $x \to \pm 2^-$, $y = -\infty$ and as $x \to \pm 2^+$, $y \to \infty$. Because the degree of the numerator is greater than the degree of the denominator, there is no horizontal asymptote. Using long division, we see that $y = x + 1 + \dfrac{4x + 4}{x^2 - 4}$, so $y = x + 1$ is a slant asymptote.

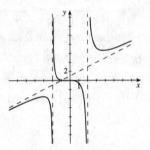

73. $f(x) = \dfrac{2x^2 + 6x + 6}{x + 3}$, $g(x) = 2x$. f has vertical asymptote $x = -3$.

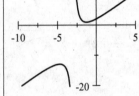 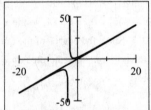

75. $f(x) = \dfrac{x^3 - 2x^2 + 16}{x - 2}$, $g(x) = x^2$. f has vertical asymptote $x = 2$.

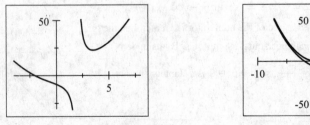

77. $f(x) = \dfrac{2x^2 - 5x}{2x + 3}$ has vertical asymptote $x = -1.5$, x-intercepts 0 and 2.5, y-intercept 0, local maximum $(-3.9, -10.4)$,

and local minimum $(0.9, -0.6)$. Using long division, we get $f(x) = x - 4 + \dfrac{12}{2x + 3}$. From the graph, we see that the end

behavior of $f(x)$ is like the end behavior of $g(x) = x - 4$.

$$
\begin{array}{r}
x - 4 \\
2x + 3 \overline{\smash{\big)}\ 2x^2 - 5x} \\
\underline{2x^2 + 3x} \\
-8x \\
\underline{-8x - 12} \\
12
\end{array}
$$

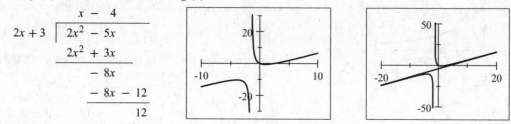

79. $f(x) = \dfrac{x^5}{x^3 - 1}$ has vertical asymptote $x = 1$, x-intercept 0, y-intercept 0, and local minimum $(1.4, 3.1)$.

Thus $y = x^2 + \dfrac{x^2}{x^3 - 1}$. From the graph we see that the end behavior of $f(x)$ is like the end behavior of $g(x) = x^2$.

$$
\begin{array}{r}
x^2 \\
x^3 - 1 \overline{\smash{\big)}\ x^5} \\
\underline{x^5 - x^2} \\
x^2
\end{array}
$$

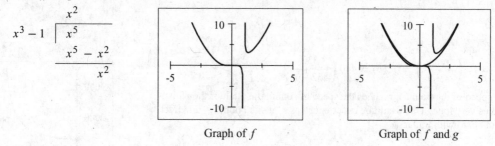

Graph of f | Graph of f and g

81. $f(x) = \dfrac{x^4 - 3x^3 + 6}{x - 3}$ has vertical asymptote $x = 3$, x-intercepts 1.6 and 2.7, y-intercept -2, local maxima $(-0.4, -1.8)$

and $(2.4, 3.8)$, and local minima $(0.6, -2.3)$ and $(3.4, 54.3)$. Thus $y = x^3 + \dfrac{6}{x - 3}$. From the graphs, we see that the end

behavior of $f(x)$ is like the end behavior of $g(x) = x^3$.

$$
\begin{array}{r}
x^3 \\
x - 3 \overline{\smash{\big)}\ x^4 - 3x^3 + 6} \\
\underline{x^4 - 3x^3} \\
6
\end{array}
$$

83. (a)

(b) $p(t) = \dfrac{3000t}{t + 1} = 3000 - \dfrac{3000}{t + 1}$. So as $t \to \infty$, we

have $p(t) \to 3000$.

85. $c(t) = \dfrac{5t}{t^2 + 1}$

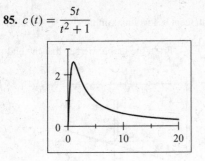

(a) The highest concentration of drug is 2.50 mg/L, and it is reached 1 hour after the drug is administered.

(b) The concentration of the drug in the bloodstream goes to 0.

(c) From the first viewing rectangle, we see that an approximate solution is near $t = 15$. Thus we graph $y = \dfrac{5t}{t^2 + 1}$ and $y = 0.3$ in the viewing rectangle $[14, 18]$ by $[0, 0.5]$. So it takes about 16.61 hours for the concentration to drop below 0.3 mg/L.

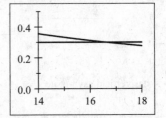

87. $P(v) = P_0 \left(\dfrac{s_0}{s_0 - v} \right) \Rightarrow P(v) = 440 \left(\dfrac{332}{332 - v} \right)$

If the speed of the train approaches the speed of sound, the pitch of the whistle becomes very loud. This would be experienced as a "sonic boom"— an effect seldom heard with trains.

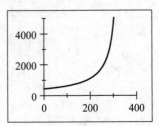

89. Vertical asymptote $x = 3$: $p(x) = \dfrac{1}{x - 3}$. Vertical asymptote $x = 3$ and horizontal asymptote $y = 2$: $r(x) = \dfrac{2x}{x - 3}$.

Vertical asymptotes $x = 1$ and $x = -1$, horizontal asymptote 0, and x-intercept 4: $q(x) = \dfrac{x - 4}{(x - 1)(x + 1)}$. Of course, other answers are possible.

91. (a) $r(x) = \dfrac{3x^2 - 3x - 6}{x - 2} = \dfrac{3(x - 2)(x + 1)}{x - 2} = 3(x + 1)$, for $x \neq 2$. Therefore,

$r(x) = 3x + 3$, $x \neq 2$. Since $3(2) + 3 = 9$, the graph is the line $y = 3x + 3$ with the point $(2, 9)$ removed.

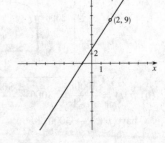

(b) $s(x) = \dfrac{x^2 + x - 20}{x + 5} = \dfrac{(x-4)(x+5)}{x+5} = x - 4$, for $x \neq -5$. Therefore, $s(x) = x - 4$, $x \neq -5$. Since

$(-5) - 4 = -9$, the graph is the line $y = x - 4$ with the point $(-5, -9)$ removed.

$t(x) = \dfrac{2x^2 - x - 1}{x - 1} = \dfrac{(2x+1)(x-1)}{x-1} = 2x + 1$, for $x \neq 1$. Therefore, $t(x) = 2x + 1$, $x \neq 1$. Since $2(1) + 1 = 3$,

the graph is the line $y = 2x + 1$ with the point $(1, 3)$ removed.

$u(x) = \dfrac{x-2}{x^2 - 2x} = \dfrac{x-2}{x(x-2)} = \dfrac{1}{x}$, for $x \neq 2$. Therefore, $u(x) = \dfrac{1}{x}$, $x \neq 2$. When $x = 2$, $\dfrac{1}{x} = \dfrac{1}{2}$, so the graph is the

curve $y = \dfrac{1}{x}$ with the point $\left(2, \frac{1}{2}\right)$ removed.

3.8 MODELING VARIATION

1. If the quantities x and y are related by the equation $y = 3x$ then we say that y is *directly proportional* to x, and the constant of *proportionality* is 3.

3. If the quantities x, y, and z are related by the equation $z = 3\dfrac{x}{y}$ then we say that z is *directly proportional* to x and *inversely proportional* to y.

5. $T = kx$, where k is constant.

7. $v = \dfrac{k}{z}$, where k is constant.

9. $y = \dfrac{ks}{t}$, where k is constant.

11. $z = k\sqrt{y}$, where k is constant.

13. $V = klwh$, where k is constant.

15. $R = \dfrac{ki}{Pt}$, where k is constant.

17. Since y is directly proportional to x, $y = kx$. Since $y = 42$ when $x = 6$, we have $42 = k(6) \Leftrightarrow k = 7$. So $y = 7x$.

19. R is inversely proportional to s, so $R = \dfrac{k}{s}$. Since $R = 3$ when $s = 4$, we have $3 = \dfrac{k}{4} \Leftrightarrow k = 12$. So $R = \dfrac{12}{s}$.

21. Since M varies directly as x and inversely as y, $M = \dfrac{kx}{y}$. Since $M = 5$ when $x = 2$ and $y = 6$, we have $5 = \dfrac{k(2)}{6} \Leftrightarrow$

$k = 15$. Therefore $M = \dfrac{15x}{y}$.

23. Since W is inversely proportional to the square of r, $W = \dfrac{k}{r^2}$. Since $W = 10$ when $r = 6$, we have $10 = \dfrac{k}{(6)^2} \Leftrightarrow k = 360$.

So $W = \dfrac{360}{r^2}$.

25. Since C is jointly proportional to l, w, and h, we have $C = klwh$. Since $C = 128$ when $l = w = h = 2$, we have

$128 = k(2)(2)(2) \Leftrightarrow 128 = 8k \Leftrightarrow k = 16$. Therefore, $C = 16lwh$.

27. Since s is inversely proportional to the square root of t, we have $s = \dfrac{k}{\sqrt{t}}$. Since $s = 100$ when $t = 25$, we have $100 = \dfrac{k}{\sqrt{25}}$

$\Leftrightarrow 100 = \dfrac{k}{5} \Leftrightarrow k = 500$. So $s = \dfrac{500}{\sqrt{t}}$.

29. (a) $z = k\dfrac{x^3}{y^2}$

(b) If we replace x with $3x$ and y with $2y$, then $z = k\dfrac{(3x)^3}{(2y)^2} = \dfrac{27}{4}\left(k\dfrac{x^3}{y^2}\right)$, so z changes by a factor of $\dfrac{27}{4}$.

31. (a) $z = kx^3y^5$

(b) If we replace x with $3x$ and y with $2y$, then $z = k(3x)^3(2y)^5 = 864kx^3y^5$, so z changes by a factor of 864.

33. (a) The force F needed is $F = kx$.

(b) Since $F = 40$ when $x = 5$, we have $40 = k(5) \Leftrightarrow k = 8$.

(c) From part (b), we have $F = 8x$. Substituting $x = 4$ into $F = 8x$ gives $F = 8(4) = 32$ N.

35. (a) $P = ks^3$.

(b) Since $P = 96$ when $s = 20$, we get $96 = k \cdot 20^3 \Leftrightarrow k = 0.012$. So $P = 0.012s^3$.

(c) Substituting $x = 30$, we get $P = 0.012 \cdot 30^3 = 324$ watts.

37. $D = ks^2$. Since $D = 240$ when $s = 50$ we have $240 = k(50)^2$ so $k = 0.096$. Thus $D = 0.096s^2$. When $D = 160$ then $160 = 0.096s^2 \Leftrightarrow s^2 = 1666.7$ so $s \approx 40$ mi/h (for safety reasons we round down).

39. $F = kAs^2$. Since $F = 220$ when $A = 40$ and $s = 5$. Solving for k we have $220 = k(40)(5)^2 \Leftrightarrow 220 = 1000k \Leftrightarrow k = 0.22$. Now when $A = 28$ and $F = 175$ we get $175 = 0.220(28)s^2 \Leftrightarrow 28.4090 = s^2$ so $s = \sqrt{28.4090} = 5.33$ mi/h.

41. (a) $P = \dfrac{kT}{V}$.

(b) Substituting $P = 33.2$, $T = 400$, and $V = 100$, we get $33.2 = \dfrac{k(400)}{100} \Leftrightarrow k = 8.3$. Thus $k = 8.3$ and the equation is $P = \dfrac{8.3T}{V}$.

(c) Substituting $T = 500$ and $V = 80$, we have $P = \dfrac{8.3(500)}{80} = 51.875$ kPa. Hence the pressure of the sample of gas is about 51.9 kPa.

43. (a) The loudness L is inversely proportional to the square of the distance d, so $L = \dfrac{k}{d^2}$.

(b) Substituting $d = 10$ and $L = 70$, we have $70 = \dfrac{k}{10^2} \Leftrightarrow k = 7000$.

(c) Substituting $2d$ for d, we have $L = \dfrac{k}{(2d)^2} = \dfrac{1}{4}\left(\dfrac{k}{d^2}\right)$, so the loudness is changed by a factor of $\dfrac{1}{4}$.

(d) Substituting $\dfrac{1}{2}d$ for d, we have $L = \dfrac{k}{\left(\frac{1}{2}d\right)^2} = 4\left(\dfrac{k}{d^2}\right)$, so the loudness is changed by a factor of 4.

45. (a) $R = \dfrac{kL}{d^2}$

(b) Since $R = 140$ when $L = 1.2$ and $d = 0.005$, we get $140 = \dfrac{k(1.2)}{(0.005)^2} \Leftrightarrow k = \dfrac{7}{2400} = 0.002916\overline{6}$.

(c) Substituting $L = 3$ and $d = 0.008$, we have $R = \dfrac{7}{2400} \cdot \dfrac{3}{(0.008)^2} = \dfrac{4375}{32} \approx 137$ Ω.

(d) If we substitute $2d$ for d and $3L$ for L, then $R = \dfrac{k(3L)}{(2d)^2} = \dfrac{3}{4}\dfrac{kL}{d^2}$, so the resistance is changed by a factor of $\dfrac{3}{4}$.

47. (a) For the sun, $E_S = k6000^4$ and for earth $E_E = k300^4$. Thus $\dfrac{E_S}{E_E} = \dfrac{k6000^4}{k300^4} = \left(\dfrac{6000}{300}\right)^4 = 20^4 = 160{,}000$. So the sun produces 160,000 times the radiation energy per unit area than the Earth.

(b) The surface area of the sun is $4\pi\,(435{,}000)^2$ and the surface area of the Earth is $4\pi\,(3{,}960)^2$. So the sun has

$$\dfrac{4\pi\,(435{,}000)^2}{4\pi\,(3{,}960)^2} = \left(\dfrac{435{,}000}{3{,}960}\right)^2$$ times the surface area of the Earth. Thus the total radiation emitted by the sun is

$$160{,}000 \times \left(\dfrac{435{,}000}{3{,}960}\right)^2 = 1{,}930{,}670{,}340$$ times the total radiation emitted by the Earth.

49. (a) Let T and l be the period and the length of the pendulum, respectively. Then $T = k\sqrt{l}$.

(b) $T = k\sqrt{l} \Rightarrow T^2 = k^2 l \Leftrightarrow l = \dfrac{T^2}{k^2}$. If the period is doubled, the new length is $\dfrac{(2T)^2}{k^2} = 4\dfrac{T^2}{k^2} = 4l$. So we would quadruple the length l to double the period T.

51. (a) Since f is inversely proportional to L, we have $f = \dfrac{k}{L}$, where k is a positive constant.

(b) If we replace L by $2L$ we have $\dfrac{k}{2L} = \dfrac{1}{2} \cdot \dfrac{k}{L} = \dfrac{1}{2}f$. So the frequency of the vibration is cut in half.

53. Using $B = k\dfrac{L}{d^2}$ with $k = 0.080$, $L = 2.5 \times 10^{26}$, and $d = 2.4 \times 10^{19}$, we have $B = 0.080\dfrac{2.5 \times 10^{26}}{(2.4 \times 10^{19})^2} \approx 3.47 \times 10^{-14}$.

The star's apparent brightness is about 3.47×10^{-14} W/m^2.

55. Examples include radioactive decay and exponential growth in biology.

CHAPTER 3 REVIEW

1. (a) $f(x) = x^2 + 4x + 1 = \left(x^2 + 4x\right) + 1$
$ = \left(x^2 + 4x + 4\right) + 1 - 4$
$ = (x + 2)^2 - 3$

(b)

3. (a) $f(x) = 1 + 8x - x^2 = -\left(x^2 - 8x\right) + 1$
$ = -\left(x^2 - 8x + 16\right) + 1 + 16$
$ = -(x - 4)^2 + 17$

(b)

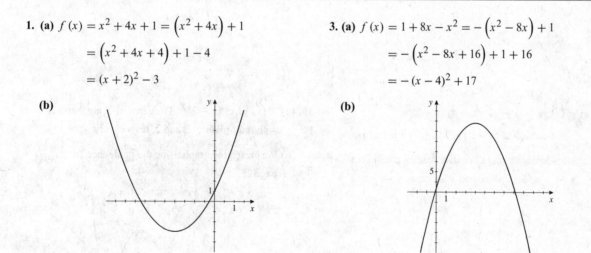

5. $f(x) = 2x^2 + 4x - 5 = 2\left(x^2 + 2x\right) - 5 = 2\left(x^2 + 2x + 1\right) - 5 - 2 = 2(x+1)^2 - 7$ has the minimum value -7 when $x = -1$.

7. We write the height function in standard form: $h(t) = -16t^2 + 48t + 32 = -16\left(t^2 - 3t\right) + 32 = -16\left(t^2 - 3t + \dfrac{9}{4}\right) + 32 + 36 = -16\left(t - \dfrac{3}{2}\right)^2 + 68$. The stone reaches a maximum height of 68 ft.

9. $P(x) = -x^3 + 64$

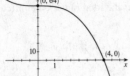

11. $P(x) = 2(x+1)^4 - 32$

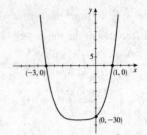

13. $P(x) = 32 + (x-1)^5$

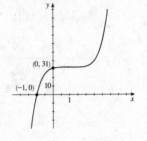

15. (a) $P(x) = (x-3)(x+1)(x-5) = x^3 - 7x^2 + 7x + 15$ has odd degree and a positive leading coefficient, so $y \to \infty$ as $x \to \infty$ and $y \to -\infty$ as $x \to -\infty$.

(b)

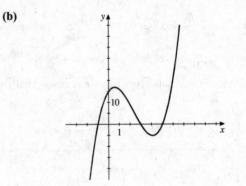

17. (a) $P(x) = -(x-1)^2(x-4)(x+2)^2$
$$= -x^5 + 2x^4 + 11x^3 - 8x^2 - 20x + 16$$
has odd degree and a negative leading coefficient, so $y \to -\infty$ as $x \to \infty$ and $y \to \infty$ as $x \to -\infty$.

(b)

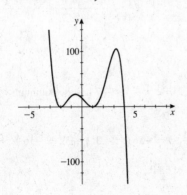

19. (a) $P(x) = x^3(x-2)^2$. The zeros of P are 0 and 2, with multiplicities 3 and 2, respectively.

(b) We sketch the graph using the guidelines on page 283.

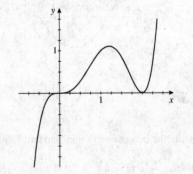

21. $P(x) = x^3 - 4x + 1$. x-intercepts: -2.1, 0.3, and 1.9. y-intercept: 1. Local maximum is $(-1.2, 4.1)$. Local minimum is $(1.2, -2.1)$. $y \to \infty$ as $x \to \infty$; $y \to -\infty$ as $x \to -\infty$.

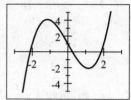

23. $P(x) = 3x^4 - 4x^3 - 10x - 1$. x-intercepts: -0.1 and 2.1. y-intercept: -1. Local maximum is $(1.4, -14.5)$. There is no local maximum. $y \to \infty$ as $x \to \pm\infty$.

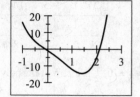

25. (a) Use the Pythagorean Theorem and solving for y^2 we have, $x^2 + y^2 = 10^2 \Leftrightarrow y^2 = 100 - x^2$. Substituting we get $S = 13.8x \left(100 - x^2\right) = 1380x - 13.8x^3$.

(b) Domain is $[0, 10]$.

(c)

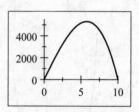

(d) The strongest beam has width 5.8 inches.

27. $\dfrac{x^2 - 3x + 5}{x - 2}$

$$
\begin{array}{r|rrr}
2 & 1 & -3 & 5 \\
  &   & 2 & -2 \\
\hline
  & 1 & -1 & 3
\end{array}
$$

Using synthetic division, we see that $Q(x) = x - 1$ and $R(x) = 3$.

29. $\dfrac{x^3 - x^2 + 11x + 2}{x - 4}$

$$
\begin{array}{r|rrrr}
4 & 1 & -1 & 11 & 2 \\
  &   & 4 & 12 & 92 \\
\hline
  & 1 & 3 & 23 & 94
\end{array}
$$

Using synthetic division, we see that $Q(x) = x^2 + 3x + 23$ and $R(x) = 94$.

31. $\dfrac{x^4 - 8x^2 + 2x + 7}{x + 5}$

$$
\begin{array}{r|rrrrr}
-5 & 1 & 0 & -8 & 2 & 7 \\
   &   & -5 & 25 & -85 & 415 \\
\hline
   & 1 & -5 & 17 & -83 & 422
\end{array}
$$

Using synthetic division, we see that $Q(x) = x^3 - 5x^2 + 17x - 83$ and $R(x) = 422$.

33. $\dfrac{2x^3 + x^2 - 8x + 15}{x^2 + 2x - 1}$

$$
\begin{array}{r}
2x - 3 \\
x^2 + 2x - 1 \overline{\smash{\big)}\ 2x^3 + x^2 - 8x + 15} \\
\underline{2x^3 + 4x^2 - 2x} \\
-3x^2 - 6x + 15 \\
\underline{-3x^2 - 6x + 3} \\
12
\end{array}
$$

Therefore, $Q(x) = 2x - 3$, and $R(x) = 12$.

35. $P(x) = 2x^3 - 9x^2 - 7x + 13$; find $P(5)$.

$$
\begin{array}{r|rrrr}
5 & 2 & -9 & -7 & 13 \\
  &   & 10 & 5 & -10 \\
\hline
  & 2 & 1 & -2 & 3
\end{array}
$$

Therefore, $P(5) = 3$.

37. $\frac{1}{2}$ is a zero of $P(x) = 2x^4 + x^3 - 5x^2 + 10x - 4$ if $P\left(\frac{1}{2}\right) = 0$.

$$
\begin{array}{r|rrrrr}
\frac{1}{2} & 2 & 1 & -5 & 10 & -4 \\
  &   & 1 & 1 & -2 & 4 \\
\hline
  & 2 & 2 & -4 & 8 & 0
\end{array}
$$

Since $P\left(\frac{1}{2}\right) = 0$, $\frac{1}{2}$ is a zero of the polynomial.

39. $P(x) = x^{500} + 6x^{201} - x^2 - 2x + 4$. The remainder from dividing $P(x)$ by $x - 1$ is

$$P(1) = (1)^{500} + 6(1)^{201} - (1)^2 - 2(1) + 4 = 8.$$

41. (a) $P(x) = x^5 - 6x^3 - x^2 + 2x + 18$ has possible rational zeros $\pm 1, \pm 2, \pm 3, \pm 6, \pm 9, \pm 18$.

(b) Since $P(x)$ has 2 variations in sign, there are either 0 or 2 positive real zeros. Since $P(-x) = -x^5 + 6x^3 - x^2 - 2x + 18$ has 3 variations in sign, there are 1 or 3 negative real zeros.

43. (a) $P(x) = x^3 - 16x = x\left(x^2 - 16\right)$

$$= x(x-4)(x+4)$$

has zeros $-4, 0, 4$ (all of multiplicity 1).

(b)

45. (a) $P(x) = x^4 + x^3 - 2x^2 = x^2\left(x^2 + x - 2\right)$

$$= x^2(x+2)(x-1)$$

The zeros are 0 (multiplicity 2), -2 (multiplicity 1), and 1 (multiplicity 1).

(b)

47. (a) $P(x) = x^4 - 2x^3 - 7x^2 + 8x + 12$. The possible rational zeros are $\pm 1, \pm 2, \pm 3, \pm 4, \pm 6, \pm 12$. P has 2 variations in sign, so it has either 2 or 0 positive real zeros.

$$
\begin{array}{r|rrrrr}
1 & 1 & -2 & -7 & 8 & 12 \\
  &   & 1 & -1 & -8 & 0 \\
\hline
  & 1 & -1 & -8 & 0 & 12
\end{array}
\qquad
\begin{array}{r|rrrrr}
2 & 1 & -2 & -7 & 8 & 12 \\
  &   & 2 & 0 & -14 & -12 \\
\hline
  & 2 & 0 & -7 & -6 & 0
\end{array}
\Rightarrow x = 2 \text{ is a root.}
$$

$P(x) = x^4 - 2x^3 - 7x^2 + 8x + 12 = (x-2)\left(x^3 - 7x - 6\right)$. Continuing:

$$
\begin{array}{r|rrrr}
2 & 1 & 0 & -6 & -6 \\
  &   & 2 & 4 & -4 \\
\hline
  & 1 & 2 & -2 & -10
\end{array}
\qquad
\begin{array}{r|rrrr}
3 & 1 & 0 & -7 & -6 \\
  &   & 3 & 9 & 6 \\
\hline
  & 1 & 3 & 2 & 0
\end{array}
$$

so $x = 3$ is a root and

$$P(x) = (x-2)(x-3)\left(x^2 + 3x + 2\right)$$

$$= (x-2)(x-3)(x+1)(x+2)$$

Therefore the real roots are $-2, -1, 2$, and 3 (all of multiplicity 1).

(b)

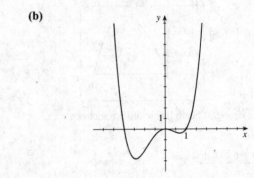

49. (a) $P(x) = 2x^4 + x^3 + 2x^2 - 3x - 2$. The possible rational roots are ± 1, ± 2, $\pm\frac{1}{2}$. P has one variation in sign, and hence 1 positive real root. $P(-x)$ has 3 variations in sign and hence either 3 or 1 negative real roots.

$$
\begin{array}{r|rrrrr}
1 & 2 & 1 & 2 & -3 & -2 \\
  &   & 2 & 3 & 5 & 2 \\
\hline
  & 2 & 3 & 5 & 2 & 0
\end{array} \Rightarrow x = 1 \text{ is a zero.}
$$

$P(x) = 2x^4 + x^3 + 2x^2 - 3x - 2 = (x - 1)\left(2x^3 + 3x^2 + 5x + 2\right)$.

Continuing: **(b)**

$$
\begin{array}{r|rrrr}
-1 & 2 & 3 & 5 & 2 \\
   &   & -2 & -1 & -4 \\
\hline
   & 2 & 1 & 4 & -2
\end{array}
\qquad
\begin{array}{r|rrrr}
-2 & 2 & 3 & 5 & 2 \\
   &   & -4 & 2 & -14 \\
\hline
   & 2 & -1 & 7 & -12
\end{array}
$$

$$
\begin{array}{r|rrrr}
-\frac{1}{2} & 2 & 3 & 5 & 2 \\
             &   & -1 & -1 & -2 \\
\hline
             & 2 & 2 & 4 & 0
\end{array} \Rightarrow x = -\tfrac{1}{2} \text{ is a zero.}
$$

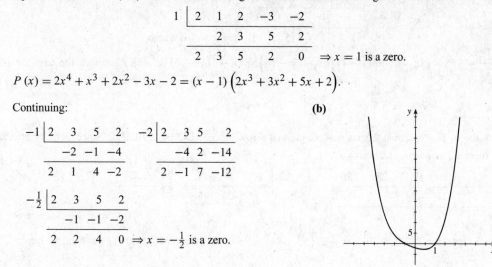

$P(x) = (x - 1)\left(x + \tfrac{1}{2}\right)\left(2x^2 + 2x + 4\right)$. The quadratic is irreducible, so the real zeros are 1 and $-\frac{1}{2}$ (each of multiplicity 1).

51. $(2 - 3i) + (1 + 4i) = (2 + 1) + (-3 + 4)i = 3 + i$

53. $(2 + i)(3 - 2i) = 6 - 4i + 3i - 2i^2 = 6 - i + 2 = 8 - i$

55. $\dfrac{4 + 2i}{2 - i} = \dfrac{4 + 2i}{2 - i} \cdot \dfrac{2 + i}{2 + i} = \dfrac{8 + 8i + 2i^2}{4 - i^2} = \dfrac{8 + 8i - 2}{4 + 1} = \dfrac{6 + 8i}{5} = \tfrac{6}{5} + \tfrac{8}{5}i$

57. $i^{25} = i^{24}i = \left(i^4\right)^6 i = (1)^6 i = i$

59. $\left(1 - \sqrt{-1}\right)\left(1 + \sqrt{-1}\right) = (1 - i)(1 + i) = 1 + i - i - i^2 = 1 + 1 = 2$

61. Since the zeros are $-\frac{1}{2}$, 2, and 3, a factorization is

$$
\begin{aligned}
P(x) &= C\left(x + \tfrac{1}{2}\right)(x - 2)(x - 3) = \tfrac{1}{2}C(2x + 1)\left(x^2 - 5x + 6\right) \\
     &= \tfrac{1}{2}C\left(2x^3 - 10x^2 + 12x + x^2 - 5x + 6\right) = \tfrac{1}{2}C\left(2x^3 - 9x^2 + 7x + 6\right)
\end{aligned}
$$

Since the constant coefficient is 12, $\frac{1}{2}C(6) = 12 \Leftrightarrow C = 4$, and so the polynomial is $P(x) = 4x^3 - 18x^2 + 14x + 12$.

63. No, there is no polynomial of degree 4 with integer coefficients that has zeros i, $2i$, $3i$ and $4i$. Since the imaginary zeros of polynomial equations with real coefficients come in complex conjugate pairs, there would have to be 8 zeros, which is impossible for a polynomial of degree 4.

65. $P(x) = x^3 - x^2 + x - 1$ has possible rational zeros ± 1.

$$
\begin{array}{r|rrrr}
1 & 1 & -1 & 1 & -1 \\
  &   & 1 & 0 & 1 \\
\hline
  & 1 & 0 & 1 & 0
\end{array} \Rightarrow x = 1 \text{ is a zero.}
$$

So $P(x) = (x - 1)\left(x^2 + 1\right)$. Therefore, the zeros are 1 and $\pm i$.

67. $P(x) = x^3 - 3x^2 - 13x + 15$ has possible rational zeros ± 1, ± 3, ± 5, ± 15.

$$
\begin{array}{r|rrrr}
1 & 1 & -3 & -13 & 15 \\
  &   & 1 & -2 & -15 \\
\hline
  & 1 & -2 & -15 & 0
\end{array}
\Rightarrow x = 1 \text{ is a zero.}
$$

So $P(x) = x^3 - 3x^2 - 13x + 15 = (x - 1)\left(x^2 - 2x - 15\right) = (x - 1)(x - 5)(x + 3)$. Therefore, the zeros are -3, 1, and 5.

69. $P(x) = x^4 + 6x^3 + 17x^2 + 28x + 20$ has possible rational zeros ± 1, ± 2, ± 4, ± 5, ± 10, ± 20. Since all of the coefficients are positive, there are no positive real zeros.

$$
\begin{array}{r|rrrrr}
-1 & 1 & 6 & 17 & 28 & 20 \\
   &   & -1 & -5 & -12 & -16 \\
\hline
   & 1 & 5 & 12 & 16 & 4
\end{array}
\qquad
\begin{array}{r|rrrrr}
-2 & 1 & 6 & 17 & 28 & 20 \\
   &   & -2 & -8 & -18 & -20 \\
\hline
   & 1 & 4 & 9 & 10 & 0
\end{array}
\Rightarrow x = -2 \text{ is a zero.}
$$

$P(x) = x^4 + 6x^3 + 17x^2 + 28x + 20 = (x + 2)\left(x^3 + 4x^2 + 9x + 10\right)$. Continuing with the quotient, we have

$$
\begin{array}{r|rrrr}
-2 & 1 & 4 & 9 & 10 \\
   &   & -2 & -4 & -10 \\
\hline
   & 1 & 2 & 5 & 0
\end{array}
\Rightarrow x = -2 \text{ is a zero.}
$$

Thus $P(x) = x^4 + 6x^3 + 17x^2 + 28x + 20 = (x + 2)^2\left(x^2 + 2x + 5\right)$. Now $x^2 + 2x + 5 = 0$ when $x = \frac{-2 \pm \sqrt{4 - 4(5)(1)}}{2} = \frac{-2 \pm 4i}{2} = -1 \pm 2i$. Thus, the zeros are -2 (multiplicity 2) and $-1 \pm 2i$.

71. $P(x) = x^5 - 3x^4 - x^3 + 11x^2 - 12x + 4$ has possible rational zeros ± 1, ± 2, ± 4.

$$
\begin{array}{r|rrrrrr}
1 & 1 & -3 & -1 & 11 & -12 & 4 \\
  &   & 1 & -2 & -3 & 8 & -4 \\
\hline
  & 1 & -2 & -3 & 8 & -4 & 0
\end{array}
\Rightarrow x = 1 \text{ is a zero.}
$$

$P(x) = x^5 - 3x^4 - x^3 + 11x^2 - 12x + 4 = (x - 1)\left(x^4 - 2x^3 - 3x^2 + 8x - 4\right)$. Continuing with the quotient, we have

$$
\begin{array}{r|rrrrr}
1 & 1 & -2 & -3 & 8 & -4 \\
  &   & 1 & -1 & -4 & 4 \\
\hline
  & 1 & -1 & -4 & 4 & 0
\end{array}
\Rightarrow x = 1 \text{ is a zero.}
$$

$$
\begin{aligned}
x^5 - 3x^4 - x^3 + 11x^2 - 12x + 4 &= (x - 1)^2\left(x^3 - x^2 - 4x + 4\right) = (x - 1)^3\left(x^2 - 4\right) \\
&= (x - 1)^3(x - 2)(x + 2)
\end{aligned}
$$

Therefore, the zeros are 1 (multiplicity 3), -2, and 2.

73. $P(x) = x^6 - 64 = \left(x^3 - 8\right)\left(x^3 + 8\right) = (x - 2)\left(x^2 + 2x + 4\right)(x + 2)\left(x^2 - 2x + 4\right)$. Now using the quadratic formula to find the zeros of $x^2 + 2x + 4$, we have $x = \frac{-2 \pm \sqrt{4 - 4(4)(1)}}{2} = \frac{-2 \pm 2i\sqrt{3}}{2} = -1 \pm i\sqrt{3}$, and using the quadratic formula to find the zeros of $x^2 - 2x + 4$, we have $x = \frac{2 \pm \sqrt{4 - 4(4)(1)}}{2} = \frac{2 \pm 2i\sqrt{3}}{2} = 1 \pm i\sqrt{3}$. Therefore, the zeros are 2, -2, $1 \pm i\sqrt{3}$, and $-1 \pm i\sqrt{3}$.

75. $P(x) = 6x^4 - 18x^3 + 6x^2 - 30x + 36 = 6\left(x^4 - 3x^3 + x^2 - 5x + 6\right)$ has possible rational zeros $\pm 1, \pm 2, \pm 3, \pm 6$.

$$
\begin{array}{r|rrrrr}
1 & 6 & -18 & 6 & -30 & 36 \\
  &   & 6 & -12 & -6 & -36 \\
\hline
  & 6 & -12 & -6 & -36 & 0
\end{array} \Rightarrow x = 1 \text{ is a zero.}
$$

So $P(x) = 6x^4 - 18x^3 + 6x^2 - 30x + 36 = (x-1)\left(6x^3 - 12x^2 - 6x - 36\right) = 6(x-1)\left(x^3 - 2x^2 - x - 6\right)$.
Continuing with the quotient we have

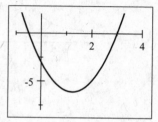

$$
\begin{array}{r|rrrr}
1 & 1 & -2 & -1 & -6 \\
  &   & 1 & -1 & -2 \\
\hline
  & 1 & -1 & -2 & -8
\end{array}
\qquad
\begin{array}{r|rrrr}
2 & 1 & -2 & -1 & -6 \\
  &   & 2 & 0 & -2 \\
\hline
  & 1 & 0 & -1 & -8
\end{array}
\qquad
\begin{array}{r|rrrr}
3 & 1 & -2 & -1 & -6 \\
  &   & 3 & 3 & 6 \\
\hline
  & 1 & 1 & 2 & 0
\end{array} \Rightarrow x = 3 \text{ is a zero.}
$$

So $P(x) = 6x^4 - 18x^3 + 6x^2 - 30x + 36 = 6(x-1)(x-3)\left(x^2 + x + 2\right)$. Now $x^2 + x + 2 = 0$ when
$x = \dfrac{-1 \pm \sqrt{1 - 4(1)(2)}}{2} = \dfrac{-1 \pm i\sqrt{7}}{2}$, and so the zeros are 1, 3, and $\dfrac{-1 \pm i\sqrt{7}}{2}$.

77. $2x^2 = 5x + 3 \Leftrightarrow 2x^2 - 5x - 3 = 0$. The solutions are $x = -0.5, 3$.

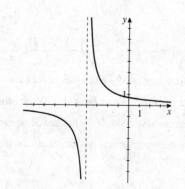

79. $x^4 - 3x^3 - 3x^2 - 9x - 2 = 0$ has solutions $x \approx -0.24$, 4.24.

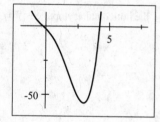

81. $P(x) = x^3 - 2x - 4$

$$
\begin{array}{r|rrrr}
1 & 1 & 0 & -2 & -4 \\
  &   & 1 & 1 & -1 \\
\hline
  & 1 & 1 & -1 & -5
\end{array}
\qquad\qquad
\begin{array}{r|rrrr}
2 & 1 & 0 & -2 & -4 \\
  &   & 2 & 4 & 4 \\
\hline
  & 1 & 2 & 2 & 0
\end{array}
$$

$P(x) = x^3 - 2x - 4 = (x-2)\left(x^2 + 2x + 2\right)$. Since $x^2 + 2x + 2 = 0$ has no real solution, the only real zero of P is $x = 2$.

83. (a) $r(x) = \dfrac{3}{x+4}$. The vertical asymptote is $x = -4$. Because the denominator has higher degree than the numerator, the horizontal asymptote is $y = 0$. When $x = 0$, $y = \frac{3}{4}$, so the y-intercept is $\frac{3}{4}$. There is no x-intercept because the numerator is never 0. The domain of r is $(-\infty, -4) \cup (-4, \infty)$ and its range is $(-\infty, 0) \cup (0, \infty)$.

(b) If $f(x) = \dfrac{1}{x}$, then $r(x) = \dfrac{3}{x+4} = 3\left(\dfrac{1}{x+4}\right) = 3f(x+4)$, so we obtain the graph of r by shifting the graph of f to the left 4 units and stretching vertically by a factor of 3.

85. (a) $r(x) = \dfrac{3x - 4}{x - 1}$. The vertical asymptote is $x = 1$. Because the

denominator has the same degree as the numerator, the horizontal

asymptote is $y = \frac{3}{1} = 3$. When $x = 0$, $y = -4$, so the y-intercept is

-4. When $y = 0$, $3x - 4 = 0 \Leftrightarrow x = \frac{4}{3}$, so the x-intercept is $\frac{4}{3}$. The

domain of r is $(-\infty, 1) \cup (1, \infty)$ and its range is $(-\infty, 3) \cup (3, \infty)$.

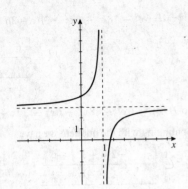

(b) If $f(x) = \dfrac{1}{x}$, then

$$r(x) = \frac{3(x - 1) - 1}{x - 1} = 3 - \frac{1}{x - 1} = 3 - f(x - 1). \text{ Thus, we}$$

obtain the graph of r by shifting the graph of f to the right 1 unit,

reflecting in the x-axis, and shifting upward 3 units.

87. $r(x) = \dfrac{3x - 12}{x + 1}$. When $x = 0$, we have $r(0) = \dfrac{-12}{1} = -12$, so the y-intercept is

-12. Since $y = 0$, when $3x - 12 = 0 \Leftrightarrow x = 4$, the x-intercept is 4. The vertical

asymptote is $x = -1$. Because the denominator has the same degree as the

numerator, the horizontal asymptote is $y = \frac{3}{1} = 3$.

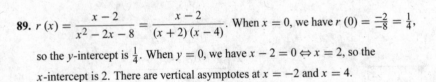

89. $r(x) = \dfrac{x - 2}{x^2 - 2x - 8} = \dfrac{x - 2}{(x + 2)(x - 4)}$. When $x = 0$, we have $r(0) = \dfrac{-2}{-8} = \frac{1}{4}$,

so the y-intercept is $\frac{1}{4}$. When $y = 0$, we have $x - 2 = 0 \Leftrightarrow x = 2$, so the

x-intercept is 2. There are vertical asymptotes at $x = -2$ and $x = 4$.

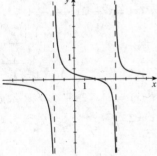

91. $r(x) = \dfrac{x^2 - 9}{2x^2 + 1} = \dfrac{(x + 3)(x - 3)}{2x^2 + 1}$. When $x = 0$, we have $r(0) = \dfrac{-9}{1}$, so the

y-intercept is -9. When $y = 0$, we have $x^2 - 9 = 0 \Leftrightarrow x = \pm 3$ so the x-intercepts

are -3 and 3. Since $2x^2 + 1 > 0$, the denominator is never zero so there are no

vertical asymptotes. The horizontal asymptote is at $y = \frac{1}{2}$ because the degree of

the denominator and numerator are the same.

93. $r(x) = \dfrac{x-3}{2x+6}$. From the graph we see that the

x-intercept is 3, the y-intercept is -0.5, there is a vertical

asymptote at $x = -3$ and a horizontal asymptote at

$y = 0.5$, and there is no local extremum.

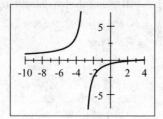

95. $r(x) = \dfrac{x^3+8}{x^2-x-2}$. From the graph we see that the x-intercept is -2, the

y-intercept is -4, there are vertical asymptotes at $x = -1$ and $x = 2$, there is no

horizontal asymptote, the local maximum is $(0.425, -3.599)$, and the local

minimum is $(4.216, 7.175)$. By using long division, we see that

$f(x) = x + 1 + \dfrac{10-x}{x^2-x-2}$, so f has a slant asymptote of $y = x + 1$.

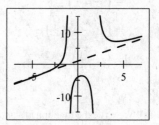

97. The graphs of $y = x^4+x^2+24x$ and $y = 6x^3+20$ intersect when $x^4+x^2+24x = 6x^3+20 \Leftrightarrow x^4-6x^3+x^2+24x-20 = 0$.

The possible rational zeros are ± 1, ± 2, ± 4, ± 5, ± 10, ± 20.

$$
\begin{array}{r|rrrrr}
1 & 1 & -6 & 1 & 24 & -20 \\
  &   & 1 & -5 & -4 & 20 \\
\hline
  & 1 & -5 & -4 & 20 & 0
\end{array}
\Rightarrow x = 1 \text{ is a zero.}
$$

So $x^4 - 6x^3 + x^2 + 24x - 20 = (x-1)\left(x^3 - 5x^2 - 4x + 20\right) = 0$. Continuing with the quotient:

$$
\begin{array}{r|rrrr}
1 & 1 & -5 & -4 & 20 \\
  &   & 1 & -4 & -8 \\
\hline
  & 1 & -4 & -8 & 12
\end{array}
\qquad
\begin{array}{r|rrrr}
2 & 1 & -5 & -4 & 20 \\
  &   & 2 & -6 & -20 \\
\hline
  & 1 & -3 & -10 & 0
\end{array}
\Rightarrow x = 2 \text{ is a zero.}
$$

So

$$x^4 - 6x^3 + x^2 + 24x - 20 = (x-1)(x-2)\left(x^2 - 3x - 10\right)$$

$$= (x-1)(x-2)(x-5)(x+2)$$

$$= 0$$

Hence, the points of intersection are $(1, 26)$, $(2, 68)$, $(5, 770)$, and $(-2, -28)$.

99. Since z is inversely proportional to y, we have $z = \dfrac{k}{y}$. Substituting $z = 12$ when $y = 16$, we find $12 = \dfrac{k}{16} \Leftrightarrow k = 192$.

Therefore $z = \dfrac{192}{y}$.

101. Let f be the frequency of the string and l be the length of the string. Since the frequency is inversely proportional to the length, we have $f = \dfrac{k}{l}$. Substituting $l = 12$ when $k = 440$, we find $440 = \dfrac{k}{12} \Leftrightarrow k = 5280$. Therefore $f = \dfrac{5280}{l}$. For $f = 660$, we must have $660 = \dfrac{5280}{l} \Leftrightarrow l = \frac{5280}{660} = 8$. So the string needs to be shortened to 8 inches.

103. Let r be the maximum range of the baseball and v be the velocity of the baseball. Since the maximum range is directly proportional to the square of the velocity, we have $r = lv^2$. Substituting $v = 60$ and $r = 242$, we find $242 = k(60)^2 \Leftrightarrow k \approx 0.0672$. If $v = 70$, then we have a maximum range of $r = 0.0672(70)^2 = 329.4$ feet.

CHAPTER 3 TEST

1. $f(x) = x^2 - x - 6$
$$= \left(x^2 - x\right) - 6$$
$$= \left(x^2 - x + \tfrac{1}{4}\right) - 6 - \tfrac{1}{4}$$
$$= \left(x - \tfrac{1}{2}\right)^2 - \tfrac{25}{4}$$

3. (a) We write the function in standard form: $h(x) = 10x - 0.01x^2 = -0.01\left(x^2 - 1000x\right) = -0.01\left(x^2 - 1000x + 500^2\right) + 0.01\left(500^2\right) = -0.01(x - 500)^2 + 2500$. Thus, the maximum height reached by the cannonball is 2500 feet.

(b) By the symmetry of the parabola, we see that the cannonball's height will be 0 again (and thus it will splash into the water) when $x = 1000$ ft.

5. (a)

$$
\begin{array}{r|rrrr}
2 & 1 & 0 & -4 & 2 & 5 \\
  &   & 2 & 4 & 0 & 4 \\
\hline
  & 1 & 2 & 0 & 2 & 9
\end{array}
$$

Therefore, the quotient is
$Q(x) = x^3 + 2x^2 + 2$, and the remainder is
$R(x) = 9$.

(b)

$$
\begin{array}{r}
x^3 + 2x^2 \qquad\quad + \tfrac{1}{2} \\
2x^2 - 1 \overline{\smash{\big)}\, 2x^5 + 4x^4 - x^3 - x^2 + 0x + 7} \\
\underline{2x^5 \qquad\quad - x^3} \\
4x^4 \qquad\quad - x^2 \\
\underline{4x^4 \qquad\quad - 2x^2} \\
x^2 \quad + 7 \\
\underline{x^2 \quad - \tfrac{1}{2}} \\
\tfrac{15}{2}
\end{array}
$$

Therefore, the quotient is $Q(x) = x^3 + 2x^2 + \tfrac{1}{2}$ and the remainder is $R(x) = \tfrac{15}{2}$.

7. (a) $(3 - 2i) + (4 + 3i) = (3 + 4) + (-2 + 3)i = 7 + i$

(b) $(3 - 2i) - (4 + 3i) = (3 - 4) + (-2 - 3)i = -1 - 5i$

(c) $(3 - 2i)(4 + 3i) = 12 + 9i - 8i - 6i^2 = 12 + i - 6(-1) = 18 + i$

(d) $\dfrac{3 - 2i}{4 + 3i} = \dfrac{(3 - 2i)(4 - 3i)}{(4 + 3i)(4 - 3i)} = \dfrac{12 - 17i - 6i^2}{16 - 9i^2} = \dfrac{6}{25} - \dfrac{17}{25}i$

(e) $i^{48} = \left(i^2\right)^{24} = (-1)^{24} = 1$

(f) $\left(\sqrt{2} - \sqrt{-2}\right)\left(\sqrt{8} + \sqrt{-2}\right) = \left(\sqrt{2} - i\sqrt{2}\right)\left(\sqrt{8} + i\sqrt{2}\right) = \sqrt{2}\sqrt{8} + i\left(\sqrt{2}\right)^2 - i\sqrt{2}\sqrt{8} - i^2\left(\sqrt{2}\right)^2$
$$= \sqrt{16} + 2i - i\sqrt{16} + 2 = 6 - 2i$$

9. $P(x) = x^4 - 2x^3 + 5x^2 - 8x + 4$. The possible rational zeros of P are: ± 1, ± 2, and ± 4. Since there are four changes in sign, P has 4, 2, or 0 positive real zeros.

$$
\begin{array}{r|rrrrr}
1 & 1 & -2 & 5 & -8 & 4 \\
  &   & 1 & -1 & 4 & -4 \\
\hline
  & 1 & -1 & 4 & -4 & 0
\end{array}
$$

So $P(x) = (x - 1)\left(x^3 - x^2 + 4x - 4\right)$. Factoring the second factor by grouping, we have
$$P(x) = (x - 1)\left[x^2(x - 1) + 4(x - 1)\right] = (x - 1)\left(x^2 + 4\right)(x - 1) = (x - 1)^2(x - 2i)(x + 2i).$$

11. $P(x) = 2x^4 - 7x^3 + x^2 - 18x + 3$.

    (a) Since $P(x)$ has 4 variations in sign, $P(x)$ can have 4, 2, or 0 positive real zeros. Since

        $P(-x) = 2x^4 + 7x^3 + x^2 + 18x + 3$ has no variations in sign, there are no negative real zeros.

    (b)

$$
\begin{array}{r|rrrrr}
4 & 2 & -7 & 1 & -18 & 3 \\
  &   & 8 & 4 & 20 & 8 \\
\hline
  & 2 & 1 & 5 & 2 & 11
\end{array}
$$

    Since the last row contains no negative entry, 4 is an upper bound for the real zeros of $P(x)$.

$$
\begin{array}{r|rrrrr}
-1 & 2 & -7 & 1 & -18 & 3 \\
   &   & -1 & 9 & -10 & 28 \\
\hline
   & 2 & -9 & 10 & -28 & 31
\end{array}
$$

    Since the last row alternates in sign, -1 is a lower bound for the real zeros of $P(x)$.

    (c) Using the upper and lower limit from part (b), we graph $P(x)$ in the viewing rectangle $[-1, 4]$ by $[-1, 1]$. The two real zeros are 0.17 and 3.93.

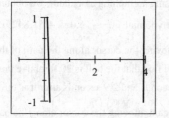

    (d) Local minimum $(2.8, -70.3)$.

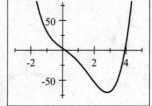

13. (a) $M = k\dfrac{wh^2}{L}$

    (b) Substituting $w = 4$, $h = 6$, $L = 12$, and $M = 4800$, we have $4800 = k\dfrac{(4)\left(6^2\right)}{12} \Leftrightarrow k = 400$. Thus $M = 400\dfrac{wh^2}{L}$.

    (c) Now if $L = 10$, $w = 3$, and $h = 10$, then $M = 400\dfrac{(3)\left(10^2\right)}{10} = 12{,}000$. So the beam can support 12,000 pounds.

FOCUS ON MODELING   Fitting Polynomial Curves to Data

1. (a) Using a graphing calculator, we obtain the quadratic polynomial

$y = -0.275428x^2 + 19.7485x - 273.5523$ (where miles are measured in thousands).

(b)

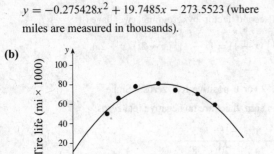

Pressure (lb/in^2)

(c) Moving the cursor along the path of the polynomial, we find that 35.85 lb/in^2 gives the longest tire life.

3. (a) Using a graphing calculator, we obtain the cubic polynomial

$y = 0.00203709x^3 - 0.104522x^2$
$+ 1.966206x + 1.45576.$

(b)

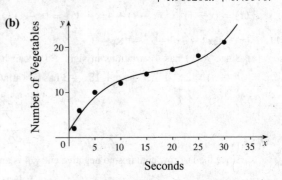

Seconds

(c) Moving the cursor along the path of the polynomial, we find that the subjects could name about 43 vegetables in 40 seconds.

(d) Moving the cursor along the path of the polynomial, we find that the subjects could name 5 vegetables in about 2.0 seconds.

(b) Using a graphing calculator, we obtain the quadratic polynomial $y = -16.0x^2 + 51.8429x + 4.20714$.

(c) Moving the cursor along the path of the polynomial, we find that the ball is 20 ft. above the ground 0.3 seconds and 2.9 seconds after it is thrown upward.

(d) Again, moving the cursor along the path of the polynomial, we find that the maximum height is 46.2 ft.

5. (a)

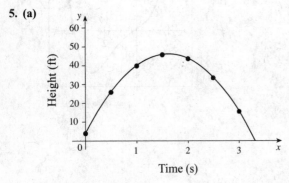

Time (s)

A quadratic model seems appropriate.

4 EXPONENTIAL AND LOGARITHMIC FUNCTIONS

4.1 EXPONENTIAL FUNCTIONS

1. The function $f(x) = 5^x$ is an exponential function with base 5; $f(-2) = 5^{-2} = \frac{1}{25}$, $f(0) = 5^0 = 1$, $f(2) = 5^2 = 25$, and $f(6) = 5^6 = 15{,}625$.

3. (a) To obtain the graph of $g(x) = 2^x - 1$ we start with the graph of $f(x) = 2^x$ and shift it *downward* 1 unit.

(b) To obtain the graph of $h(x) = 2^{x-1}$ we start with the graph of $f(x) = 2^x$ and shift it to the *right* 1 unit.

5. $f(x) = 4^x$; $f(0.5) = 2$, $f\left(\sqrt{2}\right) \approx 7.103$, $f(-1) = \frac{1}{4}$, $f\left(\frac{1}{3}\right) \approx 1.587$

7. $g(x) = \left(\frac{2}{3}\right)^{x-1}$; $g(1.3) \approx 0.885$, $g\left(\sqrt{5}\right) \approx 0.606$, $g(2\pi) \approx 0.117$, $g\left(-\frac{1}{2}\right) \approx 1.837$

9. $f(x) = 2^x$

x	y
-4	$\frac{1}{16}$
-2	$\frac{1}{4}$
0	1
2	4
4	16

11. $f(x) = \left(\frac{1}{3}\right)^x$

x	y
-2	9
-1	3
0	1
1	$\frac{1}{3}$
2	$\frac{1}{9}$

13. $g(x) = 3(1.3)^x$

x	y
-2	1.775
-1	2.308
0	3.0
1	3.9
2	5.07
3	6.591
4	8.568

15. $f(x) = 2^x$ and $g(x) = 2^{-x}$

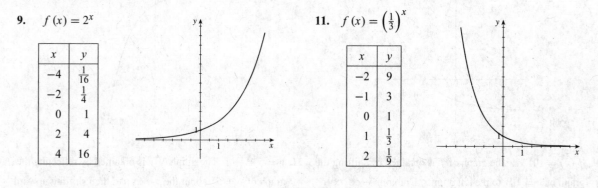

197

17. $f(x) = 4^x$ and $g(x) = 7^x$.

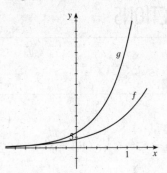

19. From the graph, $f(2) = a^2 = 9$, so $a = 3$. Thus $f(x) = 3^x$.

21. From the graph, $f(2) = a^2 = \frac{1}{16}$, so $a = \frac{1}{4}$. Thus $f(x) = \left(\frac{1}{4}\right)^x$.

23. The graph of $f(x) = 5^{x+1}$ is obtained from that of $y = 5^x$ by shifting 1 unit to the left, so it has graph II.

25. $g(x) = 2^x - 3$. The graph of g is obtained by shifting the graph of $y = 2^x$ downward 3 units. Domain: $(-\infty, \infty)$. Range: $(-3, \infty)$. Asymptote: $y = -3$.

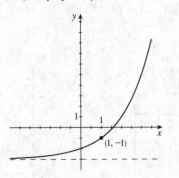

27. The graph of $f(x) = -3^x$ is obtained by reflecting the graph of $y = 3^x$ about the x-axis. Domain: $(-\infty, \infty)$. Range: $(-\infty, 0)$. Asymptote: $y = 0$.

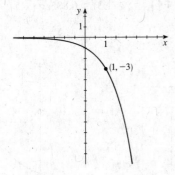

29. $f(x) = 10^{x+3}$. The graph of f is obtained by shifting the graph of $y = 10^x$ to the left 3 units. Domain: $(-\infty, \infty)$. Range: $(0, \infty)$. Asymptote: $y = 0$.

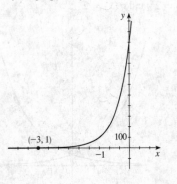

31. $y = 5^{-x} + 1$. The graph of y is obtained by reflecting the graph of $y = 5^x$ about the x-axis and then shifting upward 1 unit. Domain: $(-\infty, \infty)$. Range: $(1, \infty)$. Asymptote: $y = 1$.

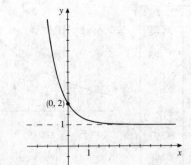

33. $h(x) = 2^{x-4} + 1$. The graph of h is obtained by shifting the graph of $y = 2^x$ to the right 4 units and upward 1 unit. Domain: $(-\infty, \infty)$. Range: $(1, \infty)$. Asymptote: $y = 1$.

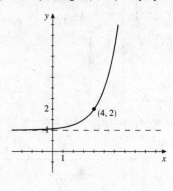

35. $g(x) = 1 - 3^{-x} = -3^{-x} + 1$. The graph of g is obtained by reflecting the graph of $y = 3^x$ about the x- and y-axes and then shifting upward 1 unit. Domain: $(-\infty, \infty)$. Range: $(-\infty, 1)$. Asymptote: $y = 1$.

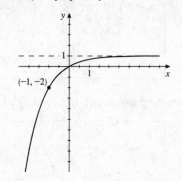

37. (a)

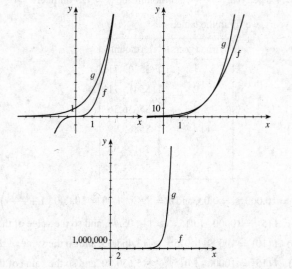

(b) Since $g(x) = 3(2^x) = 3f(x)$ and $f(x) > 0$, the height of the graph of $g(x)$ is always three times the height of the graph of $f(x) = 2^x$, so the graph of g is steeper than the graph of f.

39.

x	$f(x) = x^3$	$g(x) = 3^x$
0	0	1
1	1	3
2	8	9
3	27	27
4	64	81
5	125	243
6	216	729
7	343	2187
8	512	6561
9	729	19,683
10	1000	59,049
15	3375	14,348,907
20	8000	3,486,784,401

41. (a) From the graphs below, we see that the graph of f ultimately increases much more quickly than the graph of g.

(i) $[0, 5]$ by $[0, 20]$ **(ii)** $[0, 25]$ by $\left[0, 10^7\right]$ **(iii)** $[0, 50]$ by $\left[0, 10^8\right]$

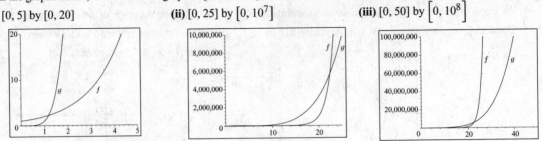

(b) From the graphs in parts (a)(i) and (a)(ii), we see that the approximate solutions are $x \approx 1.2$ and $x \approx 22.4$.

43.

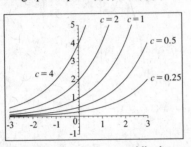

The larger the value of c, the more rapidly the graph of $f(x) = c2^x$ increases. Also notice that the graphs are just shifted horizontally 1 unit. This is because of our choice of c; each c in this exercise is of the form 2^k. So $f(x) = 2^k \cdot 2^x = 2^{x+k}$.

45. $y = 10^{x - x^2}$

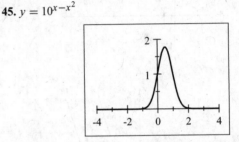

(a) From the graph, we see that the function is increasing on $(-\infty, 0.50]$ and decreasing on $[0.50, \infty)$.

(b) From the graph, we see that the range is approximately $(0, 1.78]$.

47. (a) After 1 hour, there are $1500 \cdot 2 = 3000$ bacteria. After 2 hours, there are $(1500 \cdot 2) \cdot 2 = 6000$ bacteria. After 3 hours, there are $(1500 \cdot 2 \cdot 2) \cdot 2 = 12{,}000$ bacteria. We see that after t hours, there are $N(t) = 1500 \cdot 2^t$ bacteria.

(b) After 24 hours, there are $N(24) = 1500 \cdot 2^{24} = 25{,}165{,}824{,}000$ bacteria.

49. Using the formula $A(t) = P(1 + i)^k$ with $P = 5000$, $i = 4\%$ per year $= \dfrac{0.04}{12}$ per month, and $k = 12 \cdot$ number of years, we fill in the table:

Time (years)	Amount
1	\$5203.71
2	\$5415.71
3	\$5636.36
4	\$5865.99
5	\$6104.98
6	\$6353.71

51. $P = 10{,}000$, $r = 0.03$, and $n = 2$. So $A(t) = 10{,}000\left(1 + \dfrac{0.03}{2}\right)^{2t} = 10{,}000 \cdot 1.015^{2t}$.

(a) $A(5) = 10000 \cdot 1.015^{10} \approx 11{,}605.41$, and so the value of the investment is \$11,605.41.

(b) $A(10) = 10000 \cdot 1.015^{20} \approx 13{,}468.55$, and so the value of the investment is \$13,468.55.

(c) $A(15) = 10000 \cdot 1.015^{30} \approx 15{,}630.80$, and so the value of the investment is \$15,630.80.

53. $P = 500, r = 0.0375$, and $n = 4$. So $A(t) = 500 \left(1 + \frac{0.0375}{4}\right)^{4t}$.

(a) $A(1) = 500 \left(1 + \frac{0.0375}{4}\right)^{4} \approx 519.02$, and so the value of the investment is \$519.02.

(b) $A(2) = 500 \left(1 + \frac{0.0375}{4}\right)^{8} \approx 538.75$, and so the value of the investment is \$538.75.

(c) $A(10) = 500 \left(1 + \frac{0.0375}{4}\right)^{40} \approx 726.23$, and so the value of the investment is \$726.23.

55. We must solve for P in the equation $10000 = P \left(1 + \frac{0.09}{2}\right)^{2(3)} = P (1.045)^6 \Leftrightarrow 10000 = 1.3023 P \Leftrightarrow P = 7678.96$.
Thus, the present value is \$7,678.96.

57. $r_{\text{APY}} = \left(1 + \frac{r}{n}\right)^n - 1$. Here $r = 0.08$ and $n = 12$, so $r_{\text{APY}} = \left(1 + \frac{0.08}{12}\right)^{12} - 1 \approx (1.0066667)^{12} - 1 \approx 0.083000$.
Thus, the annual percentage yield is about 8.3%.

59. (a) In this case the payment is \$1 million.

(b) In this case the total pay is $2 + 2^2 + 2^3 + \cdots + 2^{30} > 2^{30}$ cents $= \$10,737,418.24$. Since this is much more than method (a), method (b) is more profitable.

4.2   THE NATURAL EXPONENTIAL FUNCTION

1. The function $f(x) = e^x$ is called the *natural* exponential function. The number e is approximately equal to 2.71828.

3. $h(x) = e^x$; $h(3) = 20.086$, $h(0.23) = 1.259$, $h(1) = 2.718$, $h(-2) = 0.135$

5. $f(x) = 3e^x$

x	y
-2	0.41
-1	1.10
-0.5	1.82
0	3
0.5	4.95
1	8.15
2	22.17

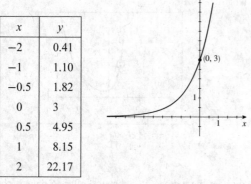

7. $y = -e^x$. The graph of $y = -e^x$ is obtained from the graph of $y = e^x$ by reflecting it about the x-axis. Domain: $(-\infty, \infty)$. Range: $(-\infty, 0)$. Asymptote: $y = 0$.

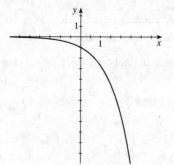

9. $y = e^{-x} - 1$. The graph of $y = e^{-x} - 1$ is obtained from the graph of $y = e^x$ by reflecting it about the y-axis then shifting downward 1 unit. Domain: $(-\infty, \infty)$. Range: $(-1, \infty)$. Asymptote: $y = -1$.

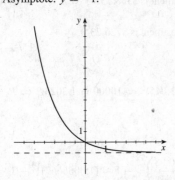

11. $y = e^{x-2}$. The graph of $y = e^{x-2}$ is obtained from the graph of $y = e^x$ by shifting it to the right 2 units. Domain: $(-\infty, \infty)$. Range: $(0, \infty)$. Asymptote: $y = 0$.

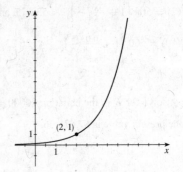

13. $h(x) = e^{x+1} - 3$. The graph of h is obtained from the graph of $y = e^x$ by shifting it to the left 1 unit and downward 3 units. Domain: $(-\infty, \infty)$. Range: $(-3, \infty)$. Asymptote: $y = -3$.

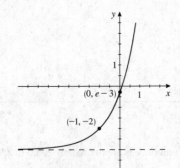

15. (a)

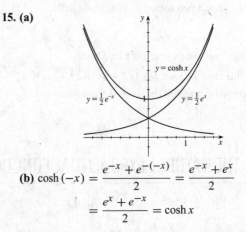

(b) $\cosh(-x) = \dfrac{e^{-x} + e^{-(-x)}}{2} = \dfrac{e^{-x} + e^x}{2}$

$= \dfrac{e^x + e^{-x}}{2} = \cosh x$

17. (a)

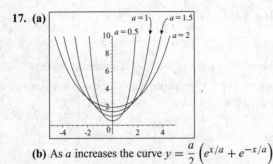

(b) As a increases the curve $y = \dfrac{a}{2}\left(e^{x/a} + e^{-x/a}\right)$ flattens out and the y intercept increases.

19. $g(x) = e^x + e^{-3x}$. The graph of $g(x)$ is shown in the viewing rectangle $[-4, 4]$ by $[0, 20]$. From the graph, we see that there is a local minimum of approximately 1.75 when $x \approx 0.27$.

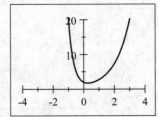

21. $m(t) = 13e^{-0.015t}$

(a) $m(0) = 13$ kg.

(b) $m(45) = 13e^{-0.015(45)} = 13e^{-0.675} = 6.619$ kg. Thus the mass of the radioactive substance after 45 days is about 6.6 kg.

23. $v(t) = 180\left(1 - e^{-0.2t}\right)$

(a) $v(0) = 180\left(1 - e^0\right) = 180(1 - 1) = 0.$

(b) $v(5) = 180\left(1 - e^{-0.2(5)}\right) \approx 180(0.632) = 113.76$ ft/s. So the

velocity after 5 s is about 113.8 ft/s.

$v(10) = 180\left(1 - e^{-0.2(10)}\right) \approx 180(0.865) = 155.7$ ft/s. So the

velocity after 10 s is about 155.7 ft/s.

(d) The terminal velocity is 180 ft/s.

(c)

25. $P(t) = \dfrac{1200}{1 + 11e^{-0.2t}}$

(a) $P(0) = \dfrac{1200}{1 + 11e^{-0.2(0)}} = \dfrac{1200}{1 + 11} = 100.$

(b) $P(10) = \dfrac{1200}{1 + 11e^{-0.2(10)}} \approx 482.$ $P(20) = \dfrac{1200}{1 + 11e^{-0.2(20)}} \approx 999.$ $P(30) = \dfrac{1200}{1 + 11e^{-0.2(30)}} \approx 1168.$

(c) As $t \to \infty$ we have $e^{-0.2t} \to 0$, so $P(t) \to \dfrac{1200}{1 + 0} = 1200$. The graph shown confirms this.

27. $P(t) = \dfrac{73.2}{6.1 + 5.9e^{-0.02t}}$

(a) In the year 2200, $t = 2200 - 2000 = 200$, and the population is

predicted to be $P(200) = \dfrac{73.2}{6.1 + 5.9e^{-0.02(200)}} \approx 11.79$ billion. In

2300, $t = 300$, and $P(300) = \dfrac{73.2}{6.1 + 5.9e^{-0.02(300)}} \approx 11.97$ billion.

(c) As t increases, the denominator approaches 6.1, so according to this

model, the world population approaches $\frac{73.2}{6.1} = 12$ billion people.

(b)

29. Using the formula $A(t) = Pe^{rt}$ with $P = 7000$ and

$r = 3\% = 0.03$, we fill in the table:

Time (years)	Amount
1	$7213.18
2	$7432.86
3	$7659.22
4	$7892.48
5	$8132.84
6	$8380.52

31. We use the formula $A(t) = Pe^{rt}$ with $P = 2000$ and $r = 3.5\% = 0.035$.

(a) $A(2) = 2000e^{0.035 \cdot 2} \approx \$2145.02$ **(b)** $A(42) = 2000e^{0.035 \cdot 4} \approx \$2300.55$ **(c)** $A(12) = 2000e^{0.035 \cdot 12} \approx \$3043.92$

33. (a) Using the formula $A(t) = P(1 + i)^k$ with $P = 600$, $i = 2.5\%$ per year $= 0.025$, and $k = 10$, we calculate

$A(10) = 600(1.025)^{10} \approx \$768.05.$

(b) Here $i = \dfrac{0.025}{2}$ semiannually and $k = 10 \cdot 2 = 20$, so $A(10) = 600\left(1 + \dfrac{0.025}{2}\right)^{20} \approx \$769.22.$

(c) Here $i = 2.5\%$ per year $= \dfrac{0.025}{4}$ quarterly and $k = 10 \cdot 4 = 40$, so $A(10) = 600\left(1 + \dfrac{0.025}{4}\right)^{40} \approx \$769.82.$

(d) Using the formula $A(t) = Pe^{rt}$ with $P = 600$, $r = 2.5\% = 0.025$, and $t = 10$, we have
$$A(10) = 600e^{0.025 \cdot 10} \approx \$770.42.$$

35. *Investment 1:* After 1 year, a \$100 investment grows to $A(1) = 100\left(1 + \frac{0.025}{2}\right)^2 \approx 102.52$.

Investment 2: After 1 year, a \$100 investment grows to $A(1) = 100\left(1 + \frac{0.0225}{4}\right)^4 = 102.27$.

Investment 3: After 1 year, a \$100 investment grows to $A(1) = 100e^{0.02} \approx 102.02$.

We see that Investment 1 yields the highest return.

37. (a) $A(t) = Pe^{rt} = 5000e^{0.09t}$

(b)

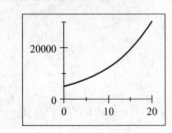

(c) $A(t) = 25{,}000$ when $t \approx 17.88$ years.

4.3   LOGARITHMIC FUNCTIONS

1. $\log x$ is the exponent to which the base 10 must be raised in order to get x.

x	10^3	10^2	10^1	10^0	10^{-1}	10^{-2}	10^{-3}	$10^{1/2}$
$\log x$	3	2	1	0	-1	-2	-3	$1/2$

3. (a) $5^3 = 125$, so $\log_5 125 = 3$. **(b)** $\log_5 25 = 2$, so $5^2 = 25$.

5.

Logarithmic form	Exponential form
$\log_8 8 = 1$	$8^1 = 8$
$\log_8 64 = 2$	$8^2 = 64$
$\log_8 4 = \frac{2}{3}$	$8^{2/3} = 4$
$\log_8 512 = 3$	$8^3 = 512$
$\log_8 \frac{1}{8} = -1$	$8^{-1} = \frac{1}{8}$
$\log_8 \frac{1}{64} = -2$	$8^{-2} = \frac{1}{64}$

7. (a) $5^2 = 25$

(b) $5^0 = 1$

9. (a) $8^{1/3} = 2$

(b) $2^{-3} = \frac{1}{8}$

11. (a) $3^x = 5$

(b) $7^2 = 3y$

13. (a) $5 = e^{3y}$

(b) $t + 1 = e^{-1}$

15. (a) $\log_5 125 = 3$

(b) $\log_{10} 0.0001 = -4$

17. (a) $\log_8 \frac{1}{8} = -1$

(b) $\log_2 \left(\frac{1}{8}\right) = -3$

19. (a) $x = \log_5 3$

(b) $5 = \log_4 z$

21. (a) $\ln 2 = x$

(b) $\ln y = 3$

23. (a) $\log_3 3 = 1$

(b) $\log_3 1 = \log_3 3^0 = 0$

(c) $\log_3 3^2 = 2$

25. (a) $\log_6 36 = \log_6 6^2 = 2$

(b) $\log_9 81 = \log_9 9^2 = 2$

(c) $\log_7 7^{10} = 10$

27. (a) $\log_3 \left(\frac{1}{27}\right) = \log_3 3^{-3} = -3$

(b) $\log_{10} \sqrt{10} = \log_{10} 10^{1/2} = \frac{1}{2}$

(c) $\log_5 0.2 = \log_5 \left(\frac{1}{5}\right) = \log_5 5^{-1} = -1$

29. (a) $2^{\log_2 37} = 37$

(b) $3^{\log_3 8} = 8$

(c) $e^{\ln \sqrt{5}} = \sqrt{5}$

31. (a) $\log_8 0.25 = \log_8 8^{-2/3} = -\frac{2}{3}$

(b) $\ln e^4 = 4$

(c) $\ln\left(\frac{1}{e}\right) = \ln e^{-1} = -1$

33. (a) $\log_2 x = 5 \Leftrightarrow x = 2^5 = 32$

(b) $x = \log_2 16 = \log_2 2^4 = 4$

35. (a) $\ln x = 3 \Leftrightarrow x = e^3$

(b) $\ln e^2 = x \Leftrightarrow x = 2\ln e = 2$

37. (a) $x = \log_3 243 = \log_3 3^5 = 5$

(b) $\log_3 x = 3 \Leftrightarrow x = 3^3 = 27$

39. (a) $\log_{10} x = 2 \Leftrightarrow x = 10^2 = 100$

(b) $\log_5 x = 2 \Leftrightarrow x = 5^2 = 25$

41. (a) $\log_x 16 = 4 \Leftrightarrow x^4 = 16 \Leftrightarrow x = 2$

(b) $\log_x 8 = \frac{3}{2} \Leftrightarrow x^{3/2} = 8 \Leftrightarrow x = 8^{2/3} = 4$

43. (a) $\log 2 \approx 0.3010$

(b) $\log 35.2 \approx 1.5465$

(c) $\log\left(\frac{2}{3}\right) \approx -0.1761$

45. (a) $\ln 5 \approx 1.6094$

(b) $\ln 25.3 \approx 3.2308$

(c) $\ln\left(1 + \sqrt{3}\right) \approx 1.0051$

47.

x	$f(x)$
$\frac{1}{3^3}$	-3
$\frac{1}{3^2}$	-2
$\frac{1}{3}$	-1
1	0
3	1
3^2	2

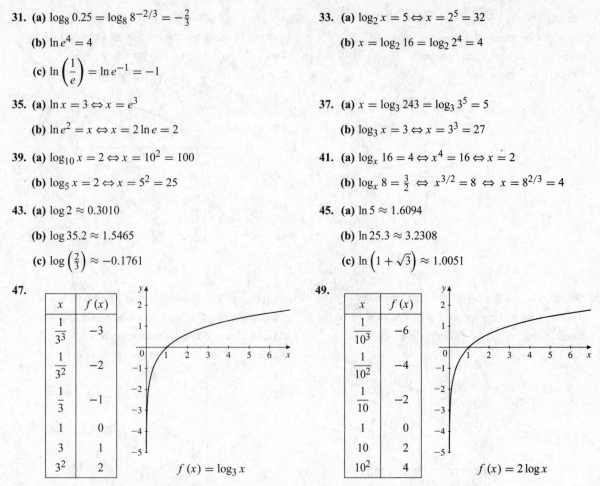

$f(x) = \log_3 x$

49.

x	$f(x)$
$\frac{1}{10^3}$	-6
$\frac{1}{10^2}$	-4
$\frac{1}{10}$	-2
1	0
10	2
10^2	4

$f(x) = 2\log x$

51. Since the point $(5, 1)$ is on the graph, we have $1 = \log_a 5 \Leftrightarrow a^1 = 5$. Thus the function is $y = \log_5 x$.

53. Since the point $\left(3, \frac{1}{2}\right)$ is on the graph, we have $\frac{1}{2} = \log_a 3 \Leftrightarrow a^{1/2} = 3 \Leftrightarrow a = 9$. Thus the function is $y = \log_9 x$.

55. I

57. The graph of $y = \log_4 x$ is obtained from that of $y = 4^x$ by reflecting it in the line $y = x$.

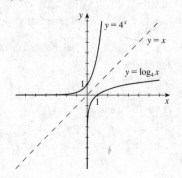

59. The graph of $f(x) = \log_2(x - 4)$ is obtained from that of $y = \log_2 x$ by shifting it to the right 4 units.
Domain: $(4, \infty)$. Range: $(-\infty, \infty)$.
Vertical asymptote: $x = 4$.

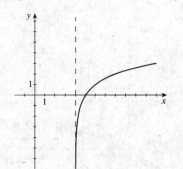

61. The graph of $g(x) = \log_5(-x)$ is obtained from that of $y = \log_5 x$ by reflecting it about the y-axis. Domain: $(-\infty, 0)$. Range: $(-\infty, \infty)$. Vertical asymptote: $x = 0$.

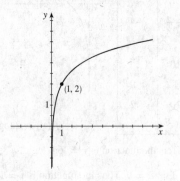

63. The graph of $h(x) = \ln(x + 5)$ is obtained from that of $y = \ln x$ by shifting to the left 5 units. Domain: $(-5, \infty)$. Range: $(-\infty, \infty)$. Vertical asymptote: $x = -5$.

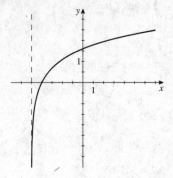

65. The graph of $y = 2 + \log_3 x$ is obtained from that of $y = \log_3 x$ by shifting upward 2 units. Domain: $(0, \infty)$. Range: $(-\infty, \infty)$. Vertical asymptote: $x = 0$.

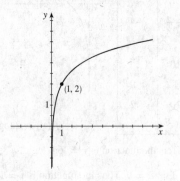

67. The graph of $y = 1 - \log_{10} x$ is obtained from that of $y = \log_{10} x$ by reflecting it about the x-axis, and then shifting it upward 1 unit. Domain: $(0, \infty)$. Range: $(-\infty, \infty)$. Vertical asymptote: $x = 0$.

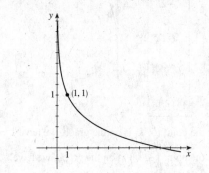

69. The graph of $y = |\ln x|$ is obtained from that of $y = \ln x$ by reflecting the part of the graph for $0 < x < 1$ about the x-axis. Domain: $(0, \infty)$. Range: $[0, \infty)$. Vertical asymptote: $x = 0$.

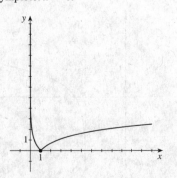

71. $f(x) = \log_{10}(x + 3)$. We require that $x + 3 > 0 \Leftrightarrow$ $x > -3$, so the domain is $(-3, \infty)$.

73. $g(x) = \log_3\left(x^2 - 1\right)$. We require that $x^2 - 1 > 0 \Leftrightarrow$ $x^2 > 1 \Rightarrow x < -1$ or $x > 1$, so the domain is $(-\infty, -1) \cup (1, \infty)$.

75. $h(x) = \ln x + \ln(2 - x)$. We require that $x > 0$ and $2 - x > 0 \Leftrightarrow x > 0$ and $x < 2 \Leftrightarrow 0 < x < 2$, so the domain is $(0, 2)$.

77. $y = \log_{10}\left(1 - x^2\right)$ has domain $(-1, 1)$, vertical asymptotes $x = -1$ and $x = 1$, and local maximum $y = 0$ at $x = 0$.

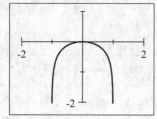

79. $y = x + \ln x$ has domain $(0, \infty)$, vertical asymptote $x = 0$, and no local maximum or minimum.

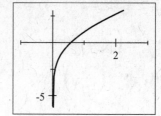

81. $y = \dfrac{\ln x}{x}$ has domain $(0, \infty)$, vertical asymptote $x = 0$, horizontal asymptote $y = 0$, and local maximum $y \approx 0.37$ at $x \approx 2.72$.

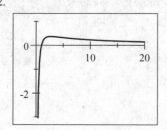

83. $f(x) = 2^x$ and $g(x) = x + 1$ both have domain $(-\infty, \infty)$, so $(f \circ g)(x) = f(g(x)) = 2^{g(x)} = 2^{x+1}$ with domain $(-\infty, \infty)$ and $(g \circ f)(x) = g(f(x)) = 2^x + 1$ with domain $(-\infty, \infty)$.

85. $f(x) = \log_2 x$ has domain $(0, \infty)$ and $g(x) = x - 2$ has domain $(-\infty, \infty)$, so $(f \circ g)(x) = f(g(x)) = \log_2(x - 2)$ with domain $(2, \infty)$ and $(g \circ f)(x) = g(f(x)) = \log_2 x - 2$ with domain $(0, \infty)$.

87. The graph of $g(x) = \sqrt{x}$ grows faster than the graph of $f(x) = \ln x$.

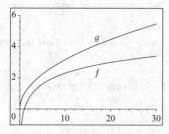

89. (a)

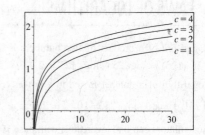

(b) Notice that $f(x) = \log(cx) = \log c + \log x$, so as c increases, the graph of $f(x) = \log(cx)$ is shifted upward $\log c$ units.

91. (a) $f(x) = \log_2\left(\log_{10} x\right)$. Since the domain of $\log_2 x$ is the positive real numbers, we have: $\log_{10} x > 0 \Leftrightarrow x > 10^0 = 1$. Thus the domain of $f(x)$ is $(1, \infty)$.

(b) $y = \log_2\left(\log_{10} x\right) \Leftrightarrow 2^y = \log_{10} x \Leftrightarrow 10^{2^y} = x$. Thus $f^{-1}(x) = 10^{2^x}$.

93. (a) $f(x) = \dfrac{2^x}{1 + 2^x}$. $y = \dfrac{2^x}{1 + 2^x} \Leftrightarrow y + y2^x = 2^x$

$\Leftrightarrow y = 2^x - y2^x = 2^x (1 - y) \Leftrightarrow 2^x = \dfrac{y}{1 - y}$

$\Leftrightarrow x = \log_2 \left(\dfrac{y}{1 - y} \right)$. Thus

$f^{-1}(x) = \log_2 \left(\dfrac{x}{1 - x} \right)$.

(b) $\dfrac{x}{1 - x} > 0$. Solving this using the methods from Chapter 1, we start with the endpoints, 0 and 1.

Interval	$(-\infty, 0)$	$(0, 1)$	$(1, \infty)$
Sign of x	$-$	$+$	$+$
Sign of $1 - x$	$+$	$+$	$-$
Sign of $\dfrac{x}{1 - x}$	$-$	$+$	$-$

Thus the domain of $f^{-1}(x)$ is $(0, 1)$.

95. Using $D = 0.73D_0$ we have $A = -8267 \ln \left(\dfrac{D}{D_0} \right) = -8267 \ln 0.73 \approx 2601$ years.

97. When $r = 6\%$ we have $t = \dfrac{\ln 2}{0.06} \approx 11.6$ years. When $r = 7\%$ we have $t = \dfrac{\ln 2}{0.07} \approx 9.9$ years. And when $r = 8\%$ we have $t = \dfrac{\ln 2}{0.08} \approx 8.7$ years.

99. Using $A = 100$ and $W = 5$ we find the ID to be $\dfrac{\log (2A/W)}{\log 2} = \dfrac{\log (2 \cdot 100/5)}{\log 2} = \dfrac{\log 40}{\log 2} \approx 5.32$. Using $A = 100$ and $W = 10$ we find the ID to be $\dfrac{\log (2A/W)}{\log 2} = \dfrac{\log (2 \cdot 100/10)}{\log 2} = \dfrac{\log 20}{\log 2} \approx 4.32$. So the smaller icon is $\dfrac{5.23}{4.32} \approx 1.23$ times harder.

101. $\log \left(\log 10^{100} \right) = \log 100 = 2$

$\log \left(\log \left(\log 10^{\text{googol}} \right) \right) = \log (\log (\text{googol})) = \log \left(\log 10^{100} \right) = \log (100) = 2$

103. The numbers between 1000 and 9999 (inclusive) each have 4 digits, while $\log 1000 = 3$ and $\log 10,000 = 4$. Since $[\![\log x]\!] = 3$ for all integers x where $1000 \leq x < 10,000$, the number of digits is $[\![\log x]\!] + 1$. Likewise, if x is an integer where $10^{n-1} \leq x < 10^n$, then x has n digits and $[\![\log x]\!] = n - 1$. Since $[\![\log x]\!] = n - 1 \Leftrightarrow n = [\![\log x]\!] + 1$, the number of digits in x is $[\![\log x]\!] + 1$.

4.4   LAWS OF LOGARITHMS

1. The logarithm of a product of two numbers is the same as the *sum* of the logarithms of these numbers. So $\log_5 (25 \cdot 125) = \log_5 25 + \log_5 125 = 2 + 3 = 5$.

3. The logarithm of a number raised to a power is the same as the power *times* the logarithm of the number. So $\log_5 \left(25^{10} \right) = 10 \cdot \log_5 25 = 10 \cdot 2 = 20$.

5. Most calculators can find logarithms with base 10 and base e. To find logarithms with different bases we use the *change of base* formula. To find $\log_7 12$ we write

$$\log_7 12 = \frac{\log 12}{\log 7} \approx \frac{1.079}{0.845} \approx 1.277$$

7. $\log 4 + \log 25 = \log (4 \cdot 25) = \log 100 = 2$

9. $\log_4 192 - \log_4 3 = \log_4 \frac{192}{3} = \log_4 64 = \log_4 4^3 = 3$ **11.** $-\frac{1}{2} \log 64 = \log 64^{-1/2} = \log \frac{1}{8} \approx -0.903$

13. $\log_3 \sqrt{27} = \log_3 3^{3/2} = \frac{3}{2}$

15. $\log_2 6 - \log_2 15 + \log_2 20 = \log_2 \frac{6}{15} + \log_2 20 = \log_2 \left(\frac{2}{5} \cdot 20 \right) = \log_2 8 = \log_2 2^3 = 3$

17. $\log_4 16^{100} = \log_4 \left(4^2\right)^{100} = \log_4 4^{200} = 200$

19. $\log\left(\log 10^{10,000}\right) = \log\left(10,000 \log 10\right) = \log\left(10,000 \cdot 1\right) = \log\left(10,000\right) = \log 10^4 = 4 \log 10 = 4$

21. $\log_2 2x = \log_2 2 + \log_2 x = 1 + \log_2 x$

23. $\log_2\left(x\left(x-1\right)\right) = \log_2 x + \log_2\left(x-1\right)$

25. $\log 6^{10} = 10 \log 6$

27. $\ln\left(\sqrt{z}\right) = \ln\left(z^{1/2}\right) = \frac{1}{2}\ln z$

29. $\log_5 \sqrt[3]{x^2+1} = \frac{1}{3}\log_5\left(x^2+1\right)$

31. $\log_2\left(xy\right)^{10} = 10 \log_2\left(xy\right) = 10\left(\log_2 x + \log_2 y\right)$

33. $\log_2\left(AB^2\right) = \log_2 A + \log_2 B^2 = \log_2 A + 2 \log_2 B$

35. $\log\left(\dfrac{x^3 y^4}{z^6}\right) = \log\left(x^3 y^4\right) - \log z^6 = 3 \log x + 4 \log y - 6 \log z$

37. $\log_2\left(\dfrac{x\left(x^2+1\right)}{\sqrt{x^2-1}}\right) = \log_2 x + \log_2\left(x^2+1\right) - \frac{1}{2}\log_2\left(x^2-1\right)$

39. $\ln\left(x\sqrt{\dfrac{y}{z}}\right) = \ln x + \frac{1}{2}\ln\left(\dfrac{y}{z}\right) = \ln x + \frac{1}{2}\left(\ln y - \ln z\right)$

41. $\log\sqrt[4]{x^2+y^2} = \frac{1}{4}\log\left(x^2+y^2\right)$

43. $\log\sqrt{\dfrac{x^2+4}{\left(x^2+1\right)\left(x^3-7\right)^2}} = \frac{1}{2}\log\dfrac{x^2+4}{\left(x^2+1\right)\left(x^3-7\right)^2} = \frac{1}{2}\left[\log\left(x^2+4\right) - \log\left(x^2+1\right)\left(x^3-7\right)^2\right]$

$$= \frac{1}{2}\left[\log\left(x^2+4\right) - \log\left(x^2+1\right) - 2\log\left(x^3-7\right)\right]$$

45. $\ln\dfrac{x^3\sqrt{x-1}}{3x+4} = \ln\left(x^3\sqrt{x-1}\right) - \ln\left(3x+4\right) = 3 \ln x + \frac{1}{2}\ln\left(x-1\right) - \ln\left(3x+4\right)$

47. $\log_3 5 + 5 \log_3 2 = \log_3 5 + \log_3 2^5 = \log_3\left(5 \cdot 2^5\right) = \log_3 160$

49. $\ln 5 + 2 \ln x + 3 \ln\left(x^2+5\right) = \ln\left(5x^2\right) + \ln\left(x^2+5\right)^3 = \ln\left[5x^2\left(x^2+5\right)^3\right]$

51. $4 \log x - \frac{1}{3}\log\left(x^2+1\right) + 2 \log\left(x-1\right) = \log x^4 - \log\sqrt[3]{x^2+1} + \log\left(x-1\right)^2$

$$= \log\left(\dfrac{x^4}{\sqrt[3]{x^2+1}}\right) + \log\left(x-1\right)^2 = \log\left(\dfrac{x^4\left(x-1\right)^2}{\sqrt[3]{x^2+1}}\right)$$

53. $\ln\left(a+b\right) + \ln\left(a-b\right) - 2 \ln c = \ln\left(\left(a+b\right)\left(a-b\right)\right) - \ln\left(c^2\right) = \ln\dfrac{a^2-b^2}{c^2}$

55. $\frac{1}{3}\log\left(x+2\right)^3 + \frac{1}{2}\left[\log x^4 - \log\left(x^2-x-6\right)^2\right] = 3 \cdot \frac{1}{3}\log\left(x+2\right) + \frac{1}{2}\log\dfrac{x^4}{\left(x^2-x-6\right)^2}$

$$= \log\left(x+2\right) + \log\left(\dfrac{x^4}{\left[\left(x-3\right)\left(x+2\right)\right]^2}\right)^{1/2} = \log\left(x+2\right) + \log\dfrac{x^2}{\left(x-3\right)\left(x+2\right)} = \log\dfrac{x^2\left(x+2\right)}{\left(x-3\right)\left(x+2\right)} = \log\dfrac{x^2}{x-3}$$

57. $\log_2 5 = \dfrac{\log 5}{\log 2} \approx 2.321928$

59. $\log_3 16 = \dfrac{\log 16}{\log 3} \approx 2.523719$

61. $\log_7 2.61 = \dfrac{\log 2.61}{\log 7} \approx 0.493008$

63. $\log_4 125 = \dfrac{\log 125}{\log 4} \approx 3.482892$

65. $\log_3 x = \dfrac{\log_e x}{\log_e 3} = \dfrac{\ln x}{\ln 3} = \dfrac{1}{\ln 3}\ln x$. The graph of $y = \dfrac{1}{\ln 3}\ln x$ is

shown in the viewing rectangle $[-1, 4]$ by $[-3, 2]$.

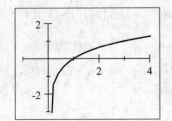

67. $\log e = \dfrac{\ln e}{\ln 10} = \dfrac{1}{\ln 10}$

69. $-\ln\left(x - \sqrt{x^2 - 1}\right) = \ln\left(\dfrac{1}{x - \sqrt{x^2 - 1}}\right) = \ln\left(\dfrac{1}{x - \sqrt{x^2 - 1}} \cdot \dfrac{x + \sqrt{x^2 - 1}}{x + \sqrt{x^2 - 1}}\right) = \ln\left(\dfrac{x + \sqrt{x^2 - 1}}{x^2 - (x^2 - 1)}\right)$

$\qquad = \ln\left(x + \sqrt{x^2 - 1}\right)$

71. (a) $\log P = \log c - k \log W \Leftrightarrow \log P = \log c - \log W^k \Leftrightarrow \log P = \log\left(\dfrac{c}{W^k}\right) \Leftrightarrow P = \dfrac{c}{W^k}$.

(b) Using $k = 2.1$ and $c = 8000$, when $W = 2$ we have $P = \dfrac{8000}{2^{2.1}} \approx 1866$ and when $W = 10$ we have $P = \dfrac{8000}{10^{2.1}} \approx 64$.

73. (a) $M = -2.5 \log(B/B_0) = -2.5 \log B + 2.5 \log B_0$.

(b) Suppose B_1 and B_2 are the brightness of two stars such that $B_1 < B_2$ and let M_1 and M_2 be their respective magnitudes. Since $\log$ is an increasing function, we have $\log B_1 < \log B_2$. Then $\log B_1 < \log B_2 \Leftrightarrow$ $\log B_1 - \log B_0 < \log B_2 - \log B_0 \Leftrightarrow \log(B_1/B_0) < \log(B_2/B_0) \Leftrightarrow -2.5 \log(B_1/B_0) > -2.5 \log(B_2/B_0) \Leftrightarrow$ $M_1 > M_2$. Thus the brighter star has less magnitudes.

(c) Let B_1 be the brightness of the star Albiero. Then $100 B_1$ is the brightness of Betelgeuse, and its magnitude is

$M = -2.5 \log(100 B_1/B_0) = -2.5\left[\log 100 + \log(B_1/B_0)\right] = -2.5\left[2 + \log(B_1/B_0)\right] = -5 - 2.5 \log(B_1/B_0)$

$\qquad = -5 + \text{magnitude of Albiero}$

75. The error is on the first line: $\log 0.1 < 0$, so $2 \log 0.1 < \log 0.1$.

4.5   EXPONENTIAL AND LOGARITHMIC EQUATIONS

1. (a) First we isolate e^x to get the equivalent equation $e^x = 25$.

(b) Next, we take the natural logarithm of each side to get the equivalent equation $x = \ln 25$.

(c) Now we use a calculator to find $x \approx 3.219$.

3. Because the function 4^x is one-to-one, $4^{x+1} = 64 \Leftrightarrow 4^{x+1} = 4^3 \Leftrightarrow x + 1 = 3 \Leftrightarrow x = 2$.

5. Because the function 5^x is one-to-one, $5^{2x-3} = 1 \Leftrightarrow 5^{2x-3} = 5^0 \Leftrightarrow 2x - 3 = 0 \Leftrightarrow x = \frac{3}{2}$.

7. Because the function 3^x is one-to-one, $3^{2x-8} = 3^{5x+1} \Leftrightarrow 2x - 8 = 5x + 1 \Leftrightarrow x = -3$.

9. Because the function 6^x is one-to-one, $6^{x^2-1} = 6^{1-x^2} \Leftrightarrow x^2 - 1 = 1 - x^2 \Leftrightarrow 2x^2 = 2 \Leftrightarrow x = \pm 1$.

11. (a) $10^x = 25 \Leftrightarrow \log 10^x = \log 25 \Leftrightarrow x \log 10 = \log 25 \Leftrightarrow x = \log 25 = 2 \log 5$

(b) $x \approx 1.397940$

13. (a) $e^{-2x} = 7 \Leftrightarrow \ln e^{-2x} = \ln 7 \Leftrightarrow -2x \ln e = \ln 7 \Leftrightarrow -2x = \ln 7 \Leftrightarrow x = -\frac{1}{2}\ln 7$

(b) $x \approx -0.972955$

15. (a) $2^{1-x} = 3 \Leftrightarrow \log 2^{1-x} = \log 3 \Leftrightarrow (1-x) \log 2 = \log 3 \Leftrightarrow 1 - x = \frac{\log 3}{\log 2} \Leftrightarrow x = 1 - \frac{\log 3}{\log 2}$

(b) $x \approx -0.584963$

17. (a) $3e^x = 10 \Leftrightarrow e^x = \frac{10}{3} \Leftrightarrow x = \ln\left(\frac{10}{3}\right)$

(b) $x \approx 1.203973$

19. (a) $100\,(1.04)^{2t} = 300 \Leftrightarrow 1.04^{2t} = 3 \Leftrightarrow \log 1.04^{2t} = \log 3 \Leftrightarrow 2t \log 1.04 = \log 3 \Leftrightarrow t = \frac{\log 3}{2 \log 1.04}$

(b) $x \approx 14.005511$

21. (a) $e^{1-4x} = 2 \Leftrightarrow 1 - 4x = \ln 2 \Leftrightarrow -4x = -1 + \ln 2 \Leftrightarrow x = \frac{1}{4}\,(1 - \ln 2)$

(b) $x \approx 0.0767132$

23. (a) $e^{2x+1} = 200 \Leftrightarrow 2x + 1 = \ln 200 \Leftrightarrow 2x = -1 + \ln 200 \Leftrightarrow x = \frac{-1 + \ln 200}{2}$

(b) $x \approx 2.149159$

25. (a) $8^{0.4x} = 5 \Leftrightarrow \log 8^{0.4x} = \log 5 \Leftrightarrow 0.4x \log 8 = \log 5 \Leftrightarrow 0.4x = \frac{\log 5}{\log 8} \Leftrightarrow x = \frac{\log 5}{0.4 \log 8}$

(b) $x \approx 1.934940$

27. (a) $3^{x/14} = 0.1 \Leftrightarrow \log 3^{x/14} = \log 0.1 \Leftrightarrow \left(\frac{x}{14}\right) \log 3 = \log 0.1 \Leftrightarrow x = \frac{14 \log 0.1}{\log 3}$

(b) $x \approx -29.342646$

29. (a) $4\left(1 + 10^{5x}\right) = 9 \Leftrightarrow 1 + 10^{5x} = \frac{9}{4} \Leftrightarrow 10^{5x} = \frac{5}{4} \Leftrightarrow 5x = \log\left(\frac{5}{4}\right) \Leftrightarrow x = \frac{1}{5}\,(\log 5 - \log 4)$

(b) $x \approx 0.0193820$

31. (a) $5^x = 4^{x+1} \Leftrightarrow \log 5^x = \log 4^{x+1} \Leftrightarrow x \log 5 = (x+1) \log 4 = x \log 4 + \log 4 \Leftrightarrow x \log 5 - x \log 4 = \log 4 \Leftrightarrow$

$x\,(\log 5 - \log 4) = \log 4 \Leftrightarrow x = \frac{\log 4}{\log 5 - \log 4}$

(b) $x \approx 6.212567$

33. (a) $2^{3x+1} = 3^{x-2} \Leftrightarrow \log 2^{3x+1} = \log 3^{x-2} \Leftrightarrow (3x + 1) \log 2 = (x - 2) \log 3 \Leftrightarrow 3x \log 2 + \log 2 = x \log 3 - 2 \log 3 \Leftrightarrow$

$3x \log 2 - x \log 3 = -\log 2 - 2 \log 3 \Leftrightarrow x\,(3 \log 2 - \log 3) = -(\log 2 + 2 \log 3) \Leftrightarrow x = -\frac{\log 2 + 2 \log 3}{3 \log 2 - \log 3}$

(b) $x \approx -2.946865$

35. (a) $\dfrac{50}{1 + e^{-x}} = 4 \Leftrightarrow 50 = 4 + 4e^{-x} \Leftrightarrow 46 = 4e^{-x} \Leftrightarrow 11.5 = e^{-x} \Leftrightarrow \ln 11.5 = -x \Leftrightarrow x = -\ln 11.5$

(b) $x \approx -2.442347$

37. $e^{2x} - 3e^x + 2 = 0 \Leftrightarrow \left(e^x - 1\right)\left(e^x - 2\right) = 0 \Rightarrow e^x - 1 = 0$ or $e^x - 2 = 0$. If $e^x - 1 = 0$, then $e^x = 1 \Leftrightarrow x = \ln 1 = 0$. If $e^x - 2 = 0$, then $e^x = 2 \Leftrightarrow x = \ln 2 \approx 0.6931$. So the solutions are $x = 0$ and $x \approx 0.6931$.

39. $e^{4x} + 4e^{2x} - 21 = 0 \Leftrightarrow \left(e^{2x} + 7\right)\left(e^{2x} - 3\right) = 0 \Rightarrow e^{2x} = -7$ or $e^{2x} = 3$. Now $e^{2x} = -7$ has no solution, since $e^{2x} > 0$ for all x. But we can solve $e^{2x} = 3 \Leftrightarrow 2x = \ln 3 \Leftrightarrow x = \frac{1}{2} \ln 3 \approx 0.5493$. So the only solution is $x \approx 0.5493$.

41. $x^2 2^x - 2^x = 0 \Leftrightarrow 2^x\left(x^2 - 1\right) = 0 \Rightarrow 2^x = 0$ (never) or $x^2 - 1 = 0$. If $x^2 - 1 = 0$, then $x^2 = 1 \Rightarrow x = \pm 1$. So the only solutions are $x = \pm 1$.

43. $4x^3 e^{-3x} - 3x^4 e^{-3x} = 0 \Leftrightarrow x^3 e^{-3x}\,(4 - 3x) = 0 \Rightarrow x = 0$ or $e^{-3x} = 0$ (never) or $4 - 3x = 0$. If $4 - 3x = 0$, then $3x = 4 \Leftrightarrow x = \frac{4}{3}$. So the solutions are $x = 0$ and $x = \frac{4}{3}$.

45. $\log x + \log\,(x - 1) = \log\,(4x) \Leftrightarrow \log\,[x\,(x - 1)] = \log\,(4x) \Leftrightarrow x^2 - x = 4x \Leftrightarrow x^2 - 5x = 0 \Leftrightarrow x\,(x - 5) = 0 \Rightarrow x = 0$ or $x = 5$. So the possible solutions are $x = 0$ and $x = 5$. However, when $x = 0$, $\log x$ is undefined. Thus the only solution is $x = 5$.

47. $2 \log x = \log 2 + \log\,(3x - 4) \Leftrightarrow \log\left(x^2\right) = \log\,(6x - 8) \Leftrightarrow x^2 = 6x - 8 \Leftrightarrow x^2 - 6x + 8 = 0 \Leftrightarrow (x - 4)\,(x - 2) = 0 \Leftrightarrow x = 4$ or $x = 2$. Thus the solutions are $x = 4$ and $x = 2$.

49. $\log_2 3 + \log_2 x = \log_2 5 + \log_2\,(x - 2) \Leftrightarrow \log_2\,(3x) = \log_2\,(5x - 10) \Leftrightarrow 3x = 5x - 10 \Leftrightarrow 2x = 10 \Leftrightarrow x = 5$

51. $\ln x = 10 \Leftrightarrow x = e^{10} \approx 22{,}026$

53. $\log x = -2 \Leftrightarrow x = 10^{-2} = 0.01$

55. $\log\,(3x + 5) = 2 \Leftrightarrow 3x + 5 = 10^2 = 100 \Leftrightarrow 3x = 95 \Leftrightarrow x = \frac{95}{3} \approx 31.6667$

57. $4 - \log(3 - x) = 3 \Leftrightarrow \log(3 - x) = 1 \Leftrightarrow 3 - x = 10 \Leftrightarrow x = -7$

59. $\log_2 x + \log_2(x - 3) = 2 \Leftrightarrow \log_2[x(x - 3)] = 2 \Leftrightarrow x^2 - 3x = 2^2 \Leftrightarrow x^2 - 3x - 4 = 0 \Leftrightarrow (x - 4)(x + 1) \Leftrightarrow x = -1$ or $x = 4$. Since $\log(-1 - 3) = \log(-4)$ is undefined, the only solution is $x = 4$.

61. $\log_9(x - 5) + \log_9(x + 3) = 1 \Leftrightarrow \log_9[(x - 5)(x + 3)] = 1 \Leftrightarrow (x - 5)(x + 3) = 9^1 \Leftrightarrow x^2 - 2x - 24 = 0 \Leftrightarrow (x - 6)(x + 4) = 0 \Rightarrow x = 6$ or -4. However, $x = -4$ is inadmissible, so $x = 6$ is the only solution.

63. $\log_5(x + 1) - \log_5(x - 1) = 2 \Leftrightarrow \log_5\left(\dfrac{x + 1}{x - 1}\right) = 2 \Leftrightarrow \dfrac{x + 1}{x - 1} = 5^2 \Leftrightarrow x + 1 = 25x - 25 \Leftrightarrow 24x = 26 \Leftrightarrow x = \frac{13}{12}$

65. $\log(x + 3) = \log x + \log 3 \Leftrightarrow \log(x + 3) = \log(3x) \Leftrightarrow x + 3 = 3x \Leftrightarrow 2x = 3 \Leftrightarrow x = \frac{3}{2}$

67. $2^{2/\log_5 x} = \frac{1}{16} \Leftrightarrow \log_2 2^{2/\log_5 x} = \log_2\left(\frac{1}{16}\right) \Leftrightarrow \dfrac{2}{\log_5 x} = -4 \Leftrightarrow \log_5 x = -\frac{1}{2} \Leftrightarrow x = 5^{-1/2} = \frac{1}{\sqrt{5}} \approx 0.4472$

69. $\ln x = 3 - x \Leftrightarrow \ln x + x - 3 = 0$. Let $f(x) = \ln x + x - 3$. We need to solve the equation $f(x) = 0$. From the graph of f, we get $x \approx 2.21$.

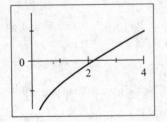

71. $x^3 - x = \log_{10}(x + 1) \Leftrightarrow x^3 - x - \log_{10}(x + 1) = 0$. Let $f(x) = x^3 - x - \log_{10}(x + 1)$. We need to solve the equation $f(x) = 0$. From the graph of f, we get $x = 0$ or $x \approx 1.14$.

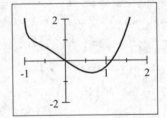

73. $e^x = -x \Leftrightarrow e^x + x = 0$. Let $f(x) = e^x + x$. We need to solve the equation $f(x) = 0$. From the graph of f, we get $x \approx -0.57$.

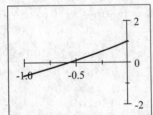

75. $4^{-x} = \sqrt{x} \Leftrightarrow 4^{-x} - \sqrt{x} = 0$. Let $f(x) = 4^{-x} - \sqrt{x}$. We need to solve the equation $f(x) = 0$. From the graph of f, we get $x \approx 0.36$.

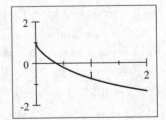

77. $\log(x-2) + \log(9-x) < 1 \Leftrightarrow \log[(x-2)(9-x)] < 1 \Leftrightarrow \log\left(-x^2 + 11x - 18\right) < 1 \Rightarrow -x^2 + 11x - 18 < 10^1$

$\Leftrightarrow 0 < x^2 - 11x + 28 \Leftrightarrow 0 < (x-7)(x-4)$. Also, since the domain of a logarithm is positive we must have

$0 < -x^2 + 11x - 18 \Leftrightarrow 0 < (x-2)(9-x)$. Using the methods from Chapter 1 with the endpoints 2, 4, 7, 9 for the

intervals, we make the following table:

Interval	$(-\infty, 2)$	$(2, 4)$	$(4, 7)$	$(7, 9)$	$(9, \infty)$
Sign of $x - 7$	$-$	$-$	$-$	$+$	$+$
Sign of $x - 4$	$-$	$-$	$+$	$+$	$+$
Sign of $x - 2$	$-$	$+$	$+$	$+$	$+$
Sign of $9 - x$	$+$	$+$	$+$	$+$	$-$
Sign of $(x-7)(x-4)$	$+$	$+$	$-$	$+$	$+$
Sign of $(x-2)(9-x)$	$-$	$+$	$+$	$+$	$-$

Thus the solution is $(2, 4) \cup (7, 9)$.

79. $2 < 10^x < 5 \Leftrightarrow \log 2 < x < \log 5 \Leftrightarrow 0.3010 < x < 0.6990$. Hence the solution to the inequality is approximately the interval $(0.3010, 0.6990)$.

81. To find the inverse of $f(x) = 2^{2x}$, we set $y = f(x)$ and solve for x. $y = 2^{2x} \Leftrightarrow \ln y = \ln\left(2^{2x}\right) = 2x \ln 2 \Leftrightarrow x = \dfrac{\ln y}{2 \ln 2}$.

Interchange x and y: $y = \dfrac{\ln x}{2 \ln 2}$. Thus, $f^{-1}(x) = \dfrac{\ln x}{2 \ln 2}$.

83. To find the inverse of $f(x) = \log_2(x-1)$, we set $y = f(x)$ and solve for x. $y = \log_2(x-1) \Leftrightarrow 2^y = 2^{\log_2(x-1)} = x - 1$

$\Leftrightarrow x = 2^y + 1$. Interchange x and y: $y = 2^x + 1$. Thus, $f^{-1}(x) = 2^x + 1$.

85. (a) $A(3) = 5000\left(1 + \dfrac{0.085}{4}\right)^{4(3)} = 5000\left(1.02125^{12}\right) = 6435.09$. Thus the amount after 3 years is $6,435.09.

(b) $10000 = 5000\left(1 + \dfrac{0.085}{4}\right)^{4t} = 5000\left(1.02125^{4t}\right) \Leftrightarrow 2 = 1.02125^{4t} \Leftrightarrow \log 2 = 4t \log 1.02125 \Leftrightarrow$

$t = \dfrac{\log 2}{4 \log 1.02125} \approx 8.24$ years. Thus the investment will double in about 8.24 years.

87. $8000 = 5000\left(1 + \dfrac{0.075}{4}\right)^{4t} = 5000\left(1.01875^{4t}\right) \Leftrightarrow 1.6 = 1.01875^{4t} \Leftrightarrow \log 1.6 = 4t \log 1.01875 \Leftrightarrow$

$t = \dfrac{\log 1.6}{4 \log 1.01875} \approx 6.33$ years. The investment will increase to $8000 in approximately 6 years and 4 months.

89. $2 = e^{0.085t} \Leftrightarrow \ln 2 = 0.085t \Leftrightarrow t = \dfrac{\ln 2}{0.085} \approx 8.15$ years. Thus the investment will double in about 8.15 years.

91. $15e^{-0.087t} = 5 \Leftrightarrow e^{-0.087t} = \frac{1}{3} \Leftrightarrow -0.087t = \ln\left(\frac{1}{3}\right) = -\ln 3 \Leftrightarrow t = \dfrac{\ln 3}{0.087} \approx 12.6277$. So only 5 grams remain after approximately 13 days.

93. (a) $P(3) = \dfrac{10}{1 + 4e^{-0.8(3)}} = 7.337$, so there are approximately 7337 fish after 3 years.

(b) We solve for t. $\dfrac{10}{1 + 4e^{-0.8t}} = 5 \Leftrightarrow 1 + 4e^{-0.8t} = \frac{10}{5} = 2 \Leftrightarrow 4e^{-0.8t} = 1 \Leftrightarrow e^{-0.8t} = 0.25 \Leftrightarrow -0.8t = \ln 0.25 \Leftrightarrow$

$t = \dfrac{\ln 0.25}{-0.8} = 1.73$. So the population will reach 5000 fish in about 1 year and 9 months.

95. (a) $\ln\left(\dfrac{P}{P_0}\right) = -\dfrac{h}{k} \Leftrightarrow \dfrac{P}{P_0} = e^{-h/k} \Leftrightarrow P = P_0 e^{-h/k}$. Substituting $k = 7$ and $P_0 = 100$ we get $P = 100e^{-h/7}$.

(b) When $h = 4$ we have $P = 100e^{-4/7} \approx 56.47$ kPa.

97. (a) $I = \frac{60}{13}\left(1 - e^{-13t/5}\right) \Leftrightarrow \frac{13}{60}I = 1 - e^{-13t/5} \Leftrightarrow e^{-13t/5} = 1 - \frac{13}{60}I \Leftrightarrow -\frac{13}{5}t = \ln\left(1 - \frac{13}{60}I\right) \Leftrightarrow$

$t = -\frac{5}{13}\ln\left(1 - \frac{13}{60}I\right)$.

(b) Substituting $I = 2$, we have $t = -\frac{5}{13}\ln\left[1 - \frac{13}{60}(2)\right] \approx 0.218$ seconds.

99. Since $9^1 = 9$, $9^2 = 81$, and $9^3 = 729$, the solution of $9^x = 20$ must be between 1 and 2 (because 20 is between 9 and 81), whereas the solution to $9^x = 100$ must be between 2 and 3 (because 100 is between 81 and 729).

101. (a) $(x-1)^{\log(x-1)} = 100(x-1) \Leftrightarrow \log\left((x-1)^{\log(x-1)}\right) = \log(100(x-1)) \Leftrightarrow$

$\left[\log(x-1)\right]\log(x-1) = \log 100 + \log(x-1) \Leftrightarrow \left[\log(x-1)\right]^2 - \log(x-1) - 2 = 0 \Leftrightarrow$

$\left[\log(x-1) - 2\right]\left[\log(x-1) + 1\right] = 0$. Thus either $\log(x-1) = 2 \Leftrightarrow x = 101$ or $\log(x-1) = -1 \Leftrightarrow x = \frac{11}{10}$.

(b) $\log_2 x + \log_4 x + \log_8 x = 11 \Leftrightarrow \log_2 x + \log_2 \sqrt{x} + \log_2 \sqrt[3]{x} = 11 \Leftrightarrow \log_2\left(x\sqrt{x}\sqrt[3]{x}\right) = 11 \Leftrightarrow \log_2\left(x^{11/6}\right) = 11$

$\Leftrightarrow \frac{11}{6}\log_2 x = 11 \Leftrightarrow \log_2 x = 6 \Leftrightarrow x = 2^6 = 64$

(c) $4^x - 2^{x+1} = 3 \Leftrightarrow \left(2^x\right)^2 - 2\left(2^x\right) - 3 = 0 \Leftrightarrow \left(2^x - 3\right)\left(2^x + 1\right) = 0 \Leftrightarrow$ either $2^x = 3 \Leftrightarrow x = \frac{\ln 3}{\ln 2}$ or $2^x = -1$, which

has no real solution. So $x = \frac{\ln 3}{\ln 2}$ is the only real solution.

4.6 MODELING WITH EXPONENTIAL AND LOGARITHMIC FUNCTIONS

1. (a) Here $n_0 = 10$ and $a = 1.5$ hours, so $n(t) = 10 \cdot 2^{t/1.5} = 10 \cdot 2^{2t/3}$.

(b) After 35 hours, there will be $n(35) = 10 \cdot 2^{2(35)/3} \approx 1.06 \times 10^8$ bacteria.

(c) $n(t) = 10 \cdot 2^{2t/3} = 10{,}000 \Leftrightarrow 2^{2t/3} = 1000 \Leftrightarrow \ln\left(2^{2t/3}\right) = \ln 1000 \Leftrightarrow \frac{2t}{3}\ln 2 = \ln 1000 \Leftrightarrow t = \frac{3}{2}\frac{\ln 1000}{\ln 2} \approx 14.9$,

so the bacteria count will reach 10,000 in about 14.9 hours.

3. (a) A model for the squirrel population is $n(t) = n_0 \cdot 2^{t/6}$. We are given

that $n(30) = 100{,}000$, so $n_0 \cdot 2^{30/6} = 100{,}000 \Leftrightarrow$

$n_0 = \frac{100{,}000}{2^5} = 3125$. Initially, there were approximately

3125 squirrels.

(b) In 10 years, we will have $t = 40$, so the population will be

$n(40) = 3125 \cdot 2^{40/6} \approx 317{,}480$ squirrels.

(c)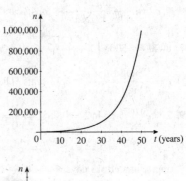

5. (a) $r = 0.08$ and $n(0) = 18{,}000$. Thus the population is given by the

formula $n(t) = 18{,}000e^{0.08t}$.

(b) $t = 2013 - 2005 = 8$. Then we have

$n(8) = 18{,}000e^{0.08(8)} = 18000e^{0.64} \approx 34{,}137$. Thus there should be

34,137 foxes in the region by the year 2013.

(d)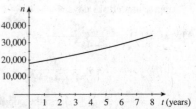

(c) Solving $n(t) = 25{,}000$, we get $18{,}000e^{0.08t} = 25{,}000 \Leftrightarrow$

$18e^{0.08t} = 25 \Leftrightarrow \ln\left(18e^{0.08t}\right) = \ln 25 \Leftrightarrow \ln 18 + 0.08t = \ln 25 \Leftrightarrow t = \frac{1}{0.08}(\ln 25 - \ln 18) \approx 4.1$, so the fox

population will reach 25,000 after about 4.1 years.

7. $n(t) = n_0 e^{rt}$; $n_0 = 110$ million, $t = 2020 - 1995 = 25$.

    (a) $r = 0.03$; $n(25) = 110{,}000{,}000 e^{0.03(25)} = 110{,}000{,}000 e^{0.75} \approx 232{,}870{,}000$. Thus at a 3% growth rate, the projected population will be approximately 233 million people by the year 2020.

    (b) $r = 0.02$; $n(25) = 110{,}000{,}000 e^{0.02(25)} = 110{,}000{,}000 e^{0.50} \approx 181{,}359{,}340$. Thus at a 2% growth rate, the projected population will be approximately 181 million people by the year 2020.

9. (a) The doubling time is 18 years and the initial population is 112,000, so a model is $n(t) = 112{,}000 \cdot 2^{t/18}$.

    (b) We need to find the relative growth rate r. Since the population is $2 \cdot 112{,}000 = 224{,}000$ when $t = 18$, we have $224{,}000 = 112{,}000 e^{18r}$

    $\Leftrightarrow 2 = e^{18r} \Leftrightarrow \ln 2 = 18r \Leftrightarrow r = \frac{\ln 2}{18} \approx 0.0385$. Thus, a model is

    $n(t) = 112{,}000 e^{0.0385t}$.

(c)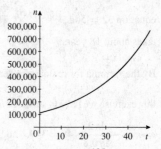

    (d) Using the model in part (a), we solve the equation $n(t) = 112{,}000 \cdot 2^{t/18} = 500{,}000 \Leftrightarrow 2^{t/18} = \frac{125}{28} \Leftrightarrow$

    $\ln 2^{t/18} = \ln \frac{125}{28} \Leftrightarrow \frac{t}{18} \ln 2 = \ln \frac{125}{28} \Leftrightarrow t = \dfrac{18 \ln \frac{125}{28}}{\ln 2} \approx 38.85$. Therefore, it takes about 38.85 years for the population to reach 500,000.

11. (a) The deer population in 2003 was 20,000.

    (b) Using the model $n(t) = 20{,}000 e^{rt}$ and the point $(4, 31000)$, we have $31{,}000 = 20{,}000 e^{4r} \Leftrightarrow 1.55 = e^{4r} \Leftrightarrow 4r = \ln 1.55 \Leftrightarrow r = \frac{1}{4} \ln 1.55 \approx 0.1096$. Thus $n(t) = 20{,}000 e^{0.1096t}$

    (c) $n(8) = 20{,}000 e^{0.1096(8)} \approx 48{,}218$, so the projected deer population in 2011 is about 48,000.

    (d) $100{,}000 = 20{,}000 e^{0.1096t} \Leftrightarrow 5 = e^{0.1096t} \Leftrightarrow 0.1096t = \ln 5 \Leftrightarrow t = \dfrac{\ln 5}{0.1096} \approx 14.68$. Thus, it takes about 14.68 years for the deer population to reach 100,000.

13. (a) Using the formula $n(t) = n_0 e^{rt}$ with $n_0 = 8600$ and $n(1) = 10000$, we solve for r, giving $10000 = n(1) = 8600 e^{r}$

    $\Leftrightarrow \frac{50}{43} = e^r \Leftrightarrow r = \ln\left(\frac{50}{43}\right) \approx 0.1508$. Thus $n(t) = 8600 e^{0.1508t}$.

    (b) $n(2) = 8600 e^{0.1508(2)} \approx 11627$. Thus the number of bacteria after two hours is about 11,600.

    (c) $17200 = 8600 e^{0.1508t} \Leftrightarrow 2 = e^{0.1508t} \Leftrightarrow 0.1508t = \ln 2 \Leftrightarrow t = \dfrac{\ln 2}{0.1508} \approx 4.596$. Thus the number of bacteria will double in about 4.6 hours.

15. (a) Calculating dates relative to 1990 gives $n_0 = 29.76$ and $n(10) = 33.87$. Then $n(10) = 29.76 e^{10r} = 33.87 \Leftrightarrow$

    $e^{10r} = \frac{33.87}{29.76} \approx 1.1381 \Leftrightarrow 10r = \ln 1.1381 \Leftrightarrow r = \frac{1}{10} \ln 1.1381 \approx 0.012936$. Thus $n(t) = 29.76 e^{0.012936t}$ million people.

    (b) $2(29.76) = 29.76 e^{0.012936t} \Leftrightarrow 2 = e^{0.012936t} \Leftrightarrow \ln 2 = 0.012936t \Leftrightarrow t = \dfrac{\ln 2}{0.012936} \approx 53.58$, so the population doubles in about 54 years.

    (c) $t = 2010 - 1990 = 20$, so our model gives the 2010 population as $n(20) \approx 29.76 e^{0.012936(20)} \approx 38.55$ million. The actual population was estimated at 36.96 million in 2009.

17. (a) Because the half-life is 1600 years and the sample weighs 22 mg initially, a suitable model is $m(t) = 22 \cdot 2^{-t/1600}$.

    (b) From the formula for radioactive decay, we have $m(t) = m_0 e^{-rt}$, where $m_0 = 22$ and $r = \dfrac{\ln 2}{h} = \dfrac{\ln 2}{1600} \approx 0.000433$. Thus, the amount after t years is given by $m(t) = 22 e^{-0.000433t}$.

    (c) $m(4000) = 22 e^{-0.000433(4000)} \approx 3.89$, so the amount after 4000 years is about 4 mg.

(d) We have to solve for t in the equation $18 = 22\,e^{-0.000433t}$. This gives $18 = 22e^{-0.000433t} \Leftrightarrow \frac{9}{11} = e^{-0.000433t} \Leftrightarrow$

$-0.000433t = \ln\left(\frac{9}{11}\right) \Leftrightarrow t = \dfrac{\ln\left(\frac{9}{11}\right)}{-0.000433} \approx 463.4$, so it takes about 463 years.

19. By the formula in the text, $m\,(t) = m_0 e^{-rt}$ where $r = \dfrac{\ln 2}{h}$, so $m\,(t) = 50e^{-[(\ln 2)/28]t}$. We need to solve for t in the

equation $32 = 50e^{-[(\ln 2)/28]t}$. This gives $e^{-(\ln 2)/28]t} = \frac{32}{50} \Leftrightarrow -\frac{\ln 2}{28}t = \ln\left(\frac{32}{50}\right) \Leftrightarrow t = -\frac{28}{\ln 2} \cdot \ln\left(\frac{32}{50}\right) \approx 18.03$, so it

takes about 18 years.

21. By the formula for radioactive decay, we have $m\,(t) = m_0 e^{-rt}$, where $r = \dfrac{\ln 2}{h}$, in other words $m\,(t) = m_0 e^{-[(\ln 2)/h]t}$. In

this exercise we have to solve for h in the equation $200 = 250e^{-[(\ln 2)/h]\cdot 48} \Leftrightarrow 0.8 = e^{-[(\ln 2)/h]\cdot 48} \Leftrightarrow \ln(0.8) = -\dfrac{\ln 2}{h}\cdot 48$

$\Leftrightarrow h = -\dfrac{\ln 2}{\ln 0.8}\cdot 48 \approx 149.1$ hours. So the half-life is approximately 149 hours.

23. By the formula in the text, $m\,(t) = m_0 e^{-[(\ln 2)/h]\cdot t}$, so we have $0.65 = 1\cdot e^{-[(\ln 2)/5730]\cdot t} \Leftrightarrow \ln(0.65) = -\dfrac{\ln 2}{5730}t \Leftrightarrow$

$t = -\dfrac{5730\ln 0.65}{\ln 2} \approx 3561$. Thus the artifact is about 3560 years old.

25. (a) $T\,(0) = 65 + 145e^{-0.05(0)} = 65 + 145 = 210°$ F.

(b) $T\,(10) = 65 + 145e^{-0.05(10)} \approx 152.9$. Thus the temperature after 10 minutes is about $153°$ F.

(c) $100 = 65 + 145e^{-0.05t} \Leftrightarrow 35 = 145e^{-0.05t} \Leftrightarrow 0.2414 = e^{-0.05t} \Leftrightarrow \ln 0.2414 = -0.05t \Leftrightarrow t = -\dfrac{\ln 0.2414}{0.05} \approx 28.4$.

Thus the temperature will be $100°$ F in about 28 minutes.

27. Using Newton's Law of Cooling, $T\,(t) = T_s + D_0 e^{-kt}$ with $T_s = 75$ and $D_0 = 185 - 75 = 110$. So $T\,(t) = 75 + 110e^{-kt}$.

(a) Since $T\,(30) = 150$, we have $T\,(30) = 75 + 110e^{-30k} = 150 \Leftrightarrow 110e^{-30k} = 75 \Leftrightarrow e^{-30k} = \frac{15}{22} \Leftrightarrow -30k = \ln\left(\frac{15}{22}\right)$

$\Leftrightarrow k = -\frac{1}{30}\ln\left(\frac{15}{22}\right)$. Thus we have $T\,(45) = 75 + 110e^{(45/30)\ln(15/22)} \approx 136.9$, and so the temperature of the turkey

after 45 minutes is about $137°$ F.

(b) The temperature will be $100°$F when $75 + 110e^{(t/30)\ln(15/22)} = 100 \Leftrightarrow e^{(t/30)\ln(15/22)} = \dfrac{25}{110} = \dfrac{5}{22} \Leftrightarrow$

$\left(\dfrac{t}{30}\right)\ln\left(\frac{15}{22}\right) = \ln\left(\frac{5}{22}\right) \Leftrightarrow t = 30\dfrac{\ln\left(\frac{5}{22}\right)}{\ln\left(\frac{15}{22}\right)} \approx 116.1$. So the temperature will be $100°$ F after about 2 hours.

29. (a) $\text{pH} = -\log\left[\text{H}^+\right] = -\log\left(5.0 \times 10^{-3}\right) \approx 2.3$

(b) $\text{pH} = -\log\left[\text{H}^+\right] = -\log\left(3.2 \times 10^{-4}\right) \approx 3.5$

(c) $\text{pH} = -\log\left[\text{H}^+\right] = -\log\left(5.0 \times 10^{-9}\right) \approx 8.3$

31. (a) $\text{pH} = -\log\left[\text{H}^+\right] = 3.0 \Leftrightarrow \left[\text{H}^+\right] = 10^{-3}$ M

(b) $\text{pH} = -\log\left[\text{H}^+\right] = 6.5 \Leftrightarrow \left[\text{H}^+\right] = 10^{-6.5} \approx 3.2 \times 10^{-7}$ M

33. $4.0 \times 10^{-7} \leq \left[\text{H}^+\right] \leq 1.6 \times 10^{-5} \Leftrightarrow \log\left(4.0 \times 10^{-7}\right) \leq \log\left[\text{H}^+\right] \leq \log\left(1.6 \times 10^{-5}\right) \Leftrightarrow$

$-\log\left(4.0 \times 10^{-7}\right) \geq \text{pH} \geq -\log\left(1.6 \times 10^{-5}\right) \Leftrightarrow 6.4 \geq \text{pH} \geq 4.8$. Therefore the range of pH readings for cheese is

approximately 4.8 to 6.4.

35. Let I_0 be the intensity of the smaller earthquake and I_1 the intensity of the larger earthquake. Then $I_1 = 20I_0$.

Notice that $M_0 = \log\left(\dfrac{I_0}{S}\right) = \log I_0 - \log S$ and $M_1 = \log\left(\dfrac{I_1}{S}\right) = \log\left(\dfrac{20I_0}{S}\right) = \log 20 + \log I_0 - \log S$. Then

$M_1 - M_0 = \log 20 + \log I_0 - \log S - \log I_0 + \log S = \log 20 \approx 1.3$. Therefore the magnitude is 1.3 times larger.

37. Let the subscript J represent the Japan earthquake and S represent the San Francisco earthquake. Then

$M_J = \log\left(\dfrac{I_J}{S}\right) = 9.1 \Leftrightarrow I_J = S \cdot 10^{9.1}$ and $M_S = \log\left(\dfrac{I_S}{S}\right) = 8.3 \Leftrightarrow I_S = S \cdot 10^{8.3}$. So $\dfrac{I_J}{I_S} = \dfrac{S \cdot 10^{9.1}}{S \cdot 10^{8.3}} = 10^{0.8} \approx 6.3$,

and hence the Japan earthquake was about six times more intense than the San Francisco earthquake.

39. $\beta = 10 \log\left(\dfrac{I}{I_0}\right) = 10 \log\left(\dfrac{2.0 \times 10^{-5}}{1.0 \times 10^{-12}}\right) = 10 \log\left(2 \times 10^7\right) = 10\left(\log 2 + \log 10^7\right) = 10\left(\log 2 + 7\right) \approx 73$. Therefore

the intensity level was 73 dB.

41. Let the subscript M represent the power mower and C the rock concert. Then $106 = 10 \log\left(\dfrac{I_M}{10^{-12}}\right) \Leftrightarrow$

$\log\left(I_M \cdot 10^{12}\right) = 10.6 \Leftrightarrow I_M \cdot 10^{12} = 10^{10.6}$. Also $120 = 10 \log\left(\dfrac{I_C}{10^{-12}}\right) \Leftrightarrow \log\left(I_C \cdot 10^{12}\right) = 12.0 \Leftrightarrow$

$I_C \cdot 10^{12} = 10^{12.0}$. So $\dfrac{I_C}{I_M} = \dfrac{10^{12}}{10^{10.6}} = 10^{1.4} \approx 25.12$, and so the ratio of intensity is roughly 25.

CHAPTER 4 REVIEW

1. $f(x) = 5^x$; $f(-1.5) \approx 0.0894$, $f\left(\sqrt{2}\right) \approx 9.739$, $f(2.5) \approx 55.902$

3. $g(x) = 4e^{x-2}$; $g(-0.7) \approx 0.269$, $g(1) \approx 1.472$, $g(\pi) \approx 12.527$

5. $f(x) = 3^{x-2}$. domain $(-\infty, \infty)$, range $(0, \infty)$, asymptote $y = 0$.

7. $g(x) = 3 + 2^x$. Domain $(-\infty, \infty)$, range $(3, \infty)$, asymptote $y = 3$.

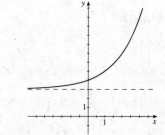

9. $F(x) = e^{x-1} + 1$. Domain $(-\infty, \infty)$, range $(1, \infty)$, asymptote $y = 1$.

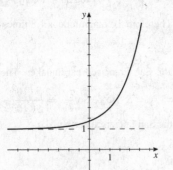

11. $f(x) = \log_3(x - 1)$. Domain $(1, \infty)$, range $(-\infty, \infty)$, asymptote $x = 1$.

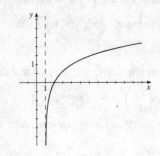

13. $f(x) = 2 - \log_2 x$. Domain $(0, \infty)$, range $(-\infty, \infty)$, asymptote $x = 0$.

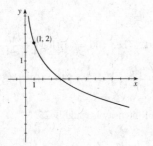

15. $g(x) = 2 \ln x$. Domain $(0, \infty)$, range $(-\infty, \infty)$, asymptote $x = 0$.

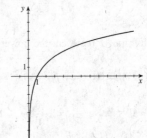

17. $f(x) = 10^{x^2} + \log(1 - 2x)$. Since $\log u$ is defined only for $u > 0$, we require $1 - 2x > 0 \Leftrightarrow -2x > -1 \Leftrightarrow x < \frac{1}{2}$, and so the domain is $\left(-\infty, \frac{1}{2}\right)$.

19. $h(x) = \ln\left(x^2 - 4\right)$. We must have $x^2 - 4 > 0$ (since $\ln y$ is defined only for $y > 0$) $\Leftrightarrow x^2 - 4 > 0 \Leftrightarrow (x - 2)(x + 2) > 0$. The endpoints of the intervals are -2 and 2.

Interval	$(-\infty, -2)$	$(-2, 2)$	$(2, \infty)$
Sign of $x - 2$	$-$	$-$	$+$
Sign of $x + 2$	$-$	$+$	$+$
Sign of $(x - 2)(x + 2)$	$+$	$-$	$+$

Thus the domain is $(-\infty, -2) \cup (2, \infty)$.

21. $\log_2 1024 = 10 \Leftrightarrow 2^{10} = 1024$

23. $\log x = y \Leftrightarrow 10^y = x$

25. $2^6 = 64 \Leftrightarrow \log_2 64 = 6$

27. $10^x = 74 \Leftrightarrow \log_{10} 74 = x \Leftrightarrow \log 74 = x$

29. $\log_2 128 = \log_2\left(2^7\right) = 7$

31. $10^{\log 45} = 45$

33. $\ln\left(e^6\right) = 6$

35. $\log_3 \frac{1}{27} = \log_3 3^{-3} = -3$

37. $\log_5 \sqrt{5} = \log_5 5^{1/2} = \frac{1}{2}$

39. $\log 25 + \log 4 = \log(25 \cdot 4) = \log 10^2 = 2$

41. $\log_2\left(16^{23}\right) = \log_2\left(2^4\right)^{23} = \log_2 2^{92} = 92$

43. $\log_8 6 - \log_8 3 + \log_8 2 = \log_8\left(\frac{6}{3} \cdot 2\right) = \log_8 4 = \log_8 8^{2/3} = \frac{2}{3}$

45. $\log\left(AB^2C^3\right) = \log A + 2\log B + 3\log C$

47. $\ln\sqrt{\dfrac{x^2-1}{x^2+1}} = \tfrac{1}{2}\ln\left(\dfrac{x^2-1}{x^2+1}\right) = \tfrac{1}{2}\left[\ln\left(x^2-1\right) - \ln\left(x^2+1\right)\right]$

49. $\log_5\left(\dfrac{x^2\left(1-5x\right)^{3/2}}{\sqrt{x^3-x}}\right) = \log_5 x^2\left(1-5x\right)^{3/2} - \log_5\sqrt{x\left(x^2-1\right)} = 2\log_5 x + \dfrac{3}{2}\log_5\left(1-5x\right) - \tfrac{1}{2}\log_5\left(x^3-x\right)$

51. $\log 6 + 4\log 2 = \log 6 + \log 2^4 = \log\left(6\cdot 2^4\right) = \log 96$

53. $\tfrac{3}{2}\log_2\left(x-y\right) - 2\log_2\left(x^2+y^2\right) = \log_2\left(x-y\right)^{3/2} - \log_2\left(x^2+y^2\right)^2 = \log_2\left(\dfrac{(x-y)^{3/2}}{\left(x^2+y^2\right)^2}\right)$

55. $\log\left(x-2\right) + \log\left(x+2\right) - \tfrac{1}{2}\log\left(x^2+4\right) = \log\left[(x-2)\left(x+2\right)\right] - \log\sqrt{x^2+4} = \log\left(\dfrac{x^2-4}{\sqrt{x^2+4}}\right)$

57. $3^{2x-7} = 27 \Leftrightarrow 3^{2x-7} = 3^3 \Leftrightarrow 2x-7 = 3 \Leftrightarrow 2x = 10 \Leftrightarrow x = 5$

59. $2^{3x-5} = 7 \Leftrightarrow \log_2\left(2^{3x-5}\right) = \log_2 7 \Leftrightarrow 3x-5 = \log_2 7 \Leftrightarrow x = \tfrac{1}{3}\left(\log_2 7 + 5\right)$. Using the Change of Base Formula, we have $\log_2 7 = \dfrac{\log 7}{\log 2} \approx 2.807$, so $x \approx \tfrac{1}{3}\left(2.807 + 5\right) \approx 2.602$.

61. $4^{1-x} = 3^{2x+5} \Leftrightarrow \log 4^{1-x} = \log 3^{2x+5} \Leftrightarrow \left(1-x\right)\log 4 = \left(2x+5\right)\log 3 \Leftrightarrow \log 4 - 5\log 3 = 2x\log 3 + x\log 4 \Leftrightarrow$
$x\left(\log 3 + \log 4\right) = \log 4 - 5\log 3 \Leftrightarrow x = \dfrac{\log 4 - 5\log 3}{2\log 3 + \log 4} \approx -1.146$

63. $x^2 e^{2x} + 2x e^{2x} = 8e^{2x} \Leftrightarrow e^{2x}\left(x^2 + 2x - 8\right) = 0 \Leftrightarrow x^2 + 2x - 8 = 0$ (since $e^{2x} \neq 0$) $\Leftrightarrow \left(x+4\right)\left(x-2\right) = 0 \Leftrightarrow x = -4$
or $x = 2$

65. $\log x + \log\left(x+1\right) = \log 12 \Leftrightarrow \log\left(x\left(x+1\right)\right) = \log 12 \Leftrightarrow x\left(x+1\right) = 12 \Leftrightarrow x^2 + x - 12 = 0 \Leftrightarrow \left(x-3\right)\left(x+4\right) = 0 \Leftrightarrow$
$x = -4$ or 3. Since $\log\left(-4\right)$ is undefined, the only solution is $x = 3$.

67. $\log_2\left(1-x\right) = 4 \Leftrightarrow 1-x = 2^4 \Leftrightarrow x = 1 - 16 = -15$

69. $\log_3\left(x-8\right) + \log_3 x = 2 \Leftrightarrow \log_3\left(x\left(x-8\right)\right) = 2 \Leftrightarrow x\left(x-8\right) = 9 \Leftrightarrow x^2 - 8x - 9 = 0 \Leftrightarrow \left(x-9\right)\left(x+1\right) = 0 \Leftrightarrow$
$x = -1$ or 9. We reject -1 because it does not satisfy the original equation, so the only solution is $x = 9$.

71. $5^{-2x/3} = 0.63 \Leftrightarrow \dfrac{-2x}{3}\log 5 = \log 0.63 \Leftrightarrow x = -\dfrac{3\log 0.63}{2\log 5} \approx 0.430618$

73. $5^{2x+1} = 3^{4x-1} \Leftrightarrow \left(2x+1\right)\log 5 = \left(4x-1\right)\log 3 \Leftrightarrow 2x\log 5 + \log 5 = 4x\log 3 - \log 3 \Leftrightarrow$
$x\left(2\log 5 - 4\log 3\right) = -\log 3 - \log 5 \Leftrightarrow x = \dfrac{\log 3 + \log 5}{4\log 3 - 2\log 5} \approx 2.303600$

75. $y = e^{x/(x+2)}$. Vertical asymptote $x = -2$, horizontal asymptote $y = 2.72$, no maximum or minimum.

77. $y = \log\left(x^3 - x\right)$. Vertical asymptotes $x = -1$, $x = 0$,
$x = 1$, no horizontal asymptote, local maximum of about
-0.41 when $x \approx -0.58$.

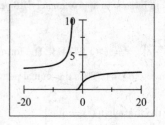

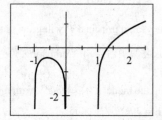

79. $3 \log x = 6 - 2x$. We graph $y = 3 \log x$ and $y = 6 - 2x$ in the same viewing rectangle. The solution occurs where the two graphs intersect. From the graphs, we see that the solution is $x \approx 2.42$.

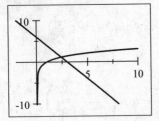

81. $\ln x > x - 2$. We graph the function $f(x) = \ln x - x + 2$, and we see that the graph lies above the x-axis for $0.16 < x < 3.15$. So the approximate solution of the given inequality is $0.16 < x < 3.15$.

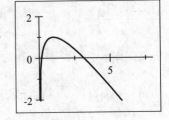

83. $f(x) = e^x - 3e^{-x} - 4x$. We graph the function $f(x)$, and we see that the function is increasing on $(-\infty, 0]$ and $[1.10, \infty)$ and that it is decreasing on $[0, 1.10]$.

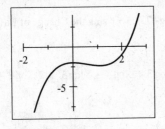

85. $\log_4 15 = \dfrac{\log 15}{\log 4} = 1.953445$

87. $\log_9 0.28 = \dfrac{\log 0.28}{\log 9} \approx -0.579352$

89. Notice that $\log_4 258 > \log_4 256 = \log_4 4^4 = 4$ and so $\log_4 258 > 4$. Also $\log_5 620 < \log_5 625 = \log_5 5^4 = 4$ and so $\log_5 620 < 4$. Then $\log_4 258 > 4 > \log_5 620$ and so $\log_4 258$ is larger.

91. $P = 12,000$, $r = 0.10$, and $t = 3$. Then $A = P \left(1 + \dfrac{r}{n}\right)^{nt}$.

    (a) For $n = 2$, $A = 12,000 \left(1 + \dfrac{0.10}{2}\right)^{2(3)} = 12,000 \left(1.05^6\right) \approx \$16,081.15$.

    (b) For $n = 12$, $A = 12,000 \left(1 + \dfrac{0.10}{12}\right)^{12(3)} \approx \$16,178.18$.

    (c) For $n = 365$, $A = 12,000 \left(1 + \dfrac{0.10}{365}\right)^{365(3)} \approx \$16,197.64$.

    (d) For $n = \infty$, $A = Pe^{rt} = 12,000e^{0.10(3)} \approx \$16,198.31$.

93. We use the formula $A = P \left(1 + \dfrac{r}{n}\right)^{nt}$ with $P = 100,000$, $r = 0.052$, $n = 365$, and $A = 100,000 + 10,000 = 110,000$, and solve for t: $110,000 = 100,000 \left(1 + \dfrac{0.052}{365}\right)^{365t} \Leftrightarrow 1.1 = \left(1 + \dfrac{0.052}{365}\right)^{365t} \Leftrightarrow \log 1.1 = 365t \log \left(1 + \dfrac{0.052}{365}\right) \Leftrightarrow$

$t = \dfrac{\log 1.1}{365 \log \left(1 + \dfrac{0.052}{365}\right)} \approx 1.833$. The account will accumulate $\$10,000$ in interest in approximately 1.8 years.

95. After one year, a principal P will grow to the amount $A = P \left(1 + \dfrac{0.0425}{365}\right)^{365} = P (1.04341)$. The formula for simple interest is $A = P (1 + r)$. Comparing, we see that $1 + r = 1.04341$, so $r = 0.04341$. Thus the annual percentage yield is 4.341%.

97. **(a)** Using the model $n(t) = n_0 e^{rt}$, with $n_0 = 30$ and $r = 0.15$, we have the formula $n(t) = 30e^{0.15t}$.

    (b) $n(4) = 30e^{0.15(4)} \approx 55$.

(c) $500 = 30e^{0.15t} \Leftrightarrow \frac{50}{3} = e^{0.15t} \Leftrightarrow 0.15t = \ln\left(\frac{50}{3}\right) \Leftrightarrow t = \frac{1}{0.15}\ln\left(\frac{50}{3}\right) \approx 18.76$. So the stray cat population will reach 500 in about 19 years.

99. (a) From the formula for radioactive decay, we have $m(t) = 10e^{-rt}$, where $r = -\dfrac{\ln 2}{2.7 \times 10^5}$. So after 1000 years

the amount remaining is $m(1000) = 10 \cdot e^{\left[-\ln 2/(2.7\times 10^5)\right]\cdot 1000} = 10e^{-(\ln 2)/(2.7\times 10^2)} = 10e^{-(\ln 2)/270} \approx 9.97$. Therefore the amount remaining is about 9.97 mg.

(b) We solve for t in the equation $7 = 10e^{-\left[\ln 2/(2.7\times 10^5)\right]\cdot t}$. We have $7 = 10e^{-\left[\ln 2/(2.7\times 10^5)\right]\cdot t} \Leftrightarrow$

$0.7 = e^{-\left[\ln 2/(2.7\times 10^5)\right]\cdot t} \Leftrightarrow \ln 0.7 = -\dfrac{\ln 2}{2.7 \times 10^5}\cdot t \Leftrightarrow t = -\dfrac{\ln 0.7}{\ln 2}\cdot 2.7 \times 10^5 \approx 138{,}934.75$. Thus it takes about 139,000 years.

101. (a) From the formula for radioactive decay, $r = \dfrac{\ln 2}{1590} \approx 0.0004359$ and $n(t) = 150 \cdot e^{-0.0004359t}$.

(b) $n(1000) = 150 \cdot e^{-0.0004359 \cdot 1000} \approx 97.00$, and so the amount remaining is about 97.00 mg.

(c) Find t so that $50 = 150 \cdot e^{-0.0004359t}$. We have $50 = 150 \cdot e^{-0.0004359t} \Leftrightarrow \frac{1}{3} = e^{-0.0004359t} \Leftrightarrow$

$t = -\dfrac{1}{0.0004359}\ln\left(\frac{1}{3}\right) \approx 2520$. Thus only 50 mg remain after about 2520 years.

103. (a) Using $n_0 = 1500$ and $n(5) = 3200$ in the formula $n(t) = n_0 e^{rt}$, we have $3200 = n(5) = 1500e^{5r} \Leftrightarrow e^{5r} = \frac{32}{15} \Leftrightarrow$

$5r = \ln\left(\frac{32}{15}\right) \Leftrightarrow r = \frac{1}{5}\ln\left(\frac{32}{15}\right) \approx 0.1515$. Thus $n(t) = 1500 \cdot e^{0.1515t}$.

(b) We have $t = 1999 - 1988 = 11$ so $n(11) = 1500e^{0.1515\cdot 11} \approx 7940$. Thus in 1999 the bird population should be about 7940.

105. $\left[H^+\right] = 1.3 \times 10^{-8}$ M. Then pH $= -\log\left[H^+\right] = -\log\left(1.3 \times 10^{-8}\right) \approx 7.9$, and so fresh egg whites are basic.

107. Let I_0 be the intensity of the smaller earthquake and I_1 be the intensity of the larger earthquake. Then $I_1 = 35I_0$. Since

$M = \log\left(\dfrac{I}{S}\right)$, we have $M_0 = \log\left(\dfrac{I_0}{S}\right) = 6.5$ and $M_1 = \log\left(\dfrac{I_1}{S}\right) = \log\left(\dfrac{35I_0}{S}\right) = \log 35 + \log\left(\dfrac{I_0}{S}\right) = \log 35 +$

$M_0 = \log 35 + 6.5 \approx 8.04$. So the magnitude on the Richter scale of the larger earthquake is approximately 8.0.

CHAPTER 4 TEST

1. (a)

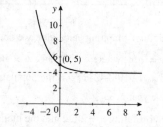

$f(x) = 2^{-x} + 4$ has domain $(-\infty, \infty)$, range $(4, \infty)$, and horizontal asymptote $y = 4$.

(b)

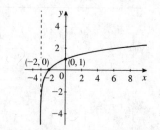

$g(x) = \log_3(x + 3)$ has domain $(-3, \infty)$, range $(-\infty, \infty)$, and vertical asymptote $x = -3$.

3. (a) $6^{2x} = 25 \Leftrightarrow \log_6 6^{2x} = \log_6 25 \Leftrightarrow 2x = \log_6 25$ (b) $\ln A = 3 \Leftrightarrow e^{\ln A} = e^3 \Leftrightarrow A = e^3$

5. (a) $\log\left(\dfrac{xy^3}{z^2}\right) = \log x + \log y^3 - \log z^2 = \log x + 3\log y - 2\log z$

(b) $\ln\sqrt{\dfrac{x}{y}} = \ln\left(\left(\dfrac{x}{y}\right)^{1/2}\right) = \frac{1}{2}\ln\left(\dfrac{x}{y}\right) = \frac{1}{2}\ln x - \frac{1}{2}\ln y$

(c) $\log \sqrt[3]{\dfrac{x+2}{x^4 \left(x^2+4\right)}} = \frac{1}{3} \log \left(\dfrac{x+2}{x^4 \left(x^2+4\right)}\right) = \frac{1}{3}\left[\log\left(x+2\right) - \left(4\log x + \log\left(x^2+4\right)\right)\right]$

$\qquad\qquad\qquad\qquad = \frac{1}{3}\log\left(x+2\right) - \frac{4}{3}\log x - \frac{1}{3}\log\left(x^2+4\right)$

7. (a) $3^{4x} = 3^{100} \Leftrightarrow 4x = 100 \Leftrightarrow x = 25$

(b) $e^{3x-2} = e^{x^2} \Leftrightarrow 3x - 2 = x^2 \Leftrightarrow x^2 - 3x + 2 = 0 \Leftrightarrow (x-1)(x-2) = 0 \Leftrightarrow x = 1 \text{ or } x = 2$

(c) $5^{x/10} + 1 = 7 \Leftrightarrow 5^{x/10} = 6 \Leftrightarrow \log_5 5^{x/10} = \log_5 6 \Leftrightarrow x/10 = \log_5 6 \Leftrightarrow x = 10\log_5 6 \approx 11.13$

(d) $10^{x+3} = 6^{2x} \Leftrightarrow \log 10^{x+3} = \log 6^{2x} \Leftrightarrow x + 3 = \dfrac{\log_6 6^{2x}}{\log_6 10} \Leftrightarrow \left(\log_6 10\right)(x+3) = 2x \Leftrightarrow \left(2 - \log_6 10\right)x = 3\log_6 10$

$\qquad \Leftrightarrow x = \dfrac{3\log_6 10}{2 - \log_6 10} \approx 5.39$

9. Using the Change of Base Formula, we have $\log_{12} 27 = \dfrac{\log 27}{\log 12} \approx 1.326$.

11. (a) $A(t) = 12{,}000\left(1 + \dfrac{0.056}{12}\right)^{12t}$, where t is in years.

(b) $A(t) = 12{,}000\left(1 + \dfrac{0.056}{365}\right)^{365t}$. So $A(3) = 12{,}000\left(1 + \dfrac{0.056}{365}\right)^{365(3)} = \$14{,}195.06$.

(c) $A(t) = 12{,}000e^{0.056t}$. So $20{,}000 = 12{,}000e^{0.056t} \Leftrightarrow 5 = 3e^{0.056t} \Leftrightarrow \ln 5 = \ln\left(3e^{0.056t}\right) \Leftrightarrow \ln 5 = \ln 3 + 0.056t \Leftrightarrow$

$\qquad t = \dfrac{1}{0.056}\left(\ln 5 - \ln 3\right) \approx 9.12$. Thus, the amount will grow to \$20,000 in approximately 9.12 years.

13. Let the subscripts J and P represent the two earthquakes. Then we have $M_J = \log\left(\dfrac{I_J}{S}\right) = 6.4 \Leftrightarrow 10^{6.4} = \dfrac{I_J}{S} \Leftrightarrow$

$10^{6.4}S = I_J$. Similarly, $M_P = \log\left(\dfrac{I_P}{S}\right) = 3.1 \Leftrightarrow 10^{3.1} = \dfrac{I_P}{S} \Leftrightarrow 10^{3.1}S = I_P$. So $\dfrac{I_J}{I_P} = \dfrac{10^{6.4}S}{10^{3.1}S} = 10^{3.3} \approx 1995.3$,

and so the Japan earthquake was about 1995 times more intense than the Pennsylvania earthquake.

FOCUS ON MODELING   Fitting Exponential and Power Curves to Data

1. (a)

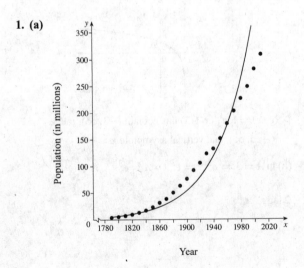

Year

(b) Using a graphing calculator, we obtain the model $y = ab^t$, where $a = 3.334926 \times 10^{-15}$ and $b = 1.019844$, and y is the population (in millions) in the year t.

(c) Substituting $t = 2020$ into the model of part (b), we get $y = ab^{2020} \approx 577.5$ million.

(d) According to the model, the population in 1995 should have been about $y = ab^{1995} \approx 353.1$ million.

(e) The values given by the model are clearly much too large. This means that an exponential model is *not* appropriate for these data.

3. (a) Yes.

(b)

Year t	Health Expenditures E ($bn)	$\ln E$
1970	74.3	4.30811
1980	251.1	5.52585
1985	434.5	6.07420
1987	506.2	6.22693
1990	696.6	6.54621
1992	820.3	6.70967
1994	937.2	6.84290
1996	1039.4	6.94640
1998	1150.0	7.04752
2000	1310.0	7.17778
2001	1424.5	7.26158
2008	2339.5	7.75769

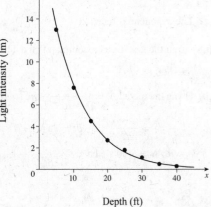

Year since 1970

Yes, the scatter plot appears to be roughly linear.

(c) Let t be the number of years elapsed since 1970 . Then $\ln E = 4.618612 + 0.0881283t$, where E is expenditure in billions of dollars.

(d) $E = e^{4.618612+0.0881283t} = 101.353256e^{0.0881283t}$

(e) In 2015 we have $t = 2015 - 1970 = 45$, so the estimated 2015 health-care expenditures are $101.353256e^{0.0881283(45)} \approx \$5347.50$ billion

5. (a) Using a graphing calculator, we find that $I_0 = 22.7586444$ and $k = 0.1062398$.

(b)

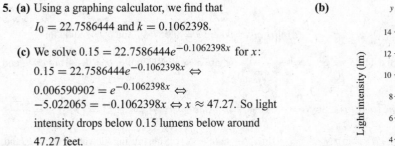

Depth (ft)

(c) We solve $0.15 = 22.7586444e^{-0.1062398x}$ for x:

$0.15 = 22.7586444e^{-0.1062398x} \Leftrightarrow$

$0.006590902 = e^{-0.1062398x} \Leftrightarrow$

$-5.022065 = -0.1062398x \Leftrightarrow x \approx 47.27$. So light intensity drops below 0.15 lumens below around 47.27 feet.

7. (a) Let A be the area of the cave and S the number of species of bat. Using a graphing calculator, we obtain the power function model $S = 0.14A^{0.64}$.

(c) According to the model, there are

$S = 0.14 (205)^{0.64} \approx 4$ species of bat living in the El Sapo cave.

(b)

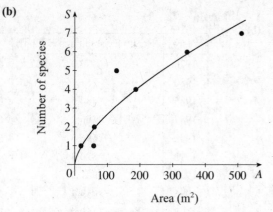

The model fits the data reasonably well.

9. (a)

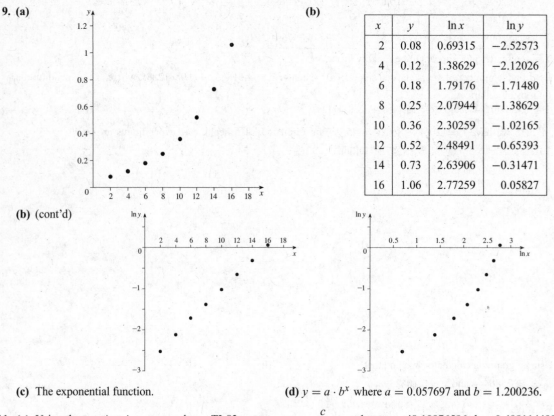

(b)

x	y	$\ln x$	$\ln y$
2	0.08	0.69315	−2.52573
4	0.12	1.38629	−2.12026
6	0.18	1.79176	−1.71480
8	0.25	2.07944	−1.38629
10	0.36	2.30259	−1.02165
12	0.52	2.48491	−0.65393
14	0.73	2.63906	−0.31471
16	1.06	2.77259	0.05827

(b) (cont'd)

(c) The exponential function.

(d) $y = a \cdot b^x$ where $a = 0.057697$ and $b = 1.200236$.

11. (a) Using the `Logistic` command on a TI-83 we get $y = \dfrac{c}{1 + ae^{-bx}}$ where $a = 49.10976596$, $b = 0.4981144989$, and $c = 500.855793$.

(b) Using the model $N = \dfrac{c}{1 + ae^{-bt}}$ we solve for t. So $N = \dfrac{c}{1 + ae^{-bt}} \Leftrightarrow 1 + ae^{-bt} = \dfrac{c}{N} \Leftrightarrow ae^{-bt} = \left(\dfrac{c}{N}\right) - 1 = \dfrac{c - N}{N}$

$\Leftrightarrow e^{-bt} = \dfrac{c - N}{aN} \Leftrightarrow -bt = \ln(c - N) - \ln aN \Leftrightarrow t = \dfrac{1}{b}[\ln aN - \ln(c - N)]$. Substituting the values for a, b, and c, with $N = 400$ we have $t = \dfrac{1}{0.4981144989}(\ln 19643.90638 - \ln 100.855793) \approx 10.58$ days.

CUMULATIVE REVIEW TEST: CHAPTERS 2, 3, and 4

1. $f(x) = x^2 - 4x$, $g(x) = \sqrt{x+4}$

  (a) The domain of f is $(-\infty, \infty)$.

  (b) The domain of g is the set of all x for which $x + 4 \geq 0 \Leftrightarrow x \geq -4$, that is, $[-4, \infty)$.

  (c) $f(-2) = (-2)^2 - 4(-2) = 12$, $f(0) = 0^2 - 4(0) = 0$, $f(4) = 4^2 - 4(4) = 0$, $g(0) = \sqrt{0+4} = 2$,

    $g(8) = \sqrt{8+4} = 2\sqrt{3}$, $g(-6) = \sqrt{-6+4} = \sqrt{-2}$, which is undefined.

  (d) $f(x+2) = (x+2)^2 - 4(x+2) = x^2 + 4x + 4 - 4x - 8 = x^2 - 4$, $g(x+2) = \sqrt{(x+2)+4} = \sqrt{x+6}$,

    $f(2+h) = (2+h)^2 - 4(2+h) = 4 + 4h + h^2 - 8 - 4h = h^2 - 4$

  (e) $\dfrac{g(21) - g(5)}{21 - 5} = \dfrac{\sqrt{21+4} - \sqrt{5+4}}{16} = \dfrac{5 - 3}{16} = \dfrac{1}{8}$

  (f) $f \circ g(x) = f(g(x)) = \left(\sqrt{x+4}\right)^2 - 4\left(\sqrt{x+4}\right) = x + 4 - 4\sqrt{x+4}$,

    $g \circ f(x) = g(f(x)) = \sqrt{x^2 - 4x + 4} = \sqrt{(x-2)^2} = |x - 2|$, $f(g(12)) = 12 + 4 - 4\sqrt{12+4} = 16 - 4\sqrt{16} = 0$,

    $g(f(12)) = 12 - 2 = 10$

  (g) $y = \sqrt{x+4} \Rightarrow y^2 = x + 4 \Leftrightarrow x = y^2 - 4$. Reverse x and y: $y = x^2 - 4$. Thus, the inverse of g is $g^{-1}(x) = x^2 - 4$,

    $x \geq 0$.

3. $f(x) = -2x^2 + 8x + 5$

  (a) $f(x) = -2\left(x^2 - 4x\right) + 5 = -2\left(x^2 - 4x + 4\right) + 8 +$

    $5 = -2(x-2)^2 + 13$

  (b) Because $a = -2 < 0$, f has a maximum value of 13 at $x = 2$.

  (d) f is increasing on $(-\infty, 2]$ and decreasing on $[2, \infty)$.

  (e) $g(x) = -2x^2 + 8x + 10 = f(x) + 5$, so its graph is obtained by

    shifting that of f upward 5 units.

  (f) $h(x) = -2(x+3)^2 + 8(x+3) + 5 = f(x+3)$, so its graph is

    obtained by shifting that of f to the left 3 units.

  (c)

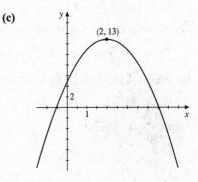

5. (a) $P(x) = 2x^3 - 11x^2 + 10x + 8$ has possible rational zeros $\pm\frac{1}{2}, \pm1,$ $\pm2, \pm4, \pm8.$

(d)

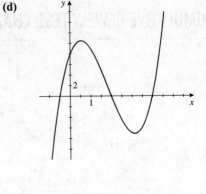

(b) P has 2 variations in sign and hence 0 or 2 positive real zeros.

$P(-x) = -x^3 - x^2 + 8x + 12$ has 1 variation in sign and hence 1 negative real zero.

$$
\begin{array}{r|rrrr}
2 & 2 & -11 & 10 & 8 \\
  &   & 4   & -14 & -8 \\
\hline
  & 2 & -7 & -4 & 0
\end{array} \Rightarrow x = 2 \text{ is a zero.}
$$

$P(x) = 2x^3 - 11x^2 + 10x + 8 = (x-2)\left(2x^2 - 7x - 4\right)$

$= (x-2)(2x+1)(x-4)$

Therefore, the zeros are $-\frac{1}{2}$, 2, and 4.

(c) See part (b).

7. $r(x) = \dfrac{3x^2 + 6x}{x^2 - x - 2} = \dfrac{3x(x+2)}{(x+1)(x-2)}$ has x-intercepts 0 and -2, y-intercept 0, horizontal asymptote $y = 3$, and vertical asymptotes -1 and 2.

9.

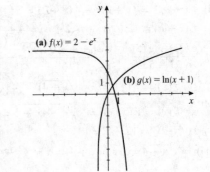

(a) $f(x) = 2 - e^x$

(b) $g(x) = \ln(x+1)$

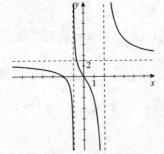

11. (a) $\log_2 x + \log_2 (x-2) = 3 \Leftrightarrow \log_2 (x(x-2)) = 3 \Rightarrow x(x-2) = 2^3 \Leftrightarrow x^2 - 2x - 8 = 0 \Leftrightarrow (x-4)(x+2) = 0 \Leftrightarrow$ $x = -2$ or $x = 4$, but $\log_2(-2)$ is undefined, so the only solution is $x = 4$.

(b) From Problem 5, $P(x) = 2x^3 - 11x^2 + 10x + 8$ has zeros $-\frac{1}{2}$, 2, and 4, so the equation $2e^{3x} - 11e^{2x} + 10e^x + 8 = 2\left(e^x\right)^3 - 11\left(e^x\right)^2 + 10e^x + 8 = 0$ has solutions $e^x = 2$ and $e^x = 4$ (since $e^x = -\frac{1}{2}$ has no solution). If $e^x = 2$, then $x = \ln 2$, and if $e^x = 4$, then $x = \ln 4 = 2\ln 2$, so the two solutions are $x = \ln 2$ and $x = 2\ln 2$.

13. (a) The rat population can be modeled by $P(t) = A_0 e^{kt}$, where t is measured in months, $A_0 = 120$, and $P(15) = 280$. Thus, $120e^{15k} = 280 \Leftrightarrow 3e^{15k} = 7 \Leftrightarrow k = \dfrac{\ln 7 - \ln 3}{15} \approx 0.0565$. So $P(t) = 120e^{0.0565t}$.

(b) After 3 years, the population will be approximately $120e^{36(\ln 7 - \ln 3)/15} \approx 917$ rats.

(c) We solve $P(t) = 2000 \Leftrightarrow 120e^{kt} = 2000 \Leftrightarrow t = \dfrac{15(\ln 50 - \ln 3)}{\ln 7 - \ln 3} \approx 49.8$. Thus, the rat population will reach 2000 after about 4 years 2 months.

5 SYSTEMS OF EQUATIONS AND INEQUALITIES

5.1 SYSTEMS OF LINEAR EQUATIONS IN TWO VARIABLES

1. The given system is a system of two equations in the two variables x and y. To check if $(5, -1)$ is a solution of this system, we check if $x = 5$ and $y = -1$ satisfy each *equation* in the system. The only solution of the given system is $(2, 1)$.

3. A system of two linear equations in two variables can have one solution, *no* solution, or *infinitely many* solutions.

5. $\begin{cases} x - y = 1 \\ 4x + 3y = 18 \end{cases}$ Solving the first equation for x, we get $x = y + 1$, and substituting this into the second equation gives

$4(y + 1) + 3y = 18 \Leftrightarrow 7y + 4 = 18 \Leftrightarrow 7y = 14 \Leftrightarrow y = 2$. Substituting for y we get $x = y + 1 = 2 + 1 = 3$. Thus, the solution is $(3, 2)$.

7. $\begin{cases} x - y = 2 \\ 2x + 3y = 9 \end{cases}$ Solving the first equation for x, we get $x = y + 2$, and substituting this into the second equation gives

$2(y + 2) + 3y = 9 \Leftrightarrow 5y + 4 = 9 \Leftrightarrow 5y = 5 \Leftrightarrow y = 1$. Substituting for y we get $x = y + 2 = (1) + 2 = 3$. Thus, the solution is $(3, 1)$.

9. $\begin{cases} 3x + 4y = 10 \\ x - 4y = -2 \end{cases}$ Adding the two equations, we get $4x = 8 \Leftrightarrow x = 2$, and substituting into the first equation in the

original system gives $3(2) + 4y = 10 \Leftrightarrow 4y = 4 \Leftrightarrow y = 1$. Thus, the solution is $(2, 1)$.

11. $\begin{cases} x + 2y = 5 \\ 2x + 3y = 8 \end{cases}$ Multiplying the first equation by 2 and the second by -1 gives the system $\begin{cases} 2x + 4y = 10 \\ -2x - 3y = -8 \end{cases}$ Adding,

we get $y = 2$, and substituting into the first equation in the original system gives $x + 2(2) = 5 \Leftrightarrow x + 4 = 5 \Leftrightarrow x = 1$. The solution is $(1, 2)$.

13. $\begin{cases} 2x + y = -1 \\ x - 2y = -8 \end{cases}$ By inspection of the graph, it appears that $(-2, 3)$ is the solution to the system. We check this in both

equations to verify that it is a solution. $2(-2) + 3 = -4 + 3 = -1$ and $-2 - 2(3) = -2 - 6 = -8$. Since both equations are satisfied, the solution is $(-2, 3)$.

15. $\begin{cases} x - y = 4 \\ 2x + y = 2 \end{cases}$

The solution is $x = 2$, $y = -2$.

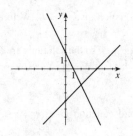

17. $\begin{cases} 2x - 3y = 12 \\ -x + \frac{3}{2}y = 4 \end{cases}$

The lines are parallel, so there is no intersection and hence no solution.

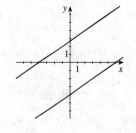

19. $\begin{cases} -x + \frac{1}{2}y = -5 \\ 2x - y = 10 \end{cases}$

There are infinitely many solutions.

21. $\begin{cases} x + y = 4 \\ -x + y = 0 \end{cases}$ Adding the two equations gives

$2y = 4 \Leftrightarrow y = 2$. Substituting for y in the first equation gives $x + 2 = 4 \Leftrightarrow x = 2$. Hence, the solution is $(2, 2)$.

23. $\begin{cases} 2x - 3y = 9 \\ 4x + 3y = 9 \end{cases}$ Adding the two equations gives

$6x = 18 \Leftrightarrow x = 3$. Substituting for x in the second equation gives $4(3) + 3y = 9 \Leftrightarrow 12 + 3y = 9 \Leftrightarrow 3y = -3 \Leftrightarrow x = -1$. Hence, the solution is $(3, -1)$.

25. $\begin{cases} x + 3y = 5 \\ 2x - y = 3 \end{cases}$ Solving the first equation for x gives $x = -3y + 5$. Substituting for x in the second equation gives $2(-3y + 5) - y = 3 \Leftrightarrow -6y + 10 - y = 3 \Leftrightarrow -7y = -7 \Leftrightarrow y = 1$. Then $x = -3(1) + 5 = 2$. Hence, the solution is $(2, 1)$.

27. $-x + y = 2 \Leftrightarrow y = x + 2$. Substituting for y into $4x - 3y = -3$ gives $4x - 3(x + 2) = -3 \Leftrightarrow 4x - 3x - 6 = -3 \Leftrightarrow x = 3$, and so $y = (3) + 2 = 5$. Hence, the solution is $(3, 5)$.

29. $x + 2y = 7 \Leftrightarrow x = 7 - 2y$. Substituting for x into $5x - y = 2$ gives $5(7 - 2y) - y = 2 \Leftrightarrow 35 - 10y - y = 2 \Leftrightarrow -11y = -33 \Leftrightarrow y = 3$, and so $x = 7 - 2(3) = 1$. Hence, the solution is $(1, 3)$.

31. $\frac{1}{2}x + \frac{1}{3}y = 2 \Leftrightarrow x + \frac{2}{3}y = 4 \Leftrightarrow x = 4 - \frac{2}{3}y$. Substituting for x into $\frac{1}{5}x - \frac{2}{3}y = 8$ gives $\frac{1}{5}\left(4 - \frac{2}{3}y\right) - \frac{2}{3}y = 8 \Leftrightarrow \frac{4}{5} - \frac{2}{15}y - \frac{10}{15}y = 8 \Leftrightarrow 12 - 2y - 10y = 120 \Leftrightarrow y = -9$, and so $x = 4 - \frac{2}{3}(-9) = 10$. Hence, the solution is $(10, -9)$.

33. $\begin{cases} 3x + 2y = 8 \\ x - 2y = 0 \end{cases}$ Multiplying the second equation by 3 gives the system $\begin{cases} 3x + 2y = 8 \\ 3x - 6y = 0 \end{cases}$ Subtracting the second equation from the first gives $8y = 8 \Leftrightarrow y = 1$. Substituting into the first equation we get $3x + 2(1) = 8 \Leftrightarrow 3x = 6 \Leftrightarrow x = 2$. Thus, the solution is $(2, 1)$.

35. $\begin{cases} x + 4y = 8 \\ 3x + 12y = 2 \end{cases}$ Adding -3 times the first equation to the second equation gives $0 = -22$, which is never true. Thus, the system has no solution.

37. $\begin{cases} 2x - 6y = 10 \\ -3x + 9y = -15 \end{cases}$ Adding 3 times the first equation to 2 times the second equation gives $0 = 0$. Writing the equation in slope-intercept form, we have $2x - 6y = 10 \Leftrightarrow -6y = -2x + 10 \Leftrightarrow y = \frac{1}{3}x - \frac{5}{3}$, so the solutions are all pairs of the form $\left(x, \frac{1}{3}x - \frac{5}{3}\right)$ where x is a real number.

39. $\begin{cases} 6x + 4y = 12 \\ 9x + 6y = 18 \end{cases}$ Adding 3 times the first equation to -2 times the second equation gives $0 = 0$. Writing the equation in slope-intercept form, we have $6x + 4y = 12 \Leftrightarrow 4y = -6x + 12 \Leftrightarrow y = -\frac{3}{2}x + 3$, so the solutions are all pairs of the form $\left(x, -\frac{3}{2}x + 3\right)$ where x is a real number.

41. $\begin{cases} 8s - 3t = -3 \\ 5s - 2t = -1 \end{cases}$ Adding 2 times the first equation to 3 times the second equation gives $s = -3$, so $8(-3) - 3t = -3 \Leftrightarrow -24 - 3t = -3 \Leftrightarrow t = -7$. Thus, the solution is $(-3, -7)$.

43. $\begin{cases} \frac{1}{2}x + \frac{3}{5}y = 3 \\ \frac{5}{3}x + 2y = 10 \end{cases}$ Adding 10 times the first equation to -3 times the second equation gives $0 = 0$. Writing the equation

in slope-intercept form, we have $\frac{1}{2}x + \frac{3}{5}y = 3 \Leftrightarrow \frac{3}{5}y = -\frac{1}{2}x + 3 \Leftrightarrow y = -\frac{5}{6}x + 5$, so the solutions are all pairs of the

form $\left(x, -\frac{5}{6}x + 5\right)$ where x is a real number.

45. $\begin{cases} 0.4x + 1.2y = 14 \\ 12x - 5y = 10 \end{cases}$ Adding 30 times the first equation to -1 times the second equation gives $41y = 410 \Leftrightarrow y = 10$, so

$12x - 5(10) = 10 \Leftrightarrow 12x = 60 \Leftrightarrow x = 5$. Thus, the solution is $(5, 10)$.

47. $\begin{cases} \frac{1}{3}x - \frac{1}{4}y = 2 \\ -8x + 6y = 10 \end{cases}$ Adding 24 times the first equation to the second equation gives $0 = 58$, which is never true. Thus,

the system has no solution.

49. $\begin{cases} 0.21x + 3.17y = 9.51 \\ 2.35x - 1.17y = 5.89 \end{cases}$

The solution is approximately $(3.87, 2.74)$.

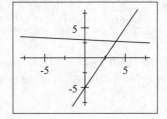

51. $\begin{cases} 2371x - 6552y = 13{,}591 \\ 9815x + 992y = 618{,}555 \end{cases}$

The solution is approximately $(61.00, 20.00)$.

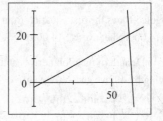

53. Subtracting the first equation from the second, we get $ay - y = 1 \Leftrightarrow y(a - 1) = 1 \Leftrightarrow y = \dfrac{1}{a - 1}, a \neq 1$. So

$x + \left(\dfrac{1}{a - 1}\right) = 0 \Leftrightarrow x = \dfrac{1}{1 - a} = -\dfrac{1}{a - 1}$. Thus, the solution is $\left(-\dfrac{1}{a - 1}, \dfrac{1}{a - 1}\right)$.

55. Subtracting b times the first equation from a times the second, we get $\left(a^2 - b^2\right)y = a - b \Leftrightarrow y = \dfrac{a - b}{a^2 - b^2} = \dfrac{1}{a + b}$,

$a^2 - b^2 \neq 0$. So $ax + \dfrac{b}{a + b} = 1 \Leftrightarrow ax = \dfrac{a}{a + b} \Leftrightarrow x = \dfrac{1}{a + b}$. Thus, the solution is $\left(\dfrac{1}{a + b}, \dfrac{1}{a + b}\right)$.

57. Let the two numbers be x and y. Then $\begin{cases} x + y = 34 \\ x - y = 10 \end{cases}$ Adding these two equations gives $2x = 44 \Leftrightarrow x = 22$. So

$22 + y = 34 \Leftrightarrow y = 12$. Therefore, the two numbers are 22 and 12.

59. Let d be the number of dimes and q be the number of quarters. This gives $\begin{cases} d + q = 14 \\ 0.10d + 0.25q = 2.75 \end{cases}$ Subtracting the

first equation from 10 times the second gives $1.5q = 13.5 \Leftrightarrow q = 9$. So $d + 9 = 14 \Leftrightarrow d = 5$. Thus, the number of dimes

is 5 and the number of quarters is 9.

61. Let r be the amount of regular gas sold and p the amount of premium gas sold. Then $\begin{cases} r + p = 280 \\ 2.20r + 3.00p = 680 \end{cases}$ Subtracting

the second equation from three times the first equation gives $3r - 2.2 = 3(280) - 680 \Leftrightarrow 0.8r = 160 \Leftrightarrow r = 200$.

Substituting this value of r into the original first equation gives $200 + p = 280 \Leftrightarrow p = 80$. Thus, 200 gallons of regular gas

and 80 gallons of premium were sold.

63. Let x be the speed of the plane in still air and y be the speed of the wind. This gives $\begin{cases} 2x - 2y = 180 \\ 1.2x + 1.2y = 180 \end{cases}$ Subtracting 6 times the first equation from 10 times the second gives $24x = 2880 \Leftrightarrow x = 120$, so $2(120) - 2y = 180 \Leftrightarrow -2y = -60 \Leftrightarrow y = 30$. Therefore, the speed of the plane is 120 mi/h and the wind speed is 30 mi/h.

65. Let a and b be the number of grams of food A and food B. Then $\begin{cases} 0.12a + 0.20b = 32 \\ 100a + 50b = 22{,}000 \end{cases}$ Subtracting 250 times the first equation from the second, we get $70a = 14{,}000 \Leftrightarrow a = 200$, so $0.12(200) + 0.20b = 32 \Leftrightarrow 0.20b = 8 \Leftrightarrow b = 40$. Thus, she should use 200 grams of food A and 40 grams of food B.

67. Let x and y be the sulfuric acid concentrations in the first and second containers. $\begin{cases} 300x + 600y = 900(0.15) \\ 100x + 500y = 600(0.125) \end{cases}$ Subtracting the first equation from 3 times the second gives $900y = 90 \Leftrightarrow y = 0.10$, so $100x + 500(0.10) = 75 \Leftrightarrow x = 0.25$. Thus, the concentrations of sulfuric acid are 25% in the first container and 10% in the second.

69. Let x be the amount invested at 5% and y the amount invested at 8%. $\begin{cases} \text{Total invested:} & x + y = 20{,}000 \\ \text{Interest earned: } 0.05x + 0.08y = 1180 \end{cases}$ Subtracting 5 times the first equation from 100 times the second gives $3y = 18{,}000 \Leftrightarrow y = 6{,}000$, so $x + 6{,}000 = 20{,}000 \Leftrightarrow x = 14{,}000$. She invests \$14,000 at 5% and \$6,000 at 8%.

71. Let x be the length of time John drives and y be the length of time Mary drives. Then $y = x + 0.25$, so $-x + y = 0.25$, and multiplying by 40, we get $-40x + 40y = 10$. Comparing the distances, we get $60x = 40y + 35$, or $60x - 40y = 35$. This gives the system $\begin{cases} -40x + 40y = 10 \\ 60x - 40y = 35 \end{cases}$ Adding, we get $20x = 45 \Leftrightarrow x = 2.25$, so $y = 2.25 + 0.25 = 2.5$. Thus, John drives for $2\frac{1}{4}$ hours and Mary drives for $2\frac{1}{2}$ hours.

73. Let x be the tens digit and y be the ones digit of the number. $\begin{cases} x + y = 7 \\ 10y + x = 27 + 10x + y \end{cases}$ Adding 9 times the first equation to the second gives $18x = 36 \Leftrightarrow x = 2$, so $2 + y = 7 \Leftrightarrow y = 5$. Thus, the number is 25.

75. $n = 5$, so $\sum_{k=1}^{n} x_k = 1 + 2 + 3 + 5 + 7 = 18$, $\sum_{k=1}^{n} y_k = 3 + 5 + 6 + 6 + 9 = 29$,
$\sum_{k=1}^{n} x_k y_k = 1(3) + 2(5) + 3(6) + 5(6) + 7(9) = 124$, and
$\sum_{k=1}^{n} x_k^2 = 1^2 + 2^2 + 3^2 + 5^2 + 7^2 = 88$. Thus we get the system

$\begin{cases} 18a + 5b = 29 \\ 88a + 18b = 124 \end{cases}$ Subtracting 18 times the first equation from 5 times the second, we get $116a = 98 \Leftrightarrow a \approx 0.845$. Then $b = \frac{1}{5}[-18(0.845) + 29] \approx 2.758$. So the regression line is $y = 0.845x + 2.758$.

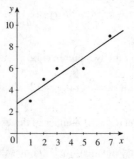

5.2   SYSTEMS OF LINEAR EQUATIONS IN SEVERAL VARIABLES

1. If we add 2 times the first equation to the second equation, the second equation becomes $x + 3z = 1$.

3. The equation $6x - \sqrt{3}y + \frac{1}{2}z = 0$ is linear.

5. The system $\begin{cases} xy - 3y + z = 5 \\ x - y^2 + 5z = 0 \\ 2x + yz = 3 \end{cases}$ is not a linear system, since the first equation contains a product of variables. In fact both the second and the third equation are not linear.

7. $\begin{cases} x - 3y + z = 0 \\ y - z = 3 \\ z = -2 \end{cases}$ Substituting $z = -2$ into the second equation gives $y - (-2) = 3 \Leftrightarrow y = 1$. Substituting $z = -2$

and $y = 1$ into the first equation gives $x - 3(1) + (-2) = 0 \Leftrightarrow x = 5$. Thus, the solution is $(5, 1, -2)$.

9. $\begin{cases} x + 2y + z = 7 \\ -y + 3z = 9 \\ 2z = 6 \end{cases}$ Solving we get $2z = 6 \Leftrightarrow z = 3$. Substituting $z = 3$ into the second equation gives

$-y + 3(3) = 9 \Leftrightarrow y = 0$. Substituting $z = 3$ and $y = 0$ into the first equation gives $x + 2(0) + 3 = 7 \Leftrightarrow x = 4$. Thus, the solution is $(4, 0, 3)$.

11. $\begin{cases} 2x - y + 6z = 5 \\ y + 4z = 0 \\ -2z = 1 \end{cases}$ Solving we get $-2z = 1 \Leftrightarrow z = -\frac{1}{2}$. Substituting $z = -\frac{1}{2}$ into the second equation gives

$y + 4\left(-\frac{1}{2}\right) = 0 \Leftrightarrow y = 2$. Substituting $z = -\frac{1}{2}$ and $y = 2$ into the first equation gives $2x - (2) + 6\left(-\frac{1}{2}\right) = 5 \Leftrightarrow x = 5$.
Thus, the solution is $\left(5, 2, -\frac{1}{2}\right)$.

13. $\begin{cases} 3x + y + z = 4 \\ -x + y + 2z = 0 \\ x - 2y - z = -1 \end{cases}$ Add the third equation to the second equation: $\begin{cases} 3x + y + z = 4 \\ -y + z = -1 \\ x - 2y - z = -1 \end{cases}$

Or, add the first equation to three times the second equation: $\begin{cases} 3x + y + z = 4 \\ 4y + 7z = 4 \\ x - 2y - z = -1 \end{cases}$

15. $\begin{cases} 2x + y - 3z = 5 \\ 2x + 3y + z = 13 \\ 6x - 5y - z = 7 \end{cases}$ Add -3 times the first equation to the third equation: $\begin{cases} 2x + y - 3z = 5 \\ 2x + 3y + z = 13 \\ -8y + 8z = -8 \end{cases}$

Or, add -3 times the second equation to the third equation: $\begin{cases} 2x + y - 3z = 5 \\ 2x + 3y + z = 13 \\ -14y - 4z = -32 \end{cases}$

17. $\begin{cases} x - y - z = 4 \\ 2y + z = -1 \\ -x + y - 2z = 5 \end{cases} \Leftrightarrow \begin{cases} x - y - z = 4 \\ 2y + z = -1 \\ -3z = 9 \quad \text{Eq. 1 + Eq. 3} \end{cases}$

So $z = -3$ and $2y + (-3) = -1 \Leftrightarrow 2y = 2 \Leftrightarrow y = 1$. Thus, $x - 1 - (-3) = 4 \Leftrightarrow x = 2$. So the solution is $(2, 1, -3)$.

19. $\begin{cases} x + 2y - z = -6 \\ y - 3z = -16 \\ x - 3y + 2z = 14 \end{cases} \Leftrightarrow \begin{cases} x + 2y - z = -6 \\ y - 3z = -16 \\ 5y - 3z = -20 \quad \text{Eq. 1} + (-1) \times \text{Eq. 3} \end{cases} \Leftrightarrow$

$\begin{cases} x + 2y - z = -6 \\ y - 3z = -16 \\ 12z = 60 \quad (-5) \times \text{Eq. 2} + \text{Eq. 3} \end{cases}$

So $z = 5$ and $y - 3(5) = -16 \Leftrightarrow y = -1$. Then $x + 2(-1) - 5 = -6 \Leftrightarrow x = 1$. So the solution is $(1, -1, 5)$.

21. $\begin{cases} x + y + z = 4 \\ x + 3y + 3z = 10 \\ 2x + y - z = 3 \end{cases} \Leftrightarrow \begin{cases} x + y + z = 4 \\ 2y + 2z = 6 \quad (-1) \times \text{Eq. 1} + \text{Eq. 2} \\ y + 3z = 5 \quad 2 \times \text{Eq. 1} + (-1) \times \text{Eq. 3} \end{cases} \Leftrightarrow \begin{cases} x + y + z = 4 \\ y + 3z = 5 \quad \text{Eq. 3} \\ 2y + 2z = 6 \quad \text{Eq. 2} \end{cases} \Leftrightarrow$

$\begin{cases} x + y + z = 4 \\ y + 3z = 5 \\ -4z = -4 \quad (-2) \times \text{Eq. 2} + \text{Eq. 3} \end{cases}$

So $z = 1$ and $y + 3(1) = 8 \Leftrightarrow y = 2$. Then $x + 2 + 1 = 4 \Leftrightarrow x = 1$. So the solution is $(1, 2, 1)$.

23. $\begin{cases} x - 4z = 1 \\ 2x - y - 6z = 4 \\ 2x + 3y - 2z = 8 \end{cases} \Leftrightarrow \begin{cases} x - 4z = 1 \\ -y + 2z = 2 \quad (-2) \times \text{Eq. 1} + \text{Eq. 2} \\ 3y + 6z = 6 \quad (-2) \times \text{Eq. 1} + \text{Eq. 3} \end{cases} \Leftrightarrow \begin{cases} x - 4z = 1 \\ -y + 2z = 2 \\ 12z = 12 \quad 3 \times \text{Eq. 2} + \text{Eq. 3} \end{cases}$

So $z = 1$ and $-y + 2(1) = 2 \Leftrightarrow y = 0$. Then $x - 4(1) = 1 \Leftrightarrow x = 5$. So the solution is $(5, 0, 1)$.

25. $\begin{cases} 2x + 4y - z = 2 \\ x + 2y - 3z = -4 \\ 3x - y + z = 1 \end{cases} \Leftrightarrow \begin{cases} 2x + 4y - z = 2 \\ 5z = 10 \quad \text{Eq. 1} + (-2) \times \text{Eq. 2} \\ 14y - 5z = 4 \quad 3 \times \text{Eq. 1} + (-2) \times \text{Eq. 3} \end{cases} \Leftrightarrow \begin{cases} x + 2y - 3z = -4 \\ 14y - 5z = 4 \quad \text{Eq. 2} \leftrightarrow \text{Eq. 3} \\ 5z = 10 \quad \text{Eq. 2} \leftrightarrow \text{Eq. 3} \end{cases}$

So $z = 2$ and $14y - 5(2) = 4 \Leftrightarrow y = 1$. Then $x + 2(1) - 3(2) = -4 \Leftrightarrow x = 0$. So the solution is $(0, 1, 2)$.

27. $\begin{cases} 2y + 4z = -1 \\ -2x + y + 2z = -1 \\ 4x - 2y = 0 \end{cases} \Leftrightarrow \begin{cases} -2x + y + 2z = -1 \quad \text{Eq. 2} \\ 2y + 4z = -1 \quad \text{Eq. 1} \\ 4z = -2 \quad 2 \times \text{Eq. 2} + \text{Eq. 3} \end{cases}$

So $z = -\frac{1}{2}$ and $2y + 4\left(-\frac{1}{2}\right) = -1 \Leftrightarrow y = \frac{1}{2}$. Then $-2x + \frac{1}{2} + 2\left(-\frac{1}{2}\right) = -1 \Leftrightarrow x = \frac{1}{4}$. So the solution is $\left(\frac{1}{4}, \frac{1}{2}, -\frac{1}{2}\right)$.

29. $\begin{cases} x + 2y - z = 1 \\ 2x + 3y - 4z = -3 \\ 3x + 6y - 3z = 4 \end{cases} \Leftrightarrow \begin{cases} x + 2y - z = 1 \\ -y - 2z = -5 \quad (-2) \times \text{Eq. 1} + \text{Eq. 2} \\ 0 = 1 \quad (-3) \times \text{Eq. 1} + \text{Eq. 3} \end{cases}$ Since $0 = 1$ is false, this system is inconsistent.

31. $\begin{cases} 2x + 3y - z = 1 \\ x + 2y = 3 \\ x + 3y + z = 4 \end{cases} \Leftrightarrow \begin{cases} x + 2y = 3 \quad \text{Eq. 2} \\ 2x + 3y - z = 1 \quad \text{Eq. 1} \\ x + 3y + z = 4 \end{cases} \Leftrightarrow \begin{cases} x + 2y = 3 \\ -y - z = -5 \quad \text{Eq. 2} + (-2) \times \text{Eq. 1} \\ y + z = 1 \quad \text{Eq. 3} - \text{Eq. 1} \end{cases} \Leftrightarrow$

$\begin{cases} x + 2y = 3 \\ -y - z = -5 \\ 0 = -4 \quad \text{Eq. 2} + \text{Eq. 3} \end{cases}$ Since $0 = -4$ is false, this system is inconsistent.

33. $\begin{cases} x + y - z = 0 \\ x + 2y - 3z = -3 \\ 2x + 3y - 4z = -3 \end{cases} \Leftrightarrow \begin{cases} x + y - z = 0 \\ y - 2z = -3 \quad \text{Eq. 2 - Eq. 1} \\ y - 2z = -3 \quad (-2) \times \text{Eq. 1 + Eq. 3} \end{cases} \Leftrightarrow \begin{cases} x + y - z = 0 \\ y - 2z = -3 \\ 0 = 0 \quad \text{Eq. 2 - Eq. 3} \end{cases}$

So $z = t$ and $y - 2t = -3 \Leftrightarrow y = 2t - 3$. Then $x + (2t - 3) - t = 0 \Leftrightarrow x = -t + 3$. So the solutions are $(-t + 3, 2t - 3, t)$, where t is any real number.

35. $\begin{cases} x + 3y - 2z = 0 \\ 2x \quad\quad + 4z = 4 \\ 4x + 6y \quad\quad = 4 \end{cases} \Leftrightarrow \begin{cases} x + 3y - 2z = 0 \\ -6y + 8z = 4 \quad \text{Eq. 2} + (-2) \times \text{Eq. 1} \\ -6y + 8z = 4 \quad \text{Eq. 3} + (-4) \times \text{Eq. 1} \end{cases} \Leftrightarrow \begin{cases} x + 3y - 2z = 0 \\ -6y + 8z = 4 \\ 0 = 0 \quad \text{Eq. 2 - Eq. 3} \end{cases}$

So $z = t$ and $-6y + 8t = 4 \Leftrightarrow -6y = -8t + 4 \Leftrightarrow y = \frac{4}{3}t - \frac{2}{3}$. Then $x + 3\left(\frac{4}{3}t - \frac{2}{3}\right) - 2t = 0 \Leftrightarrow x = -2t + 2$. So the

solutions are $\left(-2t + 2, \frac{4}{3}t - \frac{2}{3}, t\right)$, where t is any real number.

37. $\begin{cases} x \quad\quad + z + 2w = 6 \\ y - 2z \quad\quad = -3 \\ x + 2y - z \quad\quad = -2 \\ 2x + y + 3z - 2w = 0 \end{cases} \Leftrightarrow \begin{cases} x \quad\quad + z + 2w = 6 \\ y - 2z \quad\quad = -3 \\ 2y - 2z - 2w = -8 \quad \text{Eq. 3 - Eq. 1} \\ y + z - 6w = -12 \quad \text{Eq. 4} + (-2) \times \text{Eq. 1} \end{cases} \Leftrightarrow$

$\begin{cases} x \quad\quad + z + 2w = 6 \\ y - 2z \quad\quad = -3 \\ 2z - 2w = -2 \quad \text{Eq. 3} + (-2) \times \text{Eq. 2} \\ 3z - 6w = -9 \quad \text{Eq. 4 - Eq. 2} \end{cases} \Leftrightarrow \begin{cases} x \quad\quad + z + 2w = 6 \\ y - 2z \quad\quad = -3 \\ z - w = -1 \quad \frac{1}{2}\text{Eq. 3} \\ 6w = 12 \quad 3 \times \text{Eq. 3} + (-2) \times \text{Eq. 4} \end{cases}$

So $w = 2$ and $z - 2 = -1 \Leftrightarrow z = 1$. Then $y - 2(1) = -3 \Leftrightarrow y = -1$ and $x + 1 + 2(2) = 6 \Leftrightarrow x = 1$. Thus, the solution is $(1, -1, 1, 2)$.

39. Let x be the amount invested at 4%, y the amount invested at 5%, and z the amount invested at 6%. We set up a model and get the following equations: $\begin{cases} \text{Total money:} \quad\quad x + y + z = 100,000 \\ \text{Annual income:} \quad 0.04x + 0.05y + 0.06z = 0.051\,(100,000) \\ \text{Equal amounts:} \quad\quad\quad x = y \end{cases}$

$\Leftrightarrow \begin{cases} x + y + z = 100,000 \\ 4x + 5y + 6z = 510,000 \\ x - y \quad\quad = 0 \end{cases} \Leftrightarrow \begin{cases} x + y + z = 100,000 \\ y + 2z = 110,000 \quad \text{Eq. 2} + (-4) \times \text{Eq. 1} \\ -2y - z = -100,000 \quad \text{Eq. 3 - Eq. 1} \end{cases} \Leftrightarrow$

$\begin{cases} x + y + z = 100,000 \\ y + 2z = 110,000 \\ 3z = 120,000 \quad 2 \times \text{Eq. 2 + Eq. 3} \end{cases}$

So $z = 40,000$ and $y + 2(40,000) = 110,000 \Leftrightarrow y = 30,000$. Since $x = y$, $x = 30,000$. She must invest $30,000 in short-term bonds, $30,000 in intermediate-term bonds, and $40,000 in long-term bonds.

41. Let x, y, and z be the number of acres of land planted with corn, wheat, and soybeans. We set up a model and

get the following equations: $\begin{cases} \text{Total acres:} & x + y + z = 1200 \\ \text{Market demand:} & 2x = y \\ \text{Total cost:} & 45x + 60y + 50z = 63{,}750 \end{cases}$ Substituting $2x$ for y, we get

$$\begin{cases} x + 2x + z = 1200 \\ 2x = y \\ 45x + 60(2x) + 50z = 63{,}750 \end{cases} \Leftrightarrow \begin{cases} 3x + z = 1200 \\ 2x - y = 0 \\ 165x + 50z = 63{,}750 \end{cases} \Leftrightarrow \begin{cases} 3x + z = 1200 \\ 2x - y = 0 \\ 15x = 3750 \quad \text{Eq. 3} + (-50) \times \text{Eq. 1} \end{cases}$$

So $15x = 3{,}750 \Leftrightarrow x = 250$ and $y = 2(250) = 500$. Substituting into the original equation, we have $250 + 500 + z = 1200$ $\Leftrightarrow z = 450$. Thus the farmer should plant 250 acres of corn, 500 acres of wheat, and 450 acres of soybeans.

43. Let a, b, and c be the number of ounces of Type A, Type B, and Type C pellets used. The

requirements for the different vitamins gives the following system: $\begin{cases} 2a + 3b + c = 9 \\ 3a + b + 3c = 14 \\ 8a + 5b + 7c = 32 \end{cases} \Leftrightarrow$

$$\begin{cases} 2a + 3b + c = 9 \\ -7b + 3c = 1 \quad 2 \times \text{Eq. 2} + (-3) \times \text{Eq. 1} \\ -7b + 3c = -4 \quad \text{Eq. 3} + (-4) \times \text{Eq. 1} \end{cases}$$ Equations 2 and 3 are inconsistent, so there is no solution.

45. Let a, b, and c represent the number of Midnight Mango, Tropical Torrent, and Pineapple Power smoothies sold. The

given information leads to the system $\begin{cases} 8a + 6b + 2c = 820 \\ 3a + 5b + 8c = 690 \\ 3a + 3b + 4c = 450 \end{cases} \Leftrightarrow \begin{cases} 8a + 6b + 2c = 820 \\ 22b + 58c = 3060 \quad 8 \times \text{Eq. 2} + (-3) \times \text{Eq. 1} \\ 2b + 4c = 240 \quad \text{Eq. 2} - \text{Eq. 3} \end{cases} \Leftrightarrow$

$$\begin{cases} 8a + 6b + 2c = 820 \\ 22b + 58c = 3060 \\ 14c = 420 \quad \text{Eq. 2} + (-11) \times \text{Eq. 3} \end{cases}$$

Thus, $c = 30$, so $22b + 58(30) = 3060 \Leftrightarrow 22b = 1320 \Leftrightarrow b = 60$ and $8a + 6(60) + 2(30) = 820 \Leftrightarrow a = 50$. Thus, The Juice Company sold 50 Midnight Mango, 60 Tropical Torrent, and 30 Pineapple Power smoothies on that particular day.

47. Let a, b, and c be the number of shares of Stock A, Stock B, and Stock C in the investor's portfolio. Since the total value remains unchanged, we get the following system:

$$\begin{cases} 10a + 25b + 29c = 74{,}000 \\ 12a + 20b + 32c = 74{,}000 \\ 16a + 15b + 32c = 74{,}000 \end{cases} \Leftrightarrow \begin{cases} 10a + 25b + 29c = 74{,}000 \\ 50b + 14c = 74{,}000 \quad 6 \times \text{Eq. 1} + (-5) \times \text{Eq. 2} \\ 125b + 72c = 222{,}000 \quad 8 \times \text{Eq. 1} + (-5) \times \text{Eq. 3} \end{cases} \Leftrightarrow$$

$$\begin{cases} 10a + 25b + 29c = 74{,}000 \\ 50b + 14c = 74{,}000 \\ 74c = 74{,}000 \quad (-5) \times \text{Eq. 2} + 2 \times \text{Eq. 3} \end{cases}$$

So $c = 1{,}000$. Back-substituting we have $50b + 14(1000) = 74{,}000 \Leftrightarrow 50b = 60{,}000 \Leftrightarrow b = 1{,}200$. And finally $10a + 25(1200) + 29(1000) = 74{,}000$ $10a + 30{,}000 + 29{,}000 = 74{,}000 \Leftrightarrow 10a = 15{,}000 \Leftrightarrow a = 1{,}500$. Thus the portfolio consists of 1,500 shares of Stock A, 1,200 shares of Stock B, and 1,000 shares of Stock C.

49. (a) We begin by substituting $\dfrac{x_0 + x_1}{2}$, $\dfrac{y_0 + y_1}{2}$, and $\dfrac{z_0 + z_1}{2}$ into the left-hand side of the first equation:

$$a_1 \left(\frac{x_0 + x_1}{2} \right) + b_1 \left(\frac{y_0 + y_1}{2} \right) + c_1 \left(\frac{z_0 + z_1}{2} \right) = \tfrac{1}{2} \left[(a_1 x_0 + b_1 y_0 + c_1 z_0) + (a_1 x_1 + b_1 y_1 + c_1 z_1) \right]$$
$$= \tfrac{1}{2} \left[d_1 + d_1 \right] = d_1$$

Thus the given ordered triple satisfies the first equation. We can show that it satisfies the second and the third in exactly the same way. Thus it is a solution of the system.

(b) We have shown in part (a) that if the system has two different solutions, we can find a third one by averaging the two solutions. But then we can find a fourth and a fifth solution by averaging the new one with each of the previous two. Then we can find four more by repeating this process with these new solutions, and so on. Clearly this process can continue indefinitely, so there are infinitely many solutions.

5.3   PARTIAL FRACTIONS

1. (iii): $r(x) = \dfrac{4}{x(x-2)^2} = \dfrac{A}{x} + \dfrac{B}{x-2} + \dfrac{C}{(x-2)^2}$

3. $\dfrac{1}{(x-1)(x+2)} = \dfrac{A}{x-1} + \dfrac{B}{x+2}$

5. $\dfrac{x^2 - 3x + 5}{(x-2)^2(x+4)} = \dfrac{A}{x-2} + \dfrac{B}{(x-2)^2} + \dfrac{C}{x+4}$

7. $\dfrac{x^2}{(x-3)(x^2+4)} = \dfrac{A}{x-3} + \dfrac{Bx+C}{x^2+4}$

9. $\dfrac{x^3 - 4x^2 + 2}{(x^2+1)(x^2+2)} = \dfrac{Ax+B}{x^2+1} + \dfrac{Cx+D}{x^2+2}$

11. $\dfrac{x^3 + x + 1}{x(2x-5)^3(x^2+2x+5)^2} = \dfrac{A}{x} + \dfrac{B}{2x-5} + \dfrac{C}{(2x-5)^2} + \dfrac{D}{(2x-5)^3} + \dfrac{Ex+F}{x^2+2x+5} + \dfrac{Gx+H}{(x^2+2x+5)^2}$

13. $\dfrac{2}{(x-1)(x+1)} = \dfrac{A}{x-1} + \dfrac{B}{x+1}$. Multiplying by $(x-1)(x+1)$, we get $2 = A(x+1) + B(x-1) \Leftrightarrow$

$2 = Ax + A + Bx - B$. Thus $\begin{cases} A + B = 0 \\ A - B = 2 \end{cases}$ Adding we get $2A = 2 \Leftrightarrow A = 1$. Now $A + B = 0 \Leftrightarrow B = -A$, so

$B = -1$. Thus, the required partial fraction decomposition is $\dfrac{2}{(x-1)(x+1)} = \dfrac{1}{x-1} - \dfrac{1}{x+1}$.

15. $\dfrac{5}{(x-1)(x+4)} = \dfrac{A}{x-1} + \dfrac{B}{x+4}$. Multiplying by $(x-1)(x+4)$, we get $5 = A(x+4) + B(x-1) \Leftrightarrow$

$5 = Ax + 4A + Bx - B$. Thus $\begin{cases} A + B = 0 \\ 4A - B = 5 \end{cases}$ Now $A + B = 0 \Leftrightarrow B = -A$, so substituting, we get $4A - (-A) = 5 \Leftrightarrow$

$5A = 5 \Leftrightarrow A = 1$ and $B = -1$. The required partial fraction decomposition is $\dfrac{5}{(x-1)(x+4)} = \dfrac{1}{x-1} - \dfrac{1}{x+4}$.

17. $\dfrac{12}{x^2-9} = \dfrac{12}{(x-3)(x+3)} = \dfrac{A}{x-3} + \dfrac{B}{x+3}$. Multiplying by $(x-3)(x+3)$, we get $12 = A(x+3) + B(x-3) \Leftrightarrow$

$12 = Ax + 3A + Bx - 3B$. Thus $\begin{cases} A + B = 0 \\ 3A - 3B = 12 \end{cases} \Leftrightarrow \begin{cases} A + B = 0 \\ A - B = 4 \end{cases}$ Adding, we get $2A = 4 \Leftrightarrow A = 2$. So $2 + B = 0$

$\Leftrightarrow B = -2$. The required partial fraction decomposition is $\dfrac{12}{x^2-9} = \dfrac{2}{x-3} - \dfrac{2}{x+3}$.

19. $\dfrac{4}{x^2 - 4} = \dfrac{4}{(x-2)(x+2)} = \dfrac{A}{x-2} + \dfrac{B}{x+2}$. Multiplying by $x^2 - 4$, we get

$4 = A(x+2) + B(x-2) = (A+B)x + (2A - 2B)$, and so $\begin{cases} A + B = 0 \\ 2A - 2B = 4 \end{cases} \Leftrightarrow \begin{cases} A + B = 0 \\ A - B = 2 \end{cases}$ Adding we get $2A = 2$

$\Leftrightarrow A = 1$, and $B = -1$. Therefore, $\dfrac{4}{x^2 - 4} = \dfrac{1}{x-2} - \dfrac{1}{x+2}$.

21. $\dfrac{x+14}{x^2 - 2x - 8} = \dfrac{x+14}{(x-4)(x+2)} = \dfrac{A}{x-4} + \dfrac{B}{x+2}$. Hence, $x + 14 = A(x+2) + B(x-4) = (A+B)x + (2A - 4B)$,

and so $\begin{cases} A + B = 1 \\ 2A - 4B = 14 \end{cases} \Leftrightarrow \begin{cases} 2A + 2B = 2 \\ A - 2B = 7 \end{cases}$ Adding, we get $3A = 9 \Leftrightarrow A = 3$. So $(3) + B = 1 \Leftrightarrow B = -2$.

Therefore, $\dfrac{x+14}{x^2 - 2x - 8} = \dfrac{3}{x-4} - \dfrac{2}{x+2}$.

23. $\dfrac{x}{8x^2 - 10x + 3} = \dfrac{x}{(4x-3)(2x-1)} = \dfrac{A}{4x-3} + \dfrac{B}{2x-1}$. Hence,

$x = A(2x-1) + B(4x-3) = (2A + 4B)x + (-A - 3B)$, and so $\begin{cases} 2A + 4B = 1 \\ -A - 3B = 0 \end{cases} \Leftrightarrow \begin{cases} 2A + 4B = 1 \\ -2A - 6B = 0 \end{cases}$

Adding, we get $-2B = 1 \Leftrightarrow B = -\frac{1}{2}$, and $A = \frac{3}{2}$. Therefore, $\dfrac{x}{8x^2 - 10x + 3} = \dfrac{\frac{3}{2}}{4x-3} - \dfrac{\frac{1}{2}}{2x-1}$.

25. $\dfrac{9x^2 - 9x + 6}{2x^3 - x^2 - 8x + 4} = \dfrac{9x^2 - 9x + 6}{(x-2)(x+2)(2x-1)} = \dfrac{A}{x-2} + \dfrac{B}{x+2} + \dfrac{C}{2x-1}$. Thus,

$$9x^2 - 9x + 6 = A(x+2)(2x-1) + B(x-2)(2x-1) + C(x-2)(x+2)$$
$$= A\left(2x^2 + 3x - 2\right) + B\left(2x^2 - 5x + 2\right) + C\left(x^2 - 4\right)$$
$$= (2A + 2B + C)x^2 + (3A - 5B)x + (-2A + 2B - 4C)$$

This leads to the system $\begin{cases} 2A + 2B + C = 9 & \text{Coefficients of } x^2 \\ 3A - 5B = -9 & \text{Coefficients of } x \\ -2A + 2B - 4C = 6 & \text{Constant terms} \end{cases} \Leftrightarrow \begin{cases} 2A + 2B + C = 9 \\ 16B + 3C = 45 \\ 4B - 3C = 15 \end{cases} \Leftrightarrow$

$\begin{cases} 2A + 2B + C = 9 \\ 16B + 3C = 45 \\ 15C = -15 \end{cases}$ Hence, $-15C = 15 \Leftrightarrow C = -1$; $16B - 3 = 45 \Leftrightarrow B = 3$; and $2A + 6 - 1 = 9 \Leftrightarrow A = 2$.

Therefore, $\dfrac{9x^2 - 9x + 6}{2x^3 - x^2 - 8x + 4} = \dfrac{2}{x-2} + \dfrac{3}{x+2} - \dfrac{1}{2x-1}$.

27. $\dfrac{x^2 + 1}{x^3 + x^2} = \dfrac{x^2 + 1}{x^2(x+1)} = \dfrac{A}{x} + \dfrac{B}{x^2} + \dfrac{C}{x+1}$. Hence,

$x^2 + 1 = Ax(x+1) + B(x+1) + Cx^2 = (A+C)x^2 + (A+B)x + B$, and so $B = 1$; $A + 1 = 0 \Leftrightarrow A = -1$; and

$-1 + C = 1 \Leftrightarrow C = 2$. Therefore, $\dfrac{x^2 + 1}{x^3 + x^2} = \dfrac{-1}{x} + \dfrac{1}{x^2} + \dfrac{2}{x+1}$.

29. $\dfrac{2x}{4x^2 + 12x + 9} = \dfrac{2x}{(2x+3)^2} = \dfrac{A}{2x+3} + \dfrac{B}{(2x+3)^2}$. Hence, $2x = A(2x+3) + B = 2Ax + (3A + B)$. So $2A = 2 \Leftrightarrow$

$A = 1$; and $3(1) + B = 0 \Leftrightarrow B = -3$. Therefore, $\dfrac{2x}{4x^2 + 12x + 9} = \dfrac{1}{2x+3} - \dfrac{3}{(2x+3)^2}$.

31. $\dfrac{4x^2 - x - 2}{x^4 + 2x^3} = \dfrac{4x^2 - x - 2}{x^3\,(x+2)} = \dfrac{A}{x} + \dfrac{B}{x^2} + \dfrac{C}{x^3} + \dfrac{D}{x+2}$. Hence,

$$4x^2 - x - 2 \;=\; Ax^2\,(x+2) + Bx\,(x+2) + C\,(x+2) + Dx^3$$

$$= (A+D)\,x^3 + (2A+B)\,x^2 + (2B+C)\,x + 2C$$

So $2C = -2 \Leftrightarrow C = -1$; $2B - 1 = -1 \Leftrightarrow B = 0$; $2A + 0 = 4 \Leftrightarrow A = 2$; and $2 + D = 0 \Leftrightarrow D = -2$. Therefore,

$$\dfrac{4x^2 - x - 2}{x^4 + 2x^3} = \dfrac{2}{x} - \dfrac{1}{x^3} - \dfrac{2}{x+2}.$$

33. $\dfrac{-10x^2 + 27x - 14}{(x-1)^3\,(x+2)} = \dfrac{A}{x+2} + \dfrac{B}{x-1} + \dfrac{C}{(x-1)^2} + \dfrac{D}{(x-1)^3}$. Thus,

$$-10x^2 + 27x - 14 \;=\; A\,(x-1)^3 + B\,(x+2)\,(x-1)^2 + C\,(x+2)\,(x-1) + D\,(x+2)$$

$$= A\left(x^3 - 3x^2 + 3x - 1\right) + B\,(x+2)\left(x^2 - 2x + 1\right) + C\left(x^2 + x - 2\right) + D\,(x+2)$$

$$= A\left(x^3 - 3x^2 + 3x - 1\right) + B\left(x^3 - 3x + 2\right) + C\left(x^2 + x - 2\right) + D\,(x+2)$$

$$= (A+B)\,x^3 + (-3A+C)\,x^2 + (3A - 3B + C + D)\,x + (-A + 2B - 2C + 2D)$$

which leads to the system

$$\begin{cases} A + B & = 0 \\ -3A \quad + C & = -10 \\ 3A - 3B + C + D & = 27 \\ -A + 2B - 2C + 2D & = -14 \end{cases} \begin{array}{l} \text{Coefficients of } x^3 \\ \text{Coefficients of } x^2 \\ \text{Coefficients of } x \\ \text{Constant terms} \end{array} \Leftrightarrow \begin{cases} A + B & = 0 \\ 3B + C & = -10 \\ -3B + 2C + D & = 17 \\ 3B - 5C + 7D & = -15 \end{cases} \Leftrightarrow$$

$$\begin{cases} A + B & = 0 \\ 3B + C & = -10 \\ 3C + D & = 7 \\ -3C + 8D & = 2 \end{cases} \Leftrightarrow \begin{cases} A + B & = 0 \\ 3B + C & = -10 \\ 3C + D & = 7 \\ 9D & = 9 \end{cases}$$

Hence, $9D = 9 \Leftrightarrow D = 1$, $3C + 1 = 7 \Leftrightarrow C = 2$, $3B + 2 = -10 \Leftrightarrow B = -4$, and $A - 4 = 0 \Leftrightarrow A = 4$. Therefore,

$$\dfrac{-10x^2 + 27x - 14}{(x-1)^3\,(x+2)} = \dfrac{4}{x+2} - \dfrac{4}{x-1} + \dfrac{2}{(x-1)^2} + \dfrac{1}{(x-1)^3}.$$

35. $\dfrac{3x^3 + 22x^2 + 53x + 41}{(x+2)^2 (x+3)^2} = \dfrac{A}{x+2} + \dfrac{B}{(x+2)^2} + \dfrac{C}{x+3} + \dfrac{D}{(x+3)^2}$. Thus,

$$3x^3 + 22x^2 + 53x + 41 = A(x+2)(x+3)^2 + B(x+3)^2 + C(x+2)^2(x+3) + D(x+2)^2$$

$$= A\left(x^3 + 8x^2 + 21x + 18\right) + B\left(x^2 + 6x + 9\right)$$

$$+ C\left(x^3 + 7x^2 + 16x + 12\right) + D\left(x^2 + 4x + 4\right)$$

$$= (A+C)x^3 + (8A + B + 7C + D)x^2$$

$$+ (21A + 6B + 16C + 4D)x + (18A + 9B + 12C + 4D)$$

so we must solve the system $\begin{cases} A + C = 3 \\ 8A + B + 7C + D = 22 \\ 21A + 6B + 16C + 4D = 53 \\ 18A + 9B + 12C + 4D = 41 \end{cases}$ $\begin{array}{l} \text{Coefficients of } x^3 \\ \text{Coefficients of } x^2 \\ \text{Coefficients of } x \\ \text{Constant terms} \end{array}$ $\Leftrightarrow \begin{cases} A + C = 3 \\ B - C + D = -2 \\ 6B - 5C + 4D = -10 \\ 9B - 6C + 4D = -13 \end{cases}$

$\Leftrightarrow \begin{cases} A + C = 3 \\ B - C + D = -2 \\ C - 2D = 2 \\ 3C - 5D = 5 \end{cases}$ $\Leftrightarrow \begin{cases} A + C = 3 \\ B - C + D = -2 \\ C - 2D = 2 \\ D = -1 \end{cases}$ Hence, $D = -1$, $C + 2 = 2 \Leftrightarrow C = 0$, $B - 0 - 1 = -2$

$\Leftrightarrow B = -1$, and $A + 0 = 3 \Leftrightarrow A = 3$. Therefore, $\dfrac{3x^3 + 22x^2 + 53x + 41}{(x+2)^2 (x+3)^2} = \dfrac{3}{x+2} - \dfrac{1}{(x+2)^2} - \dfrac{1}{(x+3)^2}$.

37. $\dfrac{x-3}{x^3 + 3x} = \dfrac{x-3}{x\left(x^2 + 3\right)} = \dfrac{A}{x} + \dfrac{Bx + C}{x^2 + 3}$. Hence, $x - 3 = A\left(x^2 + 3\right) + Bx^2 + Cx = (A+B)x^2 + Cx + 3A$. So

$3A = -3 \Leftrightarrow A = -1$; $C = 1$; and $-1 + B = 0 \Leftrightarrow B = 1$. Therefore, $\dfrac{x-3}{x^3 + 3x} = -\dfrac{1}{x} + \dfrac{x+1}{x^2 + 3}$.

39. $\dfrac{2x^3 + 7x + 5}{\left(x^2 + x + 2\right)\left(x^2 + 1\right)} = \dfrac{Ax + B}{x^2 + x + 2} + \dfrac{Cx + D}{x^2 + 1}$. Thus,

$$2x^3 + 7x + 5 = (Ax + B)\left(x^2 + 1\right) + (Cx + D)\left(x^2 + x + 2\right)$$

$$= Ax^3 + Ax + Bx^2 + B + Cx^3 + Cx^2 + 2Cx + Dx^2 + Dx + 2D$$

$$= (A + C)x^3 + (B + C + D)x^2 + (A + 2C + D)x + (B + 2D)$$

We must solve the system

$\begin{cases} A + C = 2 \\ B + C + D = 0 \\ A + 2C + D = 7 \\ B + 2D = 5 \end{cases}$ $\begin{array}{l} \text{Coefficients of } x^3 \\ \text{Coefficients of } x^2 \\ \text{Coefficients of } x \\ \text{Constant terms} \end{array}$ $\Leftrightarrow \begin{cases} A + C = 2 \\ B + C + D = 0 \\ C + D = 5 \\ C - D = -5 \end{cases}$ $\Leftrightarrow \begin{cases} A + C = 2 \\ B + C + D = 0 \\ C + D = 5 \\ 2D = 10 \end{cases}$

Hence, $2D = 10 \Leftrightarrow D = 5$, $C + 5 = 5 \Leftrightarrow C = 0$, $B + 0 + 5 = 0 \Leftrightarrow B = -5$, and $A + 0 = 2 \Leftrightarrow A = 2$. Therefore,

$$\frac{2x^3 + 7x + 5}{(x^2 + x + 2)(x^2 + 1)} = \frac{2x - 5}{x^2 + x + 2} + \frac{5}{x^2 + 1}.$$

41. $\dfrac{x^4 + x^3 + x^2 - x + 1}{x\left(x^2 + 1\right)^2} = \dfrac{A}{x} + \dfrac{Bx + C}{x^2 + 1} + \dfrac{Dx + E}{\left(x^2 + 1\right)^2}$. Hence,

$$
\begin{aligned}
x^4 + x^3 + x^2 - x + 1 &= A\left(x^2 + 1\right)^2 + (Bx + C)\,x\left(x^2 + 1\right) + x\,(Dx + E) \\
&= A\left(x^4 + 2x^2 + 1\right) + \left(Bx^2 + Cx\right)\left(x^2 + 1\right) + Dx^2 + Ex \\
&= A\left(x^4 + 2x^2 + 1\right) + Bx^4 + Bx^2 + Cx^3 + Cx + Dx^2 + Ex \\
&= (A + B)\,x^4 + Cx^3 + (2A + B + D)\,x^2 + (C + E)\,x + A
\end{aligned}
$$

So $A = 1$, $1 + B = 1 \Leftrightarrow B = 0$; $C = 1$; $2 + 0 + D = 1 \Leftrightarrow D = -1$; and $1 + E = -1 \Leftrightarrow E = -2$. Therefore,

$$\frac{x^4 + x^3 + x^2 - x + 1}{x\left(x^2 + 1\right)^2} = \frac{1}{x} + \frac{1}{x^2 + 1} - \frac{x + 2}{\left(x^2 + 1\right)^2}.$$

43. We must first get a proper rational function. Using long division, we find that $\dfrac{x^5 - 2x^4 + x^3 + x + 5}{x^3 - 2x^2 + x - 2} = x^2 +$

$\dfrac{2x^2 + x + 5}{x^3 - 2x^2 + x - 2} = x^2 + \dfrac{2x^2 + x + 5}{(x - 2)\left(x^2 + 1\right)} = x^2 + \dfrac{A}{x - 2} + \dfrac{Bx + C}{x^2 + 1}$. Hence,

$$
\begin{aligned}
2x^2 + x + 5 &= A\left(x^2 + 1\right) + (Bx + C)(x - 2) = Ax^2 + A + Bx^2 + Cx - 2Bx - 2C \\
&= (A + B)\,x^2 + (C - 2B)\,x + (A - 2C)
\end{aligned}
$$

Equating coefficients, we get the system

$$
\begin{cases}
A + B & = 2 & \text{Coefficients of } x^2 \\
-2B + C & = 1 & \text{Coefficients of } x \\
A - 2C & = 5 & \text{Constant terms}
\end{cases}
\Leftrightarrow
\begin{cases}
A + B & = 2 \\
-2B + C & = 1 \\
B + 2C & = -3
\end{cases}
\Leftrightarrow
\begin{cases}
A + B & = 2 \\
-2B + C & = 1 \\
5C & = -5
\end{cases}
$$

Therefore, $5C = -5 \Leftrightarrow C = -1$, $-2B - 1 = 1 \Leftrightarrow B = -1$, and $A - 1 = 2 \Leftrightarrow A = 3$, so

$$\frac{x^5 - 2x^4 + x^3 + x + 5}{x^3 - 2x^2 + x - 2} = x^2 + \frac{3}{x - 2} - \frac{x + 1}{x^2 + 1}.$$

45. $\dfrac{ax + b}{x^2 - 1} = \dfrac{A}{x - 1} + \dfrac{B}{x + 1}$. Hence, $ax + b = A\,(x + 1) + B\,(x - 1) = (A + B)\,x + (A - B)$.

So $\begin{cases} A + B = a \\ A - B = b \end{cases}$ Adding, we get $2A = a + b \Leftrightarrow A = \dfrac{a + b}{2}$.

Substituting, we get $B = a - A = \dfrac{2a}{2} - \dfrac{a + b}{2} = \dfrac{a - b}{2}$. Therefore, $A = \dfrac{a + b}{2}$ and $B = \dfrac{a - b}{2}$.

47. (a) The expression $\dfrac{x}{x^2+1}+\dfrac{1}{x+1}$ is already a partial fraction decomposition. The denominator in the first term is a quadratic which cannot be factored and the degree of the numerator is less than 2. The denominator of the second term is linear and the numerator is a constant.

(b) The term $\dfrac{x}{(x+1)^2}$ can be decomposed further, since the numerator and denominator both have linear factors.

$\dfrac{x}{(x+1)^2}=\dfrac{A}{x+1}+\dfrac{B}{(x+1)^2}$. Hence, $x=A(x+1)+B=Ax+(A+B)$. So $A=1$, $B=-1$, and

$\dfrac{x}{(x+1)^2}=\dfrac{1}{x+1}+\dfrac{-1}{(x+1)^2}$.

(c) The expression $\dfrac{1}{x+1}+\dfrac{2}{(x+1)^2}$ is already a partial fraction decomposition, since each numerator is constant.

(d) The expression $\dfrac{x+2}{\left(x^2+1\right)^2}$ is already a partial fraction decomposition, since the denominator is the square of a quadratic which cannot be factored, and the degree of the numerator is less than 2.

5.4 SYSTEMS OF NONLINEAR EQUATIONS

1. The solutions of the system are the points of intersection of the two graphs, namely $(-2,2)$ and $(4,8)$.

3. $\begin{cases} y=x^2 \\ y=x+12 \end{cases}$ Substituting $y=x^2$ into the second equation gives $x^2=x+12 \Leftrightarrow$

$0=x^2-x-12=(x-4)(x+3) \Rightarrow x=4$ or $x=-3$. So since $y=x^2$, the solutions are $(-3,9)$ and $(4,16)$.

5. $\begin{cases} x^2+y^2=8 \\ x+\ y=0 \end{cases}$ Solving the second equation for y gives $y=-x$, and substituting this into the first equation gives

$x^2+(-x)^2=8 \Leftrightarrow 2x^2=8 \Leftrightarrow x=\pm2$. So since $y=-x$, the solutions are $(2,-2)$ and $(-2,2)$.

7. $\begin{cases} x+\ y^2=\ 0 \\ 2x+5y^2=75 \end{cases}$ Solving the first equation for x gives $x=-y^2$, and substituting this into the second equation gives

$2\left(-y^2\right)+5y^2=75 \Leftrightarrow 3y^2=75 \Leftrightarrow y^2=25 \Leftrightarrow y=\pm5$. So since $x=-y^2$, the solutions are $(-25,-5)$ and $(-25,5)$.

9. $\begin{cases} x^2-2y=\ 1 \\ x^2+5y=29 \end{cases}$ Subtracting the first equation from the second equation gives $7y=28 \Rightarrow y=4$. Substituting $y=4$ into

the first equation of the original system gives $x^2-2(4)=1 \Leftrightarrow x^2=9 \Leftrightarrow x=\pm3$. The solutions are $(3,4)$ and $(-3,4)$.

11. $\begin{cases} 3x^2-\ y^2=11 \\ x^2+4y^2=\ 8 \end{cases}$ Multiplying the first equation by 4 gives the system $\begin{cases} 12x^2-4y^2=44 \\ x^2+4y^2=\ 8 \end{cases}$ Adding the equations

gives $13x^2=52 \Leftrightarrow x=\pm2$. Substituting into the first equation we get $3(4)-y^2=11 \Leftrightarrow y=\pm1$. Thus, the solutions are $(2,1)$, $(2,-1)$, $(-2,1)$, and $(-2,-1)$.

13. $\begin{cases} x-y^2+3=0 \\ 2x^2+y^2-4=0 \end{cases}$ Adding the two equations gives $2x^2+x-1=0$. Using the quadratic formula we have

$x=\dfrac{-1\pm\sqrt{1-4(2)(-1)}}{2(2)}=\dfrac{-1\pm\sqrt{9}}{4}=\dfrac{-1\pm3}{4}$. So $x=\dfrac{-1-3}{4}=-1$ or $x=\dfrac{-1+3}{4}=\dfrac{1}{2}$. Substituting $x=-1$

into the first equation gives $-1-y^2+3=0 \Leftrightarrow y^2=2 \Leftrightarrow y=\pm\sqrt{2}$. Substituting $x=\dfrac{1}{2}$ into the first equation gives

$\dfrac{1}{2}-y^2+3=0 \Leftrightarrow y^2=\dfrac{7}{2} \Leftrightarrow y=\pm\sqrt{\dfrac{7}{2}}$. Thus the solutions are $\left(-1,\pm\sqrt{2}\right)$ and $\left(\dfrac{1}{2},\pm\sqrt{\dfrac{7}{2}}\right)$.

15. $\begin{cases} x^2 + y = 8 \\ x - 2y = -6 \end{cases}$ By inspection of the graph, it appears that $(2, 4)$ is a solution, but is difficult to get accurate values

for the other point. Multiplying the first equation by 2 gives the system $\begin{cases} 2x^2 + 2y = 16 \\ x - 2y = -6 \end{cases}$ Adding the equations gives

$2x^2 + x = 10 \Leftrightarrow 2x^2 + x - 10 = 0 \Leftrightarrow (2x + 5)(x - 2) = 0$. So $x = -\frac{5}{2}$ or $x = 2$. If $x = -\frac{5}{2}$, then $-\frac{5}{2} - 2y = -6 \Leftrightarrow$

$-2y = -\frac{7}{2} \Leftrightarrow y = \frac{7}{4}$, and if $x = 2$, then $2 - 2y = -6 \Leftrightarrow -2y = -8 \Leftrightarrow y = 4$. Hence, the solutions are $\left(-\frac{5}{2}, \frac{7}{4}\right)$ and

$(2, 4)$.

17. $\begin{cases} x^2 + y = 0 \\ x^3 - 2x - y = 0 \end{cases}$ By inspection of the graph, it appears that $(-2, -4)$, $(0, 0)$, and $(1, -1)$ are solutions to the system.

We check each point in both equations to verify that it is a solution.

For $(-2, -4)$: $(-2)^2 + (-4) = 4 - 4 = 0$ and $(-2)^3 - 2(-2) - (-4) = -8 + 4 + 4 = 0$.

For $(0, 0)$: $(0)^2 + (0) = 0$ and $(0)^3 - 2(0) - (0) = 0$.

For $(1, -1)$: $(1)^2 + (-1) = 1 - 1 = 0$ and $(1)^3 - 2(1) - (-1) = 1 - 2 + 1 = 0$.

Thus, the solutions are $(-2, -4)$, $(0, 0)$, and $(1, -1)$.

19. $\begin{cases} y + x^2 = 4x \\ y + 4x = 16 \end{cases}$ Subtracting the second equation from the first equation gives $x^2 - 4x = 4x - 16 \Leftrightarrow$

$x^2 - 8x + 16 = 0 \Leftrightarrow (x - 4)^2 = 0 \Leftrightarrow x = 4$. Substituting this value for x into either of the original equations gives $y = 0$.

Therefore, the solution is $(4, 0)$.

21. $\begin{cases} x - 2y = 2 \\ y^2 - x^2 = 2x + 4 \end{cases}$ Now $x - 2y = 2 \Leftrightarrow x = 2y + 2$. Substituting for x gives $y^2 - x^2 = 2x + 4 \Leftrightarrow$

$y^2 - (2y + 2)^2 = 2(2y + 2) + 4 \Leftrightarrow y^2 - 4y^2 - 8y - 4 = 4y + 4 + 4 \Leftrightarrow y^2 + 4y + 4 = 0 \Leftrightarrow (y + 2)^2 = 0 \Leftrightarrow y = -2$.

Since $x = 2y + 2$, we have $x = 2(-2) + 2 = -2$. Thus, the solution is $(-2, -2)$.

23. $\begin{cases} x - y = 4 \\ xy = 12 \end{cases}$ Now $x - y = 4 \Leftrightarrow x = 4 + y$. Substituting for x gives $xy = 12 \Leftrightarrow (4 + y)y = 12 \Leftrightarrow y^2 + 4y - 12 = 0$

$\Leftrightarrow (y + 6)(y - 2) = 0 \Leftrightarrow y = -6, y = 2$. Since $x = 4 + y$, the solutions are $(-2, -6)$ and $(6, 2)$.

25. $\begin{cases} x^2y = 16 \\ x^2 + 4y + 16 = 0 \end{cases}$ Now $x^2y = 16 \Leftrightarrow x^2 = \dfrac{16}{y}$. Substituting for x^2 gives $\dfrac{16}{y} + 4y + 16 = 0 \Rightarrow 4y^2 + 16y + 16 = 0$

$\Leftrightarrow y^2 + 4y + 4 = 0 \Leftrightarrow (y + 2)^2 = 0 \Leftrightarrow y = -2$. Therefore, $x^2 = \dfrac{16}{-2} = -8$, which has no real solution, and so the system

has no solution.

27. $\begin{cases} x^2 + y^2 = 9 \\ x^2 - y^2 = 1 \end{cases}$ Adding the equations gives $2x^2 = 10 \Leftrightarrow x^2 = 5 \Leftrightarrow x = \pm\sqrt{5}$. Now $x = \pm\sqrt{5} \Rightarrow y^2 = 9 - 5 = 4 \Leftrightarrow$

$y = \pm 2$, and so the solutions are $\left(\sqrt{5}, 2\right)$, $\left(\sqrt{5}, -2\right)$, $\left(-\sqrt{5}, 2\right)$, and $\left(-\sqrt{5}, -2\right)$.

29. $\begin{cases} 2x^2 - 8y^3 = 19 \\ 4x^2 + 16y^3 = 34 \end{cases}$ Multiplying the first equation by 2 gives the system $\begin{cases} 4x^2 - 16y^3 = 38 \\ 4x^2 + 16y^3 = 34 \end{cases}$ Adding the two

equations gives $8x^2 = 72 \Leftrightarrow x = \pm 3$, and then substituting into the first equation we have $2(9) - 8y^3 = 19 \Leftrightarrow$

$y^3 = -\frac{1}{8} \Leftrightarrow y = -\frac{1}{2}$. Therefore, the solutions are $\left(3, -\frac{1}{2}\right)$ and $\left(-3, -\frac{1}{2}\right)$.

31. $\begin{cases} \dfrac{2}{x} - \dfrac{3}{y} = 1 \\ -\dfrac{4}{x} + \dfrac{7}{y} = 1 \end{cases}$ If we let $u = \dfrac{1}{x}$ and $v = \dfrac{1}{y}$, the system is equivalent to $\begin{cases} 2u - 3v = 1 \\ -4u + 7v = 1 \end{cases}$ Multiplying the first

equation by 4 gives the system $\begin{cases} 4u - 6v = 2 \\ -4u + 7v = 1 \end{cases}$ Adding the equations gives $v = 3$, and then substituting into the first

equation gives $2u - 9 = 1 \Leftrightarrow u = 5$. Thus, the solution is $\left(\frac{1}{5}, \frac{1}{3}\right)$.

33. $\begin{cases} y = x^2 + 8x \\ y = 2x + 16 \end{cases}$

The solutions are $(-8, 0)$ and $(2, 20)$.

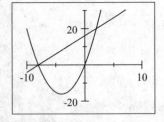

35. $\begin{cases} x^2 + y^2 = 25 \\ x + 3y = 2 \end{cases} \Leftrightarrow \begin{cases} y = \pm\sqrt{25 - x^2} \\ y = -\frac{1}{3}x + \frac{2}{3} \end{cases}$

The solutions are $(-4.51, 2.17)$ and $(4.91, -0.97)$.

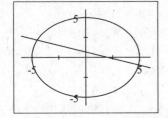

37. $\begin{cases} \dfrac{x^2}{9} + \dfrac{y^2}{18} = 1 \\ y = -x^2 + 6x - 2 \end{cases} \Leftrightarrow \begin{cases} y = \pm\sqrt{18 - 2x^2} \\ y = -x^2 + 6x - 2 \end{cases}$

The solutions are $(1.23, 3.87)$ and $(-0.35, -4.21)$.

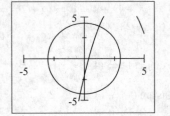

39. $\begin{cases} x^4 + 16y^4 = 32 \\ x^2 + 2x + y = 0 \end{cases} \Leftrightarrow \begin{cases} y = \pm\dfrac{\sqrt[4]{32 - x^4}}{2} \\ y = -x^2 - 2x \end{cases}$

The solutions are $(-2.30, -0.70)$ and $(0.48, -1.19)$.

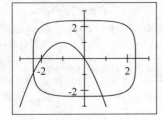

41. Let w and l be the lengths of the sides, in cm. Then we have the system $\begin{cases} lw = 180 \\ 2l + 2w = 54 \end{cases}$ We solve the second equation

for w giving, $w = 27 - l$, and substitute into the first equation to get $l(27 - l) = 180 \Leftrightarrow l^2 - 27l + 180 = 0 \Leftrightarrow$
$(l - 15)(l - 12) = 0 \Rightarrow l = 15$ or $l = 12$. If $l = 15$, then $w = 27 - 15 = 12$, and if $l = 12$, then $w = 27 - 12 = 15$.
Therefore, the dimensions of the rectangle are 12 cm by 15 cm.

43. Let l and w be the length and width, respectively, of the rectangle. Then, the system of equations is

$\begin{cases} 2l + 2w = 70 \\ \sqrt{l^2 + w^2} = 25 \end{cases}$ Solving the first equation for l, we have $l = 35 - w$, and substituting into the second gives

$\sqrt{l^2 + w^2} = 25 \Leftrightarrow l^2 + w^2 = 625 \Leftrightarrow (35 - w)^2 + w^2 = 625 \Leftrightarrow 1225 - 70w + w^2 + w^2 = 625 \Leftrightarrow 2w^2 - 70w + 600 = 0$
$\Leftrightarrow (w - 15)(w - 20) = 0 \Rightarrow w = 15$ or $w = 20$. So the dimensions of the rectangle are 15 and 20.

45. At the points where the rocket path and the hillside meet, we have $\begin{cases} y = \frac{1}{2}x \\ y = -x^2 + 401x \end{cases}$ Substituting for y in the second

equation gives $\frac{1}{2}x = -x^2 + 401x \Leftrightarrow x^2 - \frac{801}{2}x = 0 \Leftrightarrow x\left(x - \frac{801}{2}\right) = 0 \Rightarrow x = 0$, $x = \frac{801}{2}$. When $x = 0$, the rocket has

not left the pad. When $x = \frac{801}{2}$, then $y = \frac{1}{2}\left(\frac{801}{2}\right) = \frac{801}{4}$. So the rocket lands at the point $\left(\frac{801}{2}, \frac{801}{4}\right)$. The distance from

the base of the hill is $\sqrt{\left(\frac{801}{2}\right)^2 + \left(\frac{801}{4}\right)^2} \approx 447.77$ meters.

47. The point P is at an intersection of the circle of radius 26 centered at $A\,(22, 32)$

and the circle of radius 20 centered at $B\,(28, 20)$. We have the system

$\begin{cases} (x - 22)^2 + (y - 32)^2 = 26^2 \\ (x - 28)^2 + (y - 20)^2 = 20^2 \end{cases} \Leftrightarrow$

$\begin{cases} x^2 - 44x + 484 + y^2 - 64y + 1024 = 676 \\ x^2 - 56x + 784 + y^2 - 40y + 400 = 400 \end{cases} \Leftrightarrow$

$\begin{cases} x^2 - 44x + y^2 - 64y = -832 \\ x^2 - 56x + y^2 - 40y = -784 \end{cases}$ Subtracting the two equations, we get $12x - 24y = -48 \Leftrightarrow x - 2y = -4$,

which is the equation of a line. Solving for x, we have $x = 2y - 4$. Substituting into the first equation gives

$(2y - 4)^2 - 44(2y - 4) + y^2 - 64y = -832 \Leftrightarrow 4y^2 - 16y + 16 - 88y + 176 + y^2 - 64y = -832 \Leftrightarrow 5y^2 - 168y + 192 = -832$

$\Leftrightarrow 5y^2 - 168y + 1024 = 0$. Using the quadratic formula, we have $y = \frac{168 \pm \sqrt{168^2 - 4(5)(1024)}}{2(5)} = \frac{168 \pm \sqrt{7744}}{10} = \frac{168 \pm 88}{10}$

$\Leftrightarrow y = 8$ or $y = 25.60$. Since the y-coordinate of the point P must be less than that of point A, we have $y = 8$. Then

$x = 2(8) - 4 = 12$. So the coordinates of P are $(12, 8)$.

To solve graphically, we must solve each equation for y. This gives $(x - 22)^2 + (y - 32)^2 = 26^2$

$\Leftrightarrow (y - 32)^2 = 26^2 - (x - 22)^2 \Rightarrow y - 32 = \pm\sqrt{676 - (x - 22)^2} \Leftrightarrow y = 32 \pm \sqrt{676 - (x - 22)^2}$. We use the function

$y = 32 - \sqrt{676 - (x - 22)^2}$ because the intersection we at interested in is below the point A. Likewise, solving the second

equation for y, we would get the function $y = 20 - \sqrt{400 - (x - 28)^2}$. In a three-dimensional situation, you would need a

minimum of three satellites, since a point on the earth can be uniquely specified as the intersection of three spheres centered

at the satellites.

49. (a) $\begin{cases} \log x + \log y = \frac{3}{2} \\ 2\log x - \log y = 0 \end{cases}$ Adding the two equations gives $3\log x = \frac{3}{2} \Leftrightarrow \log x = \frac{1}{2} \Leftrightarrow x = \sqrt{10}$. Substituting into the

second equation we get $2\log 10^{1/2} - \log y = 0 \Leftrightarrow \log 10 - \log y = 0 \Leftrightarrow \log y = 1 \Leftrightarrow y = 10$. Thus, the solution is

$\left(\sqrt{10}, 10\right)$.

(b) $\begin{cases} 2^x + 2^y = 10 \\ 4^x + 4^y = 68 \end{cases} \Leftrightarrow \begin{cases} 2^x + 2^y = 10 \\ 2^{2x} + 2^{2y} = 68 \end{cases}$ If we let $u = 2^x$ and $v = 2^y$, the system becomes $\begin{cases} u + v = 10 \\ u^2 + v^2 = 68 \end{cases}$

Solving the first equation for u, and substituting this into the second equation gives $u + v = 10 \Leftrightarrow u = 10 - v$, so

$(10 - v)^2 + v^2 = 68 \Leftrightarrow 100 - 20v + v^2 + v^2 = 68 \Leftrightarrow v^2 - 10v + 16 = 0 \Leftrightarrow (v - 8)(v - 2) = 0 \Rightarrow v = 2$ or $v = 8$.

If $v = 2$, then $u = 8$, and so $y = 1$ and $x = 3$. If $v = 8$, then $u = 2$, and so $y = 3$ and $x = 1$. Thus, the solutions are

$(1, 3)$ and $(3, 1)$.

(c) $\begin{cases} x - y = 3 \\ x^3 - y^3 = 387 \end{cases}$ Solving the first equation for x gives $x = 3 + y$ and using the hint, $x^3 - y^3 = 387 \Leftrightarrow$

$(x - y)\left(x^2 + xy + y^2\right) = 387$. Next, substituting for x, we get $3\left[(3 + y)^2 + y(3 + y) + y^2\right] = 387 \Leftrightarrow 9 + 6y +$

$y^2 + 3y + y^2 + y^2 = 129 \Leftrightarrow 3y^2 + 9y + 9 = 129 \Leftrightarrow (y + 8)(y - 5) = 0 \Rightarrow y = -8$ or $y = 5$. If $y = -8$, then

$x = 3 + (-8) = -5$, and if $y = 5$, then $x = 3 + 5 = 8$. Thus the solutions are $(-5, -8)$ and $(8, 5)$.

(d) $\begin{cases} x^2 + xy = 1 \\ xy + y^2 = 3 \end{cases}$ Adding the equations gives $x^2 + xy + xy + y^2 = 4 \Leftrightarrow x^2 + 2xy + y^2 = 4 \Leftrightarrow$

$(x + y)^2 = 4 \Rightarrow x + y = \pm 2$. If $x + y = 2$, then from the first equation we get $x(x + y) = 1 \Rightarrow x \cdot 2 = 1 \Rightarrow x = \frac{1}{2}$,

and so $y = 2 - \frac{1}{2} = \frac{3}{2}$. If $x + y = -2$, then from the first equation we get $x(x + y) = 1 \Rightarrow x \cdot (-2) = 1 \Rightarrow x = -\frac{1}{2}$,

and so $y = -2 - \left(-\frac{1}{2}\right) = -\frac{3}{2}$. Thus the solutions are $\left(\frac{1}{2}, \frac{3}{2}\right)$ and $\left(-\frac{1}{2}, -\frac{3}{2}\right)$.

5.5 SYSTEMS OF INEQUALITIES

1. To graph an inequality we first graph the corresponding *equation*. So to graph $y \le x + 1$, we first graph the equation $y = x + 1$. To decide which side is the graph of the inequality we use *test* points.

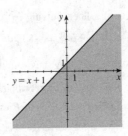

3. $y < 2x$

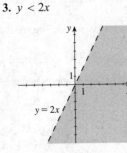

5. $y \ge 2$

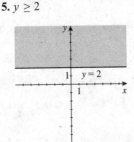

7. $x < 2$

9. $y < x - 3$

11. $-2x + y \le 4$

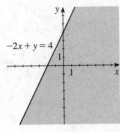

13. $-3x + 7y > 21$

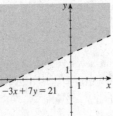

15. $2x - 3y \le 9$

17. $x^2 + y \ge 3$

19. $x^2 + y^2 \geq 100$

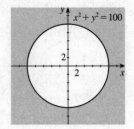

21. The boundary is a solid curve, so we have the inequality $y \leq \frac{1}{2}x - 1$. We take the test point $(0, -2)$ and verify that it satisfies the inequality: $-2 \leq \frac{1}{2}(0) - 1$.

23. The boundary is a broken curve, so we have the inequality $x^2 + y^2 > 4$. We take the test point $(0, 4)$ and verify that it satisfies the inequality: $0^2 + 4^2 > 4$.

25. $\begin{cases} x + y \leq 4 \\ y \geq x \end{cases}$ The vertices occur where $\begin{cases} x + y = 4 \\ y = x \end{cases}$ Substituting, we have

$2x = 4 \Leftrightarrow x = 2$. Since $y = x$, the vertex is $(2, 2)$, and the solution set is not bounded.

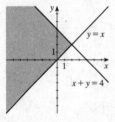

27. $\begin{cases} y < \frac{1}{4}x + 2 \\ y \geq 2x - 5 \end{cases}$ The vertex occurs where $\begin{cases} y = \frac{1}{4}x + 2 \\ y = 2x - 5 \end{cases}$ Substituting for y

gives $\frac{1}{4}x + 2 = 2x - 5 \Leftrightarrow \frac{7}{4}x = 7 \Leftrightarrow x = 4$, so $y = 3$. Hence, the vertex is $(4, 3)$, and the solution is not bounded.

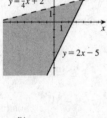

29. $\begin{cases} y \leq -2x + 8 \\ y \leq -\frac{1}{2}x + 5 \\ x \geq 0, y \geq 0 \end{cases}$ One vertex occurs where $\begin{cases} y \leq -2x + 8 \\ y \leq -\frac{1}{2}x + 5 \end{cases}$ Substituting for

y gives $-2x + 8 = -\frac{1}{2}x + 5 \Leftrightarrow -\frac{3}{2}x = -3 \Leftrightarrow x = 2$, so $y = -2(2) + 8 = 4$.

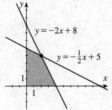

Hence, this vertex is $(2, 4)$. Another vertex occurs where $\begin{cases} y = -2x + 8 \\ y = 0 \end{cases} \Leftrightarrow$

$-2x + 8 = 0 \Leftrightarrow x = 4$; this vertex is $(4, 0)$. Another occurs where

$\begin{cases} y = -\frac{1}{2}x + 5 \\ x = 0 \end{cases} \Leftrightarrow y = 5$; this gives the vertex $(0, 5)$. The origin is another

vertex, and the solution set is bounded.

31. $\begin{cases} x \geq 0 \\ y \geq 0 \\ 3x + 5y \leq 15 \\ 3x + 2y \leq 9 \end{cases}$ From the graph, the points $(3, 0)$, $(0, 3)$ and $(0, 0)$ are vertices,

and the fourth vertex occurs where the lines $3x + 5y = 15$ and $3x + 2y = 9$

intersect. Subtracting these two equations gives $3y = 6 \Leftrightarrow y = 2$, and so $x = \frac{5}{3}$.

Thus, the fourth vertex is $\left(\frac{5}{3}, 2 \right)$, and the solution set is bounded.

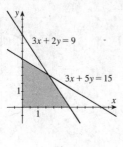

33. $\begin{cases} y \leq 9 - x^2 \\ x \geq 0, y \geq 0 \end{cases}$ From the graph, the vertices occur at $(0, 0)$, $(3, 0)$, and $(0, 9)$.

The solution set is bounded.

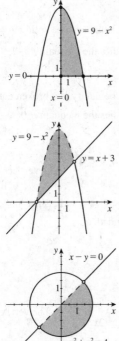

35. $\begin{cases} y < 9 - x^2 \\ y \geq x + 3 \end{cases}$ The vertices occur where $\begin{cases} y = 9 - x^2 \\ y = x + 3 \end{cases}$ Substituting for y

gives $9 - x^2 = x + 3 \Leftrightarrow x^2 + x - 6 = 0 \Leftrightarrow (x - 2)(x + 3) = 0 \Rightarrow x = -3$,
$x = 2$. Therefore, the vertices are $(-3, 0)$ and $(2, 5)$, and the solution set is
bounded.

37. $\begin{cases} x^2 + y^2 \leq 4 \\ x - y > 0 \end{cases}$ The vertices occur where $\begin{cases} x^2 + y^2 = 4 \\ x - y = 0 \end{cases}$ Since $x - y = 0$

$\Leftrightarrow x = y$, substituting for x gives $y^2 + y^2 = 4 \Leftrightarrow y^2 = 2 \Rightarrow y = \pm\sqrt{2}$, and
$x = \pm\sqrt{2}$. Therefore, the vertices are $\left(-\sqrt{2}, -\sqrt{2} \right)$ and $\left(\sqrt{2}, \sqrt{2} \right)$, and the
solution set is bounded.

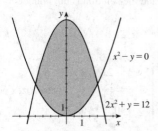

39. $\begin{cases} x^2 - y \leq 0 \\ 2x^2 + y \leq 12 \end{cases}$ The vertices occur where $\begin{cases} x^2 - y = 0 \\ 2x^2 + y = 12 \end{cases} \Leftrightarrow$

$\begin{cases} 2x^2 - 2y = 0 \\ 2x^2 + y = 12 \end{cases}$ Subtracting the equations gives $3y = 12 \Leftrightarrow y = 4$, and

$x = \pm 2$. Thus, the vertices are $(2, 4)$ and $(-2, 4)$, and the solution set is bounded.

41. $\begin{cases} x^2 + y^2 \leq 9 \\ x^2 + 2y \leq 1 \end{cases}$ We find the vertices of the region by solving pairs of the

corresponding equations: $\begin{cases} x^2 + y^2 = 9 \\ x^2 + 2y = 1 \end{cases}$ Subtracting the second equation from

the first gives $y^2 - 2y = 8 \Leftrightarrow y^2 - 2y - 8 = 0 \Leftrightarrow (y + 2)(y - 4) = 0 \Leftrightarrow y = -2$
or 4. However, if $y = 4$, the first original equation has no real solution. The
vertices are $\left(-\sqrt{5}, -2 \right)$ and $\left(\sqrt{5}, -2 \right)$. The solution set is bounded.

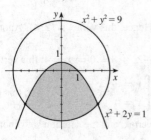

43. $\begin{cases} x + 2y \le 14 \\ 3x - y \ge 0 \\ x - y \ge 2 \end{cases}$ We find the vertices of the region by solving pairs of the

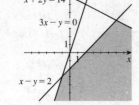

corresponding equations: $\begin{cases} x + 2y = 14 \\ x - y = 2 \end{cases} \Leftrightarrow \begin{cases} x + 2y = 14 \\ 3y = 12 \end{cases} \Leftrightarrow y = 4 \text{ and } x = 6.$

$\begin{cases} 3x - y = 0 \\ x - y = 2 \end{cases} \Leftrightarrow \begin{cases} 3x - y = 0 \\ 2x - y = -2 \end{cases} \Leftrightarrow x = -1 \text{ and } y = -3.$ Therefore, the vertices

are $(6, 4)$ and $(-1, -3)$, and the solution set is not bounded.

45. $\begin{cases} x \ge 0, y \ge 0 \\ x \le 5, x + y \le 7 \end{cases}$ The points of intersection are $(0, 7)$, $(0, 0)$, $(7, 0)$, $(5, 2)$,

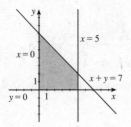

and $(5, 0)$. However, the point $(7, 0)$ is not in the solution set. Therefore, the

vertices are $(0, 7)$, $(0, 0)$, $(5, 0)$, and $(5, 2)$, and the solution set is bounded.

47. $\begin{cases} y > x + 1 \\ x + 2y \le 12 \\ x + 1 > 0 \end{cases}$ We find the vertices of the region by solving pairs of the

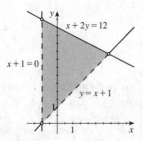

corresponding equations. Using $x = -1$ and substituting for x in the line $y = x + 1$

gives the point $(-1, 0)$. Substituting for x in the line $x + 2y = 12$ gives the point

$\left(-1, \frac{13}{2}\right)$. $\begin{cases} y = x + 1 \\ x + 2y = 12 \end{cases} \Leftrightarrow x = y - 1 \text{ and } y - 1 + 2y = 12 \Leftrightarrow 3y = 13 \Leftrightarrow$

$y = \frac{13}{3}$ and $x = \frac{10}{3}$. So the vertices are $(-1, 0)$, $\left(-1, \frac{13}{2}\right)$, and $\left(\frac{10}{3}, \frac{13}{3}\right)$, and

none of these vertices is in the solution set. The solution set is bounded.

49. $\begin{cases} x^2 + y^2 \le 8 \\ x \ge 2, y \ge 0 \end{cases}$ The intersection points are $(2, \pm 2)$, $(2, 0)$, and $\left(2\sqrt{2}, 0\right)$.

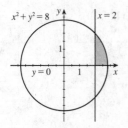

However, since $(2, -2)$ is not part of the solution set, the vertices are $(2, 2)$, $(2, 0)$,

and $\left(2\sqrt{2}, 0\right)$. The solution set is bounded.

51. $\begin{cases} x^2 + y^2 < 9 \\ x + y > 0, x \le 0 \end{cases}$ Substituting $x = 0$ into the equations $x^2 + y^2 = 9$ and

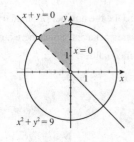

$x + y = 0$ gives the vertices $(0, \pm 3)$ and $(0, 0)$. To find the points of intersection

for the equations $x^2 + y^2 = 9$ and $x + y = 0$, we solve for $x = -y$ and substitute

into the first equation. This gives $(-y)^2 + y^2 = 9 \Rightarrow y = \pm \frac{3\sqrt{2}}{2}$. The points

$(0, -3)$ and $\left(\frac{3\sqrt{2}}{2}, -\frac{3\sqrt{2}}{2}\right)$ lie away from the solution set, so the vertices are $(0, 0)$,

$(0, 3)$, and $\left(-\frac{3\sqrt{2}}{2}, \frac{3\sqrt{2}}{2}\right)$. Note that the vertices are not solutions in this case. The solution set is bounded.

53. $\begin{cases} y \geq x - 3 \\ y \geq -2x + 6 \\ y \leq 8 \end{cases}$ Using a graphing calculator, we find the region shown. The

vertices are $(3, 0)$, $(-1, 8)$, and $(11, 8)$.

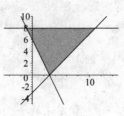

55. $\begin{cases} y \leq 6x - x^2 \\ x + y \geq 4 \end{cases}$ Using a graphing calculator, we find the region shown. The

vertices are $(0.6, 3.4)$ and $(6.4, -2.4)$.

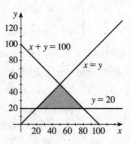

57. Let x be the number of fiction books published in a year and y the number of
nonfiction books. Then the following system of inequalities holds:

$\begin{cases} x \geq 0, y \geq 0 \\ x + y \leq 100 \\ y \geq 20, x \geq y \end{cases}$ From the graph, we see that the vertices are $(50, 50)$, $(80, 20)$

and $(20, 20)$.

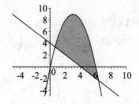

59. Let x be the number of Standard Blend packages and y be the number of Deluxe
Blend packages. Since there are 16 ounces per pound, we get the following system
of inequalities:

$\begin{cases} x \geq 0 \\ y \geq 0 \\ \frac{1}{4}x + \frac{5}{8}y \leq 80 \\ \frac{3}{4}x + \frac{3}{8}y \leq 90 \end{cases}$

From the graph, we see that the vertices are $(0, 0)$, $(120, 0)$, $(70, 100)$ and $(0, 128)$.

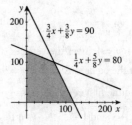

61. $x + 2y > 4$, $-x + y < 1$, $x + 3y \le 9$, $x < 3$.

Method 1:   We shade the solution to each inequality with lines perpendicular to the boundary. As you can see, as the number of inequalities in the system increases, it gets harder to locate the region where *all* of the shaded parts overlap.

Method 2: Here, if a region is shaded then it fails to satisfy at least one inequality. As a result, the region that is left unshaded satisfies each inequality, and is the solution to the system of inequalities. In this case, this method makes it easier to identify the solution set.

To finish, we find the vertices of the solution set. The line $x = 3$ intersects the line $x + 2y = 4$ at $\left(3, \frac{1}{2}\right)$ and the line $x + 3y = 9$ at $(3, 2)$. To find where the lines $-x + y = 1$ and $x + 2y = 4$ intersect, we add the two equations, which gives $3y = 5 \Leftrightarrow y = \frac{5}{3}$, and $x = \frac{2}{3}$. To find where the lines $-x + y = 1$ and $x + 3y = 9$ intersect, we add the two equations, which gives $4y = 10 \Leftrightarrow y = \frac{10}{4} = \frac{5}{2}$, and $x = \frac{3}{2}$. The vertices are $\left(3, \frac{1}{2}\right)$, $(3, 2)$, $\left(\frac{2}{3}, \frac{5}{3}\right)$, and $\left(\frac{3}{2}, \frac{5}{2}\right)$, and the solution set is bounded.

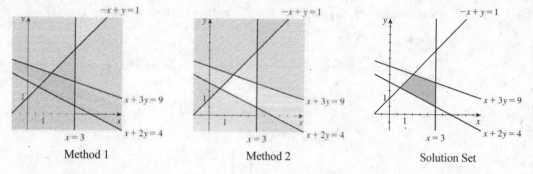

Method 1   Method 2   Solution Set

CHAPTER 5 REVIEW

1. $\begin{cases} 3x - y = 5 \\ 2x + y = 5 \end{cases}$ Adding, we get $5x = 10 \Leftrightarrow x = 2$. So $2(2) + y = 5 \Leftrightarrow y = 1$.

Thus, the solution is $(2, 1)$.

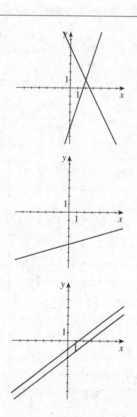

3. $\begin{cases} 2x - 7y = 28 \\ y = \frac{2}{7}x - 4 \end{cases} \Leftrightarrow \begin{cases} 2x - 7y = 28 \\ 2x - 7y = 28 \end{cases}$ Since these equations represent the same

line, any point on this line will satisfy the system. Thus the solution are

$\left(x, \frac{2}{7}x - 4\right)$, where x is any real number.

5. $\begin{cases} 6x - 8y = 16 \\ -\frac{3}{2}x + 2y = -2 \end{cases} \Leftrightarrow \begin{cases} 6x - 8y = 16 \\ 6x - 8y = 8 \end{cases}$ Subtracting gives $0 = 8$, which is

false. Hence, there is no solution. The lines are parallel.

7. $\begin{cases} 2x - y = 1 \\ x + 3y = 10 \\ 3x + 4y = 15 \end{cases}$ Solving the first equation for y, we get $y = -2x + 1$.

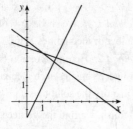

Substituting into the second equation gives $x + 3(-2x + 1) = 10 \Leftrightarrow -5x = 7 \Leftrightarrow$

$x = -\frac{7}{5}$. So $y = -\left(-\frac{7}{5}\right) + 1 = \frac{12}{5}$. Checking the point $\left(-\frac{7}{5}, \frac{12}{5}\right)$ in the third

equation we have $3\left(-\frac{7}{5}\right) + 4\left(\frac{12}{5}\right) \overset{?}{=} 15$ but $-\frac{21}{5} + \frac{48}{5} \neq 15$. Thus, there is no

solution, and the lines do not intersect at one point.

9. $\begin{cases} y = x^2 + 2x \\ y = 6 + x \end{cases}$ Substituting for y gives $6 + x = x^2 + 2x \Leftrightarrow x^2 + x - 6 = 0$. Factoring, we have $(x - 2)(x + 3) = 0$.

Thus $x = 2$ or -3. If $x = 2$, then $y = 8$, and if $x = -3$, then $y = 3$. Thus the solutions are $(-3, 3)$ and $(2, 8)$.

11. $\begin{cases} 3x + \dfrac{4}{y} = 6 \\ x - \dfrac{8}{y} = 4 \end{cases}$ Adding twice the first equation to the second gives $7x = 16 \Leftrightarrow x = \frac{16}{7}$. So $\frac{16}{7} - \frac{8}{y} = 4 \Leftrightarrow$

$16y - 56 = 28y \Leftrightarrow -12y = 56 \Leftrightarrow y = -\frac{14}{3}$. Thus, the solution is $\left(\frac{16}{7}, -\frac{14}{3}\right)$.

13. $\begin{cases} 0.32x + 0.43y = 0 \\ 7x - 12y = 341 \end{cases} \Leftrightarrow \begin{cases} y = -\dfrac{32x}{43} \\ y = \dfrac{7x - 341}{12} \end{cases}$

The solution is approximately $(21.41, -15.93)$.

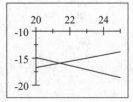

15. $\begin{cases} x - y^2 = 10 \\ x = \frac{1}{22}y + 12 \end{cases} \Leftrightarrow \begin{cases} y = \pm\sqrt{x - 10} \\ y = 22(x - 12) \end{cases}$

The solutions are $(11.94, -1.39)$ and $(12.07, 1.44)$.

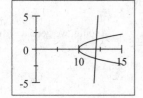

17. $\begin{cases} x + y + 2z = 6 \\ 2x + 5z = 12 \\ x + 2y + 3z = 9 \end{cases} \Leftrightarrow \begin{cases} x + y + 2z = 6 \\ 2y - z = 0 \quad \text{2 × Eq. 1 − Eq. 2} \\ 4y + z = 6 \quad \text{−Eq. 2 + 2 × Eq. 3} \end{cases} \Leftrightarrow \begin{cases} x + y + 2z = 6 \\ 2y - z = 0 \\ 3z = 6 \quad \text{(−2) × Eq. 2 + Eq. 3} \end{cases}$

Therefore, $3z = 6 \Leftrightarrow z = 2$, $2y - 2 = 0 \Leftrightarrow y = 1$, and $x + 1 + 2(2) = 6 \Leftrightarrow x = 1$. Hence, the solution is $(1, 1, 2)$.

19. $\begin{cases} x - 2y + 3z = 1 \\ 2x - y + z = 3 \\ 2x - 7y + 11z = 2 \end{cases} \Leftrightarrow \begin{cases} x - 2y + 3z = 1 \\ 3y - 5z = 1 \quad \text{(−2) × Eq. 1 + Eq. 2} \\ 6y - 10z = 1 \quad \text{Eq. 2 − Eq. 3} \end{cases} \Leftrightarrow \begin{cases} x - 2y + 3z = 1 \\ 3y - 5z = 1 \\ 0 = -1 \quad \text{(−2) × Eq. 2 + Eq. 3} \end{cases}$

which is impossible. Therefore, the system has no solution.

21. $\begin{cases} x - 3y + z = 4 \\ 4x - y + 15z = 5 \end{cases} \Leftrightarrow \begin{cases} x - 3y + z = 4 \\ y + z = -1 \quad \left(-\frac{4}{11}\right) \times \text{Eq. 1} + \frac{1}{11} \times \text{Eq. 1} \end{cases}$ Thus, the system has infinitely many

solutions given by $z = t$, $y + t = -1 \Leftrightarrow y = -1 - t$, and $x + 3(1 + t) + t = 4 \Leftrightarrow x = 1 - 4t$. Therefore, the solutions are

$(1 - 4t, -1 - t, t)$, where t is any real number.

23. $\begin{cases} -x + 4y + z = 8 \\ 2x - 6y + z = -9 \\ x - 6y - 4z = -15 \end{cases} \Leftrightarrow \begin{cases} -x + 4y + z = 8 \\ 2y + 3z = 7 \quad \text{2 × Eq. 1 + Eq. 2} \\ 6y + 9z = 21 \quad \text{Eq. 2 − 2 × Eq. 3} \end{cases} \Leftrightarrow \begin{cases} -x + 4y + z = 8 \\ 2y + 3z = 7 \\ 0 = 0 \quad \text{3 × Eq. 2 − Eq. 3} \end{cases}$

Thus, the system has infinitely many solutions. Letting $z = t$, we find $2y + 3t = 7 \Leftrightarrow y = \frac{7}{2} - \frac{3}{2}t$, and

$-x + 4\left(\frac{7}{2} - \frac{3}{2}t\right) + t = 8 \Leftrightarrow x = 6 - 5t$. Therefore, the solutions are $\left(6 - 5t, \frac{7}{2} - \frac{3}{2}t, t\right)$, where t is any real number.

25. Let k and s be be Kieran's and Siobhan's ages. From the given information, we have the system

$\begin{cases} k = s + 4 \\ k + s = 22 \end{cases}$ Subtracting the first equation from the second gives $s = 22 - s - 4 \Leftrightarrow 2s = 18 \Leftrightarrow s = 9$, so

$k = 22 - 9 = 13$. Thus, Kieran is 13 and Siobhan is 9.

27. Let n be the number of nickels, d the number of dimes, and q the number of quarter in the piggy bank. We get the following

system: $\begin{cases} n + d + q = 50 \\ 5n + 10d + 25z = 560 \\ 10d = 5(5n) \end{cases}$ Since $10d = 25n$, we have $d = \frac{5}{2}n$, so substituting into the first equation

we get $n + \frac{5}{2}n + q = 50 \Leftrightarrow \frac{7}{2}n + q = 50 \Leftrightarrow q = 50 - \frac{7}{2}n$. Now substituting this into the second equation we have

$5n + 10\left(\frac{5}{2}n\right) + 25\left(50 - \frac{7}{2}n\right) = 560 \Leftrightarrow 5n + 25n + 1250 - \frac{175}{2}n = 560 \Leftrightarrow 1250 - \frac{115}{2}n = 560 \Leftrightarrow \frac{115}{2}n = 650 \Leftrightarrow$

$n = 12$. Then $d = \frac{5}{2}(12) = 30$ and $q = 50 - n - d = 50 - 12 - 30 = 8$. Thus the piggy bank contains 12 nickels,

30 dimes, and 8 quarters.

29. $\dfrac{3x + 1}{x^2 - 2x - 15} = \dfrac{3x + 1}{(x - 5)(x + 3)} = \dfrac{A}{x - 5} + \dfrac{B}{x + 3}$. Thus, $3x + 1 = A(x + 3) + B(x - 5) = x(A + B) + (3A - 5B)$,

and so $\begin{cases} A + B = 3 \\ 3A - 5B = 1 \end{cases} \Leftrightarrow \begin{cases} -3A - 3B = -9 \\ 3A - 5B = 1 \end{cases}$ Adding, we have $-8B = -8 \Leftrightarrow B = 1$, and $A = 2$. Hence,

$\dfrac{3x + 1}{x^2 - 2x - 15} = \dfrac{2}{x - 5} + \dfrac{1}{x + 3}$.

31. $\dfrac{2x - 4}{x(x - 1)^2} = \dfrac{A}{x} + \dfrac{B}{x - 1} + \dfrac{C}{(x - 1)^2}$. Then $2x - 4 = A(x - 1)^2 + Bx(x - 1) + Cx = Ax^2 - 2Ax + A + Bx^2 - $

$Bx + Cx = x^2(A + B) + x(-2A - B + C) + A$. So $A = -4$, $-4 + B = 0 \Leftrightarrow B = 4$, and $8 - 4 + C = 2 \Leftrightarrow C = -2$.

Therefore, $\dfrac{2x - 4}{x(x - 1)^2} = -\dfrac{4}{x} + \dfrac{4}{x - 1} - \dfrac{2}{(x - 1)^2}$.

33. $\dfrac{2x - 1}{x^3 + x} = \dfrac{2x - 1}{x(x^2 + 1)} = \dfrac{A}{x} + \dfrac{Bx + C}{x^2 + 1}$. Then $2x - 1 = A(x^2 + 1) + (Bx + C)x = Ax^2 + A + Bx^2 + Cx = (A + B)x^2 + $

$Cx + A$. So $A = -1$, $C = 2$, and $A + B = 0$ gives us $B = 1$. Thus $\dfrac{2x - 1}{x^3 + x} = -\dfrac{1}{x} + \dfrac{x + 2}{x^2 + 1}$.

35. $\dfrac{3x^2 - x + 6}{(x^2 + 2)^2} = \dfrac{Ax + B}{x^2 + 2} + \dfrac{Cx + D}{(x^2 + 2)^2}$. Thus

$$3x^2 - x + 6 = (x^2 + 2)(Ax + B) + Cx + D = Ax^3 + Bx^2 + 2Ax + 2B + Cx + D$$

$$= x^3 A + x^2 B + x(2A + C) + (2B + D)$$

This leads to the system $\begin{cases} A & = 0 \\ B & = 3 \\ 2A & + C & = -1 \\ & 2B & + D = 6 \end{cases} \Leftrightarrow \begin{cases} A & = 0 \\ B & = 3 \\ C & = -1 \\ D & = 0 \end{cases}$

Thus $\dfrac{3x^2 - x + 6}{(x^2 + 2)^2} = \dfrac{3}{x^2 + 2} - \dfrac{x}{(x^2 + 2)^2}$.

37. $\begin{cases} 2x + 3y = 7 \\ x - 2y = 0 \end{cases}$ By inspection of the graph, it appears that $(2, 1)$ is the solution to the system. We check this in both equations to verify that it is the solution. $2(2) + 3(1) = 4 + 3 = 7$ and $2 - 2(1) = 2 - 2 = 0$. Since both equations are satisfied, the solution is indeed $(2, 1)$.

39. $\begin{cases} x^2 + y = 2 \\ x^2 - 3x - y = 0 \end{cases}$ By inspection of the graph, it appears that $(2, -2)$ is a solution to the system, but is difficult to get accurate values for the other point. Adding the equations, we get $2x^2 - 3x = 2 \Leftrightarrow 2x^2 - 3x - 2 = 0 \Leftrightarrow (2x + 1)(x - 2) = 0$. So $2x + 1 = 0 \Leftrightarrow x = -\frac{1}{2}$ or $x = 2$. If $x = -\frac{1}{2}$, then $\left(-\frac{1}{2}\right)^2 + y = 2 \Leftrightarrow y = \frac{7}{4}$. If $x = 2$, then $2^2 + y = 2 \Leftrightarrow y = -2$. Thus, the solutions are $\left(-\frac{1}{2}, \frac{7}{4}\right)$ and $(2, -2)$.

41. The boundary is a solid curve, so we have the inequality $x + y^2 \le 4$. We take the test point $(0, 0)$ and verify that it satisfies the inequality: $0 + 0^2 \le 4$.

43. $3x + y \le 6$ **45.** $x^2 + y^2 > 9$ **47.** $\begin{cases} y \ge x^2 - 3x \\ y \le \frac{1}{3}x - 1 \end{cases}$ **49.** $\begin{cases} x + y \ge 2 \\ y - x \le 2 \\ x \le 3 \end{cases}$

51. $\begin{cases} x^2 + y^2 < 9 \\ x + y < 0 \end{cases}$ The vertices occur where $y = -x$. By substitution, $x^2 + x^2 = 9$

$\Leftrightarrow x = \pm\frac{3}{\sqrt{2}}$, and so $y = \mp\frac{3}{\sqrt{2}}$. Therefore, the vertices are $\left(\frac{3}{\sqrt{2}}, -\frac{3}{\sqrt{2}}\right)$ and

$\left(-\frac{3}{\sqrt{2}}, \frac{3}{\sqrt{2}}\right)$ and the solution set is bounded.

53. $\begin{cases} x \ge 0, \ y \ge 0 \\ x + 2y \le 12 \\ y \le x + 4 \end{cases}$ The intersection points are $(-4, 0)$, $(0, 4)$, $\left(\frac{4}{3}, \frac{16}{3}\right)$, $(0, 6)$,

$(0, 0)$, and $(12, 0)$. Since the points $(-4, 0)$ and $(0, 6)$ are not in the solution set, the vertices are $(0, 4)$, $\left(\frac{4}{3}, \frac{16}{3}\right)$, $(12, 0)$, and $(0, 0)$. The solution set is bounded.

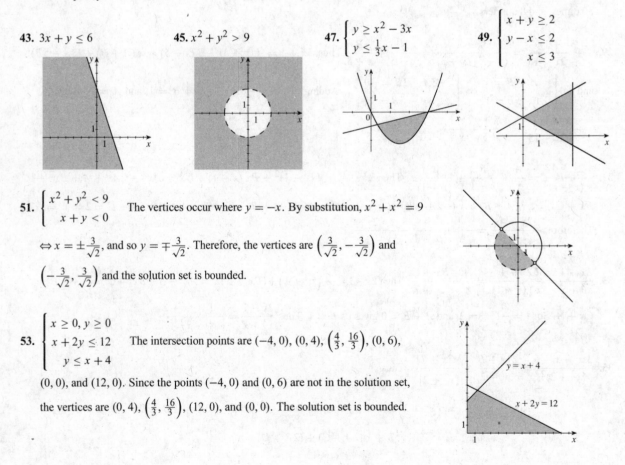

55. $\begin{cases} -x + y + z = a \\ x - y + z = b \\ x + y - z = c \end{cases} \Leftrightarrow \begin{cases} -x + y + z = a \\ \qquad\qquad 2z = a + b \\ \qquad 2y \qquad = a + c \end{cases}$ Thus, $y = \dfrac{a + c}{2}$, $z = \dfrac{a + b}{2}$, and $-x + \dfrac{a + c}{2} + \dfrac{a + b}{2} = a \Leftrightarrow$

$x = \dfrac{b + c}{2}$. The solution is $\left(\dfrac{b + c}{2}, \dfrac{a + c}{2}, \dfrac{a + b}{2} \right)$.

57. Solving the second equation for y, we have $y = kx$. Substituting for y in the first equation gives us

$x + kx = 12 \Leftrightarrow (1 + k) x = 12 \Leftrightarrow x = \dfrac{12}{k + 1}$. Substituting for y in the third equation gives us $kx - x = 2k \Leftrightarrow$

$(k - 1) x = 2k \Leftrightarrow x = \dfrac{2k}{k - 1}$. These points of intersection are the same when the x-values are equal. Thus, $\dfrac{12}{k + 1} = \dfrac{2k}{k - 1}$

$\Leftrightarrow 12 (k - 1) = 2k (k + 1) \Leftrightarrow 12k - 12 = 2k^2 + 2k \Leftrightarrow 0 = 2k^2 - 10k + 12 = 2 \left(k^2 - 5k + 6 \right) = 2 (k - 3) (k + 2)$.

Hence, $k = 2$ or $k = 3$.

CHAPTER 5 TEST

1. (a) The system is linear.

   (b) $\begin{cases} 3x + 5y = 4 \\ x - 4y = 7 \end{cases}$ Multiplying the second equation by -3 and then adding gives $17y = -17 \Leftrightarrow y = -1$, so

   $3x + 5 (-1) = 4 \Leftrightarrow 3x = 9 \Leftrightarrow x = 3$. Thus, the solution is $(3, -1)$.

3. (a) The system is nonlinear.

   (b) $\begin{cases} x^2 + y^2 = 100 \\ y = 3x \end{cases}$ Substituting $y = 3x$ into the first equation gives $x^2 + (3x)^2 = 100 \Leftrightarrow x^2 = 10 \Leftrightarrow x = \pm\sqrt{10}$.

   If $x = -\sqrt{10}$, then $y = -3\sqrt{10}$, and if $x = \sqrt{10}$, then $y = 3\sqrt{10}$. We can verify that $\left(-\sqrt{10}, -3\sqrt{10} \right)$ and

   $\left(\sqrt{10}, 3\sqrt{10} \right)$ are valid solutions to the first equation in the given system.

5. Let w be the speed of the wind and a the speed of the airplane in still air, in kilometers per hour. Then the speed of the of the

   plane flying against the wind is $a - w$ and the speed of the plane flying with the wind is $a + w$. Using distance $=$ rate $\times$ time,

   we get the system $\begin{cases} 600 = 2.5 (a - w) \\ 300 = \frac{50}{60} (a + w) \end{cases} \Leftrightarrow \begin{cases} 240 = a - w \\ 360 = a + w \end{cases}$ Adding the two equations, we get $600 = 2a \Leftrightarrow a = 300$. So

   $360 = 300 + w \Leftrightarrow w = 60$. Thus the speed of the airplane in still air is 300 km/h and the speed of the wind is 60 km/h.

7. (a) $\begin{cases} x - y + 9z = -3 \\ x \qquad - 4z = 7 \\ 3x - y + z = 5 \end{cases} \Leftrightarrow \begin{cases} x - y + 9z = -3 \\ \quad y - 13z = 10 \\ \quad 2y - 26z = 14 \end{cases} \begin{array}{l} \\ -\text{Eq. } 1 + \text{Eq. } 2 \\ (-3) \times \text{Eq. } 1 + \text{Eq. } 3 \end{array} \Leftrightarrow \begin{cases} x - y + 9z = -3 \\ \quad y - 13z = 10 \\ \qquad\qquad 0 = 6 \end{cases} \begin{array}{l} \\ \\ 2 \times \text{Eq. } 2 - \text{Eq. } 3 \end{array}$

   The last equation cannot be satisfied, so the system has no solution.

   (b) The system is inconsistent.

9. (a) $\begin{cases} x + y - 2z = 8 \\ 2x - y \qquad = 20 \\ 2x + 2y - 5z = 15 \end{cases} \Leftrightarrow \begin{cases} x + y - 2z = 8 \\ \quad 3y - 4z = -4 \\ \qquad\qquad -z = -1 \end{cases} \begin{array}{l} \\ 2 \times \text{Eq. } 1 - \text{Eq. } 2 \\ (-2) \times \text{Eq. } 1 + \text{Eq. } 3 \end{array} \Rightarrow z = 1$, so $3y - 4 (1) = -4 \Leftrightarrow y = 0$ and

   $x + 0 - 2 (1) = 8 \Rightarrow x = 10$. Thus, the solution is $(10, 0, 1)$.

   (b) The system is neither inconsistent nor dependent.

11. a. $3x + 4y < 6$ **(b)** $-x^2 + y \geq 3$

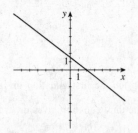
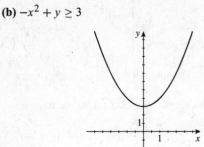

13. $\begin{cases} x^2 + y \leq 5 \\ y \geq 2x + 5 \end{cases}$ Substituting $y = 5 + 2x$ into the first equation gives

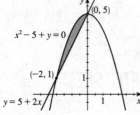

$x^2 - 5 + (5 + 2x) = 0 \Leftrightarrow x^2 + 2x = 0 \Leftrightarrow x\,(x + 2) = 0 \Leftrightarrow x = 0$ or $x = -2$. If
$x = 0$, then $y = 5 + 2\,(0) = 5$, and if $x = -2$, then $y = 5 + 2\,(-2) = 1$. Thus, the
vertices are $(0, 5)$ and $(-2, 1)$.

15. $\dfrac{2x - 3}{x^3 + 3x} = \dfrac{2x - 3}{x\,(x^2 + 3)} = \dfrac{A}{x} + \dfrac{Bx + C}{x^2 + 3}$. Then

$2x - 3 = A\left(x^2 + 3\right) + (Bx + C)\,x = Ax^2 + 3A + Bx^2 + Cx = (A + B)\,x^2 + Cx + 3A.$

So $3A = -3 \Leftrightarrow A = -1$, $C = 2$ and $A + B = 0$ gives us $B = 1$. Thus $\dfrac{2x - 3}{x^3 + x} = -\dfrac{1}{x} + \dfrac{x + 2}{x^2 + 3}$.

FOCUS ON MODELING   Linear Programming

1.

Vertex	$M = 200 - x - y$
$(0, 2)$	$200 - (0) - (2) = 198$
$(0, 5)$	$200 - (0) - (5) = 195$
$(4, 0)$	$200 - (4) - (0) = 196$

Thus, the maximum value is 198 and the minimum value
is 195.

3. $\begin{cases} x \geq 0,\ y \geq 0 \\ 2x + y \leq 10 \\ 2x + 4y \leq 28 \end{cases}$ The objective function is $P = 140 - x + 3y$. From the graph,

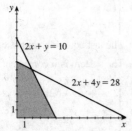

the vertices are $(0, 0)$, $(5, 0)$, $(2, 6)$, and $(0, 7)$.

Vertex	$P = 140 - x + 3y$
$(0, 0)$	$140 - (0) + 3\,(0) = 140$
$(5, 0)$	$140 - (5) + 3\,(0) = 135$
$(2, 6)$	$140 - (2) + 3\,(6) = 156$
$(0, 7)$	$140 - (0) + 3\,(7) = 161$

Thus the maximum value is 161, and the minimum value is 135.

5. Let t be the number of tables made daily and c be the number of chairs made daily. Then the data given can be summarized by the following table:

	Tables t	Chairs c	Available time
Carpentry	2 h	3 h	108 h
Finishing	1 h	$\frac{1}{2}$ h	20 h
Profit	$35	$20	

Thus we wish to maximize the total profit $P = 35t + 20c$ subject to the constraints $\begin{cases} 2t + 3c \leq 108 \\ t + \frac{1}{2}c \leq 20 \\ t \geq 0, c \geq 0 \end{cases}$

From the graph, the vertices occur at $(0, 0)$, $(20, 0)$, $(0, 36)$, and $(3, 34)$.

Vertex	$P = 35t + 20c$
$(0, 0)$	$35\,(0) + 20\,(0) = \quad 0$
$(20, 0)$	$35\,(20) + 20\,(0) = 700$
$(0, 36)$	$35\,(0) + 20\,(36) = 720$
$(3, 34)$	$35\,(3) + 20\,(34) = 785$

Hence, 3 tables and 34 chairs should be produced daily for a maximum profit of $785.

7. Let x be the number of crates of oranges and y the number of crates of grapefruit. Then the data given can be summarized by the following table:

	Oranges	Grapefruit	Available
Volume	4 ft^3	6 ft^3	300 ft^3
Weight	80 lb	100 lb	5600 lb
Profit	$2.50	$4.00	

In addition, $x \geq y$. Thus we wish to maximize the total profit $P = 2.5x + 4y$ subject to the constraints

$\begin{cases} x \geq 0, y \geq 0, x \geq y \\ 4x + 6y \leq 300 \\ 80x + 100y \leq 5600 \end{cases}$

From the graph, the vertices occur at $(0, 0)$, $(30, 30)$, $(45, 20)$, and $(70, 0)$.

Vertex	$P = 2.5x + 4y$
$(0, 0)$	$2.5\,(0) + 4\,(0) = \quad 0$
$(30, 30)$	$2.5\,(30) + 4\,(30) = 195$
$(45, 20)$	$2.5\,(45) + 4\,(20) = 192.5$
$(70, 0)$	$2.5\,(70) + 4\,(0) = 175$

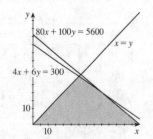

Thus, she should carry 30 crates of oranges and 30 crates of grapefruit for a maximum profit of $195.

9. Let x be the number of stereo sets shipped from Long Beach to Santa Monica and y the number of stereo sets shipped from Long Beach to El Toro. Thus, $15 - x$ sets must be shipped to Santa Monica from Pasadena and $19 - y$ sets to El Toro from Pasadena. Thus, $x \geq 0$, $y \geq 0$, $15 - x \geq 0$, $19 - y \geq 0$, $x + y \leq 24$, and $(15 - x) + (19 - y) \leq 18$. Simplifying, we get

the constraints $\begin{cases} x \geq 0, y \geq 0 \\ x \leq 15, y \leq 19 \\ x + y \leq 24 \\ x + y \geq 16 \end{cases}$

The objective function is the cost $C = 5x + 6y + 4(15 - x) + 5.5(19 - y) = x + 0.5y + 164.5$, which we wish to minimize. From the graph, the vertices occur at $(0, 16)$, $(0, 19)$, $(5, 19)$, $(15, 9)$, and $(15, 1)$.

Vertex	$C = x + 0.5y + 164.5$
$(0, 16)$	$(0) + 0.5(16) + 164.5 = 172.5$
$(0, 19)$	$(0) + 0.5(19) + 164.5 = 174$
$(5, 19)$	$(5) + 0.5(19) + 164.5 = 179$
$(15, 9)$	$(15) + 0.5(9) + 164.5 = 184$
$(15, 1)$	$(15) + 0.5(1) + 164.5 = 180$

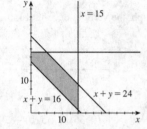

The minimum cost is \$172.50 and occurs when $x = 0$ and $y = 16$. Hence, no stereo should be shipped from Long Beach to Santa Monica, 16 from Long Beach to El Toro, 15 from Pasadena to Santa Monica, and 3 from Pasadena to El Toro.

11. Let x be the number of bags of standard mixtures and y be the number of bags of deluxe mixtures. Then the data can be summarized by the following table:

	Standard	Deluxe	Available
Cashews	100 g	150 g	15 kg
Peanuts	200 g	50 g	20 kg
Selling price	\$1.95	\$2.20	

Thus the total revenue, which we want to maximize, is given by $R = 1.95x + 2.25y$. We have the constraints

$\begin{cases} x \geq 0, y \geq 0, x \geq y \\ 0.1x + 0.15y \leq 15 \\ 0.2x + 0.05y \leq 20 \end{cases}$ $\Leftrightarrow$ $\begin{cases} x \geq 0, y \geq 0, x \geq y \\ 10x + 15y \leq 1500 \\ 20x + 5y \leq 2000 \end{cases}$

From the graph, the vertices occur at $(0, 0)$, $(60, 60)$, $(90, 40)$, and $(100, 0)$.

Vertex	$R = 1.95x + 2.25y$
$(0, 0)$	$1.96(0) + 2.25(0) = 0$
$(60, 60)$	$1.95(60) + 2.25(60) = 252$
$(90, 40)$	$1.95(90) + 2.25(40) = 265.5$
$(100, 0)$	$1.95(100) + 2.25(0) = 195$

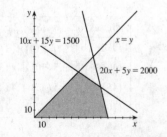

Hence, he should pack 90 bags of standard and 40 bags of deluxe mixture for a maximum revenue of \$265.50.

13. Let x be the amount in municipal bonds and y the amount in bank certificates, both in dollars. Then $12000 - x - y$ is the amount in high-risk bonds. So our constraints can be stated as

$$\begin{cases} x \geq 0,\, y \geq 0,\, x \geq 3y \\ 12{,}000 - x - y \geq 0 \\ 12{,}000 - x - y \leq 2000 \end{cases} \Leftrightarrow \begin{cases} x \geq 0,\, y \geq 0,\, x \geq 3y \\ x + y \leq 12{,}000 \\ x + y \geq 10{,}000 \end{cases}$$

From the graph, the vertices occur at $(7500, 2500)$, $(10000, 0)$, $(12000, 0)$, and $(9000, 3000)$. The objective function is

$P = 0.07x + 0.08y + 0.12(12000 - x - y) = 1440 - 0.05x - 0.04y$, which we wish to maximize.

Vertex	$P = 1440 - 0.05x - 0.04y$
$(7500, 2500)$	$1440 - 0.05(7500) - 0.04(2500) = 965$
$(10000, 0)$	$1440 - 0.05(10{,}000) - 0.04(0) = 940$
$(12000, 0)$	$1440 - 0.05(12{,}000) - 0.04(0) = 840$
$(9000, 3000)$	$1440 - 0.05(9000) - 0.04(3000) = 870$

Hence, she should invest \$7500 in municipal bonds, \$2500 in bank certificates, and the remaining \$2000 in high-risk bonds for a maximum yield of \$965.

15. Let g be the number of games published and e be the number of educational programs published. Then the number of utility programs published is $36 - g - e$. Hence we wish to maximize profit,
$P = 5000g + 8000e + 6000(36 - g - e) = 216{,}000 - 1000g + 2000e$, subject to the constraints

$$\begin{cases} g \geq 4,\, e \geq 0 \\ 36 - g - e \geq 0 \\ 36 - g - e \leq 2e \end{cases} \Leftrightarrow \begin{cases} g \geq 4,\, e \geq 0 \\ g + e \leq 36 \\ g + 3e \geq 36. \end{cases}$$

From the graph, the vertices are at $\left(4, \frac{32}{3}\right)$, $(4, 32)$, and $(36, 0)$. The objective function is $P = 216{,}000 - 1000g + 2000e$.

Vertex	$P = 216{,}000 - 1000g + 2000e$
$\left(4, \frac{32}{3}\right)$	$216{,}000 - 1000(4) + 2000\left(\frac{32}{3}\right) = 233{,}333.33$
$(4, 32)$	$216{,}000 - 1000(4) + 2000(32) = 276{,}000$
$(36, 0)$	$216{,}000 - 1000(36) + 2000(0) = 180{,}000$

So, they should publish 4 games, 32 educational programs, and no utility program for a maximum profit of \$276,000 annually.

6 MATRICES AND DETERMINANTS

6.1 MATRICES AND SYSTEMS OF LINEAR EQUATIONS

1. A system of linear equations with infinitely many solutions is called *dependent*. A system of linear equations with no solution is called *inconsistent*.

3. **(a)** The leading variables are x *and* y.

   (b) The system is dependent.

   (c) The solution of the system is $x = 3 + t$, $y = 5 - 2t$, $z = t$.

5. 3×2 7. 2×1 9. 1×3

11. $\begin{bmatrix} 3 & 1 & -1 & 2 \\ 2 & -1 & 0 & 1 \\ 1 & 0 & -1 & 3 \end{bmatrix}$

13. **(a)** Yes, this matrix is in row-echelon form.

   (b) Yes, this matrix is in reduced row-echelon form.

   (c) $\begin{cases} x = -3 \\ y = 5 \end{cases}$

15. **(a)** Yes, this matrix is in row-echelon form.

   (b) No, this matrix is not in reduced row-echelon form, since the leading 1 in the second row does not have a zero above it.

   (c) $\begin{cases} x + 2y + 8z = 0 \\ y + 3z = 2 \\ 0 = 0 \end{cases}$

17. **(a)** No, this matrix is not in row-echelon form, since the row of zeros is not at the bottom.

   (b) No, this matrix is not in reduced row-echelon form.

   (c) $\begin{cases} x = 0 \\ 0 = 0 \\ y + 5z = 1 \end{cases}$

19. **(a)** Yes, this matrix is in row-echelon form.

   (b) Yes, this matrix is in reduced row-echelon form.

   (c) $\begin{cases} x + 3y - w = 0 \\ z + 2w = 0 \\ 0 = 1 \\ 0 = 0 \end{cases}$

   Notice that this system has no solution.

21. $\begin{bmatrix} -1 & 1 & 2 & 0 \\ 3 & 1 & 1 & 4 \\ 1 & -2 & -1 & -1 \end{bmatrix} \xrightarrow{3R_1 + R_2 \to R_2} \begin{bmatrix} -1 & 1 & 2 & 0 \\ 0 & 4 & 7 & 4 \\ 1 & -2 & -1 & -1 \end{bmatrix}$

23. $\begin{bmatrix} 2 & 1 & -3 & 5 \\ 2 & 3 & 1 & 13 \\ 6 & -5 & -1 & 7 \end{bmatrix} \xrightarrow{-3R_1 + R_3 \to R_3} \begin{bmatrix} 2 & 1 & -3 & 5 \\ 2 & 3 & 1 & 13 \\ 0 & -8 & 8 & -8 \end{bmatrix}$

25. (a) $\begin{cases} x - 2y + 4z = 3 \\ \quad\quad y + 2z = 7 \\ \quad\quad\quad\quad z = 2 \end{cases}$

(b) $y + 2(2) = 7 \Leftrightarrow y = 3$, so $x - 2(3) + 4(2) = 3 \Leftrightarrow$
$x = 1$. The solution is $(1, 3, 2)$.

27. (a) $\begin{cases} x + 2y + 3z - \;w = 7 \\ \quad\quad y - 2z \quad\quad = 5 \\ \quad\quad\quad\quad z + 2w = 5 \\ \quad\quad\quad\quad\quad\quad w = 3 \end{cases}$

(b) $z + 2(3) = 5 \Leftrightarrow z = -1$, so $y - 2(-1) = 5 \Leftrightarrow$
$y = 3$ and $x + 2(3) + 3(-1) - 3 = 7 \Leftrightarrow x = 7$. The
solution is $(7, 3, -1, 3)$.

29. $\begin{bmatrix} 1 & -2 & 1 & 1 \\ 0 & 1 & 2 & 5 \\ 1 & 1 & 3 & 8 \end{bmatrix} \xrightarrow{R_3 - R_1 \to R_3} \begin{bmatrix} 1 & -2 & 1 & 1 \\ 0 & 1 & 2 & 5 \\ 0 & 3 & 2 & 7 \end{bmatrix} \xrightarrow{R_3 - 3R_2 \to R_3} \begin{bmatrix} 1 & -2 & 1 & 1 \\ 0 & 1 & 2 & 5 \\ 0 & 0 & -4 & -8 \end{bmatrix}$. Thus, $-4z = -8 \Leftrightarrow$

$z = 2$; $y + 2(2) = 5 \Leftrightarrow y = 1$; and $x - 2(1) + (2) = 1 \Leftrightarrow x = 1$. Therefore, the solution is $(1, 1, 2)$.

31. $\begin{bmatrix} 1 & 1 & 1 & 2 \\ 2 & -3 & 2 & 4 \\ 4 & 1 & -3 & 1 \end{bmatrix} \begin{array}{c} \xrightarrow{R_2 - 2R_1 \to R_2} \\ \xrightarrow{R_3 - 4R_1 \to R_3} \end{array} \begin{bmatrix} 1 & 1 & 1 & 2 \\ 0 & -5 & 0 & 0 \\ 0 & -3 & -7 & -7 \end{bmatrix} \xrightarrow{R_3 - \frac{3}{5}R_2 \to R_3} \begin{bmatrix} 1 & 1 & 1 & 2 \\ 0 & -5 & 0 & 0 \\ 0 & 0 & -7 & -7 \end{bmatrix}$.

Thus, $-7z = -7 \Leftrightarrow z = 1$; $-5y = 0 \Leftrightarrow y = 0$; and $x + 0 + 1 = 2 \Leftrightarrow x = 1$. Therefore, the solution is $(1, 0, 1)$.

33. $\begin{bmatrix} 1 & 2 & -1 & -2 \\ 1 & 0 & 1 & 0 \\ 2 & -1 & -1 & -3 \end{bmatrix} \begin{array}{c} \xrightarrow{R_2 - R_1 \to R_2} \\ \xrightarrow{R_3 - 2R_1 \to R_3} \end{array} \begin{bmatrix} 1 & 2 & -1 & -2 \\ 0 & -2 & 2 & 2 \\ 0 & -5 & 1 & 1 \end{bmatrix} \xrightarrow{-\frac{1}{2}R_2} \begin{bmatrix} 1 & 2 & -1 & -2 \\ 0 & 1 & 1 & 1 \\ 0 & -5 & 1 & 1 \end{bmatrix} \xrightarrow{R_3 + 5R_2 \to R_3}$

$\begin{bmatrix} 1 & 2 & -1 & -2 \\ 0 & 1 & 1 & 1 \\ 0 & 0 & 6 & 6 \end{bmatrix}$. Thus, $6z = 6 \Leftrightarrow z = 1$; $y + (1) = 1 \Leftrightarrow y = 0$; and $x + 2(0) - (1) = -2 \Leftrightarrow x = -1$. Therefore, the

solution is $(-1, 0, 1)$.

35. $\begin{bmatrix} 1 & 2 & -1 & 9 \\ 2 & 0 & -1 & -2 \\ 3 & 5 & 2 & 22 \end{bmatrix} \begin{array}{c} \xrightarrow{R_2 - 2R_1 \to R_2} \\ \xrightarrow{R_3 - 3R_1 \to R_3} \end{array} \begin{bmatrix} 1 & 2 & -1 & 9 \\ 0 & -4 & 1 & -20 \\ 0 & -1 & 5 & -5 \end{bmatrix} \xrightarrow{4R_3 - R_2 \to R_3} \begin{bmatrix} 1 & 2 & -1 & 9 \\ 0 & -4 & 1 & -20 \\ 0 & 0 & 19 & 0 \end{bmatrix}$.

Thus, $19x_3 = 0 \Leftrightarrow x_3 = 0$; $-4x_2 = -20 \Leftrightarrow x_2 = 5$; and $x_1 + 2(5) = 9 \Leftrightarrow x_1 = -1$. Therefore, the solution is $(-1, 5, 0)$.

37. $\begin{bmatrix} 2 & -3 & -1 & 13 \\ -1 & 2 & -5 & 6 \\ 5 & -1 & -1 & 49 \end{bmatrix} \begin{array}{c} \xrightarrow{2R_2 + R_1 \to R_2} \\ \xrightarrow{2R_3 - 5R_1 \to R_3} \end{array} \begin{bmatrix} 2 & -3 & -1 & 13 \\ 0 & 1 & -11 & 25 \\ 0 & 13 & 3 & 33 \end{bmatrix} \xrightarrow{R_3 - 13R_2 \to R_3} \begin{bmatrix} 2 & -3 & -1 & 13 \\ 0 & 1 & -11 & 25 \\ 0 & 0 & 146 & -292 \end{bmatrix}$.

Thus, $146z = -292 \Leftrightarrow z = -2$; $y - 11(-2) = 25 \Leftrightarrow y = 3$; and $2x - 3 \cdot 3 + 2 = 13 \Leftrightarrow x = 10$. Therefore, the solution is
$(10, 3, -2)$.

39. $\begin{bmatrix} 1 & 1 & 1 & 2 \\ 0 & 1 & -3 & 1 \\ 2 & 1 & 5 & 0 \end{bmatrix} \xrightarrow{R_3 - 2R_1 \to R_3} \begin{bmatrix} 1 & 1 & 1 & 2 \\ 0 & 1 & -3 & 1 \\ 0 & -1 & 3 & -4 \end{bmatrix} \xrightarrow{R_3 + R_2 \to R_3} \begin{bmatrix} 1 & 1 & 1 & 3 \\ 0 & 1 & -3 & 1 \\ 0 & 0 & 0 & -3 \end{bmatrix}$. The third row of the

matrix states $0 = -3$, which is impossible. Hence, the system is inconsistent, and there is no solution.

41.
$\begin{bmatrix} 2 & -3 & -9 & -5 \\ 1 & 0 & 3 & 2 \\ -3 & 1 & -4 & -3 \end{bmatrix} \xrightarrow{R_1 \leftrightarrow R_2} \begin{bmatrix} 1 & 0 & 3 & 2 \\ 2 & -3 & -9 & -5 \\ -3 & 1 & -4 & -3 \end{bmatrix} \xrightarrow[R_3 + 3R_1 \to R_3]{R_2 - 2R_1 \to R_2} \begin{bmatrix} 1 & 0 & 3 & 2 \\ 0 & -3 & -15 & -9 \\ 0 & 1 & 5 & 3 \end{bmatrix} \xrightarrow{-\frac{1}{3}R_2}$

$\begin{bmatrix} 1 & 0 & 3 & 2 \\ 0 & 1 & 5 & 3 \\ 0 & 1 & 5 & 3 \end{bmatrix} \xrightarrow{R_3 - R_2 \to R_3} \begin{bmatrix} 1 & 0 & 3 & 2 \\ 0 & 1 & 5 & 3 \\ 0 & 0 & 0 & 0 \end{bmatrix}$. Therefore, this system has infinitely many solutions, given by $x + 3t = 2$

$\Leftrightarrow x = 2 - 3t$, and $y + 5t = 3 \Leftrightarrow y = 3 - 5t$. Hence, the solutions are $(2 - 3t, 3 - 5t, t)$, where t is any real number.

43.
$\begin{bmatrix} 1 & -1 & 3 & 3 \\ 4 & -8 & 32 & 24 \\ 2 & -3 & 11 & 4 \end{bmatrix} \xrightarrow[R_3 - 2R_1 \to R_3]{R_2 - 4R_1 \to R_2} \begin{bmatrix} 1 & -1 & 3 & 3 \\ 0 & -4 & 20 & 12 \\ 0 & -1 & 5 & -2 \end{bmatrix} \xrightarrow[R_3 + R_2 \to R_3]{-\frac{1}{4}R_2} \begin{bmatrix} 1 & -1 & 3 & 3 \\ 0 & 1 & -5 & -3 \\ 0 & 0 & 0 & -5 \end{bmatrix}$. The third row of the

matrix states $0 = -5$, which is impossible. Hence, the system is inconsistent, and there is no solution.

45.
$\begin{bmatrix} 1 & 4 & -2 & -3 \\ 2 & -1 & 5 & 12 \\ 8 & 5 & 11 & 30 \end{bmatrix} \xrightarrow[R_3 - 8R_1 \to R_3]{R_2 - 2R_1 \to R_2} \begin{bmatrix} 1 & 4 & -2 & -3 \\ 0 & -9 & 9 & 18 \\ 0 & -27 & 27 & 54 \end{bmatrix} \xrightarrow{R_3 - 3R_2 \to R_3} \begin{bmatrix} 1 & 4 & -2 & -3 \\ 0 & -9 & 9 & 18 \\ 0 & 0 & 0 & 0 \end{bmatrix}$.

Therefore, this system has infinitely many solutions, given by $-9y + 9t = 18 \Leftrightarrow y = -2 + t$, and
$x + 4(-2 + t) - 2t = -3 \Leftrightarrow x = 5 - 2t$. Hence, the solutions are $(5 - 2t, -2 + t, t)$, where t is any real number.

47.
$\begin{bmatrix} 2 & 1 & -2 & 12 \\ -1 & -\frac{1}{2} & 1 & -6 \\ 3 & \frac{3}{2} & -3 & 18 \end{bmatrix} \xrightarrow[-R_1]{R_1 \leftrightarrow R_2} \begin{bmatrix} 1 & \frac{1}{2} & -1 & 6 \\ 2 & 1 & -2 & 12 \\ 3 & \frac{3}{2} & -3 & 18 \end{bmatrix} \xrightarrow[R_3 - 3R_1 \to R_3]{R_2 - 2R_1 \to R_2} \begin{bmatrix} 1 & \frac{1}{2} & -1 & 6 \\ 0 & 0 & 0 & 0 \\ 0 & 0 & 0 & 0 \end{bmatrix}$.

Therefore, this system has infinitely many solutions, given by $x + \frac{1}{2}s - t = 6 \Leftrightarrow x = 6 - \frac{1}{2}s + t$. Hence, the solutions are $\left(6 - \frac{1}{2}s + t, s, t\right)$, where s and t are any real numbers.

49.
$\begin{bmatrix} 4 & -3 & 1 & -8 \\ -2 & 1 & -3 & -4 \\ 1 & -1 & 2 & 3 \end{bmatrix} \xrightarrow{R_1 \leftrightarrow R_3} \begin{bmatrix} 1 & -1 & 2 & 3 \\ -2 & 1 & -3 & -4 \\ 4 & -3 & 1 & -8 \end{bmatrix} \xrightarrow[R_3 - 4R_1 \to R_3]{R_2 + 2R_1 \to R_2} \begin{bmatrix} 1 & -1 & 2 & 3 \\ 0 & -1 & 1 & 2 \\ 0 & 1 & -7 & -20 \end{bmatrix} \xrightarrow{-R_2}$

$\begin{bmatrix} 1 & -1 & 2 & 3 \\ 0 & 1 & -1 & -2 \\ 0 & 1 & -7 & -20 \end{bmatrix} \xrightarrow{R_3 - R_2 \to R_3} \begin{bmatrix} 1 & -1 & 2 & 3 \\ 0 & 1 & -1 & -2 \\ 0 & 0 & -6 & -18 \end{bmatrix}$. Therefore, $-6z = -18 \Leftrightarrow z = 3$; $y - (3) = -2 \Leftrightarrow$

$y = 1$; and $x - (1) + 2(3) = 3 \Leftrightarrow x = -2$. Hence, the solution is $(-2, 1, 3)$.

51.
$\begin{bmatrix} 2 & 1 & 3 & 9 \\ -1 & 0 & -7 & 10 \\ 3 & 2 & -1 & 4 \end{bmatrix} \xrightarrow[R_3 + 3R_2 \to R_3]{2R_2 + R_1 \to R_2} \begin{bmatrix} 2 & 1 & 3 & 9 \\ 0 & 1 & -11 & 29 \\ 0 & 2 & -22 & 34 \end{bmatrix} \xrightarrow{2R_2 - R_3 \to R_3} \begin{bmatrix} 2 & 1 & 3 & 9 \\ 0 & 1 & -11 & 29 \\ 0 & 0 & 0 & 24 \end{bmatrix}$.

Therefore, the system is inconsistent and there is no solution.

53.
$\begin{bmatrix} 1 & 2 & -3 & -5 \\ -2 & -4 & -6 & 10 \\ 3 & 7 & -2 & -13 \end{bmatrix} \xrightarrow[R_3 - 3R_1 \to R_3]{R_2 + 2R_1 \to R_2} \begin{bmatrix} 1 & 2 & -3 & -5 \\ 0 & 0 & -12 & 0 \\ 0 & 1 & 7 & 2 \end{bmatrix} \xrightarrow{R_2 \leftrightarrow R_3} \begin{bmatrix} 1 & 2 & -3 & -5 \\ 0 & 1 & 7 & 2 \\ 0 & 0 & -12 & 0 \end{bmatrix}$. Therefore,

$-12z = 0 \Leftrightarrow z = 0$; $y + 7(0) = 2 \Leftrightarrow y = 2$; and $x + 2(2) - 3(0) = -5 \Leftrightarrow x = -9$. Hence, the solution is $(-9, 2, 0)$.

55. $\begin{bmatrix} 1 & -1 & 6 & 8 \\ 1 & 0 & 1 & 5 \\ 1 & 3 & -14 & -4 \end{bmatrix} \xrightarrow[R_3 - R_1 \to R_3]{R_2 - R_1 \to R_2} \begin{bmatrix} 1 & -1 & 6 & 8 \\ 0 & 1 & -5 & -3 \\ 0 & 4 & -20 & -12 \end{bmatrix} \xrightarrow{R_3 - 4R_2 \to R_3} \begin{bmatrix} 1 & -1 & 6 & 8 \\ 0 & 1 & -5 & -3 \\ 0 & 0 & 0 & 0 \end{bmatrix}$. Therefore,

the system is dependent. Let $z = t$. Then $y - 5t = -3 \Rightarrow y = -3 + 5t$ and $x - y + 6t = 8 \Leftrightarrow x = 8 + (-3 + 5t) - 6t = 5 - t$.
The solutions are $(5 - t, -3 + 5t, t)$, where t is any real number.

57. $\begin{bmatrix} -1 & 2 & 1 & -3 & 3 \\ 3 & -4 & 1 & 1 & 9 \\ -1 & -1 & 1 & 1 & 0 \\ 2 & 1 & 4 & -2 & 3 \end{bmatrix} \xrightarrow{-R_1} \begin{bmatrix} 1 & -2 & -1 & 3 & -3 \\ 3 & -4 & 1 & 1 & 9 \\ -1 & -1 & 1 & 1 & 0 \\ 2 & 1 & 4 & -2 & 3 \end{bmatrix} \xrightarrow[\substack{R_3 + R_1 \to R_3 \\ R_4 - 2R_1 \to R_4}]{R_2 - 3R_1 \to R_2} \begin{bmatrix} 1 & -2 & -1 & 3 & -3 \\ 0 & 2 & 4 & -8 & 18 \\ 0 & -3 & 0 & 4 & -3 \\ 0 & 5 & 6 & -8 & 9 \end{bmatrix}$

$\xrightarrow{\frac{1}{2}R_2} \begin{bmatrix} 1 & -2 & -1 & 3 & -3 \\ 0 & 1 & 2 & -4 & 9 \\ 0 & -3 & 0 & 4 & -3 \\ 0 & 5 & 6 & -8 & 9 \end{bmatrix} \xrightarrow[R_4 - 5R_2 \to R_4]{R_3 + 3R_2 \to R_3} \begin{bmatrix} 1 & -2 & -1 & 3 & -3 \\ 0 & 1 & 2 & -4 & 9 \\ 0 & 0 & 6 & -8 & 24 \\ 0 & 0 & -4 & 12 & -36 \end{bmatrix} \xrightarrow{3R_4 + 2R_3 \to R_4}$

$\begin{bmatrix} 1 & -2 & -1 & 3 & -3 \\ 0 & 1 & 2 & -4 & 9 \\ 0 & 0 & 6 & -8 & 24 \\ 0 & 0 & 0 & 20 & -60 \end{bmatrix}$. Therefore, $20w = -60 \Leftrightarrow w = -3$; $6z + 24 = 24 \Leftrightarrow z = 0$. Then $y + 12 = 9 \Leftrightarrow y = -3$ and

$x + 6 - 9 = -3 \Leftrightarrow x = 0$. Hence, the solution is $(0, -3, 0, -3)$.

59. $\begin{bmatrix} 1 & 1 & 2 & -1 & -2 \\ 0 & 3 & 1 & 2 & 2 \\ 1 & 1 & 0 & 3 & 2 \\ -3 & 0 & 1 & 2 & 5 \end{bmatrix} \xrightarrow[R_4 + 3R_1 \to R_4]{R_3 - R_1 \to R_3} \begin{bmatrix} 1 & 1 & 2 & -1 & -2 \\ 0 & 3 & 1 & 2 & 2 \\ 0 & 0 & -2 & 4 & 4 \\ 0 & 3 & 7 & -1 & -1 \end{bmatrix} \xrightarrow{R_4 - R_2 \to R_4} \begin{bmatrix} 1 & 1 & 2 & -1 & -2 \\ 0 & 3 & 1 & 2 & 2 \\ 0 & 0 & -2 & 4 & 4 \\ 0 & 0 & 6 & -3 & -3 \end{bmatrix}$

$\xrightarrow{R_4 + 3R_3 \to R_4} \begin{bmatrix} 1 & 1 & 2 & -1 & -2 \\ 0 & 3 & 1 & 2 & 2 \\ 0 & 0 & -2 & 4 & 4 \\ 0 & 0 & 0 & 9 & 9 \end{bmatrix}$. Therefore, $9w = 9 \Leftrightarrow w = 1$; $-2z + 4(1) = 4 \Leftrightarrow z = 0$. Then

$3y + (0) + 2(1) = 2 \Leftrightarrow y = 0$ and $x + (0) + 2(0) - (1) = -2 \Leftrightarrow x = -1$. Hence, the solution is $(-1, 0, 0, 1)$.

61. $\begin{bmatrix} 1 & -1 & 0 & 1 & 0 \\ 3 & 0 & -1 & 2 & 0 \\ 1 & -4 & 1 & 2 & 0 \end{bmatrix} \xrightarrow[R_3 - R_1 \to R_3]{R_2 - 3R_1 \to R_2} \begin{bmatrix} 1 & -1 & 0 & 1 & 0 \\ 0 & 3 & -1 & -1 & 0 \\ 0 & -3 & 1 & 1 & 0 \end{bmatrix} \xrightarrow{R_3 + R_2 \to R_3} \begin{bmatrix} 1 & -1 & 0 & 1 & 0 \\ 0 & 3 & -1 & -1 & 0 \\ 0 & 0 & 0 & 0 & 0 \end{bmatrix}$.

Therefore, the system has infinitely many solutions, given by $3y - s - t = 0 \Leftrightarrow y = \frac{1}{3}(s + t)$ and $x - \frac{1}{3}(s + t) + t = 0$

$\Leftrightarrow x = \frac{1}{3}(s - 2t)$. So the solutions are $\left(\frac{1}{3}(s - 2t), \frac{1}{3}(s + t), s, t\right)$, where s and t are any real numbers.

63. $\begin{bmatrix} 1 & 0 & 1 & 1 & 4 \\ 0 & 1 & -1 & 0 & -4 \\ 1 & -2 & 3 & 1 & 12 \\ 2 & 0 & -2 & 5 & -1 \end{bmatrix}$ $\xrightarrow[R_4 - 2R_1 \to R_4]{R_3 - R_1 \to R_3}$ $\begin{bmatrix} 1 & 0 & 1 & 1 & 4 \\ 0 & 1 & -1 & 0 & -4 \\ 0 & -2 & 2 & 0 & 8 \\ 0 & 0 & -4 & 3 & -9 \end{bmatrix}$ $\xrightarrow{R_3 + 2R_2 \to R_3}$ $\begin{bmatrix} 1 & 0 & 1 & 1 & 4 \\ 0 & 1 & -1 & 0 & -4 \\ 0 & 0 & 0 & 0 & 0 \\ 0 & 0 & -4 & 3 & -9 \end{bmatrix}$

$\xrightarrow{R_3 \leftrightarrow -R_4}$ $\begin{bmatrix} 1 & 0 & 1 & 1 & 4 \\ 0 & 1 & -1 & 0 & -4 \\ 0 & 0 & 4 & -3 & 9 \\ 0 & 0 & 0 & 0 & 0 \end{bmatrix}$. Therefore, $4z - 3t = 9 \Leftrightarrow 4z = 9 + 3t \Leftrightarrow z = \frac{9}{4} + \frac{3}{4}t$. Then we have

$y - \left(\frac{9}{4} + \frac{3}{4}t\right) = -4 \Leftrightarrow y = \frac{-7}{4} + \frac{3}{4}t$ and $x + \left(\frac{9}{4} + \frac{3}{4}t\right) + t = 4 \Leftrightarrow x = \frac{7}{4} - \frac{7}{4}t$. Hence, the solutions are

$\left(\frac{7}{4} - \frac{7}{4}t, -\frac{7}{4} + \frac{3}{4}t, \frac{9}{4} + \frac{3}{4}t, t\right)$, where t is any real number.

65. Using `rref` on the matrix $\begin{bmatrix} 0.75 & -3.75 & 2.95 & 4.0875 \\ 0.95 & -8.75 & 0 & 3.375 \\ 1.25 & -0.15 & 2.75 & 3.6625 \end{bmatrix}$, we find that the solution is $x = 1.25$, $y = -0.25$, $z = 0.75$.

67. Using `rref` on the matrix $\begin{bmatrix} 42 & -31 & 0 & -42 & -0.4 \\ -6 & 0 & 0 & -9 & 4.5 \\ 35 & 0 & -67 & 32 & 348.8 \\ 0 & 31 & 48 & -52 & -76.6 \end{bmatrix}$, we find that the solution is $x = 1.2$, $y = 3.4$, $z = -5.2$,

$w = -1.3$.

69. Let x, y, z represent the number of VitaMax, Vitron, and VitaPlus pills taken daily. The matrix representation for the system of equations is

$\begin{bmatrix} 5 & 10 & 15 & 50 \\ 15 & 20 & 0 & 50 \\ 10 & 10 & 10 & 50 \end{bmatrix}$ $\xrightarrow[\frac{1}{5}R_3]{\substack{\frac{1}{5}R_1 \\ \frac{1}{5}R_2}}$ $\begin{bmatrix} 1 & 2 & 3 & 10 \\ 3 & 4 & 0 & 10 \\ 2 & 2 & 2 & 10 \end{bmatrix}$ $\xrightarrow[R_3 - 2R_1 \to R_3]{R_2 - 3R_1 \to R_2}$ $\begin{bmatrix} 1 & 2 & 3 & 10 \\ 0 & -2 & -9 & -20 \\ 0 & -2 & -4 & -10 \end{bmatrix}$ $\xrightarrow{R_3 - R_2 \to R_3}$ $\begin{bmatrix} 1 & 2 & 3 & 10 \\ 0 & -2 & -9 & -20 \\ 0 & 0 & 5 & 10 \end{bmatrix}$.

Thus, $5z = 10 \Leftrightarrow z = 2$; $-2y - 18 = -20 \Leftrightarrow y = 1$; and $x + 2 + 6 = 10 \Leftrightarrow x = 2$. Hence, he should take 2 VitaMax, 1 Vitron, and 2 VitaPlus pills daily.

71. Let x, y, and z represent the distance, in miles, of the run, swim, and cycle parts of the race respectively. Then, since

$\text{time} = \dfrac{\text{distance}}{\text{speed}}$, we get the following equations from the three contestants' race times:

$\begin{cases} \left(\frac{x}{10}\right) + \left(\frac{y}{4}\right) + \left(\frac{z}{20}\right) = 2.5 \\ \left(\frac{x}{7.5}\right) + \left(\frac{y}{6}\right) + \left(\frac{z}{15}\right) = 3 \\ \left(\frac{x}{15}\right) + \left(\frac{y}{3}\right) + \left(\frac{z}{40}\right) = 1.75 \end{cases}$ $\Leftrightarrow$ $\begin{cases} 2x + 5y + z = 50 \\ 4x + 5y + 2z = 90 \\ 8x + 40y + 3z = 210 \end{cases}$ which has the following matrix representation:

$\begin{bmatrix} 2 & 5 & 1 & 50 \\ 4 & 5 & 2 & 90 \\ 8 & 40 & 3 & 210 \end{bmatrix}$ $\xrightarrow[R_3 - 4R_1 \to R_3]{R_2 - 2R_1 \to R_2}$ $\begin{bmatrix} 2 & 5 & 1 & 50 \\ 0 & -5 & 0 & -10 \\ 0 & 20 & -1 & 10 \end{bmatrix}$ $\xrightarrow{R_3 + 4R_2 \to R_3}$ $\begin{bmatrix} 2 & 5 & 1 & 50 \\ 0 & -5 & 0 & -10 \\ 0 & 0 & -1 & -30 \end{bmatrix}$.

Thus, $-z = -30 \Leftrightarrow z = 30$; $-5y = -10 \Leftrightarrow y = 2$; and $2x + 10 + 30 = 50 \Leftrightarrow x = 5$. So the race has a 5 mile run, 2 mile swim, and 30 mile cycle.

73. Let t be the number of tables produced, c the number of chairs, and a the number of armoires. Then, the system of equations

is $\begin{cases} \frac{1}{2}t + c + a = 300 \\ \frac{1}{2}t + \frac{3}{2}c + a = 400 \\ t + \frac{3}{2}c + 2a = 590 \end{cases} \Leftrightarrow \begin{cases} t + 2c + 2a = 600 \\ t + 3c + 2a = 800 \\ 2t + 3c + 4a = 1180 \end{cases}$ and a matrix representation is

$\begin{bmatrix} 1 & 2 & 2 & 600 \\ 1 & 3 & 2 & 800 \\ 2 & 3 & 4 & 1180 \end{bmatrix} \xrightarrow[R_3 - 2R_1 \to R_3]{R_2 - R_1 \to R_2} \begin{bmatrix} 1 & 2 & 2 & 600 \\ 0 & 1 & 0 & 200 \\ 0 & -1 & 0 & -20 \end{bmatrix} \xrightarrow{R_3 + R_2 \to R_3} \begin{bmatrix} 1 & 2 & 2 & 600 \\ 0 & 1 & 0 & 200 \\ 0 & 0 & 0 & 180 \end{bmatrix}.$

The third row states $0 = 180$, which is impossible, and so the system is inconsistent. Therefore, it is impossible to use all of the available labor-hours.

75. *Line containing the points* $(0, 0)$ *and* $(1, 12)$: Using the general form of a line, $y = ax + b$, we substitute for x and y and solve for a and b. The point $(0, 0)$ gives $0 = a(0) + b \Rightarrow b = 0$; the point $(1, 12)$ gives $12 = a(1) + b \Rightarrow a = 12$. Since $a = 12$ and $b = 0$, the equation of the line is $y = 12x$.

Quadratic containing the points $(0, 0)$, $(1, 12)$, *and* $(3, 6)$: Using the general form of a quadratic, $y = ax^2 + bx + c$, we substitute for x and y and solve for a, b, and c. The point $(0, 0)$ gives $0 = a(0)^2 + b(0) + c \Rightarrow c = 0$; the point $(1, 12)$ gives $12 = a(1)^2 + b(1) + c \Rightarrow a + b = 12$; the point $(3, 6)$ gives $6 = a(3)^2 + b(3) + c \Rightarrow 9a + 3b = 6$. Subtracting the third equation from -3 times the third gives $6a = -30 \Leftrightarrow a = -5$. So $a + b = 12 \Leftrightarrow b = 12 - a \Rightarrow b = 17$. Since $a = -5$, $b = 17$, and $c = 0$, the equation of the quadratic is $y = -5x^2 + 17x$.

Cubic containing the points $(0, 0)$, $(1, 12)$, $(2, 40)$, *and* $(3, 6)$: Using the general form of a cubic, $y = ax^3 + bx^2 + cx + d$, we substitute for x and y and solve for a, b, c, and d. The point $(0, 0)$ gives $0 = a(0)^3 + b(0)^2 + c(0) + d \Rightarrow d = 0$; the point the point $(1, 12)$ gives $12 = a(1)^3 + b(1)^2 + c(1) + d \Rightarrow a + b + c + d = 12$; the point $(2, 40)$ gives $40 = a(2)^3 + b(2)^2 + c(2) + d \Rightarrow 8a + 4b + 2c + d = 40$; the point $(3, 6)$ gives $6 = a(3)^3 + b(3)^2 + c(3) + d \Rightarrow$

$27a + 9b + 3c + d = 6$. Since $d = 0$, the system reduces to $\begin{cases} a + b + c = 12 \\ 8a + 4b + 2c = 40 \\ 27a + 9b + 3c = 6 \end{cases}$ which has representation

$\begin{bmatrix} 1 & 1 & 1 & 12 \\ 8 & 4 & 2 & 40 \\ 27 & 9 & 3 & 6 \end{bmatrix} \xrightarrow[R_3 - 27R_1 \to R_3]{R_2 - 8R_1 \to R_2} \begin{bmatrix} 1 & 1 & 1 & 12 \\ 0 & -4 & -6 & -56 \\ 0 & -18 & -24 & -318 \end{bmatrix} \xrightarrow[-\frac{1}{6}R_3]{-\frac{1}{2}R_2} \begin{bmatrix} 1 & 1 & 1 & 12 \\ 0 & 2 & 3 & 28 \\ 0 & 3 & 4 & 53 \end{bmatrix} \xrightarrow{2R_3 - 3R_2 \to R_3} \begin{bmatrix} 1 & 1 & 1 & 12 \\ 0 & 2 & 3 & 28 \\ 0 & 0 & -1 & 22 \end{bmatrix}.$

So $c = -22$ and back-substituting we have $2b + 3(-22) = 28 \Leftrightarrow b = 47$ and $a + 47 + (-22) = 0 \Leftrightarrow a = -13$. So the cubic is $y = -13x^3 + 47x^2 - 22x$.

Fourth-degree polynomial containing the points $(0, 0)$, $(1, 12)$, $(2, 40)$, $(3, 6)$, *and* $(-1, -14)$: Using the general form of a fourth-degree polynomial, $y = ax^4 + bx^3 + cx^2 + dx + e$, we substitute for x and y and solve for a, b, c, d, and e. The point $(0, 0)$ gives $0 = a(0)^4 + b(0)^3 + c(0)^2 + d(0) + e \Rightarrow e = 0$; the point $(1, 12)$ gives $12 = a(1)^4 + b(1)^3 + c(1)^2 + d(1) + e$; the point $(2, 40)$ gives $40 = a(2)^4 + b(2)^3 + c(2)^2 + d(2) + e$; the point $(3, 6)$ gives $6 = a(3)^4 + b(3)^3 + c(3)^2 + d(3) + e$; the point $(-1, -14)$ gives $-14 = a(-1)^4 + b(-1)^3 + c(-1)^2 + d(-1) + e$.

Because the first equation is $e = 0$, we eliminate e from the other equations to get

$$\begin{cases} a + b + c + d = 12 \\ 16a + 8b + 4c + 2d = 40 \\ 81a + 27b + 9c + 3d = 6 \\ a - b + c - d = -14 \end{cases} \Leftrightarrow \begin{bmatrix} 1 & 1 & 1 & 1 & 12 \\ 16 & 8 & 4 & 2 & 40 \\ 81 & 27 & 9 & 3 & 6 \\ 1 & -1 & 1 & -1 & -14 \end{bmatrix} \xrightarrow[\substack{R_2 - 16R_1 \to R_2 \\ R_3 - 81R_1 \to R_3 \\ R_4 - R_1 \to R_4}]{} \begin{bmatrix} 1 & 1 & 1 & 1 & 12 \\ 0 & -8 & -12 & -14 & -152 \\ 0 & -54 & -72 & -78 & -966 \\ 0 & -2 & 0 & -2 & -26 \end{bmatrix} \xrightarrow[\substack{-\frac{1}{2}R_4 \to R_2 \\ R_2 \to R_3 \\ R_3 \to R_4}]{}$$

$$\begin{bmatrix} 1 & 1 & 1 & 1 & 12 \\ 0 & 1 & 0 & 1 & 13 \\ 0 & -8 & -12 & -14 & -152 \\ 0 & -54 & -72 & -78 & -966 \end{bmatrix} \xrightarrow[\substack{R_3 + 8R_2 \to R_3 \\ R_4 + 54R_2 \to R_4}]{} \begin{bmatrix} 1 & 1 & 1 & 1 & 12 \\ 0 & 1 & 0 & 1 & 13 \\ 0 & 0 & -12 & -6 & -48 \\ 0 & 0 & -72 & -24 & -264 \end{bmatrix} \xrightarrow[R_4 - 6R_3 \to R_4]{} \begin{bmatrix} 1 & 1 & 1 & 1 & 12 \\ 0 & 1 & 0 & 1 & 13 \\ 0 & 0 & -12 & -6 & -48 \\ 0 & 0 & 0 & 12 & 24 \end{bmatrix}.$$

So $d = 2$. Then $-12c - 6(2) = -48 \Leftrightarrow c = 3$ and $b + 2 = 13 \Leftrightarrow b = 11$. Finally, $a + 11 + 3 + 2 = 12 \Leftrightarrow a = -4$. So the fourth-degree polynomial containing these points is $y = -4x^4 + 11x^3 + 3x^2 + 2x$.

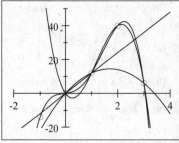

6.2   THE ALGEBRA OF MATRICES

1. We can add (or subtract) two matrices only if they have the same *dimension*.

3. (i) $A + A$ and (ii) $2A$ exist for all matrices A, but (iii) $A \cdot A$ is not defined when A is not square.

5. The matrices have different dimensions, so they cannot be equal.

7. All corresponding entries must be equal, so $a = -5$ and $b = 3$.

9. $\begin{bmatrix} 2 & 6 \\ -5 & 3 \end{bmatrix} + \begin{bmatrix} -1 & -3 \\ 6 & 2 \end{bmatrix} = \begin{bmatrix} 1 & 3 \\ 1 & 5 \end{bmatrix}$

11. $3 \begin{bmatrix} 1 & 2 \\ 4 & -1 \\ 1 & 0 \end{bmatrix} = \begin{bmatrix} 3 & 6 \\ 12 & -3 \\ 3 & 0 \end{bmatrix}$

13. $\begin{bmatrix} 2 & 6 \\ 1 & 3 \\ 2 & 4 \end{bmatrix} \begin{bmatrix} 1 & -2 \\ 3 & 6 \\ -2 & 0 \end{bmatrix}$ is undefined because these matrices have incompatible dimensions.

15. $\begin{bmatrix} 1 & 2 \\ -1 & 4 \end{bmatrix} \begin{bmatrix} 1 & -2 & 3 \\ 2 & 2 & -1 \end{bmatrix} = \begin{bmatrix} 5 & 2 & 1 \\ 7 & 10 & -7 \end{bmatrix}$

17. $2X + A = B \Leftrightarrow X = \frac{1}{2}(B - A) = \frac{1}{2}\left(\begin{bmatrix} 2 & 5 \\ 3 & 7 \end{bmatrix} - \begin{bmatrix} 4 & 6 \\ 1 & 3 \end{bmatrix} \right) = \frac{1}{2} \begin{bmatrix} -2 & -1 \\ 2 & 4 \end{bmatrix} = \begin{bmatrix} -1 & -\frac{1}{2} \\ 1 & 2 \end{bmatrix}.$

19. $2(B - X) = D$. Since B is a 2×2 matrix, $B - X$ is defined only when X is a 2×2 matrix, so $2(B - X)$ is a 2×2 matrix. But D is a 3×2 matrix. Thus, there is no solution.

21. $\frac{1}{5}(X + D) = C \Leftrightarrow X + D = 5C \Leftrightarrow$

$$X = 5C - D = 5 \begin{bmatrix} 2 & 3 \\ 1 & 0 \\ 0 & 2 \end{bmatrix} - \begin{bmatrix} 10 & 20 \\ 30 & 20 \\ 10 & 0 \end{bmatrix} = \begin{bmatrix} 10 & 15 \\ 5 & 0 \\ 0 & 10 \end{bmatrix} - \begin{bmatrix} 10 & 20 \\ 30 & 20 \\ 10 & 0 \end{bmatrix} = \begin{bmatrix} 0 & -5 \\ -25 & -20 \\ -10 & 10 \end{bmatrix}.$$

In Solutions 23–35, the matrices $A, B, C, D, E, F, G,$ **and** H **are defined as follows:**

$$A = \begin{bmatrix} 2 & -5 \\ 0 & 7 \end{bmatrix} \qquad B = \begin{bmatrix} 3 & \frac{1}{2} & 5 \\ 1 & -1 & 3 \end{bmatrix} \qquad C = \begin{bmatrix} 2 & -\frac{5}{2} & 0 \\ 0 & 2 & -3 \end{bmatrix} \qquad D = \begin{bmatrix} 7 & 3 \end{bmatrix}$$

$$E = \begin{bmatrix} 1 \\ 2 \\ 0 \end{bmatrix} \qquad F = \begin{bmatrix} 1 & 0 & 0 \\ 0 & 1 & 0 \\ 0 & 0 & 1 \end{bmatrix} \qquad G = \begin{bmatrix} 5 & -3 & 10 \\ 6 & 1 & 0 \\ -5 & 2 & 2 \end{bmatrix} \qquad H = \begin{bmatrix} 3 & 1 \\ 2 & -1 \end{bmatrix}$$

23. (a) $B + C = \begin{bmatrix} 3 & \frac{1}{2} & 5 \\ 1 & -1 & 3 \end{bmatrix} + \begin{bmatrix} 2 & -\frac{5}{2} & 0 \\ 0 & 2 & -3 \end{bmatrix} = \begin{bmatrix} 5 & -2 & 5 \\ 1 & 1 & 0 \end{bmatrix}$

(b) $B + F$ is undefined because B (2×3) and F (3×3) don't have the same dimensions.

25. (a) $5A = 5 \begin{bmatrix} 2 & -5 \\ 0 & 7 \end{bmatrix} = \begin{bmatrix} 10 & -25 \\ 0 & 35 \end{bmatrix}$

(b) $C - 5A$ is undefined because C (2×3) and A (2×2) don't have the same dimensions.

27. (a) AD is undefined because A (2×2) and D (1×2) have incompatible dimensions.

(b) $DA = \begin{bmatrix} 7 & 3 \end{bmatrix} \begin{bmatrix} 2 & -5 \\ 0 & 7 \end{bmatrix} = \begin{bmatrix} 14 & -14 \end{bmatrix}$

29. (a) $AH = \begin{bmatrix} 2 & -5 \\ 0 & 7 \end{bmatrix} \begin{bmatrix} 3 & 1 \\ 2 & -1 \end{bmatrix} = \begin{bmatrix} -4 & 7 \\ 14 & -7 \end{bmatrix}$

(b) $\begin{bmatrix} 3 & 1 \\ 2 & -1 \end{bmatrix} \begin{bmatrix} 2 & -5 \\ 0 & 7 \end{bmatrix} = \begin{bmatrix} 6 & -8 \\ 4 & -17 \end{bmatrix}$

31. (a) $GF = \begin{bmatrix} 5 & -3 & 10 \\ 6 & 1 & 0 \\ -5 & 2 & 2 \end{bmatrix} \begin{bmatrix} 1 & 0 & 0 \\ 0 & 1 & 0 \\ 0 & 0 & 1 \end{bmatrix} = \begin{bmatrix} 5 & -3 & 10 \\ 6 & 1 & 0 \\ -5 & 2 & 2 \end{bmatrix}$

(b) $GE = \begin{bmatrix} 5 & -3 & 10 \\ 6 & 1 & 0 \\ -5 & 2 & 2 \end{bmatrix} \begin{bmatrix} 1 \\ 2 \\ 0 \end{bmatrix} = \begin{bmatrix} -1 \\ 8 \\ -1 \end{bmatrix}$

33. (a) $A^2 = \begin{bmatrix} 2 & -5 \\ 0 & 7 \end{bmatrix} \begin{bmatrix} 2 & -5 \\ 0 & 7 \end{bmatrix} = \begin{bmatrix} 4 & -45 \\ 0 & 49 \end{bmatrix}$

(b) $A^3 = \begin{bmatrix} 2 & -5 \\ 0 & 7 \end{bmatrix} \begin{bmatrix} 2 & -5 \\ 0 & 7 \end{bmatrix} \begin{bmatrix} 2 & -5 \\ 0 & 7 \end{bmatrix} = \begin{bmatrix} 4 & -45 \\ 0 & 49 \end{bmatrix} \begin{bmatrix} 2 & -5 \\ 0 & 7 \end{bmatrix} = \begin{bmatrix} 8 & -335 \\ 0 & 343 \end{bmatrix}$

35. (a) $ABE = \begin{bmatrix} 2 & -5 \\ 0 & 7 \end{bmatrix} \begin{bmatrix} 3 & \frac{1}{2} & 5 \\ 1 & -1 & 3 \end{bmatrix} \begin{bmatrix} 1 \\ 2 \\ 0 \end{bmatrix} = \begin{bmatrix} 1 & 6 & -5 \\ 7 & -7 & 21 \end{bmatrix} \begin{bmatrix} 1 \\ 2 \\ 0 \end{bmatrix} = \begin{bmatrix} 13 \\ -7 \end{bmatrix}$

(b) AHE is undefined because the dimensions of AH (2×2) and E (3×1) are incompatible.

In Solutions 37–41, the matrices A, B, and C are defined as follows:

$$A = \begin{bmatrix} 0.3 & 1.1 & 2.4 \\ 0.9 & -0.1 & 0.4 \\ -0.7 & 0.3 & -0.5 \end{bmatrix} \qquad B = \begin{bmatrix} 1.2 & -0.1 \\ 0 & -0.5 \\ 0.5 & -2.1 \end{bmatrix} \qquad C = \begin{bmatrix} -0.2 & 0.2 & 0.1 \\ 1.1 & 2.1 & -2.1 \end{bmatrix}$$

37. $AB = \begin{bmatrix} 1.56 & -5.62 \\ 1.28 & -0.88 \\ -1.09 & 0.97 \end{bmatrix}$

39. $BC = \begin{bmatrix} -0.35 & 0.03 & 0.33 \\ -0.55 & -1.05 & 1.05 \\ -2.41 & -4.31 & 4.46 \end{bmatrix}$

41. B and C have different dimensions, so $B + C$ is undefined.

43. $\begin{bmatrix} x & 2y \\ 4 & 6 \end{bmatrix} = \begin{bmatrix} 2 & -2 \\ 2x & -6y \end{bmatrix}$. Thus we must solve the system $\begin{cases} x = 2 \\ 2y = -2 \\ 4 = 2x \\ 6 = -6y \end{cases}$ So $x = 2$ and $2y = -2 \Leftrightarrow y = -1$. Since

these values for x and y also satisfy the last two equations, the solution is $x = 2$, $y = -1$.

45. $2\begin{bmatrix} x & y \\ x+y & x-y \end{bmatrix} = \begin{bmatrix} 2 & -4 \\ -2 & 6 \end{bmatrix}$. Since $2\begin{bmatrix} x & y \\ x+y & x-y \end{bmatrix} = \begin{bmatrix} 2x & 2y \\ 2(x+y) & 2(x-y) \end{bmatrix}$, Thus we must solve the

system $\begin{cases} 2x = 2 \\ 2y = -4 \\ 2(x+y) = -2 \\ 2(x-y) = 6 \end{cases}$ So $x = 1$ and $y = -2$. Since these values for x and y also satisfy the last two equations, the

solution is $x = 1$, $y = -2$.

47. $\begin{cases} 2x - 5y = 7 \\ 3x + 2y = 4 \end{cases}$ written as a matrix equation is $\begin{bmatrix} 2 & -5 \\ 3 & 2 \end{bmatrix}\begin{bmatrix} x \\ y \end{bmatrix} = \begin{bmatrix} 7 \\ 4 \end{bmatrix}$.

49. $\begin{cases} 3x_1 + 2x_2 - x_3 + x_4 = 0 \\ x_1 \quad\quad - x_3 \quad\quad = 5 \\ 3x_2 + x_3 - x_4 = 4 \end{cases}$ written as a matrix equation is $\begin{bmatrix} 3 & 2 & -1 & 1 \\ 1 & 0 & -1 & 0 \\ 0 & 3 & 1 & -1 \end{bmatrix}\begin{bmatrix} x_1 \\ x_2 \\ x_3 \\ x_4 \end{bmatrix} = \begin{bmatrix} 0 \\ 5 \\ 4 \end{bmatrix}$.

51. $A = \begin{bmatrix} 1 & 0 & 6 & -1 \\ 2 & \frac{1}{2} & 4 & 0 \end{bmatrix}$, $B = \begin{bmatrix} 1 & 7 & -9 & 2 \end{bmatrix}$, and $C = \begin{bmatrix} 1 \\ 0 \\ -1 \\ -2 \end{bmatrix}$. ABC is undefined because the dimensions of A (2×4)

and B (1×4) are not compatible. $ACB = \begin{bmatrix} -3 \\ -2 \end{bmatrix}\begin{bmatrix} 1 & 7 & -9 & 2 \end{bmatrix} = \begin{bmatrix} -3 & -21 & 27 & -6 \\ -2 & -14 & 18 & -4 \end{bmatrix}$. BAC is undefined because

the dimensions of B (1×4) and A (2×4) are not compatible. BCA is undefined because the dimensions of C (4×1) and

A (2×4) are not compatible. CAB is undefined because the dimensions of C (4×1) and A (2×4) are not compatible.

CBA is undefined because the dimensions of B (1×4) and A (2×4) are not compatible.

53. (a) $BA = \begin{bmatrix} \$0.90 & \$0.80 & \$1.10 \end{bmatrix} \begin{bmatrix} 4000 & 1000 & 3500 \\ 400 & 300 & 200 \\ 700 & 500 & 9000 \end{bmatrix} = \begin{bmatrix} \$4690 & \$1690 & \$13,210 \end{bmatrix}$

(b) The entries in the product matrix represent the total food sales in Santa Monica, Long Beach, and Anaheim, respectively.

55. (a) $AB = \begin{bmatrix} 6 & 10 & 14 & 28 \end{bmatrix} \begin{bmatrix} 2000 & 2500 \\ 3000 & 1500 \\ 2500 & 1000 \\ 1000 & 500 \end{bmatrix} = \begin{bmatrix} 105,000 & 58,000 \end{bmatrix}$

(b) That day they canned 105,000 ounces of tomato sauce and 58,000 ounces of tomato paste.

57.

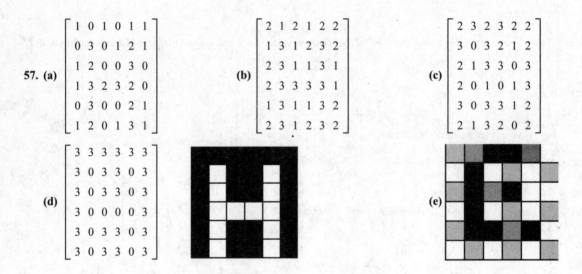

59. $A = \begin{bmatrix} 1 & 1 \\ 0 & 1 \end{bmatrix}$; $A^2 = \begin{bmatrix} 1 & 1 \\ 0 & 1 \end{bmatrix}\begin{bmatrix} 1 & 1 \\ 0 & 1 \end{bmatrix} = \begin{bmatrix} 1 & 2 \\ 0 & 1 \end{bmatrix}$; $A^3 = A \cdot A^2 = \begin{bmatrix} 1 & 1 \\ 0 & 1 \end{bmatrix}\begin{bmatrix} 1 & 2 \\ 0 & 1 \end{bmatrix} = \begin{bmatrix} 1 & 3 \\ 0 & 1 \end{bmatrix}$;

$A^4 = A \cdot A^3 = \begin{bmatrix} 1 & 1 \\ 0 & 1 \end{bmatrix}\begin{bmatrix} 1 & 3 \\ 0 & 1 \end{bmatrix} = \begin{bmatrix} 1 & 4 \\ 0 & 1 \end{bmatrix}$. Therefore, it seems that $A^n = \begin{bmatrix} 1 & n \\ 0 & 1 \end{bmatrix}$.

61. Let $A = \begin{bmatrix} a & b \\ c & d \end{bmatrix}$. For the first matrix, we have $A^2 = \begin{bmatrix} a & b \\ c & d \end{bmatrix}$.

$$\begin{bmatrix} a & b \\ c & d \end{bmatrix} = \begin{bmatrix} a^2 + bc & ab + bd \\ ac + cd & bc + d^2 \end{bmatrix} = \begin{bmatrix} a^2 + bc & b(a+d) \\ c(a+d) & bc + d^2 \end{bmatrix}. \text{ So } A^2 = \begin{bmatrix} 4 & 0 \\ 0 & 9 \end{bmatrix} \Leftrightarrow \begin{cases} a^2 + bc = 4 \\ b(a+d) = 0 \\ c(a+d) = 0 \\ bc + d^2 = 9 \end{cases}$$

If $a + d = 0$, then $a = -d$, so $4 = a^2 + bc = (-d)^2 + bc = d^2 + bc = 9$, which is a contradiction. Thus $a + d \neq 0$. Since $b(a+d) = 0$ and $c(a+d) = 0$, we must have $b = 0$ and $c = 0$. So the first equation becomes $a^2 = 4 \Rightarrow a = \pm 2$, and the fourth equation becomes $d^2 = 9 \Rightarrow d = \pm 3$.

Thus the square roots of $\begin{bmatrix} 4 & 0 \\ 0 & 9 \end{bmatrix}$ are $A_1 = \begin{bmatrix} 2 & 0 \\ 0 & 3 \end{bmatrix}$, $A_2 = \begin{bmatrix} 2 & 0 \\ 0 & -3 \end{bmatrix}$, $A_3 = \begin{bmatrix} -2 & 0 \\ 0 & 3 \end{bmatrix}$, and $A_4 = \begin{bmatrix} -2 & 0 \\ 0 & -3 \end{bmatrix}$.

For the second matrix, we have $A^2 = \begin{bmatrix} 1 & 5 \\ 0 & 9 \end{bmatrix} \Leftrightarrow \begin{cases} a^2 + bc = 1 \\ b(a+d) = 5 \\ c(a+d) = 0 \\ bc + d^2 = 9 \end{cases}$ Since $a + d \neq 0$ and $c(a+d) = 0$, we must have

$c = 0$. The equations then simplify into the system $\begin{cases} a^2 = 1 \\ b(a+d) = 5 \\ d^2 = 9 \end{cases} \Rightarrow \begin{cases} a = \pm 1 \\ b(a+d) = 5 \\ d = \pm 3 \end{cases}$

We consider the four possible values of a and d. If $a = 1$ and $d = 3$, then $b(a+d) = 5 \Rightarrow b(4) = 5 \Rightarrow b = \frac{5}{4}$. If $a = 1$ and $d = -3$, then $b(a+d) = 5 \Rightarrow b(-2) = 5 \Rightarrow b = -\frac{5}{2}$. If $a = -1$ and $d = 3$, then $b(a+d) = 5 \Rightarrow b(2) = 5 \Rightarrow b = \frac{5}{2}$. If $a = -1$ and $d = -3$, then $b(a+d) = 5 \Rightarrow b(-4) = 5 \Rightarrow b = -\frac{5}{4}$. Thus, the square roots of $\begin{bmatrix} 1 & 5 \\ 0 & 9 \end{bmatrix}$ are

$A_1 = \begin{bmatrix} 1 & \frac{5}{4} \\ 0 & 3 \end{bmatrix}$, $A_2 = \begin{bmatrix} 1 & -\frac{5}{2} \\ 0 & -3 \end{bmatrix}$, $A_3 = \begin{bmatrix} -1 & \frac{5}{2} \\ 0 & 3 \end{bmatrix}$, and $A_4 = \begin{bmatrix} -1 & -\frac{5}{4} \\ 0 & -3 \end{bmatrix}$.

6.3 INVERSES OF MATRICES AND MATRIX EQUATIONS

1. (a) The matrix $I = \begin{bmatrix} 1 & 0 \\ 0 & 1 \end{bmatrix}$ is called an *identity* matrix.

(b) If A is a 2×2 matrix then $A \times I = A$ and $I \times A = A$.

(c) If A and B are 2×2 matrices with $AB = I$ then B is the *inverse* of A.

3. $A = \begin{bmatrix} 4 & 1 \\ 7 & 2 \end{bmatrix}$; $B = \begin{bmatrix} 2 & -1 \\ -7 & 4 \end{bmatrix}$.

$AB = \begin{bmatrix} 4 & 1 \\ 7 & 2 \end{bmatrix}\begin{bmatrix} 2 & -1 \\ -7 & 4 \end{bmatrix} = \begin{bmatrix} 1 & 0 \\ 0 & 1 \end{bmatrix}$ and $BA = \begin{bmatrix} 2 & -1 \\ -7 & 4 \end{bmatrix}\begin{bmatrix} 4 & 1 \\ 7 & 2 \end{bmatrix} = \begin{bmatrix} 1 & 0 \\ 0 & 1 \end{bmatrix}$.

5. $A = \begin{bmatrix} 1 & 3 & -1 \\ 1 & 4 & 0 \\ -1 & -3 & 2 \end{bmatrix}$; $B = \begin{bmatrix} 8 & -3 & 4 \\ -2 & 1 & -1 \\ 1 & 0 & 1 \end{bmatrix}$. $AB = \begin{bmatrix} 1 & 3 & -1 \\ 1 & 4 & 0 \\ -1 & -3 & 2 \end{bmatrix} \begin{bmatrix} 8 & -3 & 4 \\ -2 & 1 & -1 \\ 1 & 0 & 1 \end{bmatrix} = \begin{bmatrix} 1 & 0 & 0 \\ 0 & 1 & 0 \\ 0 & 0 & 1 \end{bmatrix}$ and

$BA = \begin{bmatrix} 8 & -3 & 4 \\ -2 & 1 & -1 \\ 1 & 0 & 1 \end{bmatrix} \begin{bmatrix} 1 & 3 & -1 \\ 1 & 4 & 0 \\ -1 & -3 & 2 \end{bmatrix} = \begin{bmatrix} 1 & 0 & 0 \\ 0 & 1 & 0 \\ 0 & 0 & 1 \end{bmatrix}$

7. $A = \begin{bmatrix} 7 & 4 \\ 3 & 2 \end{bmatrix} \Leftrightarrow A^{-1} = \frac{1}{14-12} \begin{bmatrix} 2 & -4 \\ -3 & 7 \end{bmatrix} = \begin{bmatrix} 1 & -2 \\ -\frac{3}{2} & \frac{7}{2} \end{bmatrix}$. Then, $AA^{-1} = \begin{bmatrix} 7 & 4 \\ 3 & 2 \end{bmatrix} \begin{bmatrix} 1 & -2 \\ -\frac{3}{2} & \frac{7}{2} \end{bmatrix} = \begin{bmatrix} 1 & 0 \\ 0 & 1 \end{bmatrix}$

and $A^{-1}A = \begin{bmatrix} 1 & -2 \\ -\frac{3}{2} & \frac{7}{2} \end{bmatrix} \begin{bmatrix} 7 & 4 \\ 3 & 2 \end{bmatrix} = \begin{bmatrix} 1 & 0 \\ 0 & 1 \end{bmatrix}$.

9. Using a calculator, we find $A^{-1} = \begin{bmatrix} \frac{1}{3} & -\frac{1}{2} \\ 2 & 2 \end{bmatrix}$ and verify that $A^{-1}A = AA^{-1} = I_2$.

11. $\begin{bmatrix} -3 & -5 \\ 2 & 3 \end{bmatrix}^{-1} = \frac{1}{-9+10} \begin{bmatrix} 3 & 5 \\ -2 & -3 \end{bmatrix} = \begin{bmatrix} 3 & 5 \\ -2 & -3 \end{bmatrix}$

13. $\begin{bmatrix} 2 & 5 \\ -5 & -13 \end{bmatrix}^{-1} = \frac{1}{-26+25} \begin{bmatrix} -13 & -5 \\ 5 & 2 \end{bmatrix} = \begin{bmatrix} 13 & 5 \\ -5 & -2 \end{bmatrix}$

15. $\begin{bmatrix} 6 & -3 \\ -8 & 4 \end{bmatrix}^{-1} = \frac{1}{24-24} \begin{bmatrix} 4 & 3 \\ 8 & 6 \end{bmatrix}$, which is not defined, and so there is no inverse.

17. $\begin{bmatrix} 0.4 & -1.2 \\ 0.3 & 0.6 \end{bmatrix}^{-1} = \frac{1}{0.24+0.36} \begin{bmatrix} 0.6 & 1.2 \\ -0.3 & 0.4 \end{bmatrix} = \begin{bmatrix} 1 & 2 \\ -\frac{1}{2} & \frac{2}{3} \end{bmatrix}$

19. $\begin{bmatrix} 2 & 4 & 1 & 1 & 0 & 0 \\ -1 & 1 & -1 & 0 & 1 & 0 \\ 1 & 4 & 0 & 0 & 0 & 1 \end{bmatrix} \xrightarrow[2R_3 - R_1 \to R_3]{2R_2 + R_1 \to R_2} \begin{bmatrix} 2 & 4 & 1 & 1 & 0 & 0 \\ 0 & 6 & -1 & 1 & 2 & 0 \\ 0 & 4 & -1 & -1 & 0 & 2 \end{bmatrix} \xrightarrow[3R_1 - 2R_2 \to R_1]{3R_3 - 2R_2 \to R_3} \begin{bmatrix} 6 & 0 & 5 & 1 & -4 & 0 \\ 0 & 6 & -1 & 1 & 2 & 0 \\ 0 & 0 & -1 & -5 & -4 & 6 \end{bmatrix}$

$\xrightarrow[R_2 - R_3 \to R_2]{R_1 + 5R_3 \to R_1} \begin{bmatrix} 6 & 0 & 0 & -24 & -24 & 30 \\ 0 & 6 & 0 & 6 & 6 & -6 \\ 0 & 0 & -1 & -5 & -4 & 6 \end{bmatrix} \xrightarrow[\frac{1}{6}R_2, -R_3]{\frac{1}{6}R_1} \begin{bmatrix} 1 & 0 & 0 & -4 & -4 & 5 \\ 0 & 1 & 0 & 1 & 1 & -1 \\ 0 & 0 & 1 & 5 & 4 & -6 \end{bmatrix}$.

Therefore, the inverse matrix is $\begin{bmatrix} -4 & -4 & 5 \\ 1 & 1 & -1 \\ 5 & 4 & -6 \end{bmatrix}$.

21. $\begin{bmatrix} 1 & 2 & 3 & 1 & 0 & 0 \\ 4 & 5 & -1 & 0 & 1 & 0 \\ 1 & -1 & -10 & 0 & 0 & 1 \end{bmatrix} \xrightarrow[R_3 - R_1 \to R_3]{R_2 - 4R_1 \to R_2} \begin{bmatrix} 1 & 2 & 3 & 1 & 0 & 0 \\ 0 & -3 & -13 & -4 & 1 & 0 \\ 0 & -3 & -13 & -1 & 0 & 1 \end{bmatrix} \xrightarrow{R_3 - R_2 \to R_3} \begin{bmatrix} 1 & 2 & 3 & 1 & 0 & 0 \\ 0 & -3 & -13 & -4 & 1 & 0 \\ 0 & 0 & 0 & 3 & -1 & 1 \end{bmatrix}$.

Since the left half of the last row consists entirely of zeros, there is no inverse matrix.

23.
$$\begin{bmatrix} 0 & -2 & 2 & 1 & 0 & 0 \\ 3 & 1 & 3 & 0 & 1 & 0 \\ 1 & -2 & 3 & 0 & 0 & 1 \end{bmatrix} \xrightarrow{R_1 \leftrightarrow R_3} \begin{bmatrix} 1 & -2 & 3 & 0 & 0 & 1 \\ 3 & 1 & 3 & 0 & 1 & 0 \\ 0 & -2 & 2 & 1 & 0 & 0 \end{bmatrix} \xrightarrow{R_2 - 3R_1 \to R_2} \begin{bmatrix} 1 & -2 & 3 & 0 & 0 & 1 \\ 0 & 7 & -6 & 0 & 1 & -3 \\ 0 & -2 & 2 & 1 & 0 & 0 \end{bmatrix}$$

$$\xrightarrow[R_2 + 3R_3 \to R_2]{R_1 - R_3 \to R_1} \begin{bmatrix} 1 & 0 & 1 & -1 & 0 & 1 \\ 0 & 1 & 0 & 3 & 1 & -3 \\ 0 & -2 & 2 & 1 & 0 & 0 \end{bmatrix} \xrightarrow{R_3 + 2R_2 \to R_3} \begin{bmatrix} 1 & 0 & 1 & -1 & 0 & 1 \\ 0 & 1 & 0 & 3 & 1 & -3 \\ 0 & 0 & 2 & 7 & 2 & -6 \end{bmatrix} \xrightarrow{\frac{1}{2}R_3}$$

$$\begin{bmatrix} 1 & 0 & 1 & -1 & 0 & 1 \\ 0 & 1 & 0 & 3 & 1 & -3 \\ 0 & 0 & 1 & \frac{7}{2} & 1 & -3 \end{bmatrix} \xrightarrow{R_1 - R_3 \to R_1} \begin{bmatrix} 1 & 0 & 0 & -\frac{9}{2} & -1 & 4 \\ 0 & 1 & 0 & 3 & 1 & -3 \\ 0 & 0 & 1 & \frac{7}{2} & 1 & -3 \end{bmatrix}.$$ Therefore, the inverse matrix is

$$\begin{bmatrix} -\frac{9}{2} & -1 & 4 \\ 3 & 1 & -3 \\ \frac{7}{2} & 1 & -3 \end{bmatrix}.$$

25.
$$\begin{bmatrix} 1 & 2 & 0 & 3 & 1 & 0 & 0 & 0 \\ 0 & 1 & 1 & 1 & 0 & 1 & 0 & 0 \\ 0 & 1 & 0 & 1 & 0 & 0 & 1 & 0 \\ 1 & 2 & 0 & 2 & 0 & 0 & 0 & 1 \end{bmatrix} \xrightarrow[R_4 - R_1 \to R_4]{R_3 - R_2 \to R_3} \begin{bmatrix} 1 & 2 & 0 & 3 & 1 & 0 & 0 & 0 \\ 0 & 1 & 1 & 1 & 0 & 1 & 0 & 0 \\ 0 & 0 & -1 & 0 & 0 & -1 & 1 & 0 \\ 0 & 0 & 0 & -1 & -1 & 0 & 0 & 1 \end{bmatrix} \xrightarrow[-R_4]{-R_3}$$

$$\begin{bmatrix} 1 & 2 & 0 & 3 & 1 & 0 & 0 & 0 \\ 0 & 1 & 1 & 1 & 0 & 1 & 0 & 0 \\ 0 & 0 & 1 & 0 & 0 & 1 & -1 & 0 \\ 0 & 0 & 0 & 1 & 1 & 0 & 0 & -1 \end{bmatrix} \xrightarrow[R_2 - R_3 \to R_2]{R_1 - 2R_2 \to R_1} \begin{bmatrix} 1 & 0 & -2 & 1 & 1 & -2 & 0 & 0 \\ 0 & 1 & 0 & 1 & 0 & 0 & 1 & 0 \\ 0 & 0 & 1 & 0 & 0 & 1 & -1 & 0 \\ 0 & 0 & 0 & 1 & 1 & 0 & 0 & -1 \end{bmatrix} \xrightarrow[R_2 - R_4 \to R_2]{R_1 + 2R_3 \to R_1}$$

$$\begin{bmatrix} 1 & 0 & 0 & 1 & 1 & 0 & -2 & 0 \\ 0 & 1 & 0 & 0 & -1 & 0 & 1 & 1 \\ 0 & 0 & 1 & 0 & 0 & 1 & -1 & 0 \\ 0 & 0 & 0 & 1 & 1 & 0 & 0 & -1 \end{bmatrix} \xrightarrow{R_1 \to R_1 - R_4} \begin{bmatrix} 1 & 0 & 0 & 0 & 0 & 0 & -2 & 1 \\ 0 & 1 & 0 & 0 & -1 & 0 & 1 & 1 \\ 0 & 0 & 1 & 0 & 0 & 1 & -1 & 0 \\ 0 & 0 & 0 & 1 & 1 & 0 & 0 & -1 \end{bmatrix}.$$ Therefore, the inverse matrix is

$$\begin{bmatrix} 0 & 0 & -2 & 1 \\ -1 & 0 & 1 & 1 \\ 0 & 1 & -1 & 0 \\ 1 & 0 & 0 & -1 \end{bmatrix}.$$

27.
$$\begin{bmatrix} -3 & 2 & 3 \\ 0 & -1 & 3 \\ 1 & 0 & -2 \end{bmatrix}^{-1} = \begin{bmatrix} \frac{2}{3} & \frac{4}{3} & 3 \\ 1 & 1 & 3 \\ \frac{1}{3} & \frac{2}{3} & 1 \end{bmatrix}$$

29.
$$\begin{bmatrix} -1 & -4 & 0 & 1 \\ 1 & 0 & -1 & 0 \\ 0 & 4 & 1 & -2 \\ 2 & 2 & -2 & 0 \end{bmatrix}^{-1} = \begin{bmatrix} -2 & 3 & -1 & -2 \\ 0 & -1 & 0 & \frac{1}{2} \\ -2 & 2 & -1 & -2 \\ -1 & -1 & -1 & 0 \end{bmatrix}$$

31.
$$\begin{bmatrix} 1 & 7 & 3 \\ 0 & 2 & 1 \\ 0 & 0 & 3 \end{bmatrix}^{-1} = \begin{bmatrix} 1 & -\frac{7}{2} & \frac{1}{6} \\ 0 & \frac{1}{2} & -\frac{1}{6} \\ 0 & 0 & \frac{1}{3} \end{bmatrix}$$

33.
$$\begin{bmatrix} 1 & 0 & 0 & 0 \\ 0 & 2 & 0 & 0 \\ 0 & 0 & 4 & 0 \\ 0 & 0 & 0 & 7 \end{bmatrix}^{-1} = \begin{bmatrix} 1 & 0 & 0 & 0 \\ 0 & \frac{1}{2} & 0 & 0 \\ 0 & 0 & \frac{1}{4} & 0 \\ 0 & 0 & 0 & \frac{1}{7} \end{bmatrix}$$

In Solutions 35 and 37, the matrices A and B are defined as follows:

$$A = \begin{bmatrix} -1 & 0 & 2 \\ 0 & -2 & -1 \\ 4 & 2 & 1 \end{bmatrix} \qquad B = \begin{bmatrix} 2 & -1 & -2 \\ 0 & 3 & 1 \\ -1 & 0 & 2 \end{bmatrix}$$

35. $A^{-1}B = \begin{bmatrix} -\frac{1}{4} & \frac{3}{4} & \frac{3}{4} \\ -\frac{7}{16} & -\frac{23}{16} & -\frac{3}{16} \\ \frac{7}{8} & -\frac{1}{8} & -\frac{5}{8} \end{bmatrix}$ \qquad **37.** $BAB^{-1} = \begin{bmatrix} -7 & -3 & -4 \\ \frac{22}{7} & -\frac{2}{7} & \frac{16}{7} \\ \frac{50}{7} & \frac{26}{7} & \frac{37}{7} \end{bmatrix}$

39. $\begin{cases} -3x - 5y = 4 \\ 2x + 3y = 0 \end{cases}$ is equivalent to the matrix equation $\begin{bmatrix} -3 & -5 \\ 2 & 3 \end{bmatrix} \begin{bmatrix} x \\ y \end{bmatrix} = \begin{bmatrix} 4 \\ 0 \end{bmatrix}$. Using the inverse from Exercise 11,

$\begin{bmatrix} x \\ y \end{bmatrix} = \begin{bmatrix} 3 & 5 \\ -2 & -3 \end{bmatrix} \begin{bmatrix} 4 \\ 0 \end{bmatrix} = \begin{bmatrix} 12 \\ -8 \end{bmatrix}$. Therefore, $x = 12$ and $y = -8$.

41. $\begin{cases} 2x + 5y = 2 \\ -5x - 13y = 20 \end{cases}$ is equivalent to the matrix equation $\begin{bmatrix} 2 & 5 \\ -5 & -13 \end{bmatrix} \begin{bmatrix} x \\ y \end{bmatrix} = \begin{bmatrix} 2 \\ 20 \end{bmatrix}$. Using the inverse from

Exercise 13, $\begin{bmatrix} x \\ y \end{bmatrix} = \begin{bmatrix} 13 & 5 \\ -5 & -2 \end{bmatrix} \begin{bmatrix} 2 \\ 20 \end{bmatrix} = \begin{bmatrix} 126 \\ -50 \end{bmatrix}$. Therefore, $x = 126$ and $y = -50$.

43. $\begin{cases} 2x + 4y + z = 7 \\ -x + y - z = 0 \\ x + 4y = -2 \end{cases}$ is equivalent to the matrix equation $\begin{bmatrix} 2 & 4 & 1 \\ -1 & 1 & -1 \\ 1 & 4 & 0 \end{bmatrix} \begin{bmatrix} x \\ y \\ z \end{bmatrix} = \begin{bmatrix} 7 \\ 0 \\ -2 \end{bmatrix}$. Using the inverse from

Exercise 19, $\begin{bmatrix} x \\ y \\ z \end{bmatrix} = \begin{bmatrix} -4 & -4 & 5 \\ 1 & 1 & -1 \\ 5 & 4 & -6 \end{bmatrix} \begin{bmatrix} 7 \\ 0 \\ -2 \end{bmatrix} = \begin{bmatrix} -38 \\ 9 \\ 47 \end{bmatrix}$. Therefore, $x = -38$, $y = 9$, and $z = 47$.

45. $\begin{cases} -2y + 2z = 12 \\ 3x + y + 3z = -2 \\ x - 2y + 3z = 8 \end{cases}$ is equivalent to the matrix equation $\begin{bmatrix} 0 & -2 & 2 \\ 3 & 1 & 3 \\ 1 & -2 & 3 \end{bmatrix} \begin{bmatrix} x \\ y \\ z \end{bmatrix} = \begin{bmatrix} 12 \\ -2 \\ 8 \end{bmatrix}$. Using the inverse from

Exercise 23, $\begin{bmatrix} x \\ y \\ z \end{bmatrix} = \begin{bmatrix} -\frac{9}{2} & -1 & 4 \\ 3 & 1 & -3 \\ \frac{7}{2} & 1 & -3 \end{bmatrix} \begin{bmatrix} 12 \\ -2 \\ 8 \end{bmatrix} = \begin{bmatrix} -20 \\ 10 \\ 16 \end{bmatrix}$. Therefore, $x = -20$, $y = 10$, and $z = 16$.

47. Using a calculator, we get the result $(3, 2, 1)$. \qquad **49.** Using a calculator, we get the result $(3, -2, 2)$.

51. Using a calculator, we get the result $(8, 1, 0, 3)$.

53. This has the form $MX = C$, so $M^{-1}(MX) = M^{-1}C$ and $M^{-1}(MX) = (M^{-1}M)X = X$.

Now $M^{-1} = \begin{bmatrix} 3 & -2 \\ -4 & 3 \end{bmatrix}^{-1} = \frac{1}{9-8} \begin{bmatrix} 3 & 2 \\ 4 & 3 \end{bmatrix} = \begin{bmatrix} 3 & 2 \\ 4 & 3 \end{bmatrix}$. Since $X = M^{-1}C$, we get

$\begin{bmatrix} x & y & z \\ u & v & w \end{bmatrix} = \begin{bmatrix} 3 & 2 \\ 4 & 3 \end{bmatrix} \begin{bmatrix} 1 & 0 & -1 \\ 2 & 1 & 3 \end{bmatrix} = \begin{bmatrix} 7 & 2 & 3 \\ 10 & 3 & 5 \end{bmatrix}$.

55. $\begin{bmatrix} a & -a \\ a & a \end{bmatrix}^{-1} = \dfrac{1}{a^2 - (-a^2)} \begin{bmatrix} a & a \\ -a & a \end{bmatrix} = \dfrac{1}{2a^2} \begin{bmatrix} a & a \\ -a & a \end{bmatrix} = \dfrac{1}{2a} \begin{bmatrix} 1 & 1 \\ -1 & 1 \end{bmatrix}$

57. $\begin{bmatrix} 2 & x \\ x & x^2 \end{bmatrix}^{-1} = \dfrac{1}{2x^2 - x^2} \begin{bmatrix} x^2 & -x \\ -x & 2 \end{bmatrix} = \dfrac{1}{x^2} \begin{bmatrix} x^2 & -x \\ -x & 2 \end{bmatrix} = \begin{bmatrix} 1 & -1/x \\ -1/x & 2/x^2 \end{bmatrix}$.

The inverse does not exist when $x = 0$.

59. $\begin{bmatrix} 1 & e^x & 0 & 1 & 0 & 0 \\ e^x & -e^{2x} & 0 & 0 & 1 & 0 \\ 0 & 0 & 2 & 0 & 0 & 1 \end{bmatrix} \xrightarrow{R_2 - e^x R_1 \to R_2} \begin{bmatrix} 1 & e^x & 0 & 1 & 0 & 0 \\ 0 & -2e^{2x} & 0 & -e^x & 1 & 0 \\ 0 & 0 & 2 & 0 & 0 & 1 \end{bmatrix} \xrightarrow[\frac{1}{2}R_3]{-\frac{1}{2}e^{-2x}R_2}$

$\begin{bmatrix} 1 & e^x & 0 & 1 & 0 & 0 \\ 0 & 1 & 0 & \frac{1}{2}e^{-x} & -\frac{1}{2}e^{-2x} & 0 \\ 0 & 0 & 1 & 0 & 0 & \frac{1}{2} \end{bmatrix} \xrightarrow{R_1 - e^x R_2 \to R_1} \begin{bmatrix} 1 & 0 & 0 & \frac{1}{2} & \frac{1}{2}e^{-x} & 0 \\ 0 & 1 & 0 & \frac{1}{2}e^{-x} & -\frac{1}{2}e^{-2x} & 0 \\ 0 & 0 & 1 & 0 & 0 & \frac{1}{2} \end{bmatrix}$.

Therefore, the inverse matrix is $\begin{bmatrix} \frac{1}{2} & \frac{1}{2}e^{-x} & 0 \\ \frac{1}{2}e^{-x} & -\frac{1}{2}e^{-2x} & 0 \\ 0 & 0 & \frac{1}{2} \end{bmatrix}$. The inverse exists for all x.

61. (a) $\begin{bmatrix} 3 & 1 & 3 & 1 & 0 & 0 \\ 4 & 2 & 4 & 0 & 1 & 0 \\ 3 & 2 & 4 & 0 & 0 & 1 \end{bmatrix} \xrightarrow[R_1 \leftrightarrow R_2]{R_3 - R_1 \to R_3} \begin{bmatrix} 4 & 2 & 4 & 0 & 1 & 0 \\ 3 & 1 & 3 & 1 & 0 & 0 \\ 0 & 1 & 1 & -1 & 0 & 1 \end{bmatrix} \xrightarrow{R_1 - R_2 \to R_1} \begin{bmatrix} 1 & 1 & 1 & -1 & 1 & 0 \\ 3 & 1 & 3 & 1 & 0 & 0 \\ 0 & 1 & 1 & -1 & 0 & 1 \end{bmatrix}$

$\xrightarrow{R_2 - 3R_1 \to R_2} \begin{bmatrix} 1 & 1 & 1 & -1 & 1 & 0 \\ 0 & -2 & 0 & 4 & -3 & 0 \\ 0 & 1 & 1 & -1 & 0 & 1 \end{bmatrix} \xrightarrow[\substack{-\frac{1}{2}R_2 \\ R_3 + \frac{1}{2}R_2 \to R_3}]{R_1 + \frac{1}{2}R_2 \to R_1} \begin{bmatrix} 1 & 0 & 1 & 1 & -\frac{1}{2} & 0 \\ 0 & 1 & 0 & -2 & \frac{3}{2} & 0 \\ 0 & 0 & 1 & 1 & -\frac{3}{2} & 1 \end{bmatrix} \xrightarrow{R_1 - R_3 \to R_1}$

$\begin{bmatrix} 1 & 0 & 0 & 0 & 1 & -1 \\ 0 & 1 & 0 & -2 & \frac{3}{2} & 0 \\ 0 & 0 & 1 & 1 & -\frac{3}{2} & 1 \end{bmatrix}$. Therefore, the inverse of the matrix is $\begin{bmatrix} 0 & 1 & -1 \\ -2 & \frac{3}{2} & 0 \\ 1 & -\frac{3}{2} & 1 \end{bmatrix}$.

(b) $\begin{bmatrix} A \\ B \\ C \end{bmatrix} = \begin{bmatrix} 0 & 1 & -1 \\ -2 & \frac{3}{2} & 0 \\ 1 & -\frac{3}{2} & 1 \end{bmatrix} \begin{bmatrix} 10 \\ 14 \\ 13 \end{bmatrix} = \begin{bmatrix} 1 \\ 1 \\ 2 \end{bmatrix}$.

Therefore, he should feed the rats 1 oz of food A, 1 oz of food B, and 2 oz of food C.

(c) $\begin{bmatrix} A \\ B \\ C \end{bmatrix} = \begin{bmatrix} 0 & 1 & -1 \\ -2 & \frac{3}{2} & 0 \\ 1 & -\frac{3}{2} & 1 \end{bmatrix} \begin{bmatrix} 9 \\ 12 \\ 10 \end{bmatrix} = \begin{bmatrix} 2 \\ 0 \\ 1 \end{bmatrix}$.

Therefore, he should feed the rats 2 oz of food A, no food B, and 1 oz of food C.

(d) $\begin{bmatrix} A \\ B \\ C \end{bmatrix} = \begin{bmatrix} 0 & 1 & -1 \\ -2 & \frac{3}{2} & 0 \\ 1 & -\frac{3}{2} & 1 \end{bmatrix} \begin{bmatrix} 2 \\ 4 \\ 11 \end{bmatrix} = \begin{bmatrix} -7 \\ 2 \\ 7 \end{bmatrix}$.

Since $A < 0$, there is no combination of foods giving the required supply.

63. (a) $\begin{cases} 9x + 11y + 8z = 740 \\ 13x + 15y + 16z = 1204 \\ 8x + 7y + 14z = 828 \end{cases}$

(b) $\begin{bmatrix} 9 & 11 & 8 \\ 13 & 15 & 16 \\ 8 & 7 & 14 \end{bmatrix} \begin{bmatrix} x \\ y \\ z \end{bmatrix} = \begin{bmatrix} 740 \\ 1204 \\ 828 \end{bmatrix}$

(c) $\begin{bmatrix} 9 & 11 & 8 & 1 & 0 & 0 \\ 13 & 15 & 16 & 0 & 1 & 0 \\ 8 & 7 & 14 & 0 & 0 & 1 \end{bmatrix}$ $\xrightarrow[9R_3 - 8R_1 \to R_3]{9R_2 - 13R_1 \to R_2}$ $\begin{bmatrix} 9 & 11 & 8 & 1 & 0 & 0 \\ 0 & -8 & 40 & -13 & 9 & 0 \\ 0 & -25 & 62 & -8 & 0 & 9 \end{bmatrix}$ $\xrightarrow{20R_3 - 31R_2 \to R_3}$

$\begin{bmatrix} 9 & 11 & 8 & 1 & 0 & 0 \\ 0 & -8 & 40 & -13 & 9 & 0 \\ 0 & -252 & 0 & 243 & -279 & 180 \end{bmatrix}$ $\xrightarrow{R_2 \leftrightarrow R_3}$ $\begin{bmatrix} 9 & 11 & 8 & 1 & 0 & 0 \\ 0 & -252 & 0 & 243 & -279 & 180 \\ 0 & -8 & 40 & -13 & 9 & 0 \end{bmatrix}$ $\xrightarrow[63R_3 - 2R_2 \to R_3]{-\frac{1}{252}R_2}$

$\begin{bmatrix} 9 & 11 & 8 & 1 & 0 & 0 \\ 0 & 1 & 0 & -\frac{27}{28} & \frac{31}{28} & -\frac{5}{7} \\ 0 & 0 & 2520 & -1305 & 1125 & -360 \end{bmatrix}$ $\xrightarrow{\frac{1}{2520}R_3}$ $\begin{bmatrix} 9 & 11 & 8 & 1 & 0 & 0 \\ 0 & 1 & 0 & -\frac{27}{28} & \frac{31}{28} & -\frac{5}{7} \\ 0 & 0 & 1 & -\frac{29}{56} & \frac{25}{56} & -\frac{1}{7} \end{bmatrix}$ $\xrightarrow{R_1 - 11R_2 - 8R_3 \to R_1}$

$\begin{bmatrix} 9 & 0 & 0 & \frac{63}{4} & -\frac{63}{4} & 9 \\ 0 & 1 & 0 & -\frac{27}{28} & \frac{31}{28} & -\frac{5}{7} \\ 0 & 0 & 1 & -\frac{29}{56} & \frac{25}{56} & -\frac{1}{7} \end{bmatrix}$ $\xrightarrow{\frac{1}{9}R_1}$ $\begin{bmatrix} 1 & 0 & 0 & \frac{7}{4} & -\frac{7}{4} & 1 \\ 0 & 1 & 0 & -\frac{27}{28} & \frac{31}{28} & -\frac{5}{7} \\ 0 & 0 & 1 & -\frac{29}{56} & \frac{25}{56} & -\frac{1}{7} \end{bmatrix}$. Thus, the inverse of the

matrix is $\begin{bmatrix} \frac{7}{4} & -\frac{7}{4} & 1 \\ -\frac{27}{28} & \frac{31}{28} & -\frac{5}{7} \\ -\frac{29}{56} & \frac{25}{56} & -\frac{1}{7} \end{bmatrix}$. (You could also use a calculator to find the inverse.) Therefore,

$\begin{bmatrix} x \\ y \\ z \end{bmatrix} = \begin{bmatrix} \frac{7}{4} & -\frac{7}{4} & 1 \\ -\frac{27}{28} & \frac{31}{28} & -\frac{5}{7} \\ -\frac{29}{56} & \frac{25}{56} & -\frac{1}{7} \end{bmatrix} \begin{bmatrix} 740 \\ 1204 \\ 828 \end{bmatrix} = \begin{bmatrix} 16 \\ 28 \\ 36 \end{bmatrix}$. She earns \$16 on a standard phone, \$28 on a deluxe phone, and

\$36 on a super deluxe phone.

6.4 DETERMINANTS AND CRAMER'S RULE

1. True. det (A) is defined only for a square matrix A.

3. True. If det $(A) = 0$ then A is not invertible.

5. The matrix $\begin{bmatrix} 2 & 0 \\ 0 & 3 \end{bmatrix}$ has determinant $|D| = (2)(3) - (0)(0) = 6$.

7. The matrix $\begin{bmatrix} \frac{3}{2} & 1 \\ -1 & -\frac{2}{3} \end{bmatrix}$ has determinant $|D| = \frac{3}{2}\left(-\frac{2}{3}\right) - 1(-1) = 0$.

9. The matrix $\begin{bmatrix} 4 & 5 \\ 0 & -1 \end{bmatrix}$ has determinant $|D| = (4)(-1) - (5)(0) = -4$.

11. The matrix $\begin{bmatrix} 2 & 5 \end{bmatrix}$ does not have a determinant because it is not square.

13. The matrix $\begin{bmatrix} \frac{1}{2} & \frac{1}{8} \\ 1 & \frac{1}{2} \end{bmatrix}$ has determinant $|D| = \frac{1}{2} \cdot \frac{1}{2} - 1 \cdot \frac{1}{8} = \frac{1}{4} - \frac{1}{8} = \frac{1}{8}$.

In Solutions 15–19, $A = \begin{bmatrix} 1 & 0 & \frac{1}{2} \\ -3 & 5 & 2 \\ 0 & 0 & 4 \end{bmatrix}$.

15. $M_{11} = 5 \cdot 4 - 0 \cdot 2 = 20$, $A_{11} = (-1)^2 M_{11} = 20$ **17.** $M_{12} = -3 \cdot 4 - 0 \cdot 2 = -12$, $A_{12} = (-1)^3 M_{12} = 12$

19. $M_{23} = 1 \cdot 0 - 0 \cdot 0 = 0$, $A_{23} = (-1)^5 M_{23} = 0$

21. $M = \begin{bmatrix} 2 & 1 & 0 \\ 0 & -2 & 4 \\ 0 & 1 & -3 \end{bmatrix}$. Therefore, expanding by the first column, $|M| = 2 \begin{vmatrix} -2 & 4 \\ 1 & -3 \end{vmatrix} = 2(6-4) = 4$. Since $|M| \neq 0$, the

matrix has an inverse.

23. $M = \begin{bmatrix} 30 & 0 & 20 \\ 0 & -10 & -20 \\ 40 & 0 & 10 \end{bmatrix}$. Therefore, expanding by the first row,

$|M| = 30 \begin{vmatrix} -10 & -20 \\ 0 & 10 \end{vmatrix} + 20 \begin{vmatrix} 0 & -10 \\ 40 & 0 \end{vmatrix} = 30(-100+0) + 20(0+400) = -3000 + 8000 = 5000$, and so M^{-1} exists.

25. $M = \begin{bmatrix} 1 & 3 & 7 \\ 2 & 0 & 8 \\ 0 & 2 & 2 \end{bmatrix}$. Therefore, expanding by the second row, $|M| = -2 \begin{vmatrix} 3 & 7 \\ 2 & 2 \end{vmatrix} - 8 \begin{vmatrix} 1 & 3 \\ 0 & 2 \end{vmatrix} = -2(6-14) - 16 = 0$. Since

$|M| = 0$, the matrix does not have an inverse.

27. $M = \begin{bmatrix} 1 & 3 & 3 & 0 \\ 0 & 2 & 0 & 1 \\ -1 & 0 & 0 & 2 \\ 1 & 6 & 4 & 1 \end{bmatrix}$. Therefore, expanding by the third row,

$|M| = -1 \begin{vmatrix} 3 & 3 & 0 \\ 2 & 0 & 1 \\ 6 & 4 & 1 \end{vmatrix} - 2 \begin{vmatrix} 1 & 3 & 3 \\ 0 & 2 & 0 \\ 1 & 6 & 4 \end{vmatrix} = 1 \begin{vmatrix} 3 & 3 \\ 6 & 4 \end{vmatrix} - 1 \begin{vmatrix} 3 & 3 \\ 2 & 0 \end{vmatrix} - 4 \begin{vmatrix} 1 & 3 \\ 1 & 4 \end{vmatrix} = -6 + 6 - 4 = -4$, and so M^{-1} exists.

29. $\begin{vmatrix} 1 & 2 & -1 \\ 2 & 2 & 1 \\ 1 & 2 & 2 \end{vmatrix} = -6$. The matrix has an inverse.   **31.** $\begin{vmatrix} 1 & 10 & 2 & 7 \\ 2 & 18 & 18 & 13 \\ -3 & -30 & -4 & -24 \\ 1 & 10 & 2 & 10 \end{vmatrix} = -12$. The matrix has an inverse.

33. $\begin{vmatrix} 4 & 3 & -2 & 10 \\ -8 & -6 & 24 & -1 \\ 20 & 15 & 3 & 27 \\ 12 & 9 & -6 & -1 \end{vmatrix} = 0$. The matrix has no inverse.

35. $|M| = \begin{vmatrix} 0 & 0 & 4 & 6 \\ 2 & 1 & 1 & 3 \\ 2 & 1 & 2 & 3 \\ 3 & 0 & 1 & 7 \end{vmatrix} = \begin{vmatrix} 0 & 0 & 4 & 6 \\ 2 & 1 & 1 & 3 \\ 0 & 0 & 1 & 0 \\ 3 & 0 & 1 & 7 \end{vmatrix}$, by replacing R_3 with $R_3 - R_2$. Then, expanding by the third row,

$$|M| = 1 \begin{vmatrix} 0 & 0 & 6 \\ 2 & 1 & 3 \\ 3 & 0 & 7 \end{vmatrix} = 6 \begin{vmatrix} 2 & 1 \\ 3 & 0 \end{vmatrix} = 6\,(2 \cdot 0 - 3 \cdot 1) = -18.$$

37. $M = \begin{bmatrix} 1 & 2 & 3 & 4 & 5 \\ 0 & 2 & 4 & 6 & 8 \\ 0 & 0 & 3 & 6 & 9 \\ 0 & 0 & 0 & 4 & 8 \\ 0 & 0 & 0 & 0 & 5 \end{bmatrix}$, so $|M| = 5 \begin{vmatrix} 1 & 2 & 3 & 4 \\ 0 & 2 & 4 & 6 \\ 0 & 0 & 3 & 6 \\ 0 & 0 & 0 & 4 \end{vmatrix} = 5 \cdot 4 \begin{vmatrix} 1 & 2 & 3 \\ 0 & 2 & 4 \\ 0 & 0 & 3 \end{vmatrix} = 20 \cdot 3 \begin{vmatrix} 1 & 2 \\ 0 & 2 \end{vmatrix} = 60 \cdot 2 = 120.$

39. $B = \begin{bmatrix} 4 & 1 & 0 \\ -2 & -1 & 1 \\ 4 & 0 & 3 \end{bmatrix}$

   (a) $|B| = 2 \begin{vmatrix} 1 & 0 \\ 0 & 3 \end{vmatrix} - 1 \begin{vmatrix} 4 & 0 \\ 4 & 3 \end{vmatrix} - 1 \begin{vmatrix} 4 & 1 \\ 4 & 0 \end{vmatrix} = 6 - 12 + 4 = -2$

   (b) $|B| = -1 \begin{vmatrix} 4 & 1 \\ 4 & 0 \end{vmatrix} + 3 \begin{vmatrix} 4 & 1 \\ -2 & -1 \end{vmatrix} = 4 - 6 = -2$

   (c) Yes, as expected, the results agree.

41. $\begin{cases} 2x - y = -9 \\ x + 2y = 8 \end{cases}$ Then $|D| = \begin{vmatrix} 2 & -1 \\ 1 & 2 \end{vmatrix} = 5$, $|D_x| = \begin{vmatrix} -9 & -1 \\ 8 & 2 \end{vmatrix} = -10$, and $|D_y| = \begin{vmatrix} 2 & -9 \\ 1 & 8 \end{vmatrix} = 25.$

   Hence, $x = \dfrac{|D_x|}{|D|} = \dfrac{-10}{5} = -2$, $y = \dfrac{|D_y|}{|D|} = \dfrac{25}{5} = 5$, and so the solution is $(-2, 5)$.

43. $\begin{cases} x - 6y = 3 \\ 3x + 2y = 1 \end{cases}$ Then, $|D| = \begin{vmatrix} 1 & -6 \\ 3 & 2 \end{vmatrix} = 20$, $|D_x| = \begin{vmatrix} 3 & -6 \\ 1 & 2 \end{vmatrix} = 12$, and $|D_y| = \begin{vmatrix} 1 & 3 \\ 3 & 1 \end{vmatrix} = -8.$

   Hence, $x = \dfrac{|D_x|}{|D|} = \dfrac{12}{20} = 0.6$, $y = \dfrac{|D_y|}{|D|} = \dfrac{-8}{20} = -0.4$, and so the solution is $(0.6, -0.4)$.

45. $\begin{cases} 0.4x + 1.2y = 0.4 \\ 1.2x + 1.6y = 3.2 \end{cases}$ Then, $|D| = \begin{vmatrix} 0.4 & 1.2 \\ 1.2 & 1.6 \end{vmatrix} = -0.8$, $|D_x| = \begin{vmatrix} 0.4 & 1.2 \\ 3.2 & 1.6 \end{vmatrix} = -3.2$, and $|D_y| = \begin{vmatrix} 0.4 & 0.4 \\ 1.2 & 3.2 \end{vmatrix} = 0.8.$

   Hence, $x = \dfrac{|D_x|}{|D|} = \dfrac{-3.2}{-0.8} = 4$, $y = \dfrac{|D_y|}{|D|} = \dfrac{0.8}{-0.8} = -1$, and so the solution is $(4, -1)$.

47. $\begin{cases} x - y + 2z = 0 \\ 3x + z = 11 \\ -x + 2y = 0 \end{cases}$ Then expanding by the second row,

$$|D| = \begin{vmatrix} 1 & -1 & 2 \\ 3 & 0 & 1 \\ -1 & 2 & 0 \end{vmatrix} = -3 \begin{vmatrix} -1 & 2 \\ 2 & 0 \end{vmatrix} - 1 \begin{vmatrix} 1 & -1 \\ -1 & 2 \end{vmatrix} = 12 - 1 = 11, \ |D_x| = \begin{vmatrix} 0 & -1 & 2 \\ 11 & 0 & 1 \\ 0 & 2 & 0 \end{vmatrix} = -11 \begin{vmatrix} -1 & 2 \\ 2 & 0 \end{vmatrix} = 44,$$

$$|D_y| = \begin{vmatrix} 1 & 0 & 2 \\ 3 & 11 & 1 \\ -1 & 0 & 0 \end{vmatrix} = 11 \begin{vmatrix} 1 & 2 \\ -1 & 0 \end{vmatrix} = 22, \text{ and } |D_z| = \begin{vmatrix} 1 & -1 & 0 \\ 3 & 0 & 11 \\ -1 & 2 & 0 \end{vmatrix} = -11 \begin{vmatrix} 1 & -1 \\ -1 & 2 \end{vmatrix} = -11.$$

Therefore, $x = \frac{44}{11} = 4$, $y = \frac{22}{11} = 2$, $z = \frac{-11}{11} = -1$, and so the solution is $(4, 2, -1)$.

49. $\begin{cases} 2x_1 + 3x_2 - 5x_3 = 1 \\ x_1 + x_2 - x_3 = 2 \\ 2x_2 + x_3 = 8 \end{cases}$

Then, expanding by the third row,

$$|D| = \begin{vmatrix} 2 & 3 & -5 \\ 1 & 1 & -1 \\ 0 & 2 & 1 \end{vmatrix} = -2 \begin{vmatrix} 2 & -5 \\ 1 & -1 \end{vmatrix} + \begin{vmatrix} 2 & 3 \\ 1 & 1 \end{vmatrix} = -6 - 1 = -7,$$

$$|D_{x_1}| = \begin{vmatrix} 1 & 3 & -5 \\ 2 & 1 & -1 \\ 8 & 2 & 1 \end{vmatrix} = \begin{vmatrix} 1 & -1 \\ 2 & 1 \end{vmatrix} - 3 \begin{vmatrix} 2 & -1 \\ 8 & 1 \end{vmatrix} - 5 \begin{vmatrix} 2 & 1 \\ 8 & 2 \end{vmatrix} = 3 - 30 + 20 = -7,$$

$$|D_{x_2}| = \begin{vmatrix} 2 & 1 & -5 \\ 1 & 2 & -1 \\ 0 & 8 & 1 \end{vmatrix} = -8 \begin{vmatrix} 2 & -5 \\ 1 & -1 \end{vmatrix} + \begin{vmatrix} 2 & 1 \\ 1 & 2 \end{vmatrix} = -24 + 3 = -21, \text{ and}$$

$$|D_{x_3}| = \begin{vmatrix} 2 & 3 & 1 \\ 1 & 1 & 2 \\ 0 & 2 & 8 \end{vmatrix} = -2 \begin{vmatrix} 2 & 1 \\ 1 & 2 \end{vmatrix} + 8 \begin{vmatrix} 2 & 3 \\ 1 & 1 \end{vmatrix} = -6 - 8 = -14.$$

Thus, $x_1 = \frac{-7}{-7} = 1$, $x_2 = \frac{-21}{-7} = 3$, $x_3 = \frac{-14}{-7} = 2$, and so the solution is $(1, 3, 2)$.

51. $\begin{cases} \frac{1}{3}x - \frac{1}{5}y + \frac{1}{2}z = \frac{7}{10} \\ -\frac{2}{3}x + \frac{2}{5}y + \frac{3}{2}z = \frac{11}{10} \\ x - \frac{4}{5}y + z = \frac{9}{5} \end{cases} \Leftrightarrow \begin{cases} 10x - 6y + 15z = 21 \\ -20x + 12y + 45z = 33 \\ 5x - 4y + 5z = 9 \end{cases}$ Then

$|D| = \begin{vmatrix} 10 & -6 & 15 \\ -20 & 12 & 45 \\ 5 & -4 & 5 \end{vmatrix} = 10 \begin{vmatrix} 12 & 45 \\ -4 & 5 \end{vmatrix} + 6 \begin{vmatrix} -20 & 45 \\ 5 & 5 \end{vmatrix} + 15 \begin{vmatrix} -20 & 12 \\ 5 & -4 \end{vmatrix} = 2400 - 1950 + 300 = 750,$

$|D_x| = \begin{vmatrix} 21 & -6 & 15 \\ 33 & 12 & 45 \\ 9 & -4 & 5 \end{vmatrix} = 21 \begin{vmatrix} 12 & 45 \\ -4 & 5 \end{vmatrix} + 6 \begin{vmatrix} 33 & 45 \\ 9 & 5 \end{vmatrix} + 15 \begin{vmatrix} 33 & 12 \\ 9 & -4 \end{vmatrix} = 5040 - 1440 - 3600 = 0,$

$|D_y| = \begin{vmatrix} 10 & 21 & 15 \\ -20 & 33 & 45 \\ 5 & 9 & 5 \end{vmatrix} = 10 \begin{vmatrix} 33 & 45 \\ 9 & 5 \end{vmatrix} - 21 \begin{vmatrix} -20 & 45 \\ 5 & 5 \end{vmatrix} + 15 \begin{vmatrix} -20 & 33 \\ 5 & 9 \end{vmatrix} = -2400 + 6825 - 5175 = -750,$ and

$|D_z| = \begin{vmatrix} 10 & -6 & 21 \\ -20 & 12 & 33 \\ 5 & -4 & 9 \end{vmatrix} = 10 \begin{vmatrix} 12 & 33 \\ -4 & 9 \end{vmatrix} + 6 \begin{vmatrix} -20 & 33 \\ 5 & 9 \end{vmatrix} + 21 \begin{vmatrix} -20 & 12 \\ 5 & -4 \end{vmatrix} = 2400 - 2070 + 420 = 750.$

Therefore, $x = 0$, $y = -1$, $z = 1$, and so the solution is $(0, -1, 1)$.

53. $\begin{cases} 3y + 5z = 4 \\ 2x - z = 10 \\ 4x + 7y = 0 \end{cases}$ Then $|D| = \begin{vmatrix} 0 & 3 & 5 \\ 2 & 0 & -1 \\ 4 & 7 & 0 \end{vmatrix} = -3 \begin{vmatrix} 2 & -1 \\ 4 & 0 \end{vmatrix} + 5 \begin{vmatrix} 2 & 0 \\ 4 & 7 \end{vmatrix} = -12 + 70 = 58,$

$|D_x| = \begin{vmatrix} 4 & 3 & 5 \\ 10 & 0 & -1 \\ 0 & 7 & 0 \end{vmatrix} = -7 \begin{vmatrix} 4 & 5 \\ 10 & -1 \end{vmatrix} = 378,$ $|D_y| = \begin{vmatrix} 0 & 4 & 5 \\ 2 & 10 & -1 \\ 4 & 0 & 0 \end{vmatrix} = 4 \begin{vmatrix} 4 & 5 \\ 10 & -1 \end{vmatrix} = -216,$ and

$|D_z| = \begin{vmatrix} 0 & 3 & 4 \\ 2 & 0 & 10 \\ 4 & 7 & 0 \end{vmatrix} = 4 \begin{vmatrix} 3 & 4 \\ 0 & 10 \end{vmatrix} - 7 \begin{vmatrix} 0 & 4 \\ 2 & 10 \end{vmatrix} = 120 + 56 = 176.$

Thus, $x = \frac{189}{29}$, $y = -\frac{108}{29}$, and $z = \frac{88}{29}$, and so the solution is $\left(\frac{189}{29}, -\frac{108}{29}, \frac{88}{29} \right)$.

55. $\begin{cases} x + y + z + w = 0 \\ 2z + w = 0 \\ y - z = 0 \\ x + 2z = 1 \end{cases}$ Then

$|D| = \begin{vmatrix} 1 & 1 & 1 & 1 \\ 2 & 0 & 0 & 1 \\ 0 & 1 & -1 & 0 \\ 1 & 0 & 2 & 0 \end{vmatrix} = -1 \begin{vmatrix} 2 & 0 & 1 \\ 0 & -1 & 0 \\ 1 & 2 & 0 \end{vmatrix} - 1 \begin{vmatrix} 1 & 1 & 1 \\ 2 & 0 & 1 \\ 1 & 2 & 0 \end{vmatrix} = -\left(2 \begin{vmatrix} -1 & 0 \\ 2 & 0 \end{vmatrix} + 1 \begin{vmatrix} 0 & 1 \\ -1 & 0 \end{vmatrix} \right) - \left(-1 \begin{vmatrix} 2 & 1 \\ 1 & 0 \end{vmatrix} - 2 \begin{vmatrix} 1 & 1 \\ 2 & 1 \end{vmatrix} \right)$

$= -2(0) - 1(1) + 1(-1) + 2(-1) = -4,$

$$|D_x| = \begin{vmatrix} 0 & 1 & 1 & 1 \\ 0 & 0 & 0 & 1 \\ 0 & 1 & -1 & 0 \\ 1 & 0 & 2 & 0 \end{vmatrix} = -1 \begin{vmatrix} 1 & 1 & 1 \\ 0 & 0 & 1 \\ 1 & -1 & 0 \end{vmatrix} = -1\,(-1) \begin{vmatrix} 1 & 1 \\ 1 & -1 \end{vmatrix} = -2,$$

$$|D_y| = \begin{vmatrix} 1 & 0 & 1 & 1 \\ 2 & 0 & 0 & 1 \\ 0 & 0 & -1 & 0 \\ 1 & 1 & 2 & 0 \end{vmatrix} = 1 \begin{vmatrix} 1 & 1 & 1 \\ 2 & 0 & 1 \\ 0 & -1 & 0 \end{vmatrix} = 1 \begin{vmatrix} 0 & 1 \\ -1 & 0 \end{vmatrix} - 2 \begin{vmatrix} 1 & 1 \\ -1 & 0 \end{vmatrix} = 1 - 2\,(1) = -1,$$

$$|D_z| = \begin{vmatrix} 1 & 1 & 0 & 1 \\ 2 & 0 & 0 & 1 \\ 0 & 1 & 0 & 0 \\ 1 & 0 & 1 & 0 \end{vmatrix} = -1 \begin{vmatrix} 1 & 1 & 1 \\ 2 & 0 & 1 \\ 0 & 1 & 0 \end{vmatrix} = -1 \begin{vmatrix} 0 & 1 \\ 1 & 0 \end{vmatrix} + 2 \begin{vmatrix} 1 & 1 \\ 1 & 0 \end{vmatrix} = -1\,(-1) + 2\,(-1) = -1, \text{ and}$$

$$|D_w| = \begin{vmatrix} 1 & 1 & 1 & 0 \\ 2 & 0 & 0 & 0 \\ 0 & 1 & -1 & 0 \\ 1 & 0 & 2 & 1 \end{vmatrix} = 1 \begin{vmatrix} 1 & 1 & 1 \\ 2 & 0 & 0 \\ 0 & 1 & -1 \end{vmatrix} = -2 \begin{vmatrix} 1 & 1 \\ 1 & -1 \end{vmatrix} = -2\,(-2) = 4. \text{ Hence, we have } x = \frac{|D_x|}{|D|} = \frac{-2}{-4} = \frac{1}{2},$$

$$y = \frac{|D_y|}{|D|} = \frac{-1}{-4} = \frac{1}{4}, z = \frac{|D_z|}{|D|} = \frac{-1}{-4} = \frac{1}{4}, \text{ and } w = \frac{|D_w|}{|D|} = \frac{4}{-4} = -1, \text{ and the solution is } \left(\frac{1}{2}, \frac{1}{4}, \frac{1}{4}, -1\right).$$

57. $\begin{vmatrix} a & 0 & 0 & 0 & 0 \\ 0 & b & 0 & 0 & 0 \\ 0 & 0 & c & 0 & 0 \\ 0 & 0 & 0 & d & 0 \\ 0 & 0 & 0 & 0 & e \end{vmatrix} = a \begin{vmatrix} b & 0 & 0 & 0 \\ 0 & c & 0 & 0 \\ 0 & 0 & d & 0 \\ 0 & 0 & 0 & e \end{vmatrix} = ab \begin{vmatrix} c & 0 & 0 \\ 0 & d & 0 \\ 0 & 0 & e \end{vmatrix} = abc \begin{vmatrix} d & 0 \\ 0 & e \end{vmatrix} = abcde$

59. $\begin{vmatrix} x & 12 & 13 \\ 0 & x-1 & 23 \\ 0 & 0 & x-2 \end{vmatrix} = 0 \Leftrightarrow (x-2) \begin{vmatrix} x & 12 \\ 0 & x-1 \end{vmatrix} = 0 \Leftrightarrow (x-2)\cdot x\,(x-1) = 0 \Leftrightarrow x = 0, 1, \text{ or } 2$

61. $\begin{vmatrix} 1 & 0 & x \\ x^2 & 1 & 0 \\ x & 0 & 1 \end{vmatrix} = 0 \Leftrightarrow 1 \begin{vmatrix} 1 & 0 \\ 0 & 1 \end{vmatrix} + x \begin{vmatrix} x^2 & 1 \\ x & 0 \end{vmatrix} = 0 \Leftrightarrow 1 - x^2 = 0 \Leftrightarrow x^2 = 1 \Leftrightarrow x = \pm 1$

63. Area $= \pm\frac{1}{2} \begin{vmatrix} 0 & 0 & 1 \\ 6 & 2 & 1 \\ 3 & 8 & 1 \end{vmatrix} = \pm\frac{1}{2} \begin{vmatrix} 6 & 2 \\ 3 & 8 \end{vmatrix} = \pm\frac{1}{2}\,(48 - 6) = \frac{1}{2}\,(42) = 21$

65. Area $= \pm\dfrac{1}{2}\begin{vmatrix} -1 & 3 & 1 \\ 2 & 9 & 1 \\ 5 & -6 & 1 \end{vmatrix} = \pm\dfrac{1}{2}\left[-1 \begin{vmatrix} 9 & 1 \\ -6 & 1 \end{vmatrix} - 3 \begin{vmatrix} 2 & 1 \\ 5 & 1 \end{vmatrix} + 1 \begin{vmatrix} 2 & 9 \\ 5 & -6 \end{vmatrix} \right]$

$\qquad = \pm\dfrac{1}{2}\left[-1(9+6) - 3(2-5) + 1(-12-45) \right]$

$\qquad = \pm\dfrac{1}{2}\left[-15 - 3(-3) + (-57) \right] = \pm\dfrac{1}{2}(-63) = \dfrac{63}{2}$

67. $\begin{vmatrix} 1 & x & x^2 \\ 1 & y & y^2 \\ 1 & z & z^2 \end{vmatrix} = 1\begin{vmatrix} y & y^2 \\ z & z^2 \end{vmatrix} - 1\begin{vmatrix} x & x^2 \\ z & z^2 \end{vmatrix} + 1\begin{vmatrix} x & x^2 \\ y & y^2 \end{vmatrix} = yz^2 - y^2z - \left(xz^2 - x^2z\right) + \left(xy^2 - xy^2\right)$

$\qquad = yz^2 - y^2z - xz^2 - x^2z + xy^2 - xy^2 + xyz - xyz = xyz - xz^2 - y^2z + yz^2 - x^2y + x^2z + zy^2 - xyz$

$\qquad = z\left(xy - xz - y^2 + yz\right) - x\left(xy - xz - y^2 + yz\right) = (z-x)\left(xy - xz - y^2 + yz\right)$

$\qquad = (z-x)\left[x(y-z) - y(y-z)\right] = (z-x)(x-y)(y-z)$

69. (a) Using the points $(10, 25)$, $(15, 33.75)$, and $(40, 40)$, we substitute for x and y and get the system

$$\begin{cases} 100a + 10b + c = 25 \\ 225a + 15b + c = 33.75 \\ 1600a + 40b + c = 40 \end{cases}$$

(b) $|D| = \begin{vmatrix} 100 & 10 & 1 \\ 225 & 15 & 1 \\ 1600 & 40 & 1 \end{vmatrix} = 1 \cdot \begin{vmatrix} 225 & 15 \\ 1600 & 40 \end{vmatrix} - 1 \cdot \begin{vmatrix} 100 & 10 \\ 1600 & 40 \end{vmatrix} + 1 \cdot \begin{vmatrix} 100 & 10 \\ 225 & 15 \end{vmatrix}$

$\qquad = (9000 - 24{,}000) - (4000 - 16{,}000) + (1500 - 2250) = -15{,}000 + 12{,}000 - 750 = -3750,$

$|D_a| = \begin{vmatrix} 25 & 10 & 1 \\ 33.75 & 15 & 1 \\ 40 & 40 & 1 \end{vmatrix} = 1 \cdot \begin{vmatrix} 33.75 & 15 \\ 40 & 40 \end{vmatrix} - 1 \cdot \begin{vmatrix} 25 & 10 \\ 40 & 40 \end{vmatrix} + 1 \cdot \begin{vmatrix} 25 & 10 \\ 33.75 & 15 \end{vmatrix}$

$\qquad = (1350 - 600) - (1000 - 400) + (375 - 337.5) = 750 - 600 + 37.5 = 187.5,$

$|D_b| = \begin{vmatrix} 100 & 25 & 1 \\ 225 & 33.75 & 1 \\ 1600 & 40 & 1 \end{vmatrix} = 1 \cdot \begin{vmatrix} 225 & 33.75 \\ 1600 & 40 \end{vmatrix} - 1 \cdot \begin{vmatrix} 100 & 25 \\ 1600 & 40 \end{vmatrix} + 1 \cdot \begin{vmatrix} 100 & 25 \\ 225 & 33.75 \end{vmatrix}$

$\qquad = (9000 - 54{,}000) - (4000 - 40{,}000) + (3375 - 5625) = -45{,}000 + 36{,}000 - 2250 = -11{,}250,$ and

$|D_c| = \begin{vmatrix} 100 & 10 & 25 \\ 225 & 15 & 33.75 \\ 1600 & 40 & 40 \end{vmatrix} = 25 \cdot \begin{vmatrix} 225 & 15 \\ 1600 & 40 \end{vmatrix} - 33.75 \cdot \begin{vmatrix} 100 & 10 \\ 1600 & 40 \end{vmatrix} + 40 \cdot \begin{vmatrix} 100 & 10 \\ 225 & 15 \end{vmatrix}$

$\qquad = 25 \cdot (9{,}000 - 24{,}000) - 33.75 \cdot (4{,}000 - 16{,}000) + 40 \cdot (1{,}500 - 2{,}250)$

$\qquad = 25 \cdot (-15{,}000) + 33.75 \cdot 12{,}000 + 40 \cdot (-750) = -375{,}000 + 405{,}000 - 30{,}000 = 0.$

Thus, $a = \dfrac{|D_a|}{|D|} = \dfrac{187.5}{-3750} = 0.05$, $b = \dfrac{|D_b|}{|D|} = \dfrac{-11{,}250}{-3{,}750} = 3$, and $c = \dfrac{|D_c|}{|D|} = \dfrac{0}{-3{,}750} = 0$. Thus, the model is $y = 0.05x^2 + 3x$.

71. (a) The coordinates of the vertices of the surrounding rectangle are (a_1, b_1), (a_2, b_1), (a_2, b_3), and (a_1, b_3). The area of the surrounding rectangle is given by $(a_2 - a_1) \cdot (b_3 - b_1) = a_2 b_3 + a_1 b_1 - a_2 b_1 - a_1 b_3 = a_1 b_1 + a_2 b_3 - a_1 b_3 - a_2 b_1$.

(b) The area of the three blue triangles are as follows:

Area of $\triangle\left((a_1, b_1), (a_2, b_1), (a_2, b_2)\right)$: $\frac{1}{2}(a_2 - a_1) \cdot (b_2 - b_1) = \frac{1}{2}(a_2 b_2 + a_1 b_1 - a_2 b_1 - a_1 b_2)$

Area of $\triangle\left((a_2, b_2), (a_2, b_3), (a_3, b_3)\right)$: $\frac{1}{2}(a_2 - a_3) \cdot (b_3 - b_2) = \frac{1}{2}(a_2 b_3 + a_3 b_2 - a_2 b_2 - a_3 b_3)$

Area of $\triangle\left((a_1, b), (a_1, b_3), (a_3, b_3)\right)$: $\frac{1}{2}(a_3 - a_1) \cdot (b_3 - b_1) = \frac{1}{2}(a_3 b_3 + a_1 b_1 - a_3 b_1 - a_1 b_3)$.

Thus the sum of the areas of the blue triangles, B, is

$$B = \frac{1}{2}(a_2 b_2 + a_1 b_1 - a_2 b_1 - a_1 b_2) + \frac{1}{2}(a_2 b_3 + a_3 b_2 - a_2 b_2 - a_3 b_3) + \frac{1}{2}(a_3 b_3 + a_1 b_1 - a_3 b_1 - a_1 b_3)$$

$$= \frac{1}{2}(a_1 b_1 + a_1 b_1 + a_2 b_2 + a_2 b_3 + a_3 b_2 + a_3 b_3) - \frac{1}{2}(a_1 b_2 + a_1 b_3 + a_2 b_1 + a_2 b_2 + a_3 b_1 + a_3 b_3)$$

$$= a_1 b_1 + \frac{1}{2}(a_2 b_3 + a_3 b_2) - \frac{1}{2}(a_1 b_2 + a_1 b_3 + a_2 b_1 + a_3 b_1)$$

So the area of the red triangle A is the area of the rectangle minus the sum of the areas of the blue triangles, that is,

$$A = (a_1 b_1 + a_2 b_3 - a_1 b_3 - a_2 b_1) - \left[a_1 b_1 + \frac{1}{2}(a_2 b_3 + a_3 b_2) - \frac{1}{2}(a_1 b_2 + a_1 b_3 + a_2 b_1 + a_3 b_1)\right]$$

$$= a_1 b_1 + a_2 b_3 - a_1 b_3 - a_2 b_1 - a_1 b_1 - \frac{1}{2}(a_2 b_3 + a_3 b_2) + \frac{1}{2}(a_1 b_2 + a_1 b_3 + a_2 b_1 + a_3 b_1)$$

$$= \frac{1}{2}(a_1 b_2 + a_2 b_3 + a_3 b_1) - \frac{1}{2}(a_1 b_3 + a_2 b_1 + a_3 b_2)$$

(c) We first find $Q = \begin{vmatrix} a_1 & b_1 & 1 \\ a_2 & b_2 & 1 \\ a_3 & b_3 & 1 \end{vmatrix}$ by expanding about the third column.

$$Q = 1\begin{vmatrix} a_2 & b_2 \\ a_3 & b_3 \end{vmatrix} - 1\begin{vmatrix} a_1 & b_1 \\ a_3 & b_3 \end{vmatrix} + 1\begin{vmatrix} a_1 & b_1 \\ a_2 & b_2 \end{vmatrix} = a_2 b_3 - a_3 b_2 - (a_1 b_3 - a_3 b_1) + a_1 b_2 - a_2 b_1$$

$$= a_1 b_2 + a_2 b_3 + a_3 b_1 - a_1 b_3 - a_2 b_1 - a_3 b_2$$

So $\frac{1}{2}Q = \frac{1}{2}(a_1 b_2 + a_2 b_3 + a_3 b_1) - \frac{1}{2}(a_1 b_3 - a_2 b_1 - a_3 b_2)$, the area of the red triangle. Since $\frac{1}{2}Q$ is not always positive, the area is $\pm\frac{1}{2}Q$.

73. (a) Let $|M| = \begin{vmatrix} x & y & 1 \\ x_1 & y_1 & 1 \\ x_2 & y_2 & 1 \end{vmatrix}$. Then, expanding by the third column,

$$|M| = \begin{vmatrix} x_1 & y_1 \\ x_2 & y_2 \end{vmatrix} - \begin{vmatrix} x & y \\ x_2 & y_2 \end{vmatrix} + \begin{vmatrix} x & y \\ x_1 & y_1 \end{vmatrix} = (x_1 y_2 - x_2 y_1) - (x y_2 - x_2 y) + (x y_1 - x_1 y)$$

$$= x_1 y_2 - x_2 y_1 - x y_2 + x_2 y + x y_1 - x_1 y = x_2 y - x_1 y - x y_2 + x y_1 + x_1 y_2 - x_2 y_1$$

$$= (x_2 - x_1) y - (y_2 - y_1) x + x_1 y_2 - x_2 y_1$$

So $|M| = 0 \Leftrightarrow (x_2 - x_1) y - (y_2 - y_1) x + x_1 y_2 - x_2 y_1 = 0 \Leftrightarrow (x_2 - x_1) y = (y_2 - y_1) x - x_1 y_2 + x_2 y_1 \Leftrightarrow$

$(x_2 - x_1) y = (y_2 - y_1) x - x_1 y_2 + x_1 y_1 - x_1 y_1 + x_2 y_1 \Leftrightarrow y = \dfrac{y_2 - y_1}{x_2 - x_1} x - \dfrac{x_1(y_2 - y_1)}{x_2 - x_1} + \dfrac{y_1(x_2 - x_1)}{x_2 - x_1} \Leftrightarrow$

$y = \dfrac{y_2 - y_1}{x_2 - x_1}(x - x_1) + y_1 \Leftrightarrow y - y_1 = \dfrac{y_2 - y_1}{x_2 - x_1}(x - x_1)$, which is the "two-point" form of the equation for the line passing through the points (x_1, y_1) and (x_2, y_2).

(b) Using the result of part (a), the line has equation

$$\begin{vmatrix} x & y & 1 \\ 20 & 50 & 1 \\ -10 & 25 & 1 \end{vmatrix} = 0 \Leftrightarrow \begin{vmatrix} 20 & 50 \\ -10 & 25 \end{vmatrix} - \begin{vmatrix} x & y \\ -10 & 25 \end{vmatrix} + \begin{vmatrix} x & y \\ 20 & 50 \end{vmatrix} = 0 \Leftrightarrow$$

$$(500 + 500) - (25x + 10y) + (50x - 20y) = 0 \Leftrightarrow 25x - 30y + 1000 = 0 \Leftrightarrow 5x - 6y + 200 = 0.$$

75. Gaussian elimination is superior, since it takes much longer to evaluate six 5×5 determinants than it does to perform one five-equation Gaussian elimination.

CHAPTER 6 REVIEW

1. (a) 2×3

   (b) Yes, this matrix is in row-echelon form.

   (c) No, this matrix is not in reduced row-echelon form, since the leading 1 in the second row does not have a 0 above it.

   (d) $\begin{cases} x + 2y = -5 \\ y = 3 \end{cases}$

3. (a) 3×4

   (b) Yes, this matrix is in row-echelon form.

   (c) Yes, this matrix is in reduced row-echelon form.

   (d) $\begin{cases} x + 8z = 0 \\ y + 5z = -1 \\ 0 = 0 \end{cases}$

5. (a) 3×4

   (b) No, this matrix is not in row-echelon form. The leading 1 in the second row is not to the left of the one above it.

   (c) No, this matrix is not in reduced row-echelon form.

   (d) $\begin{cases} y - 3z = 4 \\ x + y = 7 \\ x + 2y + z = 2 \end{cases}$

7. $\begin{bmatrix} 1 & 2 & 2 & 6 \\ 1 & -1 & 0 & -1 \\ 2 & 1 & 3 & 7 \end{bmatrix} \xrightarrow{R_2 \leftrightarrow R_1} \begin{bmatrix} 1 & -1 & 0 & -1 \\ 1 & 2 & 2 & 6 \\ 2 & 1 & 3 & 7 \end{bmatrix} \xrightarrow[R_3 - 2R_1 \to R_3]{R_2 - R_1 \to R_2} \begin{bmatrix} 1 & -1 & 0 & -1 \\ 0 & 3 & 2 & 7 \\ 0 & 3 & 3 & 9 \end{bmatrix} \xrightarrow{R_3 - R_2 \to R_3} \begin{bmatrix} 1 & -1 & 0 & -1 \\ 0 & 3 & 2 & 7 \\ 0 & 0 & 1 & 2 \end{bmatrix}.$

Thus, $z = 2$, $3y + 2(2) = 7 \Leftrightarrow 3y = 3 \Leftrightarrow y = 1$, and $x - (1) = -1 \Leftrightarrow x = 0$, and so the solution is $(0, 1, 2)$.

9. $\begin{bmatrix} 1 & -2 & 3 & -2 \\ 2 & -1 & 1 & 2 \\ 2 & -7 & 11 & -9 \end{bmatrix} \xrightarrow[R_3 - 2R_1 \to R_3]{R_2 - 2R_1 \to R_2} \begin{bmatrix} 1 & -2 & 3 & -2 \\ 0 & 3 & -5 & 6 \\ 0 & -3 & 5 & -5 \end{bmatrix} \xrightarrow{R_3 + R_2 \to R_3} \begin{bmatrix} 1 & -2 & 3 & -2 \\ 0 & 3 & -5 & 6 \\ 0 & 0 & 0 & 1 \end{bmatrix}.$

The last row corresponds to the equation $0 = 1$, which is always false. Thus, there is no solution.

11.
$$\begin{bmatrix} 1 & 1 & 1 & 1 & 0 \\ 1 & -1 & -4 & -1 & -1 \\ 1 & -2 & 0 & 4 & -7 \\ 2 & 2 & 3 & 4 & -3 \end{bmatrix} \xrightarrow[\begin{subarray}{l} R_2 - R_1 \to R_2 \\ R_3 - R_1 \to R_3 \\ R_4 - 2R_1 \to R_4 \end{subarray}]{} \begin{bmatrix} 1 & 1 & 1 & 1 & 0 \\ 0 & -2 & -5 & -2 & -1 \\ 0 & -3 & -1 & 3 & -7 \\ 0 & 0 & 1 & 2 & -3 \end{bmatrix} \xrightarrow{-R_3 + R_2 \to R_3} \begin{bmatrix} 1 & 1 & 1 & 1 & 0 \\ 0 & 1 & -4 & -5 & 6 \\ 0 & -3 & -1 & 3 & -7 \\ 0 & 0 & 1 & 2 & -3 \end{bmatrix}$$

$$\xrightarrow{R_3 + 3R_2 \to R_3} \begin{bmatrix} 1 & 1 & 1 & 1 & 0 \\ 0 & 1 & -4 & -5 & 6 \\ 0 & 0 & -13 & -12 & 11 \\ 0 & 0 & 1 & 2 & -3 \end{bmatrix} \xrightarrow{R_3 \leftrightarrow R_4} \begin{bmatrix} 1 & 1 & 1 & 1 & 0 \\ 0 & 1 & -4 & -5 & 6 \\ 0 & 0 & 1 & 2 & -3 \\ 0 & 0 & -13 & -12 & 11 \end{bmatrix} \xrightarrow{R_4 + 13R_3 \to R_4} \begin{bmatrix} 1 & 1 & 1 & 1 & 0 \\ 0 & 1 & -4 & -5 & 6 \\ 0 & 0 & 1 & 2 & -3 \\ 0 & 0 & 0 & 14 & -28 \end{bmatrix}.$$

Therefore, $14w = -28 \Leftrightarrow w = -2$, $z + 2(-2) = -3 \Leftrightarrow z = 1$, $y - 4(1) - 5(-2) = 6 \Leftrightarrow y = 0$, and
$x + 0 + 1 + (-2) = 0 \quad \Leftrightarrow \quad x = 1$. So the solution is $(1, 0, 1, -2)$.

13.
$$\begin{bmatrix} 1 & -1 & 3 & 2 \\ 2 & 1 & 1 & 2 \\ 3 & 0 & 4 & 4 \end{bmatrix} \xrightarrow[\begin{subarray}{l} R_2 - 2R_1 \to R_2 \\ R_3 - 3R_1 \to R_3 \end{subarray}]{} \begin{bmatrix} 1 & -1 & 3 & 2 \\ 0 & 3 & -5 & -2 \\ 0 & 3 & -5 & -2 \end{bmatrix} \xrightarrow{R_3 - R_2 \to R_3} \begin{bmatrix} 1 & -1 & 3 & 2 \\ 0 & 3 & -5 & -2 \\ 0 & 0 & 0 & 0 \end{bmatrix} \xrightarrow{\frac{1}{3}R_2}$$

$$\begin{bmatrix} 1 & -1 & 3 & 2 \\ 0 & 1 & -\frac{5}{3} & -\frac{2}{3} \\ 0 & 0 & 0 & 0 \end{bmatrix} \xrightarrow{R_1 + R_2 \to R_1} \begin{bmatrix} 1 & 0 & \frac{4}{3} & \frac{4}{3} \\ 0 & 1 & -\frac{5}{3} & -\frac{2}{3} \\ 0 & 0 & 0 & 0 \end{bmatrix}.$$ The system is dependent, so let $z = t$: $y - \frac{5}{3}t = -\frac{2}{3} \Leftrightarrow$

$y = \frac{5}{3}t - \frac{2}{3}$ and $x + \frac{4}{3}t = \frac{4}{3} \Leftrightarrow x = -\frac{4}{3}t + \frac{4}{3}$. So the solution is $\left(-\frac{4}{3}t + \frac{4}{3}, \frac{5}{3}t - \frac{2}{3}, t \right)$, where t is any real number.

15.
$$\begin{bmatrix} 1 & -1 & 1 & -1 & 0 \\ 3 & -1 & -1 & -1 & 2 \end{bmatrix} \xrightarrow{R_2 - 3R_1 \to R_2} \begin{bmatrix} 1 & -1 & 1 & -1 & 0 \\ 0 & 2 & -4 & 2 & 2 \end{bmatrix} \xrightarrow{\frac{1}{2}R_4} \begin{bmatrix} 1 & -1 & 1 & -1 & 0 \\ 0 & 1 & -2 & 1 & 1 \end{bmatrix} \xrightarrow{R_1 + R_2 \to R_1}$$

$$\begin{bmatrix} 1 & 0 & -1 & 0 & 1 \\ 0 & 1 & -2 & 1 & 1 \end{bmatrix}.$$ Since the system is dependent, Let $z = s$ and $w = t$. Then $y - 2s + t = 1 \Leftrightarrow y = 2s - t + 1$ and

$x - s = 1 \Leftrightarrow x = s + 1$. So the solution is $(s + 1, 2s - t + 1, s, t)$, where s and t are any real numbers.

17.
$$\begin{bmatrix} 1 & -1 & 1 & 0 \\ 3 & 2 & -1 & 6 \\ 1 & 4 & -3 & 3 \end{bmatrix} \xrightarrow[\begin{subarray}{l} R_2 - 3R_1 \to R_2 \\ R_3 - R_1 \to R_3 \end{subarray}]{} \begin{bmatrix} 1 & -1 & 1 & 0 \\ 0 & 5 & -4 & 6 \\ 0 & 5 & -4 & 3 \end{bmatrix} \xrightarrow{R_3 - R_2 \to R_3} \begin{bmatrix} 1 & -1 & 1 & 0 \\ 0 & 5 & -4 & 6 \\ 0 & 0 & 0 & 3 \end{bmatrix}.$$ The last row of this

matrix corresponds to the equation $0 = 3$, which is always false. Hence there is no solution.

19. $\begin{bmatrix} 1 & 1 & -1 & -1 & 2 \\ 1 & -1 & 1 & -1 & 0 \\ 2 & 0 & 0 & 2 & 2 \\ 2 & 4 & -4 & -2 & 6 \end{bmatrix} \xrightarrow{R_1 \leftrightarrow \frac{1}{2}R_3} \begin{bmatrix} 1 & 0 & 0 & 1 & 1 \\ 1 & -1 & 1 & -1 & 0 \\ 1 & 1 & -1 & -1 & 2 \\ 2 & 4 & -4 & -2 & 6 \end{bmatrix} \xrightarrow[\substack{R_3 - R_1 \to R_3 \\ R_4 - 2R_1 \to R_4}]{R_2 - R_1 \to R_2} \begin{bmatrix} 1 & 0 & 0 & 1 & 1 \\ 0 & -1 & 1 & -2 & -1 \\ 0 & 1 & -1 & -2 & 1 \\ 0 & 4 & -4 & -4 & 4 \end{bmatrix}$

$\xrightarrow[\substack{R_4 + 4R_2 \to R_4}]{R_3 + R_2 \to R_3} \begin{bmatrix} 1 & 0 & 0 & 1 & 1 \\ 0 & -1 & 1 & -2 & -1 \\ 0 & 0 & 0 & -4 & 0 \\ 0 & 0 & 0 & -12 & 0 \end{bmatrix} \xrightarrow[\substack{-\frac{1}{12}R_4}]{-\frac{1}{4}R_3} \begin{bmatrix} 1 & 0 & 0 & 1 & 1 \\ 0 & -1 & 1 & -2 & -1 \\ 0 & 0 & 0 & 0 & 0 \\ 0 & 0 & 0 & 1 & 0 \end{bmatrix} \xrightarrow[\substack{R_2 + 2R_3 \to R_2 \\ R_4 - R_3 \to R_4}]{R_1 - R_3 \to R_1}$

$\begin{bmatrix} 1 & 0 & 0 & 0 & 1 \\ 0 & -1 & 1 & 0 & -1 \\ 0 & 0 & 0 & 1 & 0 \\ 0 & 0 & 0 & 0 & 0 \end{bmatrix}$. This system is dependent. Let $z = t$, so $-y + t = -1 \Leftrightarrow y = t + 1$; $x = 1 \quad \Leftrightarrow \quad x = 1$. So the

solution is $(1, t + 1, t, 0)$, where t is any real number.

21. A (3×3) and B (2×3) have different dimensions, so they are not equal.

In Solutions 23–33, the matrices A, B, C, D, E, F, and G are defined as follows:

$$A = \begin{bmatrix} 2 & 0 & -1 \end{bmatrix} \qquad B = \begin{bmatrix} 1 & 2 & 4 \\ -2 & 1 & 0 \end{bmatrix} \qquad C = \begin{bmatrix} \frac{1}{2} & 3 \\ 2 & \frac{3}{2} \\ -2 & 1 \end{bmatrix}$$

$$D = \begin{bmatrix} 1 & 4 \\ 0 & -1 \\ 2 & 0 \end{bmatrix} \qquad E = \begin{bmatrix} 2 & -1 \\ -\frac{1}{2} & 1 \end{bmatrix} \qquad F = \begin{bmatrix} 4 & 0 & 2 \\ -1 & 1 & 0 \\ 7 & 5 & 0 \end{bmatrix} \qquad G = \begin{bmatrix} 5 \end{bmatrix}$$

23. $A + B$ is not defined because the matrix dimensions 1×3 and 2×3 are not compatible.

25. $2C + 3D = 2\begin{bmatrix} \frac{1}{2} & 3 \\ 2 & \frac{3}{2} \\ -2 & 1 \end{bmatrix} + 3\begin{bmatrix} 1 & 4 \\ 0 & -1 \\ 2 & 0 \end{bmatrix} = \begin{bmatrix} 1 & 6 \\ 4 & 3 \\ -4 & 2 \end{bmatrix} + \begin{bmatrix} 3 & 12 \\ 0 & -3 \\ 6 & 0 \end{bmatrix} = \begin{bmatrix} 4 & 18 \\ 4 & 0 \\ 2 & 2 \end{bmatrix}$

27. $GA = \begin{bmatrix} 5 \end{bmatrix}\begin{bmatrix} 2 & 0 & -1 \end{bmatrix} = \begin{bmatrix} 10 & 0 & -5 \end{bmatrix}$

29. $BC = \begin{bmatrix} 1 & 2 & 4 \\ -2 & 1 & 0 \end{bmatrix}\begin{bmatrix} \frac{1}{2} & 3 \\ 2 & \frac{3}{2} \\ -2 & 1 \end{bmatrix} = \begin{bmatrix} -\frac{7}{2} & 10 \\ 1 & -\frac{9}{2} \end{bmatrix}$ **31.** $BF = \begin{bmatrix} 1 & 2 & 4 \\ -2 & 1 & 0 \end{bmatrix}\begin{bmatrix} 4 & 0 & 2 \\ -1 & 1 & 0 \\ 7 & 5 & 0 \end{bmatrix} = \begin{bmatrix} 30 & 22 & 2 \\ -9 & 1 & -4 \end{bmatrix}$

33. $(C + D)E = \left(\begin{bmatrix} \frac{1}{2} & 3 \\ 2 & \frac{3}{2} \\ -2 & 1 \end{bmatrix} + \begin{bmatrix} 1 & 4 \\ 0 & -1 \\ 2 & 0 \end{bmatrix}\right)\begin{bmatrix} 2 & -1 \\ -\frac{1}{2} & 1 \end{bmatrix} = \begin{bmatrix} \frac{3}{2} & 7 \\ 2 & \frac{1}{2} \\ 0 & 1 \end{bmatrix}\begin{bmatrix} 2 & -1 \\ -\frac{1}{2} & 1 \end{bmatrix} = \begin{bmatrix} -\frac{1}{2} & \frac{11}{2} \\ \frac{15}{4} & -\frac{3}{2} \\ -\frac{1}{2} & 1 \end{bmatrix}$

In Solutions 35–43, the matrices A and B are defined as follows:

$$A = \begin{bmatrix} 3 & 0 & -3 \\ -2 & 1 & 2 \\ 1 & 6 & 0 \end{bmatrix} \qquad\qquad B = \begin{bmatrix} -1 & 4 & -1 \\ 1 & -1 & 0 \\ -2 & 0 & 2 \end{bmatrix}$$

35. $AB^2 = \begin{bmatrix} 27 & 0 & -21 \\ -20 & 5 & 13 \\ -5 & 22 & -7 \end{bmatrix}$

37. $A^{-1}BA = \begin{bmatrix} 14 & 26 & -8 \\ -3 & -\frac{7}{3} & \frac{7}{3} \\ 18 & \frac{80}{3} & -\frac{35}{3} \end{bmatrix}$

39. $|AB| = -12$

41. $\left| A^{-1} \right| = \dfrac{1}{3}$

43. $\left| A^{-1}BA \right| = -4$

45. $AB = \begin{bmatrix} 2 & -5 \\ -2 & 6 \end{bmatrix} \begin{bmatrix} 3 & \frac{5}{2} \\ 1 & 1 \end{bmatrix} = \begin{bmatrix} 1 & 0 \\ 0 & 1 \end{bmatrix}$ and $BA = \begin{bmatrix} 3 & \frac{5}{2} \\ 1 & 1 \end{bmatrix} \begin{bmatrix} 2 & -5 \\ -2 & 6 \end{bmatrix} = \begin{bmatrix} 1 & 0 \\ 0 & 1 \end{bmatrix}$.

In Solutions 47–51, the matrices A, B, and C are defined as follows:

$$A = \begin{bmatrix} 2 & 1 \\ 3 & 2 \end{bmatrix} \qquad B = \begin{bmatrix} 1 & -2 \\ -2 & 4 \end{bmatrix} \qquad C = \begin{bmatrix} 0 & 1 & 3 \\ -2 & 4 & 0 \end{bmatrix}$$

47. $A + 3X = B \iff 3X = B - A \iff X = \frac{1}{3}(B - A)$. Thus, $X = \frac{1}{3}\left(\begin{bmatrix} 1 & -2 \\ -2 & 4 \end{bmatrix} - \begin{bmatrix} 2 & 1 \\ 3 & 2 \end{bmatrix} \right) = \frac{1}{3} \begin{bmatrix} -1 & -3 \\ -5 & 2 \end{bmatrix}$.

49. $2(X - A) = 3B \iff X - A = \frac{3}{2}B \iff X = A + \frac{3}{2}B$. Thus,

$$X = \begin{bmatrix} 2 & 1 \\ 3 & 2 \end{bmatrix} + \frac{3}{2} \begin{bmatrix} 1 & -2 \\ -2 & 4 \end{bmatrix} = \begin{bmatrix} 2 & 1 \\ 3 & 2 \end{bmatrix} + \begin{bmatrix} \frac{3}{2} & -3 \\ -3 & 6 \end{bmatrix} = \begin{bmatrix} \frac{7}{2} & -2 \\ 0 & 8 \end{bmatrix}.$$

51. $AX = C \iff A^{-1}AX = X = A^{-1}C$. Now

$$A^{-1} = \frac{1}{4 - 3} \begin{bmatrix} 2 & -1 \\ -3 & 2 \end{bmatrix} = \begin{bmatrix} 2 & -1 \\ -3 & 2 \end{bmatrix}. \text{ Thus, } X = A^{-1}C = \begin{bmatrix} 2 & -1 \\ -3 & 2 \end{bmatrix} \begin{bmatrix} 0 & 1 & 3 \\ -2 & 4 & 0 \end{bmatrix} = \begin{bmatrix} 2 & -2 & 6 \\ -4 & 5 & -9 \end{bmatrix}.$$

53. $D = \begin{bmatrix} 1 & 4 \\ 2 & 9 \end{bmatrix}$. Then $|D| = 1(9) - 2(4) = 1$, and so $D^{-1} = \begin{bmatrix} 9 & -4 \\ -2 & 1 \end{bmatrix}$.

55. $D = \begin{bmatrix} 4 & -12 \\ -2 & 6 \end{bmatrix}$. Then $|D| = 4(6) - 2(12) = 0$, and so D has no inverse.

57. $D = \begin{bmatrix} 3 & 0 & 1 \\ 2 & -3 & 0 \\ 4 & -2 & 1 \end{bmatrix}$. Then, $|D| = 1 \begin{vmatrix} 2 & -3 \\ 4 & -2 \end{vmatrix} + 1 \begin{vmatrix} 3 & 0 \\ 2 & -3 \end{vmatrix} = -4 + 12 - 9 = -1$. So D^{-1} exists.

$\begin{bmatrix} 3 & 0 & 1 & 1 & 0 & 0 \\ 2 & -3 & 0 & 0 & 1 & 0 \\ 4 & -2 & 1 & 0 & 0 & 1 \end{bmatrix} \xrightarrow{R_1 - R_2 \to R_1} \begin{bmatrix} 1 & 3 & 1 & 1 & -1 & 0 \\ 2 & -3 & 0 & 0 & 1 & 0 \\ 4 & -2 & 1 & 0 & 0 & 1 \end{bmatrix} \xrightarrow[R_3 - 4R_1 \to R_3]{R_2 - 2R_1 \to R_2} \begin{bmatrix} 1 & 3 & 1 & 1 & -1 & 0 \\ 0 & -9 & -2 & -2 & 3 & 0 \\ 0 & -14 & -3 & -4 & 4 & 1 \end{bmatrix} \xrightarrow[-2R_3]{-3R_2}$

$\begin{bmatrix} 1 & 3 & 1 & 1 & -1 & 0 \\ 0 & 27 & 6 & 6 & -9 & 0 \\ 0 & 28 & 6 & 8 & -8 & -2 \end{bmatrix} \xrightarrow{R_3 - R_2 \to R_3} \begin{bmatrix} 1 & 3 & 1 & 1 & -1 & 0 \\ 0 & 27 & 6 & 6 & -9 & 0 \\ 0 & 1 & 0 & 2 & 1 & -2 \end{bmatrix} \xrightarrow[\frac{1}{3}R_3]{R_3 \leftrightarrow R_2} \begin{bmatrix} 1 & 3 & 1 & 1 & -1 & 0 \\ 0 & 1 & 0 & 2 & 1 & -2 \\ 0 & 9 & 2 & 2 & -3 & 0 \end{bmatrix} \xrightarrow[R_1 - 3R_2 \to R_1]{R_3 - 9R_2 \to R_3}$

$\begin{bmatrix} 1 & 0 & 1 & -5 & -4 & 6 \\ 0 & 1 & 0 & 2 & 1 & -2 \\ 0 & 0 & 2 & -16 & -12 & 18 \end{bmatrix} \xrightarrow[R_1 - R_3 \to R_1]{\frac{1}{2}R_3} \begin{bmatrix} 1 & 0 & 0 & 3 & 2 & -3 \\ 0 & 1 & 0 & 2 & 1 & -2 \\ 0 & 0 & 1 & -8 & -6 & 9 \end{bmatrix}$. Thus, $D^{-1} = \begin{bmatrix} 3 & 2 & -3 \\ 2 & 1 & -2 \\ -8 & -6 & 9 \end{bmatrix}$.

59. $D = \begin{bmatrix} 1 & 0 & 0 & 1 \\ 0 & 2 & 0 & 2 \\ 0 & 0 & 3 & 3 \\ 0 & 0 & 0 & 4 \end{bmatrix}$. Thus, $|D| = \begin{vmatrix} 2 & 0 & 2 \\ 0 & 3 & 3 \\ 0 & 0 & 4 \end{vmatrix} = 2 \begin{vmatrix} 3 & 3 \\ 0 & 4 \end{vmatrix} = 24$ and D^{-1} exists. $\begin{bmatrix} 1 & 0 & 0 & 1 & 1 & 0 & 0 & 0 \\ 0 & 2 & 0 & 2 & 0 & 1 & 0 & 0 \\ 0 & 0 & 3 & 3 & 0 & 0 & 1 & 0 \\ 0 & 0 & 0 & 4 & 0 & 0 & 0 & 1 \end{bmatrix} \xrightarrow[\frac{1}{4}R_4]{\overset{\frac{1}{2}R_2}{\frac{1}{3}R_3}}$

$\begin{bmatrix} 1 & 0 & 0 & 1 & 1 & 0 & 0 & 0 \\ 0 & 1 & 0 & 1 & 0 & \frac{1}{2} & 0 & 0 \\ 0 & 0 & 1 & 1 & 0 & 0 & \frac{1}{3} & 0 \\ 0 & 0 & 0 & 1 & 0 & 0 & 0 & \frac{1}{4} \end{bmatrix} \xrightarrow[R_3 - R_4 \to R_3]{\overset{R_1 - R_4 \to R_1}{R_2 - R_4 \to R_2}} \begin{bmatrix} 1 & 0 & 0 & 0 & 1 & 0 & 0 & -\frac{1}{4} \\ 0 & 1 & 0 & 0 & 0 & \frac{1}{2} & 0 & -\frac{1}{4} \\ 0 & 0 & 1 & 0 & 0 & 0 & \frac{1}{3} & -\frac{1}{4} \\ 0 & 0 & 0 & 1 & 0 & 0 & 0 & \frac{1}{4} \end{bmatrix}$. Therefore, $D^{-1} = \begin{bmatrix} 1 & 0 & 0 & -\frac{1}{4} \\ 0 & \frac{1}{2} & 0 & -\frac{1}{4} \\ 0 & 0 & \frac{1}{3} & -\frac{1}{4} \\ 0 & 0 & 0 & \frac{1}{4} \end{bmatrix}$.

61. $\begin{bmatrix} 12 & -5 \\ 5 & -2 \end{bmatrix} \begin{bmatrix} x \\ y \end{bmatrix} = \begin{bmatrix} 10 \\ 17 \end{bmatrix}$. If we let $A = \begin{bmatrix} 12 & -5 \\ 5 & -2 \end{bmatrix}$, then $A^{-1} = \frac{1}{-24 + 25} \begin{bmatrix} -2 & 5 \\ -5 & 12 \end{bmatrix} = \begin{bmatrix} -2 & 5 \\ -5 & 12 \end{bmatrix}$, and so

$\begin{bmatrix} x \\ y \end{bmatrix} = \begin{bmatrix} -2 & 5 \\ -5 & 12 \end{bmatrix} \begin{bmatrix} 10 \\ 17 \end{bmatrix} = \begin{bmatrix} 65 \\ 154 \end{bmatrix}$. Therefore, the solution is $(65, 154)$.

63. $\begin{bmatrix} 2 & 1 & 5 \\ 1 & 2 & 2 \\ 1 & 0 & 3 \end{bmatrix} \begin{bmatrix} x \\ y \\ z \end{bmatrix} = \begin{bmatrix} \frac{1}{3} \\ \frac{1}{4} \\ \frac{1}{6} \end{bmatrix}$. Let $A = \begin{bmatrix} 2 & 1 & 5 \\ 1 & 2 & 2 \\ 1 & 0 & 3 \end{bmatrix}$. Then $\begin{bmatrix} 2 & 1 & 5 & 1 & 0 & 0 \\ 1 & 2 & 2 & 0 & 1 & 0 \\ 1 & 0 & 3 & 0 & 0 & 1 \end{bmatrix} \xrightarrow{R_1 \leftrightarrow R_2} \begin{bmatrix} 1 & 2 & 2 & 0 & 1 & 0 \\ 2 & 1 & 5 & 1 & 0 & 0 \\ 1 & 0 & 3 & 0 & 0 & 1 \end{bmatrix}$

$\xrightarrow[R_3 - R_1 \to R_3]{R_2 - 2R_1 \to R_2} \begin{bmatrix} 1 & 2 & 2 & 0 & 1 & 0 \\ 0 & -3 & 1 & 1 & -2 & 0 \\ 0 & -2 & 1 & 0 & -1 & 1 \end{bmatrix} \xrightarrow{R_2 - 2R_3 \to R_2} \begin{bmatrix} 1 & 2 & 2 & 0 & 1 & 0 \\ 0 & 1 & -1 & 1 & 0 & -2 \\ 0 & -2 & 1 & 0 & -1 & 1 \end{bmatrix} \xrightarrow[R_3 \to R_3 + 2R_2]{R_1 - 2R_2 \to R_1}$

$\begin{bmatrix} 1 & 0 & 4 & -2 & 1 & 4 \\ 0 & 1 & -1 & 1 & 0 & -2 \\ 0 & 0 & -1 & 2 & -1 & -3 \end{bmatrix} \xrightarrow{-R_3} \begin{bmatrix} 1 & 0 & 4 & -2 & 1 & 4 \\ 0 & 1 & -1 & 1 & 0 & -2 \\ 0 & 0 & 1 & -2 & 1 & 3 \end{bmatrix} \xrightarrow[R_2 + R_3 \to R_2]{R_1 - 4R_3 \to R_1} \begin{bmatrix} 1 & 0 & 0 & 6 & -3 & -8 \\ 0 & 1 & 0 & -1 & 1 & 1 \\ 0 & 0 & 1 & -2 & 1 & 3 \end{bmatrix}$.

Hence, $A^{-1} = \begin{bmatrix} 6 & -3 & -8 \\ -1 & 1 & 1 \\ -2 & 1 & 3 \end{bmatrix}$ and $\begin{bmatrix} x \\ y \\ z \end{bmatrix} = \begin{bmatrix} 6 & -3 & -8 \\ -1 & 1 & 1 \\ -2 & 1 & 3 \end{bmatrix} \begin{bmatrix} \frac{1}{3} \\ \frac{1}{4} \\ \frac{1}{6} \end{bmatrix} = \begin{bmatrix} -\frac{1}{12} \\ \frac{1}{12} \\ \frac{1}{12} \end{bmatrix}$, and so the solution is

$\left(-\frac{1}{12}, \frac{1}{12}, \frac{1}{12} \right)$.

65. (a) The (i, j)th entry of A represents how many pounds of vegetable j were sold on day i, and the ith entry of B represents the price of vegetable i.

(b) $AB = \begin{bmatrix} 25 & 16 & 30 \\ 14 & 12 & 16 \end{bmatrix} \begin{bmatrix} 1.50 \\ 1.00 \\ 0.50 \end{bmatrix} = \begin{bmatrix} 68.5 \\ 41.0 \end{bmatrix}$. The jth entry of AB represents the total revenue on day j.

67. $|D| = \begin{vmatrix} 2 & 7 \\ 6 & 16 \end{vmatrix} = 32 - 42 = -10$, $|D_x| = \begin{vmatrix} 13 & 7 \\ 30 & 16 \end{vmatrix} = 208 - 210 = -2$, and $|D_y| = \begin{vmatrix} 2 & 13 \\ 6 & 30 \end{vmatrix} = 60 - 78 = -18$.

Therefore, $x = \frac{-2}{-10} = \frac{1}{5}$ and $y = \frac{-18}{-10} = \frac{9}{5}$, and so the solution is $\left(\frac{1}{5}, \frac{9}{5} \right)$.

69. $|D| = \begin{vmatrix} 2 & -1 & 5 \\ -1 & 7 & 0 \\ 5 & 4 & 3 \end{vmatrix} = 5 \begin{vmatrix} -1 & 7 \\ 5 & 4 \end{vmatrix} + 3 \begin{vmatrix} 2 & -1 \\ -1 & 7 \end{vmatrix} = -195 + 39 = -156$,

$|D_x| = \begin{vmatrix} 0 & -1 & 5 \\ 9 & 7 & 0 \\ -9 & 4 & 3 \end{vmatrix} = 5 \begin{vmatrix} 9 & 7 \\ -9 & 4 \end{vmatrix} + 3 \begin{vmatrix} 0 & -1 \\ 9 & 7 \end{vmatrix} = 495 + 27 = 522$,

$|D_y| = \begin{vmatrix} 2 & 0 & 5 \\ -1 & 9 & 0 \\ 5 & -9 & 3 \end{vmatrix} = 5 \begin{vmatrix} -1 & 9 \\ 5 & -9 \end{vmatrix} + 3 \begin{vmatrix} 2 & 0 \\ -1 & 9 \end{vmatrix} = -180 + 54 = -126$, and

$|D_z| = \begin{vmatrix} 2 & -1 & 0 \\ -1 & 7 & 9 \\ 5 & 4 & -9 \end{vmatrix} = -9 \begin{vmatrix} 2 & -1 \\ 5 & 4 \end{vmatrix} - 9 \begin{vmatrix} 2 & -1 \\ -1 & 7 \end{vmatrix} = -117 - 117 = -234$.

Therefore, $x = \frac{522}{-156} = -\frac{87}{26}$, $y = \frac{-126}{-156} = \frac{21}{26}$, and $z = \frac{-234}{-156} = \frac{3}{2}$, and so the solution is $\left(-\frac{87}{26}, \frac{21}{26}, \frac{3}{2} \right)$.

71. The area is $\pm \frac{1}{2} \begin{vmatrix} -1 & 3 & 1 \\ 3 & 1 & 1 \\ -2 & -2 & 1 \end{vmatrix} = \pm \frac{1}{2} \left(\begin{vmatrix} 3 & 1 \\ -2 & -2 \end{vmatrix} - \begin{vmatrix} -1 & 3 \\ -2 & -2 \end{vmatrix} + \begin{vmatrix} -1 & 3 \\ 3 & 1 \end{vmatrix} \right) = \pm \frac{1}{2} (-4 - 8 - 10) = 11$.

73. Let x be the amount invested in Bank A, y the amount invested in Bank B, and z the amount invested in Bank C.

We get the following system: $\begin{cases} x + y + z = 60,000 \\ 0.02x + 0.025y + 0.03z = 1575 \\ 2x + 2z = y \end{cases} \Leftrightarrow \begin{cases} x + y + z = 60,000 \\ 2x + 2.5y + 3z = 157,500 \\ 2x - y + 2z = 0 \end{cases}$ which

has matrix representation $\begin{bmatrix} 1 & 1 & 1 & 60,000 \\ 2 & 2.5 & 3 & 157,500 \\ 2 & -1 & 2 & 0 \end{bmatrix} \xrightarrow[R_3 - 2R_1 \to R_3]{R_2 - 2R_1 \to R_2} \begin{bmatrix} 1 & 1 & 1 & 60,000 \\ 0 & 0.5 & 1 & 37,500 \\ 0 & -3 & 0 & -120,000 \end{bmatrix} \xrightarrow{R_2 \leftrightarrow -\frac{1}{3}R_3}$

$\begin{bmatrix} 1 & 1 & 1 & 60,000 \\ 0 & 1 & 0 & 40,000 \\ 0 & 0.5 & 1 & 37,500 \end{bmatrix} \xrightarrow[R_3 - 0.5R_2 \to R_3]{R_1 - R_2 \to R_1} \begin{bmatrix} 1 & 0 & 1 & 20,000 \\ 0 & 1 & 0 & 40,000 \\ 0 & 0 & 1 & 17,500 \end{bmatrix} \xrightarrow{R_1 - R_3 \to R_1} \begin{bmatrix} 1 & 0 & 1 & 2,500 \\ 0 & 1 & 0 & 40,000 \\ 0 & 0 & 1 & 17,500 \end{bmatrix}$. Thus, she invests

$2,500 in Bank A, $40,000 in Bank B, and $17,500 in Bank C.

CHAPTER 6 TEST

1. $\begin{bmatrix} 1 & 8 & 0 & 0 \\ 0 & 1 & 7 & 10 \\ 0 & 0 & 0 & 0 \end{bmatrix}$ is in row-echelon form, but not reduced row-echelon form because the 1 in the second row does not have a 0 above it.

3. $\begin{bmatrix} 1 & 0 & 0 \\ 0 & 0 & 1 \end{bmatrix}$ is in reduced row-echelon form.

5. $\begin{cases} x - y + 2z = 0 \\ 2x - 4y + 5z = -5 \\ 2y - 3z = 5 \end{cases}$ has the matrix representation $\begin{bmatrix} 1 & -1 & 2 & 0 \\ 2 & -4 & 5 & -5 \\ 0 & 2 & -3 & 5 \end{bmatrix}$ $\xrightarrow{R_2 - R_1 \to R_2}$ $\begin{bmatrix} 1 & -1 & 2 & 0 \\ 0 & -2 & 1 & -5 \\ 0 & 2 & -3 & 5 \end{bmatrix}$

$\xrightarrow{R_3 + R_2 \to R_3}$ $\begin{bmatrix} 1 & -1 & 2 & 0 \\ 0 & -2 & 1 & -5 \\ 0 & 0 & -2 & 0 \end{bmatrix}$ $\xrightarrow{-\frac{1}{2}R_3}$ $\begin{bmatrix} 1 & -1 & 2 & 0 \\ 0 & -2 & 1 & -5 \\ 0 & 0 & 1 & 0 \end{bmatrix}$. Thus $z = 0$, $-2y + 0 = -5 \Leftrightarrow y = \frac{5}{2}$, and

$x - \frac{5}{2} + 2(0) = 0 \Leftrightarrow x = \frac{5}{2}$. Thus, the solution is $\left(\frac{5}{2}, \frac{5}{2}, 0\right)$.

7. $\begin{cases} x + 3y - z = 0 \\ 3x + 4y - 2z = -1 \\ -x + 2y = 1 \end{cases}$ has the matrix representation $\begin{bmatrix} 1 & 3 & -1 & 0 \\ 3 & 4 & -2 & -1 \\ -1 & 2 & 0 & 1 \end{bmatrix}$ $\xrightarrow[R_3 + R_1 \to R_3]{R_2 - 3R_1 \to R_2}$ $\begin{bmatrix} 1 & 3 & -1 & 0 \\ 0 & -5 & 1 & -1 \\ 0 & 5 & -1 & 1 \end{bmatrix}$

$\xrightarrow{R_3 - R_2 \to R_3}$ $\begin{bmatrix} 1 & 3 & -1 & 0 \\ 0 & -5 & 1 & -1 \\ 0 & 0 & 0 & 0 \end{bmatrix}$ $\xrightarrow{-\frac{1}{5}R_2}$ $\begin{bmatrix} 1 & 3 & -1 & 0 \\ 0 & 1 & -\frac{1}{5} & \frac{1}{5} \\ 0 & 0 & 0 & 0 \end{bmatrix}$ $\xrightarrow{R_1 - 3R_2 \to R_1}$ $\begin{bmatrix} 1 & 0 & -\frac{2}{5} & -\frac{3}{5} \\ 0 & 1 & -\frac{1}{5} & \frac{1}{5} \\ 0 & 0 & 0 & 0 \end{bmatrix}$.

Since this system is dependent, let $z = t$. Then $y - \frac{1}{5}t = \frac{1}{5} \Leftrightarrow y = \frac{1}{5}t + \frac{1}{5}$ and $x - \frac{2}{5}t = -\frac{3}{5} \Leftrightarrow x = \frac{2}{5}t - \frac{3}{5}$. Thus, the

solution is $\left(\frac{2}{5}t - \frac{3}{5}, \frac{1}{5}t + \frac{1}{5}, t\right)$.

In Solutions 9–15, the matrices A, B, and C are defined as follows:

$$A = \begin{bmatrix} 2 & 3 \\ 2 & 4 \end{bmatrix} \qquad B = \begin{bmatrix} 2 & 4 \\ -1 & 1 \\ 3 & 0 \end{bmatrix} \qquad C = \begin{bmatrix} 1 & 0 & 4 \\ -1 & 1 & 2 \\ 0 & 1 & 3 \end{bmatrix}$$

9. AB is undefined because A is 2×2 and B is 3×2, so they have incompatible dimensions.

11. $CBA = \begin{bmatrix} 1 & 0 & 4 \\ -1 & 1 & 2 \\ 0 & 1 & 3 \end{bmatrix} \begin{bmatrix} 2 & 4 \\ -1 & 1 \\ 3 & 0 \end{bmatrix} \begin{bmatrix} 2 & 3 \\ 2 & 4 \end{bmatrix} = \begin{bmatrix} 14 & 4 \\ 3 & -3 \\ 8 & 1 \end{bmatrix} \begin{bmatrix} 2 & 3 \\ 2 & 4 \end{bmatrix} = \begin{bmatrix} 36 & 58 \\ 0 & -3 \\ 18 & 28 \end{bmatrix}$

13. B^{-1} does not exist because B is not a square matrix.

15. $\det(C) = \begin{vmatrix} 1 & 0 & 4 \\ -1 & 1 & 2 \\ 0 & 1 & 3 \end{vmatrix} = 1 \begin{vmatrix} 1 & 2 \\ 1 & 3 \end{vmatrix} + 4 \begin{vmatrix} -1 & 1 \\ 0 & 1 \end{vmatrix} = 1 - 4 = -3$

17. $|A| = \begin{vmatrix} 1 & 4 & 1 \\ 0 & 2 & 0 \\ 1 & 0 & 1 \end{vmatrix} = 2 \begin{vmatrix} 1 & 1 \\ 1 & 1 \end{vmatrix} = 0$, $|B| = \begin{vmatrix} 1 & 4 & 0 \\ 0 & 2 & 0 \\ -3 & 0 & 1 \end{vmatrix} = 2 \begin{vmatrix} 1 & 0 \\ -3 & 1 \end{vmatrix} = 2$. Since $|A| = 0$, A does not have an inverse, and

since $|B| \neq 0$, B does have an inverse. $\begin{bmatrix} 1 & 4 & 0 & 1 & 0 & 0 \\ 0 & 2 & 0 & 0 & 1 & 0 \\ -3 & 0 & 1 & 0 & 0 & 1 \end{bmatrix}$ $\xrightarrow{R_3 + 3R_1 \to R_3}$ $\begin{bmatrix} 1 & 4 & 0 & 1 & 0 & 0 \\ 0 & 2 & 0 & 0 & 1 & 0 \\ 0 & 12 & 1 & 3 & 0 & 1 \end{bmatrix}$ $\xrightarrow[R_3 - 6R_2 \to R_3]{R_1 - 2R_2 \to R_1}$

$\begin{bmatrix} 1 & 0 & 0 & 1 & -2 & 0 \\ 0 & 2 & 0 & 0 & 1 & 0 \\ 0 & 0 & 1 & 3 & -6 & 1 \end{bmatrix}$ $\xrightarrow{\frac{1}{2}R_2}$ $\begin{bmatrix} 1 & 0 & 0 & 1 & -2 & 0 \\ 0 & 1 & 0 & 0 & \frac{1}{2} & 0 \\ 0 & 0 & 1 & 3 & -6 & 1 \end{bmatrix}$. Therefore, $B^{-1} = \begin{bmatrix} 1 & -2 & 0 \\ 0 & \frac{1}{2} & 0 \\ 3 & -6 & 1 \end{bmatrix}$.

19. Let x and y represent the number of pounds of almonds and walnuts respectively. Then the problem is modeled by the system

of equations $\begin{cases} x + y = 3 \\ 4.75x + 3.45y = 11.91 \end{cases}$ Then $D = \begin{bmatrix} 1 & 1 \\ 4.75 & 3.45 \end{bmatrix}$, so $\det(D) = \begin{vmatrix} 1 & 1 \\ 4.75 & 3.45 \end{vmatrix} = 3.45 - 4.75 = -1.3$,

$\det(D_x) = \begin{vmatrix} 3 & 1 \\ 11.91 & 3.45 \end{vmatrix} = 10.35 - 11.91 = -1.56$, and $\det(D_y) = \begin{vmatrix} 1 & 3 \\ 4.75 & 11.91 \end{vmatrix} = 11.91 - 14.25 = -2.34$. Then

$x = \frac{|D_x|}{|D|} = \frac{-1.56}{-1.3} = 1.2$ and $y = \frac{|D_y|}{|D|} = \frac{-2.34}{-1.3} = 1.8$, so she bought 1.2 pounds of almonds and 1.8 pounds of walnuts.

FOCUS ON MODELING   Computer Graphics

1. The data matrix $D = \begin{bmatrix} 0 & 1 & 1 & 0 \\ 0 & 0 & 1 & 1 \end{bmatrix}$ represents the gray square.

Reflection using $T = \begin{bmatrix} 1 & 0 \\ 0 & -1 \end{bmatrix}$:

$TD = \begin{bmatrix} 1 & 0 \\ 0 & -1 \end{bmatrix} \begin{bmatrix} 0 & 1 & 1 & 0 \\ 0 & 0 & 1 & 1 \end{bmatrix} = \begin{bmatrix} 0 & 1 & 1 & 0 \\ 0 & 0 & -1 & -1 \end{bmatrix}$

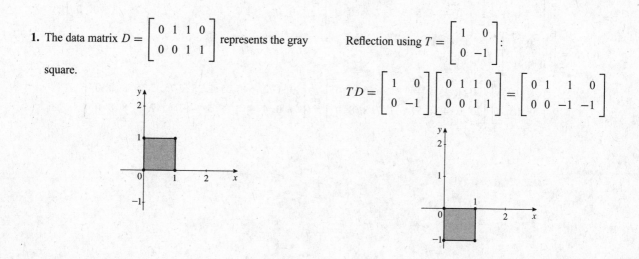

Expansion with $c = 2$ using $T = \begin{bmatrix} 2 & 0 \\ 0 & 1 \end{bmatrix}$:

$$TD = \begin{bmatrix} 2 & 0 \\ 0 & 1 \end{bmatrix} \begin{bmatrix} 0 & 1 & 1 & 0 \\ 0 & 0 & 1 & 1 \end{bmatrix} = \begin{bmatrix} 0 & 2 & 2 & 0 \\ 0 & 0 & 1 & 1 \end{bmatrix}$$

Shearing with $c = 1$ using $T = \begin{bmatrix} 1 & 1 \\ 0 & 1 \end{bmatrix}$:

$$TD = \begin{bmatrix} 1 & 1 \\ 0 & 1 \end{bmatrix} \begin{bmatrix} 0 & 1 & 1 & 0 \\ 0 & 0 & 1 & 1 \end{bmatrix} = \begin{bmatrix} 0 & 1 & 2 & 1 \\ 0 & 0 & 1 & 1 \end{bmatrix}$$

3. (a) $T = \begin{bmatrix} 1 & 1.5 \\ 0 & 1 \end{bmatrix}$ is a shear in the x-direction.

(b) $T^{-1} = \dfrac{1}{1} \begin{bmatrix} 1 & -1.5 \\ 0 & 1 \end{bmatrix} = \begin{bmatrix} 1 & -1.5 \\ 0 & 1 \end{bmatrix}$

(c) T^{-1} is a leftward shear in the x-direction.

(d) The result is the original matrix. Algebraically, $T^{-1}(TD) = \left(T^{-1}T\right)D = ID = D$ where I is the 2×2 identity

matrix: $\begin{bmatrix} 1 & -1.5 \\ 0 & 1 \end{bmatrix} \left(\begin{bmatrix} 1 & 1.5 \\ 0 & 1 \end{bmatrix} \begin{bmatrix} 0 & 1 & 1 & 0 \\ 0 & 0 & 1 & 1 \end{bmatrix} \right) = \begin{bmatrix} 1 & -1.5 \\ 0 & 1 \end{bmatrix} \begin{bmatrix} 0 & 1 & 2.5 & 1.5 \\ 0 & 0 & 1 & 1 \end{bmatrix} = \begin{bmatrix} 0 & 1 & 1 & 0 \\ 0 & 0 & 1 & 1 \end{bmatrix}$

5. (a) $D = \begin{bmatrix} 0 & 1 & 1 & 4 & 4 & 1 & 1 & 6 & 6 & 0 & 0 \\ 0 & 0 & 4 & 4 & 5 & 5 & 7 & 7 & 8 & 8 & 0 \end{bmatrix}$

(b) $T = \begin{bmatrix} 0.75 & 0 \\ 0 & 1 \end{bmatrix}$,

$TD = \begin{bmatrix} 0.75 & 0 \\ 0 & 1 \end{bmatrix} \begin{bmatrix} 0 & 1 & 1 & 4 & 4 & 1 & 1 & 6 & 6 & 0 & 0 \\ 0 & 0 & 4 & 4 & 5 & 5 & 7 & 7 & 8 & 8 & 0 \end{bmatrix} = \begin{bmatrix} 0 & 0.75 & 0.75 & 3 & 3 & 0.75 & 0.75 & 4.5 & 4.5 & 0 & 0 \\ 0 & 0 & 4 & 4 & 5 & 5 & 7 & 7 & 8 & 8 & 0 \end{bmatrix}$

(c) $S = \begin{bmatrix} 1 & 0.25 \\ 0 & 1 \end{bmatrix}$,

$SD = \begin{bmatrix} 1 & 0.25 \\ 0 & 1 \end{bmatrix} \begin{bmatrix} 0 & 1 & 1 & 4 & 4 & 1 & 1 & 6 & 6 & 0 & 0 \\ 0 & 0 & 4 & 4 & 5 & 5 & 7 & 7 & 8 & 8 & 0 \end{bmatrix} = \begin{bmatrix} 0 & 1 & 2 & 5 & 5.25 & 2.25 & 2.75 & 7.75 & 8 & 2 & 0 \\ 0 & 0 & 4 & 4 & 5 & 5 & 7 & 7 & 8 & 8 & 0 \end{bmatrix}$

7 CONIC SECTIONS

7.1 PARABOLAS

1. A parabola is the set of all points in the plane equidistant from a fixed point called the *focus* and a fixed line called the *directrix* of the parabola.

3. The graph of the equation $y^2 = 4px$ is a parabola with focus $F(p, 0)$ and directrix $x = -p$. So the graph of $y^2 = 12x$ is a parabola with focus $F(3, 0)$ and directrix $x = -3$.

5. $y^2 = 2x$ is Graph III, which opens to the right and is not as wide as the graph for Exercise 5.

7. $x^2 = -6y$ is Graph II, which opens downward and is narrower than the graph for Exercise 6.

9. $y^2 - 8x = 0$ is Graph VI, which opens to the right and is wider than the graph for Exercise 1.

11. (a) $x^2 = 8y$, so $4p = 8 \Leftrightarrow p = 2$. The focus is $(0, 2)$, the directrix is $y = -2$, and the focal diameter is 8.

(b)

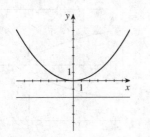

13. (a) $y^2 = -24x$, so $4p = -24 \Leftrightarrow p = -6$. The focus is $(-6, 0)$, the directrix is $x = 6$, and the focal diameter is 24.

(b)

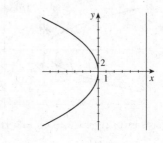

15. (a) $y = -\frac{1}{8}x^2 \Leftrightarrow x^2 = -8y$, so $4p = -8 \Leftrightarrow p = -2$. The focus is $(0, -2)$, the directrix is $y = 2$, and the focal diameter is 8.

(b)

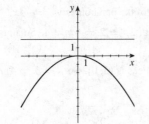

17. (a) $x = -2y^2 \Leftrightarrow y^2 = -\frac{1}{2}x$, so $4p = -\frac{1}{2} \Leftrightarrow p = -\frac{1}{8}$. The focus is $\left(-\frac{1}{8}, 0\right)$, the directrix is $x = \frac{1}{8}$, and the focal diameter is $\frac{1}{2}$.

(b)

291

19. (a) $5y = x^2$, so $4p = 5 \Leftrightarrow p = \frac{5}{4}$. The focus is $\left(0, \frac{5}{4}\right)$, the directrix is $y = -\frac{5}{4}$, and the focal diameter is 5.

(b)

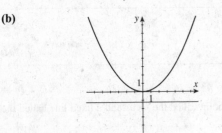

21. (a) $x^2 + 12y = 0 \Leftrightarrow x^2 = -12y$, so $4p = -12 \Leftrightarrow p = -3$. The focus is $(0, -3)$, the directrix is $y = 3$, and the focal diameter is 12.

(b)

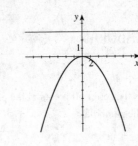

23. (a) $5x + 3y^2 = 0 \Leftrightarrow y^2 = -\frac{5}{3}x$. Then $4p = -\frac{5}{3} \Leftrightarrow p = -\frac{5}{12}$. The focus is $\left(-\frac{5}{12}, 0\right)$, the directrix is $x = \frac{5}{12}$, and the focal diameter is $\frac{5}{3}$.

(b)

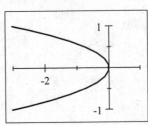

25. $x^2 = 16y$

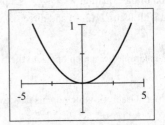

27. $y^2 = -\frac{1}{3}x$

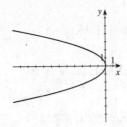

29. $4x + y^2 = 0$

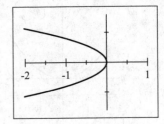

31. Since the focus is $(0, 2)$, $p = 2 \Leftrightarrow 4p = 8$. Hence, an equation of the parabola is $x^2 = 8y$.

33. Since the focus is $(-8, 0)$, $p = -8 \Leftrightarrow 4p = -32$. Hence, an equation of the parabola is $y^2 = -32x$.

35. Since the focus is $\left(0, -\frac{3}{4}\right)$, $p = -\frac{3}{4} \Leftrightarrow 4p = -3$. Hence, an equation of the parabola is $x^2 = -3y$.

37. Since the directrix is $x = 2$, $p = -2 \Leftrightarrow 4p = -8$. Hence, an equation of the parabola is $y^2 = -8x$.

39. Since the directrix is $y = -10$, $p = 10 \Leftrightarrow 4p = 40$. Hence, an equation of the parabola is $x^2 = 40y$.

41. Since the directrix is $x = \frac{1}{20}$, $p = -\frac{1}{20} \Leftrightarrow 4p = -\frac{1}{5}$. Hence, an equation of the parabola is $y^2 = -\frac{1}{5}x$.

43. The focus is on the positive x-axis, so the parabola opens horizontally with $2p = 2 \Leftrightarrow 4p = 4$. So an equation of the parabola is $y^2 = 4x$.

45. Since the parabola opens upward with focus 5 units from the vertex, the focus is $(5, 0)$. So $p = 5 \Leftrightarrow 4p = 20$. Thus an equation of the parabola is $x^2 = 20y$.

47. $p = 2 \Leftrightarrow 4p = 8$. Since the parabola opens upward, its equation is $x^2 = 8y$.

49. $p = 4 \Leftrightarrow 4p = 16$. Since the parabola opens to the left, its equation is $y^2 = -16x$.

51. The focal diameter is $4p = \frac{3}{2} + \frac{3}{2} = 3$. Since the parabola opens to the left, its equation is $y^2 = -3x$.

53. The equation of the parabola has the form $y^2 = 4px$. Since the parabola passes through the point $(4, -2)$, $(-2)^2 = 4p\,(4)$ $\Leftrightarrow 4p = 1$, and so an equation is $y^2 = x$.

55. The area of the shaded region is width × height $= 4p \cdot p = 8$, and so $p^2 = 2 \Leftrightarrow p = -\sqrt{2}$ (because the parabola opens downward). Therefore, an equation is $x^2 = 4py = -4\sqrt{2}y \Leftrightarrow x^2 = -4\sqrt{2}y$.

57. (a) A parabola with directrix $y = -p$ has equation $x^2 = 4py$. If the directrix

is $y = \frac{1}{2}$, then $p = -\frac{1}{2}$, so an equation is $x^2 = 4\left(-\frac{1}{2}\right)y \Leftrightarrow x^2 = -2y$.

If the directrix is $y = 1$, then $p = -1$, so an equation is $x^2 = 4\,(-1)\,y \Leftrightarrow$

$x^2 = -4y$. If the directrix is $y = 4$, then $p = -4$, so an equation is

$x^2 = 4\,(-4)\,y \Leftrightarrow x^2 = -16y$. If the directrix is $y = 8$, then $p = -8$, so

an equation is $x^2 = 4\,(-8)\,y \Leftrightarrow x^2 = -32y$.

(b)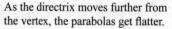

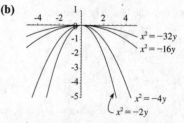

As the directrix moves further from the vertex, the parabolas get flatter.

59. (a) Since the focal diameter is 12 cm, $4p = 12$. Hence, the parabola has equation $y^2 = 12x$.

(b) At a point 20 cm horizontally from the vertex, the parabola passes through the point $(20, y)$, and hence from part (a), $y^2 = 12\,(20) \Leftrightarrow y^2 = 240 \Leftrightarrow y = \pm 4\sqrt{15}$. Thus, $|CD| = 8\sqrt{15} \approx 31$ cm.

61. With the vertex at the origin, the top of one tower will be at the point $(300, 150)$. Inserting this point into the equation $x^2 = 4py$ gives $(300)^2 = 4p\,(150) \Leftrightarrow 90000 = 600p \Leftrightarrow p = 150$. So an equation of the parabolic part of the cables is $x^2 = 4\,(150)\,y \Leftrightarrow x^2 = 600y$.

63. Many answers are possible: satellite dish TV antennas, sound surveillance equipment, solar collectors for hot water heating or electricity generation, bridge pillars, etc.

7.2   ELLIPSES

1. An ellipse is the set of all points in the plane for which the *sum* of the distances from two fixed points F_1 and F_2 is constant. The points F_1 and F_2 are called the *foci* of the ellipse.

3. The graph of the equation $\dfrac{x^2}{b^2} + \dfrac{y^2}{a^2} = 1$ with $a > b > 0$ is an ellipse with vertices $(0, a)$ and $(0, -a)$ and foci $(0, \pm c)$,

where $c = \sqrt{a^2 - b^2}$. So the graph of $\dfrac{x^2}{4^2} + \dfrac{y^2}{5^2} = 1$ is an ellipse with vertices $(0, 5)$ and $(0, -5)$ and foci $(0, 3)$ and $(0, -3)$.

5. $\dfrac{x^2}{16} + \dfrac{y^2}{4} = 1$ is Graph II. The major axis is horizontal and the vertices are $(\pm 4, 0)$.

7. $4x^2 + y^2 = 4$ is Graph I. The major axis is vertical and the vertices are $(0, \pm 2)$.

9. $\dfrac{x^2}{25} + \dfrac{y^2}{9} = 1$.

(a) This ellipse has $a = 5$, $b = 3$, and so $c^2 = a^2 - b^2 = 16 \Leftrightarrow c = 4$. The vertices

are $(\pm 5, 0)$, the foci are $(\pm 4, 0)$, and the eccentricity is $e = \dfrac{c}{a} = \dfrac{4}{5} = 0.8$.

(b) The length of the major axis is $2a = 10$, and the length of the minor axis is $2b = 6$.

(c)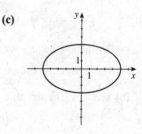

11. $\dfrac{x^2}{36} + \dfrac{y^2}{81} = 1$

(c)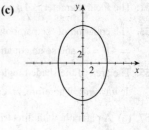

(a) This ellipse has $a = 9$, $b = 6$, and so $c^2 = 81 - 36 = 45 \Leftrightarrow c = 3\sqrt{5}$. The
vertices are $(0, \pm 9)$, the foci are $\left(0, \pm 3\sqrt{5}\right)$, and the eccentricity is

$e = \dfrac{c}{a} = \dfrac{\sqrt{5}}{3}$.

(b) The length of the major axis is $2a = 18$ and the length of the minor axis is
$2b = 12$.

13. $\dfrac{x^2}{49} + \dfrac{y^2}{25} = 1$

(c)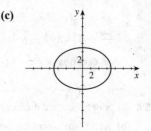

(a) This ellipse has $a = 7$, $b = 5$, and so $c^2 = 49 - 25 = 24 \Leftrightarrow c = 2\sqrt{6}$. The
vertices are $(\pm 7, 0)$, the foci are $\left(\pm 2\sqrt{6}, 0\right)$, and the eccentricity is

$e = \dfrac{c}{a} = \dfrac{2\sqrt{6}}{7}$.

(b) The length of the major axis is $2a = 14$ and the length of the minor axis is
$2b = 10$.

15. $9x^2 + 4y^2 = 36 \Leftrightarrow \dfrac{x^2}{4} + \dfrac{y^2}{9} = 1$

(c)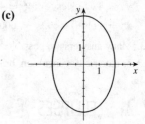

(a) This ellipse has $a = 3$, $b = 2$, and so $c^2 = 9 - 4 = 5 \Leftrightarrow c = \sqrt{5}$. The vertices
are $(0, \pm 3)$, the foci are $\left(0, \pm\sqrt{5}\right)$, and the eccentricity is $e = \dfrac{c}{a} = \dfrac{\sqrt{5}}{3}$.

(b) The length of the major axis is $2a = 6$, and the length of the minor axis is
$2b = 4$.

17. $x^2 + 4y^2 = 16 \Leftrightarrow \dfrac{x^2}{16} + \dfrac{y^2}{4} = 1$

(c)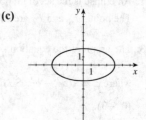

(a) This ellipse has $a = 4$, $b = 2$, and so $c^2 = 16 - 4 = 12 \Leftrightarrow c = 2\sqrt{3}$. The
vertices are $(\pm 4, 0)$, the foci are $\left(\pm 2\sqrt{3}, 0\right)$, and the eccentricity is

$e = \dfrac{c}{a} = \dfrac{2\sqrt{3}}{4} = \dfrac{\sqrt{3}}{2}$.

(b) The length of the major axis is $2a = 8$, and the length of the minor axis is
$2b = 4$.

19. $16x^2 + 25y^2 = 1600 \Leftrightarrow \dfrac{x^2}{100} + \dfrac{y^2}{64} = 1$

(c)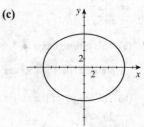

(a) This ellipse has $a = 10$, $b = 8$, and so $c^2 = 100 - 64 = 36 \Leftrightarrow c = 6$. The
vertices are $(\pm 10, 0)$, the foci are $(\pm 6, 0)$, and the eccentricity is $e = \dfrac{c}{a} = \dfrac{3}{5}$.

(b) The length of the major axis is $2a = 20$ and the length of the minor axis is
$2b = 16$.

21. $3x^2 + y^2 = 9 \Leftrightarrow \dfrac{x^2}{3} + \dfrac{y^2}{9} = 1$ **(c)**

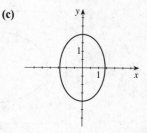

    (a) This ellipse has $a = 3$, $b = \sqrt{3}$, and so $c^2 = 9 - 3 = 6 \Leftrightarrow c = \sqrt{6}$. The

        vertices are $(0, \pm 3)$, the foci are $\left(0, \pm\sqrt{6}\right)$, and the eccentricity is

        $e = \dfrac{c}{a} = \dfrac{\sqrt{6}}{3}$.

    (b) The length of the major axis is $2a = 6$ and the length of the minor axis is

        $2b = 2\sqrt{3}$.

23. $2x^2 + y^2 = 4 \Leftrightarrow \dfrac{x^2}{2} + \dfrac{y^2}{4} = 1$ **(c)**

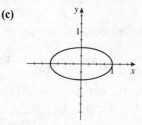

    (a) This ellipse has $a = 2$, $b = \sqrt{2}$, and so $c^2 = 4 - 2 = 2 \Leftrightarrow c = \sqrt{2}$. The

        vertices are $(0, \pm 2)$, the foci are $\left(0, \pm\sqrt{2}\right)$, and the eccentricity is

        $e = \dfrac{c}{a} = \dfrac{\sqrt{2}}{2}$.

    (b) The length of the major axis is $2a = 4$ and the length of the minor axis is

        $2b = 2\sqrt{2}$.

25. $x^2 + 4y^2 = 1 \Leftrightarrow \dfrac{x^2}{1} + \dfrac{y^2}{\frac{1}{4}} = 1$ **(c)**

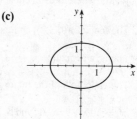

    (a) This ellipse has $a = 1$, $b = \frac{1}{2}$, and so $c^2 = 1 - \frac{1}{4} = \frac{3}{4} \Leftrightarrow c = \dfrac{\sqrt{3}}{2}$. The vertices

        are $(\pm 1, 0)$, the foci are $\left(\pm\dfrac{\sqrt{3}}{2}, 0\right)$, and the eccentricity is

        $e = \dfrac{c}{a} = \dfrac{\sqrt{3}/2}{1} = \dfrac{\sqrt{3}}{2}$.

    (b) The length of the major axis is $2a = 2$, and the length of the minor axis is

        $2b = 1$.

27. $x^2 = 4 - 2y^2 \Leftrightarrow x^2 + 2y^2 = 4 \Leftrightarrow \dfrac{x^2}{4} + \dfrac{y^2}{2} = 1$ **(c)**

    (a) This ellipse has $a = 2$, $b = \sqrt{2}$, and so $c^2 = 4 - 2 = 2 \Leftrightarrow c = \sqrt{2}$. The

        vertices are $(\pm 2, 0)$, the foci are $\left(\pm\sqrt{2}, 0\right)$, and the eccentricity is

        $e = \dfrac{c}{a} = \dfrac{\sqrt{2}}{2}$.

    (b) The length of the major axis is $2a = 4$, and the length of the minor axis is

        $2b = 2\sqrt{2}$.

29. This ellipse has a horizontal major axis with $a = 5$ and $b = 4$, so an equation is $\dfrac{x^2}{(5)^2} + \dfrac{y^2}{(4)^2} = 1 \Leftrightarrow \dfrac{x^2}{25} + \dfrac{y^2}{16} = 1$.

31. This ellipse has a vertical major axis with $c = 2$ and $b = 2$. So $a^2 = c^2 + b^2 = 2^2 + 2^2 = 8 \Leftrightarrow a = 2\sqrt{2}$. So an equation

    is $\dfrac{x^2}{(2)^2} + \dfrac{y^2}{\left(2\sqrt{2}\right)^2} = 1 \Leftrightarrow \dfrac{x^2}{4} + \dfrac{y^2}{8} = 1$.

33. This ellipse has a horizontal major axis with $a = 16$, so an equation of the ellipse is of the form $\dfrac{x^2}{16^2} + \dfrac{y^2}{b^2} = 1$. Substituting

the point $(8, 6)$ into the equation, we get $\dfrac{64}{256} + \dfrac{36}{b^2} = 1 \Leftrightarrow \dfrac{36}{b^2} = 1 - \dfrac{1}{4} \Leftrightarrow \dfrac{36}{b^2} = \dfrac{3}{4} \Leftrightarrow b^2 = \dfrac{4\,(36)}{3} = 48$. Thus, an

equation of the ellipse is $\dfrac{x^2}{256} + \dfrac{y^2}{48} = 1$.

35. $\dfrac{x^2}{25} + \dfrac{y^2}{20} = 1 \Leftrightarrow \dfrac{y^2}{20} = 1 - \dfrac{x^2}{25} \Leftrightarrow y^2 = 20 - \dfrac{4x^2}{5} \Rightarrow$ **37.** $6x^2 + y^2 = 36 \Leftrightarrow y^2 = 36 - 6x^2 \Rightarrow y = \pm\sqrt{36 - 6x^2}$.

$y = \pm\sqrt{20 - \dfrac{4x^2}{5}}$.

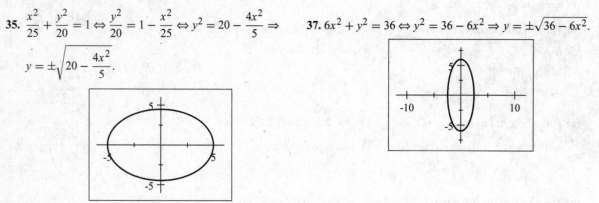

39. The foci are $(\pm 4, 0)$, and the vertices are $(\pm 5, 0)$. Thus, $c = 4$ and $a = 5$, and so $b^2 = 25 - 16 = 9$. Therefore, an equation

of the ellipse is $\dfrac{x^2}{25} + \dfrac{y^2}{9} = 1$.

41. The foci are $(\pm 1, 0)$ and the vertices are $(\pm 2, 0)$. Thus, $c = 1$ and $a = 2$, so $c^2 = a^2 - b^2 \Leftrightarrow 1 = 4 - b^2 \Leftrightarrow b^2 = 4 - 1 = 3$.

Therefore, an equation of the ellipse is $\dfrac{x^2}{4} + \dfrac{y^2}{3} = 1$.

43. The foci are $\left(0, \pm\sqrt{10}\right)$ and the vertices are $(0, \pm 7)$. Thus, $c = \sqrt{10}$ and $a = 7$, so $c^2 = a^2 - b^2 \Leftrightarrow 10 = 49 - b^2 \Leftrightarrow$

$b^2 = 49 - 10 = 39$. Therefore, an equation of the ellipse is $\dfrac{x^2}{39} + \dfrac{y^2}{49} = 1$.

45. The length of the major axis is $2a = 4 \Leftrightarrow a = 2$, the length of the minor axis is $2b = 2 \Leftrightarrow b = 1$, and the foci are on the

y-axis. Therefore, an equation of the ellipse is $x^2 + \dfrac{y^2}{4} = 1$.

47. The foci are $(0, \pm 2)$, and the length of the minor axis is $2b = 6 \Leftrightarrow b = 3$. Thus, $a^2 = 4 + 9 = 13$. Since the foci are on the

y-axis, an equation is $\dfrac{x^2}{9} + \dfrac{y^2}{13} = 1$.

49. The endpoints of the major axis are $(\pm 10, 0) \Leftrightarrow a = 10$, and the distance between the foci is $2c = 6 \Leftrightarrow c = 3$. Therefore,

$b^2 = 100 - 9 = 91$, and so an equation of the ellipse is $\dfrac{x^2}{100} + \dfrac{y^2}{91} = 1$.

51. The length of the major axis is 10, so $2a = 10 \Leftrightarrow a = 5$, and the foci are on the x-axis, so the form of the equation is

$\dfrac{x^2}{25} + \dfrac{y^2}{b^2} = 1$. Since the ellipse passes through $\left(\sqrt{5}, 2\right)$, we have $\dfrac{\left(\sqrt{5}\right)^2}{25} + \dfrac{(2)^2}{b^2} = 1 \Leftrightarrow \dfrac{5}{25} + \dfrac{4}{b^2} = 1 \Leftrightarrow \dfrac{4}{b^2} = \dfrac{4}{5} \Leftrightarrow$

$b^2 = 5$, and so an equation is $\dfrac{x^2}{25} + \dfrac{y^2}{5} = 1$.

53. The eccentricity is $\dfrac{1}{3}$, so $e = \dfrac{1}{3}$, and the foci are $(0, \pm 2)$, so $c = 2$. Thus, $e = \dfrac{c}{a} \Leftrightarrow a = \dfrac{c}{e} = \dfrac{2}{1/3} = 6$ and

$b^2 = a^2 - c^2 = 36 - 4 = 32$. The major axis lies on the y-axis, so an equation is $\dfrac{x^2}{32} + \dfrac{y^2}{36} = 1$.

55. Since the length of the major axis is $2a = 4$, we have $a = 2$. The eccentricity is $\frac{\sqrt{3}}{2} = \frac{c}{a} = \frac{c}{2}$, so $c = \sqrt{3}$. Then

$b^2 = a^2 - c^2 = 4 - 3 = 1$, and since the foci are on the y-axis, an equation of the ellipse is $x^2 + \frac{y^2}{4} = 1$.

57. $\begin{cases} 4x^2 + y^2 = 4 \\ 4x^2 + 9y^2 = 36 \end{cases}$ Subtracting the first equation from the second gives

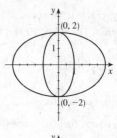

$8y^2 = 32 \Leftrightarrow y^2 = 4 \Leftrightarrow y = \pm 2$. Substituting $y = \pm 2$ in the first equation gives

$4x^2 + (\pm 2)^2 = 4 \Leftrightarrow x = 0$, and so the points of intersection are $(0, \pm 2)$.

59. $\begin{cases} 100x^2 + 25y^2 = 100 \\ x^2 + \frac{y^2}{9} = 1 \end{cases}$ Dividing the first equation by 100 gives $x^2 + \frac{y^2}{4} = 1$.

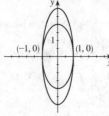

Subtracting this equation from the second equation gives $\frac{y^2}{9} - \frac{y^2}{4} = 0 \Leftrightarrow$

$\left(\frac{1}{9} - \frac{1}{4}\right) y^2 = 0 \Leftrightarrow y = 0$. Substituting $y = 0$ in the second equation gives

$x^2 + (0)^2 = 1 \Leftrightarrow x = \pm 1$, and so the points of intersection are $(\pm 1, 0)$.

61. (a) The ellipse $x^2 + 4y^2 = 16 \Leftrightarrow \frac{x^2}{16} + \frac{y^2}{4} = 1$ has $a = 4$ and $b = 2$. Thus, an equation of the ancillary circle is

$x^2 + y^2 = 4$.

(b) If (s, t) is a point on the ancillary circle, then $s^2 + t^2 = 4 \Leftrightarrow 4s^2 + 4t^2 = 16 \Leftrightarrow (2s)^2 + 4(t)^2 = 16$, which implies

that $(2s, t)$ is a point on the ellipse.

63. $\frac{x^2}{k} + \frac{y^2}{4+k} = 1$ is an ellipse for $k > 0$. Then $a^2 = 4 + k$, $b^2 = k$, and so $c^2 = 4 + k - k = 4 \Leftrightarrow c = \pm 2$. Therefore, all of

the ellipses' foci are $(0, \pm 2)$ regardless of the value of k.

65. Using the eccentricity, $e = 0.25 = \frac{c}{a} \Leftrightarrow c = 0.25a$. Using the length of the minor axis, $2b = 10{,}000{,}000{,}000 \Leftrightarrow$

$b = 5 \times 10^9$. Since $a^2 = c^2 + b^2$, $a^2 = (0.25a)^2 + 25 \times 10^{18} \Leftrightarrow \frac{15}{16}a^2 = 25 \times 10^{18} \Leftrightarrow a^2 = \frac{80}{3} \times 10^{18} \Leftrightarrow$

$a = \sqrt{\frac{80}{3}} \times 10^9 = 4\sqrt{\frac{5}{3}} \times 10^9$. Then $c = 0.25\left(4\sqrt{\frac{5}{3}} \times 10^9\right) = \sqrt{\frac{5}{3}} \times 10^9$. Since the Sun is at one focus of the ellipse,

the distance from Pluto to the Sun at perihelion is $a - c = 4\sqrt{\frac{5}{3}} \times 10^9 - \sqrt{\frac{5}{3}} \times 10^9 = 3\sqrt{\frac{5}{3}} \times 10^9 \approx 3.87 \times 10^9$ km; the

distance from Pluto to the Sun at aphelion is $a + c = 4\sqrt{\frac{5}{3}} \times 10^9 + \sqrt{\frac{5}{3}} \times 10^9 = 5\sqrt{\frac{5}{3}} \times 10^9 \approx 6.45 \times 10^9$ km.

67. Placing the origin at the center of the sheet of plywood and letting the x-axis be the long central axis, we have $2a = 8$, so

that $a = 4$, and $2b = 4$, so that $b = 2$. So $c^2 = a^2 - b^2 = 4^2 - 2^2 = 12 \Rightarrow c = 2\sqrt{3} \approx 3.46$. So the tacks should be

located $2(3.46) = 6.92$ feet apart and the string should be $2a = 8$ feet long.

69. Have each friend hold one end of the string on the blackboard. These fixed points will be the foci. Then, keeping the string

taut with the chalk, draw the ellipse.

71. The foci are $(\pm c, 0)$, where $c^2 = a^2 - b^2$. The endpoints of one *latus rectum* are the points $(c, \pm k)$, and the length is $2k$.

Substituting this point into the equation, we get $\frac{c^2}{a^2} + \frac{k^2}{b^2} = 1 \Leftrightarrow \frac{k^2}{b^2} = 1 - \frac{c^2}{a^2} = \frac{a^2 - c^2}{a^2} \Leftrightarrow k^2 = \frac{b^2\left(a^2 - c^2\right)}{a^2}$. Since

$b^2 = a^2 - c^2$, the last equation becomes $k^2 = \frac{b^4}{a^2} \Rightarrow k = \frac{b^2}{a}$. Thus, the length of the *latus rectum* is $2k = 2\left(\frac{b^2}{a}\right) = \frac{2b^2}{a}$.

7.3 HYPERBOLAS

1. A hyperbola is the set of all points in the plane for which the *difference* of the distances from two fixed point F_1 and F_2 is constant. The points F_1 and F_2 are called the *foci* of the hyperbola.

3. The graph of the equation $\dfrac{y^2}{a^2} - \dfrac{x^2}{b^2} = 1$ with $a > 0$, $b > 0$ is a hyperbola with *vertical* transverse axis, vertices $(0, a)$ and $(0, -a)$ and foci $(0, \pm c)$, where $c = \sqrt{a^2 + b^2}$. So the graph of $\dfrac{y^2}{4^2} - \dfrac{x^2}{3^2} = 1$ is a hyperbola with vertices $(0, 4)$ and $(0, -4)$ and foci $(0, 5)$ and $(0, -5)$.

5. $\dfrac{x^2}{4} - y^2 = 1$ is Graph III, which opens horizontally and has vertices at $(\pm 2, 0)$.

7. $16y^2 - x^2 = 144$ is Graph II, which pens vertically and has vertices at $(0, \pm 3)$.

9. $\dfrac{x^2}{4} - \dfrac{y^2}{16} = 1$ **(c)**

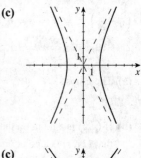

    (a) The hyperbola has $a = 2$, $b = 4$, and $c^2 = 16 + 4 \Rightarrow c = 2\sqrt{5}$. The vertices are $(\pm 2, 0)$, the foci are $\left(\pm 2\sqrt{5}, 0\right)$, and the asymptotes are $y = \pm \frac{4}{2}x \Leftrightarrow y = \pm 2x$.

    (b) The transverse axis has length $2a = 4$.

11. $\dfrac{y^2}{36} - \dfrac{x^2}{4} = 1$ **(c)**

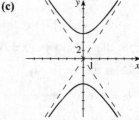

    (a) The hyperbola has $a = 6$, $b = 2$, and $c^2 = 36 + 4 = 40 \Rightarrow c = 2\sqrt{10}$. The vertices are $(0, \pm 6)$, the foci are $\left(0, \pm 2\sqrt{10}\right)$, and the asymptotes are $y = \pm 3x$.

    (b) The transverse axis has length $2a = 12$.

13. $\dfrac{y^2}{1} - \dfrac{x^2}{25} = 1$ **(c)**

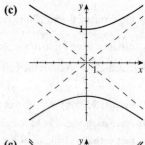

    (a) The hyperbola has $a = 1$, $b = 5$, and $c^2 = 1 + 25 = 26 \Rightarrow c = \sqrt{26}$. The vertices are $(0, \pm 1)$, the foci are $\left(0, \pm\sqrt{26}\right)$, and the asymptotes are $y = \pm\frac{1}{5}x$.

    (b) The transverse axis has length $2a = 2$.

15. $x^2 - y^2 = 1$ **(c)**

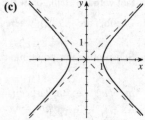

    (a) The hyperbola has $a = 1$, $b = 1$, and $c^2 = 1 + 1 = 2 \Rightarrow c = \sqrt{2}$. The vertices are $(\pm 1, 0)$, the foci are $\left(\pm\sqrt{2}, 0\right)$, and the asymptotes are $y = \pm x$.

    (b) The transverse axis has length $2a = 2$.

17. $9x^2 - 4y^2 = 36 \Leftrightarrow \dfrac{x^2}{4} - \dfrac{y^2}{9} = 1$

(c)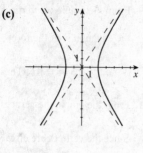

    (a) The hyperbola has $a = 2$, $b = 3$, and $c^2 = 4 + 9 = 13 \Rightarrow c = \sqrt{13}$. The

    vertices are $(\pm 2, 0)$, the foci are $\left(\pm\sqrt{13}, 0\right)$, and the asymptotes are $y = \pm\dfrac{3}{2}x$.

    (b) The transverse axis has length $2a = 4$.

19. $4y^2 - 9x^2 = 144 \Leftrightarrow \dfrac{y^2}{36} - \dfrac{x^2}{16} = 1$

(c)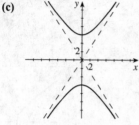

    (a) The hyperbola has $a = 6$, $b = 4$, and $c^2 = a^2 + b^2 = 52 \Rightarrow c = 2\sqrt{13}$. The

    vertices are $(0, \pm 6)$, the foci are $\left(0, \pm 2\sqrt{13}\right)$, and the asymptotes are

    $y = \pm\dfrac{3}{2}x$.

    (b) The transverse axis has length $2a = 12$.

21. $x^2 - 4y^2 - 8 = 0 \Leftrightarrow \dfrac{x^2}{8} - \dfrac{y^2}{2} = 1$

(c)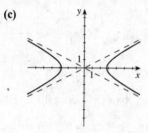

    (a) The hyperbola has $a = 2\sqrt{2}$, $b = \sqrt{2}$, and $c^2 = 8 + 2 = 10 \Rightarrow c = \sqrt{10}$. The

    vertices are $\left(\pm 2\sqrt{2}, 0\right)$, the foci are $\left(\pm\sqrt{10}, 0\right)$, and the asymptotes are

    $y = \pm\dfrac{\sqrt{2}}{\sqrt{8}}x = \pm\dfrac{1}{2}x$.

    (b) The transverse axis has length $2a = 4\sqrt{2}$.

23. $x^2 - y^2 + 4 = 0 \Leftrightarrow y^2 - x^2 = 4 \Leftrightarrow \dfrac{y^2}{4} - \dfrac{x^2}{4} = 1$

(c)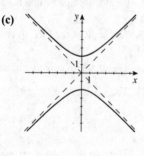

    (a) The hyperbola has $a = 2$, $b = 2$, and $c^2 = 4 + 4 = 8 = 2\sqrt{2}$. The vertices are

    $(0, \pm 2)$, the foci are $\left(0, \pm 2\sqrt{2}\right)$, and the asymptotes are $y = \pm x$.

    (b) The transverse axis has length $2a = 4$.

25. $4y^2 - x^2 = 1 \Leftrightarrow \dfrac{y^2}{\frac{1}{4}} - x^2 = 1$

(c)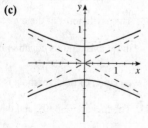

    (a) The hyperbola has $a = \dfrac{1}{2}$, $b = 1$, and $c^2 = \dfrac{1}{4} + 1 = \dfrac{5}{4} \Rightarrow c = \dfrac{\sqrt{5}}{2}$. The

    vertices are $\left(0, \pm\dfrac{1}{2}\right)$, the foci are $\left(0, \pm\dfrac{\sqrt{5}}{2}\right)$, and the asymptotes are

    $y = \pm\dfrac{1/2}{1}x = \pm\dfrac{1}{2}x$.

    (b) The transverse axis has length $2a = 1$.

27. From the graph, the foci are $(\pm 4, 0)$, and the vertices are $(\pm 2, 0)$, so $c = 4$ and $a = 2$. Thus, $b^2 = 16 - 4 = 12$, and since

the vertices are on the x-axis, an equation of the hyperbola is $\dfrac{x^2}{4} - \dfrac{y^2}{12} = 1$.

29. From the graph, the vertices are $(0, \pm 4)$, the foci are on the y-axis, and the hyperbola passes through the point $(3, -5)$. So

the equation is of the form $\dfrac{y^2}{16} - \dfrac{x^2}{b^2} = 1$. Substituting the point $(3, -5)$, we have $\dfrac{(-5)^2}{16} - \dfrac{(3)^2}{b^2} = 1 \Leftrightarrow \dfrac{25}{16} - 1 = \dfrac{9}{b^2} \Leftrightarrow$

$\dfrac{9}{16} = \dfrac{9}{b^2} \Leftrightarrow b^2 = 16$. Thus, an equation of the hyperbola is $\dfrac{y^2}{16} - \dfrac{x^2}{16} = 1$.

31. From the graph, the vertices are $(0, \pm 3)$, so $a = 3$. Since the asymptotes are $y = \pm 3x = \pm \dfrac{a}{b}x$, we have $\dfrac{3}{b} = 3 \Leftrightarrow b = 1$.

Since the vertices are on the x-axis, an equation is $\dfrac{y^2}{3^2} - \dfrac{x^2}{1^2} = 1 \Leftrightarrow \dfrac{y^2}{9} - x^2 = 1$.

33. $x^2 - 2y^2 = 8 \Leftrightarrow 2y^2 = x^2 - 8 \Leftrightarrow y^2 = \frac{1}{2}x^2 - 4 \Rightarrow$ **35.** $\dfrac{y^2}{2} - \dfrac{x^2}{6} = 1 \Leftrightarrow \dfrac{y^2}{2} = \dfrac{x^2}{6} + 1 \Leftrightarrow y^2 = \dfrac{x^2}{3} + 2 \Rightarrow$

$y = \pm\sqrt{\frac{1}{2}x^2 - 4}$ $\qquad y = \pm\sqrt{\dfrac{x^2}{3} + 2}$

37. The foci are $(\pm 5, 0)$ and the vertices are $(\pm 3, 0)$, so $c = 5$ and $a = 3$. Then $b^2 = 25 - 9 = 16$, and since the vertices are on

the x-axis, an equation of the hyperbola is $\dfrac{x^2}{9} - \dfrac{y^2}{16} = 1$.

39. The foci are $(0, \pm 2)$ and the vertices are $(0, \pm 1)$, so $c = 2$ and $a = 1$. Then $b^2 = 4 - 1 = 3$, and since the vertices are on

the y-axis, an equation is $y^2 - \dfrac{x^2}{3} = 1$.

41. The vertices are $(\pm 1, 0)$ and the asymptotes are $y = \pm 5x$, so $a = 1$. The asymptotes are $y = \pm \dfrac{b}{a}x$, so $\dfrac{b}{1} = 5 \Leftrightarrow b = 5$.

Therefore, an equation of the hyperbola is $x^2 - \dfrac{y^2}{25} = 1$.

43. The vertices are $(0, \pm 6)$, so $a = 6$. Since the vertices are on the y-axis, the hyperbola has an equation of the form

$\dfrac{y^2}{36} - \dfrac{x^2}{b^2} = 1$. Since the hyperbola passes through the point $(-5, 9)$, we have $\dfrac{81}{36} - \dfrac{25}{b^2} = 1 \Leftrightarrow \dfrac{25}{b^2} = \dfrac{45}{36} \Leftrightarrow b^2 = 20$.

Thus, an equation is $\dfrac{y^2}{36} - \dfrac{x^2}{20} = 1$.

45. The asymptotes of the hyperbola are $y = \pm x$, so $b = a$. Since the hyperbola passes through the point $(5, 3)$, its foci are on

the x-axis, and its equation has the form, $\dfrac{x^2}{a^2} - \dfrac{y^2}{a^2} = 1$, so it follows that $\dfrac{25}{a^2} - \dfrac{9}{a^2} = 1 \Leftrightarrow a^2 = 16 = b^2$. Therefore, an

equation of the hyperbola is $\dfrac{x^2}{16} - \dfrac{y^2}{16} = 1$.

47. The foci are $(\pm 5, 0)$, and the length of the transverse axis is 6, so $c = 5$ and $2a = 6 \Leftrightarrow a = 3$. Thus, $b^2 = 25 - 9 = 16$, and

an equation is $\dfrac{x^2}{9} - \dfrac{y^2}{16} = 1$.

49. (a) The hyperbola $x^2 - y^2 = 5 \Leftrightarrow \dfrac{x^2}{5} - \dfrac{y^2}{5} = 1$ has $a = \sqrt{5}$ and $b = \sqrt{5}$. Thus, the asymptotes are $y = \pm x$, and their

slopes are $m_1 = 1$ and $m_2 = -1$. Since $m_1 \cdot m_2 = -1$, the asymptotes are perpendicular.

(b) Since the asymptotes are perpendicular, they must have slopes ± 1, so $a = b$. Therefore, $c^2 = 2a^2 \Leftrightarrow a^2 = \dfrac{c^2}{2}$, and

since the vertices are on the x-axis, an equation is $\dfrac{x^2}{\frac{1}{2}c^2} - \dfrac{y^2}{\frac{1}{2}c^2} = 1 \Leftrightarrow x^2 - y^2 = \dfrac{c^2}{2}$.

51. $\sqrt{(x+c)^2 + y^2} - \sqrt{(x-c)^2 + y^2} = \pm 2a$. Let us consider the positive case only. Then

$\sqrt{(x+c)^2 + y^2} = 2a + \sqrt{(x-c)^2 + y^2}$, and squaring both sides gives $x^2 + 2cx + c^2 + y^2 = 4a^2 +$

$4a\sqrt{(x-c)^2 + y^2} + x^2 - 2cx + c^2 + y^2 \Leftrightarrow 4a\sqrt{(x-c)^2 + y^2} = 4cx - 4a^2$. Dividing by 4 and squaring both sides

gives $a^2 \left(x^2 - 2cx + c^2 + y^2\right) = c^2 x^2 - 2a^2 cx + a^4 \Leftrightarrow a^2 x^2 - 2a^2 cx + a^2 c^2 + a^2 y^2 = c^2 x^2 - 2a^2 cx + a^4$

$\Leftrightarrow a^2 x^2 + a^2 c^2 + a^2 y^2 = c^2 x^2 + a^4$. Rearranging the order, we have $c^2 x^2 - a^2 x^2 - a^2 y^2 = a^2 c^2 - a^4 \Leftrightarrow$

$\left(c^2 - a^2\right) x^2 - a^2 y^2 = a^2 \left(c^2 - a^2\right)$. The negative case gives the same result.

53. (a) From the equation, we have $a^2 = k$ and $b^2 = 16 - k$. Thus, $c^2 = a^2 + b^2 = k + 16 - k = 16 \Rightarrow c = \pm 4$. Thus the
foci of the family of hyperbolas are $(0, \pm 4)$.

(b) $\dfrac{y^2}{k} - \dfrac{x^2}{16 - k} = 1 \Leftrightarrow y^2 = k\left(1 + \dfrac{x^2}{16 - k}\right) \Rightarrow y = \pm\sqrt{k + \dfrac{kx^2}{16 - k}}$. For

the top branch, we graph $y = \sqrt{k + \dfrac{kx^2}{16 - k}}$, $k = 1, 4, 8, 12$. As k

increases, the asymptotes get steeper and the vertices move further apart.

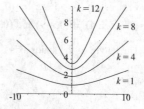

55. Since the asymptotes are perpendicular, $a = b$. Also, since the sun is a focus and the closest distance is 2×10^9, it follows

that $c - a = 2 \times 10^9$. Now $c^2 = a^2 + b^2 = 2a^2$, and so $c = \sqrt{2}a$. Thus, $\sqrt{2}a - a = 2 \times 10^9 \Rightarrow a = \dfrac{2 \times 10^9}{\sqrt{2} - 1}$ and

$a^2 = b^2 = \dfrac{4 \times 10^{18}}{3 - 2\sqrt{2}} \approx 2.3 \times 10^{19}$. Therefore, an equation of the hyperbola is $\dfrac{x^2}{2.3 \times 10^{19}} - \dfrac{y^2}{2.3 \times 10^{19}} = 1 \Leftrightarrow$

$x^2 - y^2 = 2.3 \times 10^{19}$.

57. Some possible answers are: as cross-sections of nuclear power plant cooling towers, or as reflectors for camouflaging the
location of secret installations.

7.4   SHIFTED CONICS

1. (a) If we replace x by $x - 3$ the graph of the equation is shifted to the *right* by 3 units. If we replace x by $x + 3$ the graph is
shifted to the *left* by 3 units.

(b) If we replace y by $y - 1$ the graph of the equation is shifted *upward* by 1 unit. If we replace y by $y + 1$ the graph is
shifted *downward* by 1 unit.

3. $\dfrac{x^2}{5^2} + \dfrac{y^2}{4^2} = 1$, from left to right: vertex $(-5, 0)$, focus $(-3, 0)$, focus $(3, 0)$, vertex $(5, 0)$. $\dfrac{(x-3)^2}{5^2} + \dfrac{(y-1)^2}{4^2} = 1$, from
left to right: vertex $(-2, 1)$, focus $(0, 1)$, focus $(6, 1)$, vertex $(8, 1)$.

5. $\dfrac{(x-2)^2}{9} + \dfrac{(y-1)^2}{4} = 1$

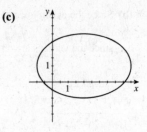

(c)

(a) This ellipse is obtained from the ellipse $\dfrac{x^2}{9} + \dfrac{y^2}{4} = 1$ by shifting it 2 units to the right and 1 unit upward. So $a = 3$, $b = 2$, and $c = \sqrt{9-4} = \sqrt{5}$. The center is $(2, 1)$, the vertices are $(2 \pm 3, 1) = (-1, 1)$ and $(5, 1)$, and the foci are $\left(2 \pm \sqrt{5}, 1\right)$.

(b) The length of the major axis is $2a = 6$ and the length of the minor axis is $2b = 4$.

7. $\dfrac{x^2}{9} + \dfrac{(y+5)^2}{25} = 1$

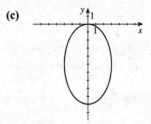

(c)

(a) This ellipse is obtained from the ellipse $\dfrac{x^2}{9} + \dfrac{y^2}{25} = 1$ by shifting it 5 units downward. So $a = 5$, $b = 3$, and $c = \sqrt{25-9} = 4$. The center is $(0, -5)$, the vertices are $(0, -5 \pm 5) = (0, -10)$ and $(0, 0)$, and the foci are $(0, -5 \pm 4) = (0, -9)$ and $(0, -1)$.

(b) The length of the major axis is $2a = 10$ and the length of the minor axis is $2b = 6$.

9. $\dfrac{(x+5)^2}{16} + \dfrac{(y-1)^2}{4} = 1$

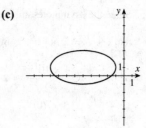

(c)

(a) This ellipse is obtained from the ellipse $\dfrac{x^2}{16} + \dfrac{y^2}{4} = 1$ by shifting it 5 units to the left and 1 units upward. So $a = 4$, $b = 2$, and $c = \sqrt{16-4} = 2\sqrt{3}$. The center is $(-5, 1)$, the vertices are $(-5 \pm 4, 1) = (-9, 1)$ and $(-1, 1)$, and the foci are $\left(-5 \pm 2\sqrt{3}, 1\right) = \left(-5 - 2\sqrt{3}, 1\right)$ and $\left(-5 + 2\sqrt{3}, 1\right)$.

(b) The length of the major axis is $2a = 8$ and the length of the minor axis is $2b = 4$.

11. $4x^2 + 25y^2 - 50y = 75 \Leftrightarrow 4x^2 + 25(y-1)^2 - 25 = 75 \Leftrightarrow \dfrac{x^2}{25} + \dfrac{(y-1)^2}{4} = 1$

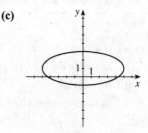

(c)

(a) This ellipse is obtained from the ellipse $\dfrac{x^2}{25} + \dfrac{y^2}{4} = 1$ by shifting it 1 unit upward. So $a = 5$, $b = 2$, and $c = \sqrt{25-4} = \sqrt{21}$. The center is $(0, 1)$, the vertices are $(\pm 5, 1) = (-5, 1)$ and $(5, 1)$, and the foci are $\left(\pm\sqrt{21}, 1\right) = \left(-\sqrt{21}, 1\right)$ and $\left(\sqrt{21}, 1\right)$.

(b) The length of the major axis is $2a = 10$ and the length of the minor axis is $2b = 4$.

13. $(x-3)^2 = 8(y+1)$ **(b)**

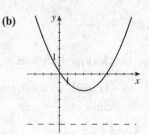

(a) This parabola is obtained from the parabola $x^2 = 8y$ by shifting it 3 units to the right and 1 unit down. So $4p = 8 \Leftrightarrow p = 2$. The vertex is $(3, -1)$, the focus is $(3, -1+2) = (3, 1)$, and the directrix is $y = -1 - 2 = -3$.

15. $(y+5)^2 = -6x + 12 = -6(x-2)$ **(b)**

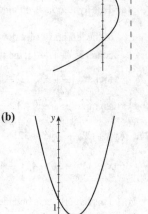

(a) This parabola is obtained from the parabola $y^2 = -6x$ by shifting to the right 2 units and down 5 units. So $4p = -6 \Leftrightarrow p = -\frac{3}{2}$. The vertex is $(2, -5)$, the focus is $\left(2 - \frac{3}{2}, -5\right) = \left(\frac{1}{2}, -5\right)$, and the directrix is $x = 2 + \frac{3}{2} = \frac{7}{2}$.

17. $2(x-1)^2 = y \Leftrightarrow (x-1)^2 = \frac{1}{2}y$ **(b)**

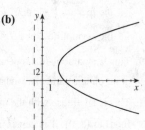

(a) This parabola is obtained from the parabola $x^2 = \frac{1}{2}y$ by shifting it 1 unit to the right. So $4p = \frac{1}{2} \Leftrightarrow p = \frac{1}{8}$. The vertex is $(1, 0)$, the focus is $\left(1, \frac{1}{8}\right)$, and the directrix is $y = -\frac{1}{8}$.

19. $y^2 - 6y - 12x + 33 = 0 \Leftrightarrow (y-3)^2 - 9 - 12x + 33 = 0 \Leftrightarrow (y-3)^2 = 12(x-2)$ **(b)**

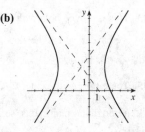

(a) This parabola is obtained from the parabola $y^2 = 12x$ by shifting it 2 units to the right and 3 units upward. So $4p = 12 \Leftrightarrow p = 3$. The vertex is $(2, 3)$, the focus is $(2+3, 3) = (5, 3)$, and the directrix is $x = 2 - 3 = -1$.

21. $\dfrac{(x+1)^2}{9} - \dfrac{(y-3)^2}{16} = 1$ **(b)**

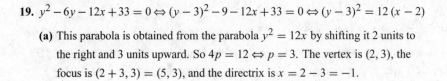

(a) This hyperbola is obtained from the hyperbola $\dfrac{x^2}{9} - \dfrac{y^2}{16} = 1$ by shifting it 1 unit to the left and 3 units up. So $a = 3$, $b = 4$, and $c = \sqrt{9+16} = 5$. The center is $(-1, 3)$, the vertices are $(-1 \pm 3, 3) = (-4, 3)$ and $(2, 3)$, the foci are $(-1 \pm 5, 3) = (-6, 3)$ and $(4, 3)$, and the asymptotes are $(y-3) = \pm\frac{4}{3}(x+1) \Leftrightarrow y = \pm\frac{4}{3}(x+1) + 3 \Leftrightarrow y = \frac{4}{3}x + \frac{13}{3}$ and $y = -\frac{4}{3}x + \frac{5}{3}$.

23. $y^2 - \dfrac{(x+1)^2}{4} = 1$

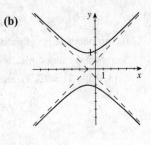

(b)

(a) This hyperbola is obtained from the hyperbola $y^2 - \dfrac{x^2}{4} = 1$ by shifting it 1 unit

to the left. So $a = 1$, $b = 2$, and $c = \sqrt{1+4} = \sqrt{5}$. The center is $(-1, 0)$, the

vertices are $(-1, \pm 1) = (-1, -1)$ and $(-1, 1)$, the foci are

$\left(-1, \pm\sqrt{5}\right) = \left(-1, -\sqrt{5}\right)$ and $\left(-1, \sqrt{5}\right)$, and the asymptotes are

$y = \pm\frac{1}{2}(x+1) \Leftrightarrow y = \frac{1}{2}x + \frac{1}{2}$ and $y = -\frac{1}{2}x - \frac{1}{2}$.

25. $\dfrac{(x+1)^2}{9} - \dfrac{(y+1)^2}{4} = 1$

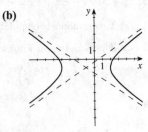

(b)

(a) This hyperbola is obtained from the hyperbola $\dfrac{x^2}{9} - \dfrac{y^2}{4} = 1$ by shifting it 1 unit

to the left and 1 unit downward, so $a = 3$, $b = 2$, and $c = \sqrt{9+4} = \sqrt{13}$. The

center is $(-1, -1)$, the vertices are $(-1 \pm 3, -1) = (-4, -1)$ and $(2, -1)$, the

foci are $\left(-1 \pm \sqrt{13}, -1\right) = \left(-1 - \sqrt{13}, -1\right)$ and $\left(\sqrt{13} - 1, -1\right)$, and the

asymptotes are $y + 1 = \pm\frac{2}{3}(x+1) \Leftrightarrow y = \frac{2}{3}x - \frac{1}{3}$ and $y = -\frac{2}{3}x - \frac{5}{3}$.

27. $36x^2 + 72x - 4y^2 + 32y + 116 = 0 \Leftrightarrow$

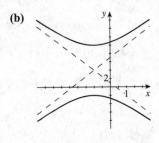

(b)

$36(x+1)^2 - 36 - 4(y-4)^2 + 64 + 116 = 0 \Leftrightarrow \dfrac{(y-4)^2}{36} - \dfrac{(x+1)^2}{4} = 1$

(a) This hyperbola is obtained from the hyperbola $\dfrac{y^2}{36} - \dfrac{x^2}{4} = 1$ by shifting it 1 unit

to the left and 4 units upward. So $a = 6$, $b = 2$, and $c = \sqrt{36+4} = 2\sqrt{10}$.

The center is $(-1, 4)$, the vertices are $(-1, 4 \pm 6) = (-1, -2)$ and $(-1, 10)$,

the foci are $\left(-1, 4 \pm 2\sqrt{10}\right) = \left(-1, -2\sqrt{10} + 4\right)$ and $\left(-1, 4 + 2\sqrt{10}\right)$, and

the asymptotes are $y - 4 = \pm 3(x+1) \Leftrightarrow y = 3x + 7$ and $y = -3x + 1$.

29. This is a parabola that opens down with its vertex at $(0, 4)$, so its equation is of the form $x^2 = a(y - 4)$. Since $(1, 0)$ is a

point on this parabola, we have $(1)^2 = a(0 - 4) \Leftrightarrow 1 = -4a \Leftrightarrow a = -\frac{1}{4}$. Thus, an equation is $x^2 = -\frac{1}{4}(y - 4)$.

31. This is an ellipse with the major axis parallel to the x-axis, with one vertex at $(0, 0)$, the other vertex at $(10, 0)$, and one

focus at $(8, 0)$. The center is at $\left(\dfrac{0+10}{2}, 0\right) = (5, 0)$, $a = 5$, and $c = 3$ (the distance from one focus to the center). So

$b^2 = a^2 - c^2 = 25 - 9 = 16$. Thus, an equation is $\dfrac{(x-5)^2}{25} + \dfrac{y^2}{16} = 1$.

33. This is a hyperbola with center $(0, 1)$ and vertices $(0, 0)$ and $(0, 2)$. Since a is the distance form the center to a vertex,

we have $a = 1$. The slope of the given asymptote is 1, so $\dfrac{a}{b} = 1 \Leftrightarrow b = 1$. Thus, an equation of the hyperbola is

$(y - 1)^2 - x^2 = 1$.

35. The ellipse with center $C(2, -3)$, vertices $V_1(-8, -3)$ and $V_2(12, -3)$, and foci $F_1(-4, -3)$ and $F_2(8, -3)$ has

a horizontal major axis, so its equation has the form $\dfrac{(x-2)^2}{a^2} + \dfrac{(y+3)^2}{b^2} = 1$. The distance between the vertices

is $2a = 12 - (-8) = 20$, so $a = 10$. Also, the distance from the center to each focus is $c = 2 - (-4) = 6$, so

$b^2 = a^2 - c^2 = 100 - 36 = 64$. Thus, an equation is $\dfrac{(x-2)^2}{100} + \dfrac{(y+3)^2}{64} = 1$.

37. The hyperbola with center $C(-1, 4)$, vertices $V_1(-1, -3)$ and $V_2(-1, 11)$, and foci $F_1(-1, -5)$ and $F_2(-1, 13)$ has a vertical transverse axis, so its equation has the form $\dfrac{(y-4)^2}{a^2} - \dfrac{(x+1)^2}{b^2} = 1$. The distance between the vertices is $2a = 11 - (-3) = 14$, so $a = 7$. Also, the distance from the center to each focus is $c = 4 - (-5) = 9$, so $b^2 = c^2 - a^2 = 81 - 49 = 32$. Thus, an equation is $\dfrac{(y-4)^2}{49} - \dfrac{(x+1)^2}{32} = 1$.

39. The parabola with vertex $V(-3, 5)$ and directrix $y = 2$ has an equation of the form $(x+3)^2 = 4p(y-5)$. The distance from the vertex to the directrix is $p = 5 - 2 = 3$, so an equation is $(x+3)^2 = 12(y-5)$.

41. The hyperbola with foci $F_1(1, -5)$ and $F_2(1, 5)$ that passes through the point $(1, 4)$ is centered midway between the foci; that is, it has $C\left(1, \frac{1}{2}(-5+5)\right) = (1, 0)$. It has a vertical transverse axis, so its equation has the form $\dfrac{y^2}{a^2} - \dfrac{(x-1)^2}{b^2} = 1$. The point $(1, 4)$ lies on the transverse axis, so it is a vertex and we have $a = 4 - 0 = 4$. Also, the distance from the center to each focus is $c = 5 - 0 = 5$, so $b^2 = c^2 - a^2 = 25 - 16 = 9$. Thus, an equation is $\dfrac{y^2}{16} - \dfrac{(x-1)^2}{9} = 1$.

43. The ellipse with foci $F_1(3, -4)$ and $F_2(3, 4)$ and x-intercepts 0 and 6 is centered midway between the foci; that is, it has $C\left(3, \frac{1}{2}(-4+4)\right) = (3, 0)$, and so its equation has the form $\dfrac{(x-3)^2}{b^2} + \dfrac{y^2}{a^2} = 1$. Because 0 is an x-intercept, we substitute $(0, 0)$ into this equation, obtaining $\dfrac{(-3)^2}{b^2} + \dfrac{0^2}{a^2} = 1 \Leftrightarrow b = 3$. We also have $c = 4 - 0 = 4$, so $a^2 = b^2 + c^2 = 25$ and an equation is $\dfrac{(x-3)^2}{9} + \dfrac{y^2}{25} = 1$.

45. $y^2 = 4(x+2y) \Leftrightarrow y^2 - 8y = 4x \Leftrightarrow y^2 - 8y + 16 = 4x + 16 \Leftrightarrow$
$(y-4)^2 = 4(x+4)$. This is a parabola with $4p = 4 \Leftrightarrow p = 1$. The vertex is $(-4, 4)$, the focus is $(-4+1, 4) = (-3, 4)$, and the directrix is $x = -4 - 1 = -5$.

47. $x^2 - 5y^2 - 2x + 20y = 44 \Leftrightarrow \left(x^2 - 2x + 1\right) - 5\left(y^2 - 4y + 4\right) = 44 + 1 - 20$
$\Leftrightarrow (x-1)^2 - 5(y-2)^2 = 25 \Leftrightarrow \dfrac{(x-1)^2}{25} - \dfrac{(y-2)^2}{5} = 1$. This is a hyperbola

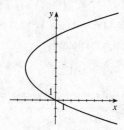

with $a = 5$, $b = \sqrt{5}$, and $c = \sqrt{25+5} = \sqrt{30}$. The center is $(1, 2)$, the foci are $\left(1 \pm \sqrt{30}, 2\right)$, the vertices are $(1 \pm 5, 2) = (-4, 2)$ and $(6, 2)$, and the asymptotes are $y - 2 = \pm\dfrac{\sqrt{5}}{5}(x-1) \Leftrightarrow y = \pm\dfrac{\sqrt{5}}{5}(x-1) + 2 \Leftrightarrow y = -\dfrac{\sqrt{5}}{5}x + 2 + \dfrac{\sqrt{5}}{5}$ and
$y = \dfrac{\sqrt{5}}{5}x + 2 - \dfrac{\sqrt{5}}{5}$.

49. $4x^2 + 25y^2 - 24x + 250y + 561 = 0 \Leftrightarrow$

$4\left(x^2 - 6x + 9\right) + 25\left(y^2 + 10y + 25\right) = -561 + 36 + 625 \Leftrightarrow$

$4\left(x - 3\right)^2 + 25\left(y + 5\right)^2 = 100 \Leftrightarrow \dfrac{(x-3)^2}{25} + \dfrac{(y+5)^2}{4} = 1$. This is an ellipse

with $a = 5$, $b = 2$, and $c = \sqrt{25 - 4} = \sqrt{21}$. The center is $(3, -5)$, the foci are

$\left(3 \pm \sqrt{21}, -5\right)$, the vertices are $(3 \pm 5, -5) = (-2, -5)$ and $(8, -5)$, the length

of the major axis is $2a = 10$, and the length of the minor axis is $2b = 4$.

51. $16x^2 - 9y^2 - 96x + 288 = 0 \Leftrightarrow 16\left(x^2 - 6x\right) - 9y^2 + 288 = 0 \Leftrightarrow$

$16\left(x^2 - 6x + 9\right) - 9y^2 = 144 - 288 \Leftrightarrow 16(x-3)^2 - 9y^2 = -144 \Leftrightarrow$

$\dfrac{y^2}{16} - \dfrac{(x-3)^2}{9} = 1$. This is a hyperbola with $a = 4$, $b = 3$, and

$c = \sqrt{16 + 9} = 5$. The center is $(3, 0)$, the foci are $(3, \pm 5)$, the vertices are

$(3, \pm 4)$, and the asymptotes are $y = \pm \tfrac{4}{3}(x - 3) \Leftrightarrow y = \tfrac{4}{3}x - 4$ and $y = 4 - \tfrac{4}{3}x$.

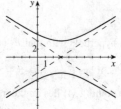

53. $x^2 + 16 = 4\left(y^2 + 2x\right) \Leftrightarrow x^2 - 8x - 4y^2 + 16 = 0 \Leftrightarrow$

$\left(x^2 - 8x + 16\right) - 4y^2 = -16 + 16 \Leftrightarrow 4y^2 = (x - 4)^2 \Leftrightarrow y = \pm \tfrac{1}{2}(x - 4)$.

Thus, the conic is degenerate, and its graph is the pair of lines $y = \tfrac{1}{2}(x - 4)$ and

$y = -\tfrac{1}{2}(x - 4)$.

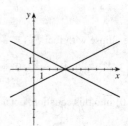

55. $3x^2 + 4y^2 - 6x - 24y + 39 = 0 \Leftrightarrow 3\left(x^2 - 2x\right) + 4\left(y^2 - 6y\right) = -39 \Leftrightarrow$

$3\left(x^2 - 2x + 1\right) + 4\left(y^2 - 6y + 9\right) = -39 + 3 + 36 \Leftrightarrow$

$3(x - 1)^2 + 4(y - 3)^2 = 0 \Leftrightarrow x = 1$ and $y = 3$. This is a degenerate conic whose

graph is the point $(1, 3)$.

57. $2x^2 - 4x + y + 5 = 0 \Leftrightarrow y = -2x^2 + 4x - 5$.

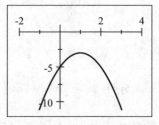

59. $9x^2 + 36 = y^2 + 36x + 6y \Leftrightarrow x^2 - 36x + 36 = y^2 + 6y \Leftrightarrow$

$9x^2 - 36x + 45 = y^2 + 6y + 9 \Leftrightarrow 9\left(x^2 - 4x + 5\right) = (y + 3)^2 \Leftrightarrow$

$y + 3 = \pm\sqrt{9\left(x^2 - 4x + 5\right)} \Leftrightarrow y = -3 \pm 3\sqrt{x^2 - 4x + 5}$

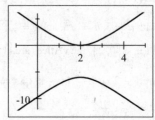

61. $4x^2 + y^2 + 4(x - 2y) + F = 0 \Leftrightarrow 4\left(x^2 + x\right) + \left(y^2 - 8y\right) = -F \Leftrightarrow$

$4\left(x^2 + x + \frac{1}{4}\right) + \left(y^2 - 8y + 16\right) = 16 + 1 - F \Leftrightarrow 4\left(x + \frac{1}{2}\right)^2 + (y - 1)^2 = 17 - F$

(a) For an ellipse, $17 - F > 0 \Leftrightarrow F < 17$.

(b) For a single point, $17 - F = 0 \Leftrightarrow F = 17$.

(c) For the empty set, $17 - F < 0 \Leftrightarrow F > 17$.

63. (a) $x^2 = 4p(y + p)$, for

$p = -2, -\frac{3}{2}, -1, -\frac{1}{2}, \frac{1}{2}, 1, \frac{3}{2}, 2$.

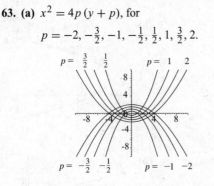

(b) The graph of $x^2 = 4p(y + p)$ is obtained by shifting the graph of $x^2 = 4py$ vertically $-p$ units so that the vertex is at $(0, -p)$. The focus of $x^2 = 4py$ is at $(0, p)$, so this point is also shifted $-p$ units vertically to the point $(0, p - p) = (0, 0)$. Thus, the focus is located at the origin.

(c) The parabolas become narrower as the vertex moves toward the origin.

65. Since the height of the satellite above the earth varies between 140 and 440, the length of the major axis is $2a = 140 + 2(3960) + 440 = 8500 \Leftrightarrow a = 4250$. Since the center of the earth is at one focus, we have

$a - c = (\text{earth radius}) + 140 = 3960 + 140 = 4100 \Leftrightarrow$

$c = a - 4100 = 4250 - 4100 = 150$. Thus, the center of the ellipse is $(-150, 0)$.

So $b^2 = a^2 - c^2 = 4250^2 - 150^2 = 18{,}062{,}500 - 22500 = 18{,}040{,}000$. Hence,

an equation is $\dfrac{(x + 150)^2}{18{,}062{,}500} + \dfrac{y^2}{18{,}040{,}000} = 1$.

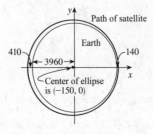

CHAPTER 7 REVIEW

1. (a) $y^2 = 4x$. This is a parabola with $4p = 4 \Leftrightarrow p = 1$. The vertex is $(0, 0)$, the focus is $(1, 0)$, and the directrix is $x = -1$.

(b)

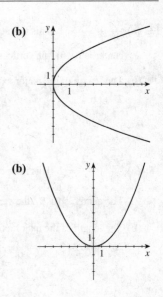

3. (a) $\frac{1}{8}x^2 = y \Leftrightarrow x^2 = 8y$. This is a parabola with $4p = 8 \Leftrightarrow p = 2$. The vertex is $(0, 0)$, the focus is $(0, 2)$, and the directrix is $y = -2$.

(b)

5. (a) $x^2 + 8y = 0 \Leftrightarrow x^2 = -8y$. This is a parabola with $4p = -8 \Leftrightarrow p = -2$. The
vertex is $(0, 0)$, the focus is $(0, -2)$, and the directrix is $y = 2$.

(b)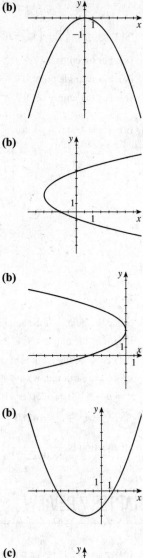

7. (a) $(y - 2)^2 = 4(x + 2)$. This is a parabola with $4p = 4 \Leftrightarrow p = 1$. The vertex is
$(-2, 2)$, the focus is $(-1, 2)$, and the directrix is $x = -3$.

(b)

9. (a) $\frac{1}{2}(y - 3)^2 + x = 0 \Leftrightarrow (y - 3)^2 = -2x$. This is a parabola with $4p = -2 \Leftrightarrow$
$p = -\frac{1}{2}$. The vertex is $(0, 3)$, the focus is $\left(-\frac{1}{2}, 3\right)$, and the directrix is $x = \frac{1}{2}$.

(b)

11. (a) $\frac{1}{2}x^2 + 2x = 2y + 4 \Leftrightarrow x^2 + 4x = 4y + 8 \Leftrightarrow x^2 + 4x + 4 = 4y + 8 \Leftrightarrow$
$(x + 2)^2 = 4(y + 3)$. This is a parabola with $4p = 4 \Leftrightarrow p = 1$. The vertex is
$(-2, -3)$, the focus is $(-2, -3 + 1) = (-2, -2)$, and the directrix is
$y = -3 - 1 = -4$.

(b)

13. (a) $\frac{x^2}{9} + \frac{y^2}{25} = 1$. This is an ellipse with $a = 5$, $b = 3$, and $c = \sqrt{25 - 9} = 4$. The
center is $(0, 0)$, the vertices are $(0, \pm 5)$, and the foci are $(0, \pm 4)$.

(b) The length of the major axis is $2a = 10$ and the length of the minor axis is
$2b = 6$.

(c)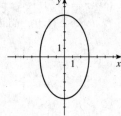

15. (a) $\frac{x^2}{49} + \frac{y^2}{4} = 1$. This is an ellipse with $a = 7$, $b = 2$, and $c = \sqrt{49 - 4} = 3\sqrt{5}$.
The center is $(0, 0)$, the vertices are $(\pm 7, 0)$, and the foci are $\left(\pm 3\sqrt{5}, 0\right)$.

(b) The length of the major axis is $2a = 14$ and the length of the minor axis is
$2b = 4$.

(c)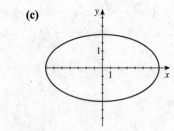

17. (a) $x^2 + 4y^2 = 16 \Leftrightarrow \dfrac{x^2}{16} + \dfrac{y^2}{4} = 1$. This is an ellipse with $a = 4$, $b = 2$, and $c = \sqrt{16 - 4} = 2\sqrt{3}$. The center is $(0, 0)$, the vertices are $(\pm 4, 0)$, and the foci are $\left(\pm 2\sqrt{3}, 0\right)$.

(c)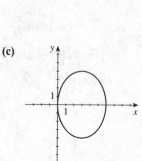

(b) The length of the major axis is $2a = 8$ and the length of the minor axis is $2b = 4$.

19. (a) $\dfrac{(x - 3)^2}{9} + \dfrac{y^2}{16} = 1$. This is an ellipse with $a = 4$, $b = 3$, and $c = \sqrt{16 - 9} = \sqrt{7}$. The center is $(3, 0)$, the vertices are $(3, \pm 4)$, and the foci are $\left(3, \pm\sqrt{7}\right)$.

(c)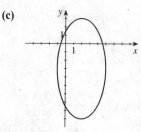

(b) The length of the major axis is $2a = 8$ and the length of the minor axis is $2b = 6$.

21. (a) $\dfrac{(x - 2)^2}{9} + \dfrac{(y + 3)^2}{36} = 1$. This is an ellipse with $a = 6$, $b = 3$, and $c = \sqrt{36 - 9} = 3\sqrt{3}$. The center is $(2, -3)$, the vertices are $(2, -3 \pm 6) = (2, -9)$ and $(2, 3)$, and the foci are $\left(2, -3 \pm 3\sqrt{3}\right)$.

(c)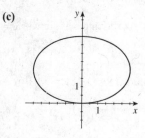

(b) The length of the major axis is $2a = 12$ and the length of the minor axis is $2b = 6$.

23. (a) $4x^2 + 9y^2 = 36y \Leftrightarrow 4x^2 + 9\left(y^2 - 4y + 4\right) = 36 \Leftrightarrow 4x^2 + 9(y - 2)^2 = 36$ $\Leftrightarrow \dfrac{x^2}{9} + \dfrac{(y - 2)^2}{4} = 1$. This is an ellipse with $a = 3$, $b = 2$, and $c = \sqrt{9 - 4} = \sqrt{5}$. The center is $(0, 2)$, the vertices are $(\pm 3, 2)$, and the foci are $\left(\pm\sqrt{5}, 2\right)$.

(c)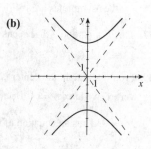

(b) The length of the major axis is $2a = 6$ and the length of the minor axis is $2b = 4$.

25. (a) $-\dfrac{x^2}{9} + \dfrac{y^2}{16} = 1 \Leftrightarrow \dfrac{y^2}{16} - \dfrac{x^2}{9} = 0$. This is a hyperbola with $a = 4$, $b = 3$, and $c = \sqrt{16 + 9} = \sqrt{25} = 5$. The center is $(0, 0)$, the vertices are $(0, \pm 4)$, the foci are $(0, \pm 5)$, and the asymptotes are $y = \pm\frac{4}{3}x$.

(b)

27. (a) $\dfrac{x^2}{4} - \dfrac{y^2}{49} = 1$. This is a hyperbola with $a = 2$, $b = 7$, and $c = \sqrt{4 + 49} = \sqrt{53}$. The center is $(0, 0)$, the vertices are $(\pm 2, 0)$, the foci are $\left(\pm\sqrt{53}, 0\right)$, and the asymptotes are $y = \pm\frac{7}{2}x$.

(b)

29. (a) $x^2 - 2y^2 = 16 \Leftrightarrow \dfrac{x^2}{16} - \dfrac{y^2}{8} = 1$. This is a hyperbola with $a = 4$, $b = 2\sqrt{2}$, and $c = \sqrt{16 + 8} = \sqrt{24} = 2\sqrt{6}$. The center is $(0, 0)$, the vertices are $(\pm 4, 0)$, the foci are $\left(\pm 2\sqrt{6}, 0\right)$, and the asymptotes are $y = \pm\frac{2\sqrt{2}}{4}x \Leftrightarrow y = \pm\frac{1}{\sqrt{2}}x$.

(b)

31. (a) $\dfrac{(x + 4)^2}{16} - \dfrac{y^2}{16} = 1$. This is a hyperbola with $a = 4$, $b = 4$ and $c = \sqrt{16 + 16} = 4\sqrt{2}$. The center is $(-4, 0)$, the vertices are $(-4 \pm 4, 0)$ which are $(-8, 0)$ and $(0, 0)$, the foci are $\left(-4 \pm 4\sqrt{2}, 0\right)$, and the asymptotes are $y = \pm(x + 4)$.

(b)

33. (a) $\dfrac{(y - 3)^2}{4} - \dfrac{(x + 1)^2}{36} = 1$. This is a hyperbola with $a = 2$, $b = 6$, and $c = \sqrt{4 + 36} = 2\sqrt{10}$. The center is $(-1, 3)$, the vertices are $(-1, 3 \pm 2) = (-1, 1)$ and $(-1, 5)$, the foci are $\left(-1, 3 \pm 2\sqrt{10}\right)$, and the asymptotes are $y - 3 = \pm\frac{1}{3}(x + 1) \Leftrightarrow y = \frac{1}{3}x + \frac{10}{3}$ and $y = -\frac{1}{3}x + \frac{8}{3}$.

(b)

35. (a) $9y^2 + 18y = x^2 + 6x + 18 \Leftrightarrow 9\left(y^2 + 2y + 1\right) = \left(x^2 + 6x + 9\right) + 9 - 9 + 18$ $\Leftrightarrow 9(y + 1)^2 - (x + 3)^2 = 18 \Leftrightarrow \dfrac{(y + 1)^2}{2} - \dfrac{(x + 3)^2}{18} = 1$. This is a hyperbola with $a = \sqrt{2}$, $b = 3\sqrt{2}$, and $c = \sqrt{2 + 18} = 2\sqrt{5}$. The center is $(-3, -1)$, the vertices are $\left(-3, -1 \pm \sqrt{2}\right)$, the foci are $\left(-3, -1 \pm 2\sqrt{5}\right)$, and the asymptotes are $y + 1 = \pm\frac{1}{3}(x + 3) \Leftrightarrow y = \frac{1}{3}x$ and $y = -\frac{1}{3}x - 2$.

(b)

37. This is a parabola that opens to the right with its vertex at $(0, 0)$ and the focus at $(2, 0)$. So $p = 2$, and the equation is $y^2 = 4(2)x \Leftrightarrow y^2 = 8x$.

39. From the graph, the center is $(0, 0)$, and the vertices are $(0, -4)$ and $(0, 4)$. Since a is the distance from the center to a vertex, we have $a = 4$. Because one focus is $(0, 5)$, we have $c = 5$, and since $c^2 = a^2 + b^2$, we have $25 = 16 + b^2 \Leftrightarrow b^2 = 9$. Thus an equation of the hyperbola is $\dfrac{y^2}{16} - \dfrac{x^2}{9} = 1$.

41. From the graph, the center of the ellipse is $(4, 2)$, and so $a = 4$ and $b = 2$. The equation is $\dfrac{(x-4)^2}{4^2} + \dfrac{(y-2)^2}{2^2} = 1 \Leftrightarrow$

$\dfrac{(x-4)^2}{16} + \dfrac{(y-2)^2}{4} = 1$.

43. $\dfrac{x^2}{12} + y = 1 \Leftrightarrow \dfrac{x^2}{12} = -(y-1) \Leftrightarrow x^2 = -12\,(y-1)$. This is a parabola with

$4p = -12 \Leftrightarrow p = -3$. The vertex is $(0, 1)$, the focus is $(0, 1-3) = (0, -2)$, and

the directrix is $y = 1 + 3 = 4$.

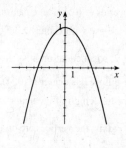

45. $x^2 - y^2 + 144 = 0 \Leftrightarrow \dfrac{y^2}{144} - \dfrac{x^2}{144} = 1$. This is a hyperbola with $a = 12$, $b = 12$,

and $c = \sqrt{144 + 144} = 12\sqrt{2}$. The center is $(0, 0)$, the foci are $\left(0, \pm 12\sqrt{2}\right)$, the

vertices are $(0, \pm 12)$, and the asymptotes are $y = \pm x$.

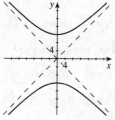

47. $4x^2 + y^2 = 8\,(x + y) \Leftrightarrow 4\left(x^2 - 2x\right) + \left(y^2 - 8y\right) = 0 \Leftrightarrow$

$4\left(x^2 - 2x + 1\right) + \left(y^2 - 8y + 16\right) = 4 + 16 \Leftrightarrow 4\,(x-1)^2 + (y-4)^2 = 20 \Leftrightarrow$

$\dfrac{(x-1)^2}{5} + \dfrac{(y-4)^2}{20} = 1$. This is an ellipse with $a = 2\sqrt{5}$, $b = \sqrt{5}$, and

$c = \sqrt{20 - 5} = \sqrt{15}$. The center is $(1, 4)$, the foci are $\left(1, 4 \pm \sqrt{15}\right)$, and the

vertices are $\left(1, 4 \pm 2\sqrt{5}\right)$.

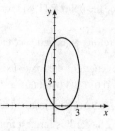

49. $x = y^2 - 16y \Leftrightarrow x + 64 = y^2 - 16y + 64 \Leftrightarrow (y-8)^2 = x + 64$. This is a

parabola with $4p = 1 \Leftrightarrow p = \frac{1}{4}$. The vertex is $(-64, 8)$, the focus is

$\left(-64 + \frac{1}{4}, 8\right) = \left(-\frac{255}{4}, 8\right)$, and the directrix is $x = -64 - \frac{1}{4} = -\frac{257}{4}$.

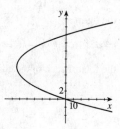

51. $2x^2 - 12x + y^2 + 6y + 26 = 0 \Leftrightarrow 2\left(x^2 - 6x\right) + \left(y^2 + 6y\right) = -26 \Leftrightarrow$

$2\left(x^2 - 6x + 9\right) + \left(y^2 + 6y + 9\right) = -26 + 18 + 9 \Leftrightarrow 2\,(x-3)^2 + (y+3)^2 = 1$

$\Leftrightarrow \dfrac{(x-3)^2}{\frac{1}{2}} + (y+3)^2 = 1$. This is an ellipse with $a = 1$, $b = \frac{\sqrt{2}}{2}$, and

$c = \sqrt{1 - \frac{1}{2}} = \frac{\sqrt{2}}{2}$. The center is $(3, -3)$, the foci are $\left(3, -3 \pm \frac{\sqrt{2}}{2}\right)$, and the

vertices are $(3, -3 \pm 1) = (3, -4)$ and $(3, -2)$.

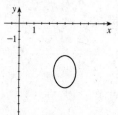

53. $9x^2 + 8y^2 - 15x + 8y + 27 = 0 \Leftrightarrow 9\left(x^2 - \frac{5}{3}x + \frac{25}{36}\right) + 8\left(y^2 + y + \frac{1}{4}\right) = -27 + \frac{25}{4} + 2 \Leftrightarrow 9\left(x - \frac{5}{6}\right)^2 + 8\left(y + \frac{1}{2}\right)^2 = -\frac{75}{4}$.

However, since the left-hand side of the equation is greater than or equal to 0, there is no point that satisfies this equation. The graph is empty.

55. The parabola has focus $(0, 1)$ and directrix $y = -1$. Therefore, $p = 1$ and so $4p = 4$. Since the focus is on the y-axis and the vertex is $(0, 0)$, an equation of the parabola is $x^2 = 4y$.

57. The ellipse with center at the origin and with x-intercepts ± 2 and y-intercepts ± 5 has a vertical major axis, $a = 5$, and $b = 2$, so an equation is $\dfrac{x^2}{4} + \dfrac{y^2}{25} = 1$.

59. The ellipse has center $C(0, 4)$, foci $F_1(0, 0)$ and $F_2(0, 8)$, and major axis of length 10. Then $2c = 8 - 0 \Leftrightarrow c = 4$. Also, since the length of the major axis is 10, $2a = 10 \Leftrightarrow a = 5$. Therefore, $b^2 = a^2 - c^2 = 25 - 16 = 9$. Since the foci are on the y-axis, the vertices are on the y-axis, and an equation of the ellipse is $\dfrac{x^2}{9} + \dfrac{(y-4)^2}{25} = 1$.

61. The ellipse has foci $F_1(1, 1)$ and $F_2(1, 3)$, and one vertex is on the x-axis. Thus, $2c = 3 - 1 = 2 \Leftrightarrow c = 1$, and so the center of the ellipse is $C(1, 2)$. Also, since one vertex is on the x-axis, $a = 2 - 0 = 2$, and thus $b^2 = 4 - 1 = 3$. So an equation of the ellipse is $\dfrac{(x-1)^2}{3} + \dfrac{(y-2)^2}{4} = 1$.

63. The ellipse has vertices $V_1(7, 12)$ and $V_2(7, -8)$ and passes through the point $P(1, 8)$. Thus, $2a = 12 - (-8) = 20 \Leftrightarrow a = 10$, and the center is $\left(7, \dfrac{-8 + 12}{2}\right) = (7, 2)$. Thus an equation of the ellipse has the form $\dfrac{(x-7)^2}{b^2} + \dfrac{(y-2)^2}{100} = 1$.

Since the point $P(1, 8)$ is on the ellipse, $\dfrac{(1-7)^2}{b^2} + \dfrac{(8-2)^2}{100} = 1 \Leftrightarrow 3600 + 36b^2 = 100b^2 \Leftrightarrow 64b^2 = 3600 \Leftrightarrow b^2 = \dfrac{225}{4}$.

Therefore, an equation of the ellipse is $\dfrac{(x-7)^2}{225/4} + \dfrac{(y-2)^2}{100} = 1 \Leftrightarrow \dfrac{4(x-7)^2}{225} + \dfrac{(y-2)^2}{100} = 1$.

65. The length of the major axis is $2a = 186{,}000{,}000 \Leftrightarrow a = 93{,}000{,}000$. The eccentricity is $e = c/a = 0.017$, and so $c = 0.017\,(93{,}000{,}000) = 1{,}581{,}000$.

(a) The earth is closest to the sun when the distance is $a - c = 93{,}000{,}000 - 1{,}581{,}000 = 91{,}419{,}000$.

(b) The earth is furthest from the sun when the distance is $a + c = 93{,}000{,}000 + 1{,}581{,}000 = 94{,}581{,}000$.

67. (a) The graphs of $\dfrac{x^2}{16 + k^2} + \dfrac{y^2}{k^2} = 1$ for $k = 1, 2, 4$, and 8 are shown in the figure.

(b) $c^2 = \left(16 + k^2\right) - k^2 = 16 \Rightarrow \quad c = \pm 4$. Since the center is $(0, 0)$, the foci of each of the ellipses are $(\pm 4, 0)$.

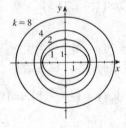

CHAPTER 7 TEST

1. $x^2 = -12y$. This is a parabola with $4p = -12 \Leftrightarrow p = -3$. The focus is $(0, -3)$ and the directrix is $y = 3$.

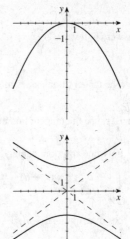

3. $\dfrac{y^2}{9} - \dfrac{x^2}{16} = 1$. This is a hyperbola with $a = 3$, $b = 4$, and $c = \sqrt{9 + 16} = 5$. The vertices are $(0, \pm 3)$, the foci are $(0, \pm 5)$, and the asymptotes are $y = \pm \frac{3}{4}x$.

5. The ellipse with foci $(\pm 3, 0)$ and vertices $(\pm 4, 0)$ has $a = 4$ and $c = 3$, so $c^2 = a^2 - b^2 \Leftrightarrow 9 = 16 - b^2 \Leftrightarrow b^2 = 16 - 9 = 7$. Thus, an equation is $\dfrac{x^2}{16} + \dfrac{y^2}{7} = 1$.

7. This is a parabola that opens to the left with its vertex at $(0, 0)$. So its equation is of the form $y^2 = 4px$ with $p < 0$. Substituting the point $(-4, 2)$, we have $2^2 = 4p(-4) \Leftrightarrow 4 = -16p \Leftrightarrow p = -\frac{1}{4}$. So an equation is $y^2 = 4\left(-\frac{1}{4}\right)x \Leftrightarrow$ $y^2 = -x$.

9. This a hyperbola with a horizontal transverse axis, vertices at $(1, 0)$ and $(3, 0)$, and foci at $(0, 0)$ and $(4, 0)$. Thus the center is $(2, 0)$, and $a = 3 - 2 = 1$ and $c = 4 - 2 = 2$. Thus $b^2 = 2^2 - 1^2 = 3$. So an equation is $\dfrac{(x - 2)^2}{1^2} - \dfrac{y^2}{3} = 1 \Leftrightarrow$ $(x - 2)^2 - \dfrac{y^2}{3} = 1$.

11. $9x^2 - 8y^2 + 36x + 64y = 164 \Leftrightarrow 9\left(x^2 + 4x\right) - 8\left(y^2 - 8y\right) = 164 \Leftrightarrow$

$9\left(x^2 + 4x + 4\right) - 8\left(y^2 - 8y + 16\right) = 164 + 36 - 128 \Leftrightarrow$

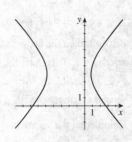

$9(x + 2)^2 - 8(y - 4)^2 = 72 \Leftrightarrow \dfrac{(x + 2)^2}{8} - \dfrac{(y - 4)^2}{9} = 1$. This is a hyperbola with $a = 2\sqrt{2}$, $b = 3$, and $c^2 = a^2 + b^2 = 17$. The center is $(-2, 4)$, the foci are $\left(-2 \pm \sqrt{17}, 4\right)$, the vertices are $\left(-2 \pm 2\sqrt{2}, 4\right)$, and the asymptotes are

$y - 4 = \pm \frac{3\sqrt{2}}{4}(x + 2) \Leftrightarrow y = \frac{3\sqrt{2}}{4}x + 4 + \frac{3\sqrt{2}}{2}$ and $y = -\frac{3\sqrt{2}}{4}x + 4 - \frac{3\sqrt{2}}{2}$.

13. The ellipse with center $(2, 0)$, foci $(2, \pm 3)$ and major axis of length 8 has a horizontal major axis with $2a = 8 \Leftrightarrow a = 4$. Also, $c = 3$, so $b^2 = a^2 - c^2 = 16 - 9 = 7$. Thus, an equation is $\dfrac{(x - 2)^2}{7} + \dfrac{y^2}{16} = 1$.

15. We place the vertex of the parabola at the origin, so the parabola contains the points $(3, \pm 3)$, and the equation is of the form $y^2 = 4px$. Substituting the point $(3, 3)$, we get $3^2 = 4p(3) \Leftrightarrow 9 = 12p \Leftrightarrow p = \frac{3}{4}$. So the focus is $\left(\frac{3}{4}, 0\right)$, and we should place the light bulb $\frac{3}{4}$ inch from the vertex.

FOCUS ON MODELING   Conics in Architecture

1. Answers will vary.

5. (a) The tangent line passes though the point $\left(a, a^2\right)$, so an equation is $y - a^2 = m\left(x - a\right)$.

(b) Because the tangent line intersects the parabola at only the one point $\left(a, a^2\right)$, the system $\begin{cases} y - a^2 = m\left(x - a\right) \\ y = x^2 \end{cases}$ has

only one solution, namely $x = a$, $y = a^2$.

(c) $\begin{cases} y - a^2 = m\left(x - a\right) \\ y = x^2 \end{cases} \Leftrightarrow \begin{cases} y = a^2 + m\left(x - a\right) \\ y = x^2 \end{cases} \Leftrightarrow a^2 + m\left(x - a\right) = x^2 \Leftrightarrow x^2 - mx + am - a^2 = 0$. This quadratic

has discriminant $(-m)^2 - 4\left(1\right)\left(am - a^2\right) = m^2 - 4am + 4a^2 = (m - 2a)^2$. Setting this equal to 0, we find $m = 2a$.

(d) An equation of the tangent line is $y - a^2 = 2a\left(x - a\right) \Leftrightarrow y = a^2 + 2ax - 2a^2 \Leftrightarrow y = 2ax - a^2$.

CUMULATIVE REVIEW TEST: CHAPTERS 5, 6, and 7

1. (a) Because some variables are raised to the power 2, the system $\begin{cases} x^2 + y^2 = 4y \\ x^2 - 2y = 0 \end{cases}$ is nonlinear.

(b) Subtracting the second equation from the first, we obtain

$y^2 + 2y = 4y \Leftrightarrow y^2 - 2y = 0 \Leftrightarrow y\left(y - 2\right) = 0 \Leftrightarrow y = 0$ or 2.

If $y = 0$, then the first equation gives $x^2 + 0 = 0 \Leftrightarrow x = 0$, and

if $y = 2$, then the first equation gives $x^2 + 2^2 = 4 \cdot 2 \Leftrightarrow x^2 = 4$

$\Leftrightarrow x = \pm 2$. Therefore, the three possible solutions are $(0, 0)$,

$(-2, 2)$, and $(2, 2)$. We can verify that all three satisfy the

second equation as well.

(c) Completing the square in y in the first equation, we have

$x^2 + y^2 = 4y \Leftrightarrow x^2 + (y - 2)^2 = 4$, a circle with radius 2

centered at $(0, 2)$. The second equation can be written as

$y = \frac{1}{2}x^2$, a parabola opening upward.

(d), (e) Taking $(0, 1)$ as a test point, we find that

$0^2 + 1^2 \le 4\left(1\right)$ and $0^2 - 2\left(1\right) \le 0$, so

the region containing this point satisfies

both inequalities.

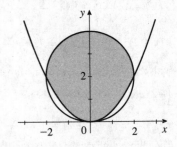

3. Let X, Y, and Z represent the number of fish caught by each of Xavier, Yolanda, and Zachary. "Yolanda catches as many fish as Xavier and Zachary put together" means that $Y = X + Z$. "Zachary catches 2 more fish than Xavier" means that $Z = X + 2$. "The total catch for all three people is 20 fish," so $X + Y + Z = 20$. From the first and third equations, we have $2Y = 20$, so $Y = 10$. From the first and second equations, we have $Y = X + (X + 2) = 2X + 2$, so $X = 4$. Finally, the third equation gives $Z = 20 - 10 - 4 = 6$. Therefore, Xavier caught 4 fish, Yolanda caught 10, and Zachary caught 6.

5. (a) A matrix equation equivalent to the system $\begin{cases} 5x - 3y = 5 \\ 6x - 4y = 0 \end{cases}$ is $AX = B$, where $A = \begin{bmatrix} 5 & -3 \\ 6 & -4 \end{bmatrix}$, $X = \begin{bmatrix} x \\ y \end{bmatrix}$, and

$B = \begin{bmatrix} 5 \\ 0 \end{bmatrix}$.

(b) Using the rule for finding the inverse of a 2×2 matrix, we get $A^{-1} = \dfrac{1}{5(-4) - (-3)6} \begin{bmatrix} -4 & -(-3) \\ -6 & 5 \end{bmatrix} = \begin{bmatrix} 2 & -\frac{3}{2} \\ 3 & -\frac{5}{2} \end{bmatrix}$.

(c) $AX = B \Leftrightarrow A^{-1}AX = A^{-1}B \Leftrightarrow X = A^{-1}B = -\dfrac{1}{2}\begin{bmatrix} -4 & 3 \\ -6 & 5 \end{bmatrix}\begin{bmatrix} 5 \\ 0 \end{bmatrix} = -\dfrac{1}{2}\begin{bmatrix} -4(5) + 3(0) \\ -6(5) + 5(0) \end{bmatrix} = \begin{bmatrix} 10 \\ 15 \end{bmatrix}$.

Thus, $x = 10$ and $y = 15$.

(d) For this system we have $|D| = |A| = -2$, $|D_x| = \begin{vmatrix} 5 & -3 \\ 0 & -4 \end{vmatrix} = -20$, and $|D_y| = \begin{vmatrix} 5 & 5 \\ 6 & 0 \end{vmatrix} = -30$. Thus, the solution is

$x = \dfrac{|D_x|}{|D|} = \dfrac{-20}{-2} = 10$ and $y = \dfrac{|D_y|}{|D|} = \dfrac{-30}{-2} = 15$, as in part (c).

7. The parabola with vertex at the origin and focus $F(0, 3)$ opens upward and has equation $x^2 = 4py$, where $p = 3$ is the y-coordinate of the focus. Thus, an equation is $x^2 = 12y$.

9. The center of the hyperbola lies halfway between its vertices; that is, at $\left(\dfrac{1+9}{2}, 0\right) = (5, 0)$. Because its transverse axis has length 8, we have $a = 4$. The foci lie 1 unit to the left and right of the center, so $c = 5$, and thus $b^2 = c^2 - a^2 = 3$. Therefore, an equation is $\dfrac{(x - 5)^2}{16} - \dfrac{y^2}{9} = 1$.

8 SEQUENCES AND SERIES

8.1 SEQUENCES AND SUMMATION NOTATION

1. A sequence is a function whose domain is *the natural numbers*.

3. $a_n = n + 1$. Then $a_1 = 1 + 1 = 2$, $a_2 = 2 + 1 = 3$, $a_3 = 3 + 1 = 4$, $a_4 = 4 + 1 = 5$, and $a_{100} = 100 + 1 = 101$.

5. $a_n = \dfrac{1}{n+1}$. Then $a_1 = \dfrac{1}{1+1} = \dfrac{1}{2}$, $a_2 = \dfrac{1}{2+1} = \dfrac{1}{3}$, $a_3 = \dfrac{1}{3+1} = \dfrac{1}{4}$, $a_4 = \dfrac{1}{4+1} = \dfrac{1}{5}$, and $a_{100} = \dfrac{1}{100+1} = \dfrac{1}{101}$.

7. $a_n = 5^n$. Then $a_1 = 5^1 = 5$, $a_2 = 5^2 = 25$, $a_3 = 5^3 = 125$, $a_4 = 5^4 = 625$, and $a_{100} = 5^{100} \approx 7.9 \times 10^{69}$.

9. $a_n = \dfrac{(-1)^n}{n^2}$. Then $a_1 = \dfrac{(-1)^1}{1^2} = -1$, $a_2 = \dfrac{(-1)^2}{2^2} = \dfrac{1}{4}$, $a_3 = \dfrac{(-1)^3}{3^2} = -\dfrac{1}{9}$, $a_4 = \dfrac{(-1)^4}{4^2} = \dfrac{1}{16}$, and

$$a_{100} = \dfrac{(-1)^{100}}{100^2} = \dfrac{1}{10,000}.$$

11. $a_n = 1 + (-1)^n$. Then $a_1 = 1 + (-1)^1 = 0$, $a_2 = 1 + (-1)^2 = 2$, $a_3 = 1 + (-1)^3 = 0$, $a_4 = 1 + (-1)^4 = 2$, and $a_{100} = 1 + (-1)^{100} = 2$.

13. $a_n = n^n$. Then $a_1 = 1^1 = 1$, $a_2 = 2^2 = 4$, $a_3 = 3^3 = 27$, $a_4 = 4^4 = 256$, and $a_{100} = 100^{100} = 10^{200}$.

15. $a_n = 2\left(a_{n-1} - 2\right)$ and $a_1 = 3$. Then $a_2 = 2\left[(3) - 2\right] = 2$, $a_3 = 2\left[(2) - 2\right] = 0$, $a_4 = 2\left[(0) - 2\right] = -4$, and $a_5 = 2\left[(-4) - 2\right] = -12$.

17. $a_n = 2a_{n-1} + 1$ and $a_1 = 1$. Then $a_2 = 2(1) + 1 = 3$, $a_3 = 2(3) + 1 = 7$, $a_4 = 2(7) + 1 = 15$, and $a_5 = 2(15) + 1 = 31$.

19. $a_n = a_{n-1} + a_{n-2}$, $a_1 = 1$, and $a_2 = 2$. Then $a_3 = 2 + 1 = 3$, $a_4 = 3 + 2 = 5$, and $a_5 = 5 + 3 = 8$.

21. **(a)** $a_1 = 7$, $a_2 = 11$, $a_3 = 15$, $a_4 = 19$, $a_5 = 23$, $a_6 = 27$, $a_7 = 31$, $a_8 = 35$, $a_9 = 39$, $a_{10} = 43$

    (b)

23. **(a)** $a_1 = \dfrac{12}{1} = 12$, $a_2 = \dfrac{12}{2} = 6$, $a_3 = \dfrac{12}{3} = 4$, $a_4 = \dfrac{12}{4} = 3$, $a_5 = \dfrac{12}{5}$, $a_6 = \dfrac{12}{6} = 2$, $a_7 = \dfrac{12}{7}$, $a_8 = \dfrac{12}{8} = \dfrac{3}{2}$, $a_9 = \dfrac{12}{9} = \dfrac{4}{3}$, $a_{10} = \dfrac{12}{10} = \dfrac{6}{5}$

    (b)

25. **(a)** $a_1 = 2$, $a_2 = 0.5$, $a_3 = 2$, $a_4 = 0.5$, $a_5 = 2$, $a_6 = 0.5$, $a_7 = 2$, $a_8 = 0.5$, $a_9 = 2$, $a_{10} = 0.5$

    (b)

27. $2, 4, 6, 8, \ldots$ All are multiples of 2, so $a_1 = 2$, $a_2 = 2 \cdot 2$, $a_3 = 3 \cdot 2$, $a_4 = 4 \cdot 2$, $\ldots$ Thus $a_n = 2n$.

29. $2, 4, 8, 16, \ldots$ All are powers of 2, so $a_1 = 2$, $a_2 = 2^2$, $a_3 = 2^3$, $a_4 = 2^4$, $\ldots$ Thus $a_n = 2^n$.

31. $1, 4, 7, 10, \ldots$ The difference between any two consecutive terms is 3, so $a_1 = 3(1) - 2$, $a_2 = 3(2) - 2$, $a_3 = 3(3) - 2$, $a_4 = 3(4) - 2$, $\ldots$ Thus $a_n = 3n - 2$.

33. $5, -25, 125, -625, \ldots$ These terms are powers of 5, and the terms alternate in sign. So $a_1 = (-1)^2 \cdot 5^1$, $a_2 = (-1)^3 \cdot 5^2$, $a_3 = (-1)^4 \cdot 5^3$, $a_4 = (-1)^5 \cdot 5^4$, $\ldots$ Thus $a_n = (-1)^{n+1} \cdot 5^n$.

35. $1, \frac{3}{4}, \frac{5}{9}, \frac{7}{16}, \frac{9}{25}, \dots$. We consider the numerator separately from the denominator. The numerators of the terms differ by 2, and the denominators are perfect squares. So $a_1 = \dfrac{2(1) - 1}{1^2}$, $a_2 = \dfrac{2(2) - 1}{2^2}$, $a_3 = \dfrac{2(3) - 1}{3^2}$, $a_4 = \dfrac{2(4) - 1}{4^2}$, $a_5 = \dfrac{2(5) - 1}{5^2}, \dots$. Thus $a_n = \dfrac{2n - 1}{n^2}$.

37. $0, 2, 0, 2, 0, 2, \dots$. These terms alternate between 0 and 2. So $a_1 = 1 - 1$, $a_2 = 1 + 1$, $a_3 = 1 - 1$, $a_4 = 1 + 1$, $a_5 = 1 - 1$, $a_6 = 1 + 1, \dots$ Thus $a_n = 1 + (-1)^n$.

39. $a_1 = 1$, $a_2 = 3$, $a_3 = 5$, $a_4 = 7, \dots$. Therefore, $a_n = 2n - 1$. So $S_1 = 1$, $S_2 = 1 + 3 = 4$, $S_3 = 1 + 3 + 5 + = 9$, $S_4 = 1 + 3 + 5 + 7 = 16$, $S_5 = 1 + 3 + 5 + 7 + 9 = 25$, and $S_6 = 1 + 3 + 5 + 7 + 9 + 11 = 36$.

41. $a_1 = \frac{1}{3}$, $a_2 = \frac{1}{3^2}$, $a_3 = \frac{1}{3^3}$, $a_4 = \frac{1}{3^4}, \dots$. Therefore, $a_n = \dfrac{1}{3^n}$. So $S_1 = \frac{1}{3}$, $S_2 = \frac{1}{3} + \frac{1}{3^2} = \frac{4}{9}$, $S_3 = \frac{1}{3} + \frac{1}{3^2} + \frac{1}{3^3} = \frac{13}{27}$, $S_4 = \frac{1}{3} + \frac{1}{3^2} + \frac{1}{3^3} + \frac{1}{3^4} = \frac{40}{81}$, and $S_5 = \frac{1}{3} + \frac{1}{3^2} + \frac{1}{3^3} + \frac{1}{3^4} + \frac{1}{3^5} = \frac{121}{243}$, $S_6 = \frac{1}{3} + \frac{1}{3^2} + \frac{1}{3^3} + \frac{1}{3^4} + \frac{1}{3^5} + \frac{1}{3^6} = \frac{364}{729}$.

43. $a_n = \dfrac{2}{3^n}$. So $S_1 = \frac{2}{3}$, $S_2 = \frac{2}{3} + \frac{2}{3^2} = \frac{8}{9}$, $S_3 = \frac{2}{3} + \frac{2}{3^2} + \frac{2}{3^3} = \frac{26}{27}$, and $S_4 = \frac{2}{3} + \frac{2}{3^2} + \frac{2}{3^3} + \frac{2}{3^4} = \frac{80}{81}$. Therefore, $S_n = \dfrac{3^n - 1}{3^n}$.

45. $a_n = \sqrt{n} - \sqrt{n+1}$. So $S_1 = \sqrt{1} - \sqrt{2} = 1 - \sqrt{2}$, $S_2 = \left(\sqrt{1} - \sqrt{2}\right) + \left(\sqrt{2} - \sqrt{3}\right) = 1 + \left(-\sqrt{2} + \sqrt{2}\right) - \sqrt{3} = 1 - \sqrt{3}$,

$S_3 = \left(\sqrt{1} - \sqrt{2}\right) + \left(\sqrt{2} - \sqrt{3}\right) + \left(\sqrt{3} - \sqrt{4}\right) = 1 + \left(-\sqrt{2} + \sqrt{2}\right) + \left(-\sqrt{3} + \sqrt{3}\right) - \sqrt{4} = 1 - \sqrt{4}$,

$S_4 = \left(\sqrt{1} - \sqrt{2}\right) + \left(\sqrt{2} - \sqrt{3}\right) + \left(\sqrt{3} - \sqrt{4}\right) + \left(\sqrt{4} - \sqrt{5}\right)$

$\quad = 1 + \left(-\sqrt{2} + \sqrt{2}\right) + \left(-\sqrt{3} + \sqrt{3}\right) + \left(-\sqrt{4} + \sqrt{4}\right) - \sqrt{5} = 1 - \sqrt{5}$

Therefore,

$S_n = \left(\sqrt{1} - \sqrt{2}\right) + \left(\sqrt{2} - \sqrt{3}\right) + \cdots + \left(\sqrt{n} - \sqrt{n+1}\right)$

$\quad = 1 + \left(-\sqrt{2} + \sqrt{2}\right) + \left(-\sqrt{3} + \sqrt{3}\right) + \cdots + \left(-\sqrt{n} + \sqrt{n}\right) - \sqrt{n+1} = 1 - \sqrt{n+1}$

47. $\sum_{k=1}^{4} k = 1 + 2 + 3 + 4 = 10$

49. $\sum_{k=1}^{3} \frac{1}{k} = 1 + \frac{1}{2} + \frac{1}{3} = \frac{6}{6} + \frac{3}{6} + \frac{2}{6} = \frac{11}{6}$

51. $\sum_{i=1}^{8} \left[1 + (-1)^i\right] = 0 + 2 + 0 + 2 + 0 + 2 + 0 + 2 = 8$

53. $\sum_{k=1}^{5} 2^{k-1} = 2^0 + 2^1 + 2^2 + 2^3 + 2^4 = 1 + 2 + 4 + 8 + 16 = 31$

55. 385   **57.** 46,438   **59.** 22

61. $\sum_{k=1}^{5} \sqrt{k} = \sqrt{1} + \sqrt{2} + \sqrt{3} + \sqrt{4} + \sqrt{5}$

63. $\sum_{k=0}^{6} \sqrt{k+4} = \sqrt{4} + \sqrt{5} + \sqrt{6} + \sqrt{7} + \sqrt{8} + \sqrt{9} + \sqrt{10}$

65. $\sum_{k=3}^{100} x^k = x^3 + x^4 + x^5 + \cdots + x^{100}$

67. $1 + 2 + 3 + 4 + \cdots + 100 = \sum_{k=1}^{100} k$

69. $1^2 + 2^2 + 3^2 + \cdots + 10^2 = \sum_{k=1}^{10} k^2$

71. $\dfrac{1}{1 \cdot 2} + \dfrac{1}{2 \cdot 3} + \dfrac{1}{3 \cdot 4} + \cdots + \dfrac{1}{999 \cdot 1000} = \sum_{k=1}^{999} \dfrac{1}{k(k+1)}$

73. $1 + x + x^2 + x^3 + \cdots + x^{100} = \sum_{k=0}^{100} x^k$

75. $\sqrt{2}, \sqrt{2\sqrt{2}}, \sqrt{2\sqrt{2\sqrt{2}}}, \sqrt{2\sqrt{2\sqrt{2\sqrt{2}}}}, \dots$. We simplify each term in an attempt to determine a formula for a_n. So $a_1 = 2^{1/2}$, $a_2 = \sqrt{2 \cdot 2^{1/2}} = \sqrt{2^{3/2}} = 2^{3/4}$, $a_3 = \sqrt{2 \cdot 2^{3/4}} = \sqrt{2^{7/4}} = 2^{7/8}$, $a_4 = \sqrt{2 \cdot 2^{7/8}} = \sqrt{2^{15/8}} = 2^{15/16}, \dots$. Thus $a_n = 2^{(2^n - 1)/2^n}$.

77. (a) $A_1 = \$2004$, $A_2 = \$2008.01$, $A_3 = \$2012.02$, $A_4 = \$2016.05$, $A_5 = \$2020.08$, $A_6 = \$2024.12$

(b) Since 3 years is 36 months, we get $A_{36} = \$2149.16$.

79. (a) $P_1 = 35{,}700$, $P_2 = 36{,}414$, $P_3 = 37{,}142$, $P_4 = 37{,}885$, $P_5 = 38{,}643$

(b) Since 2014 is 10 years after 2004, $P_{10} = 42{,}665$.

81. (a) The number of catfish at the end of the month, P_n, is the population at the start of the month, P_{n-1}, plus the increase in population, $0.08P_{n-1}$, minus the 300 catfish harvested. Thus $P_n = P_{n-1} + 0.08P_{n-1} - 300 \Leftrightarrow P_n = 1.08P_{n-1} - 300$.

(b) $P_1 = 5100$, $P_2 = 5208$, $P_3 = 5325$, $P_4 = 5451$, $P_5 = 5587$, $P_6 = 5734$, $P_7 = 5892$, $P_8 = 6064$, $P_9 = 6249$, $P_{10} = 6449$, $P_{11} = 6665$, $P_{12} = 6898$. Thus there should be 6898 catfish in the pond at the end of 12 months.

83. (a) Let S_n be his salary in the nth year. Then $S_1 = \$30{,}000$. Since his salary increase by 2000 each year, $S_n = S_{n-1} + 2000$. Thus $S_1 = \$30{,}000$ and $S_n = S_{n-1} + 2000$.

(b) $S_5 = S_4 + 2000 = (S_3 + 2000) + 2000 = (S_2 + 2000) + 4000 = (S_1 + 2000) + 6000 = \$38{,}000$.

85. Let F_n be the number of pairs of rabbits in the nth month. Clearly $F_1 = F_2 = 1$. In the nth month each pair that is two or more months old (that is, F_{n-2} pairs) will add a pair of offspring to the F_{n-1} pairs already present. Thus $F_n = F_{n-1} + F_{n-2}$. So F_n is the Fibonacci sequence.

87. $a_{n+1} = \begin{cases} \dfrac{a_n}{2} & \text{if } a_n \text{ is even} \\ 3a_n + 1 & \text{if } a_n \text{ is odd} \end{cases}$ With $a_1 = 11$, we have $a_2 = 34$, $a_3 = 17$, $a_4 = 52$, $a_5 = 26$, $a_6 = 13$, $a_7 = 40$,

$a_8 = 20$, $a_9 = 10$, $a_{10} = 5$, $a_{11} = 16$, $a_{12} = 8$, $a_{13} = 4$, $a_{14} = 2$, $a_{15} = 1$, $a_{16} = 4$, $a_{17} = 2$, $a_{18} = 1$, $\ldots$ (with 4, 2, 1 repeating). So $a_{3n+1} = 4$, $a_{3n+2} = 2$, and $a_{3n} = 1$, for $n \geq 5$. With $a_1 = 25$, we have $a_2 = 76$, $a_3 = 38$, $a_4 = 19$, $a_5 = 58$, $a_6 = 29$, $a_7 = 88$, $a_8 = 44$, $a_9 = 22$, $a_{10} = 11$, $a_{11} = 34$, $a_{12} = 17$, $a_{13} = 52$, $a_{14} = 26$, $a_{15} = 13$, $a_{16} = 40$, $a_{17} = 20$, $a_{18} = 10$, $a_{19} = 5$, $a_{20} = 16$, $a_{21} = 8$, $a_{22} = 4$, $a_{23} = 2$, $a_{24} = 1$, $a_{25} = 4$, $a_{26} = 2$, $a_{27} = 1$, $\ldots$ (with 4, 2, 1 repeating). So $a_{3n+1} = 4$, $a_{3n+2} = 2$, and $a_{3n+3} = 1$ for $n \geq 7$.

We conjecture that the sequence will always return to the numbers 4, 2, 1 repeating.

8.2   ARITHMETIC SEQUENCES

1. An arithmetic sequence is sequence where the *difference* between successive terms is constant.

3. True. The nth partial sum of an arithmetic sequence is the average of the first and last terms times n.

5. (a) $a_1 = 5 + 2(1 - 1) = 5$,
$a_2 = 5 + 2(2 - 1) = 5 + 2 = 7$,
$a_3 = 5 + 2(3 - 1) = 5 + 4 = 9$,
$a_4 = 5 + 2(4 - 1) = 5 + 6 = 11$,
$a_5 = 5 + 2(5 - 1) = 5 + 8 = 13$

(b) The common difference is 2.

(c)

7. (a) $a_1 = \frac{5}{2} - (1 - 1) = \frac{5}{2}$, $a_2 = \frac{5}{2} - (2 - 1) = \frac{3}{2}$,
$a_3 = \frac{5}{2} - (3 - 1) = \frac{1}{2}$, $a_4 = \frac{5}{2} - (4 - 1) = -\frac{1}{2}$,
$a_5 = \frac{5}{2} - (5 - 1) = -\frac{3}{2}$

(b) The common difference is -1.

(c)

9. $a = 3$, $d = 5$, $a_n = a + d(n - 1) = 3 + 5(n - 1)$. So $a_{10} = 3 + 5(10 - 1) = 48$.

11. $a = \frac{5}{2}$, $d = -\frac{1}{2}$, $a_n = a + d(n - 1) = \frac{5}{2} - \frac{1}{2}(n - 1)$. So $a_{10} = \frac{5}{2} - \frac{1}{2}(10 - 1) = -2$.

13. $a_4 - a_3 = 14 - 11 = 3, a_3 - a_2 = 11 - 8 = 3, a_2 - a_1 = 8 - 5 = 3$. This sequence is arithmetic with common difference 3.

15. $a_4 - a_3 = a_3 - a_2 = a_2 - a_1 = -25$. This sequence is arithmetic with common difference -25.

17. Since $a_2 - a_1 = 4 - 2 = 2$ and $a_4 - a_3 = 16 - 8 = 8$, the terms of the sequence do not have a common difference. This sequence is not arithmetic.

19. $a_4 - a_3 = -\frac{3}{2} - 0 = -\frac{3}{2}, a_3 - a_2 = 0 - \frac{3}{2} = -\frac{3}{2}, a_2 - a_1 = \frac{3}{2} - 3 = -\frac{3}{2}$. This sequence is arithmetic with common difference $-\frac{3}{2}$.

21. $a_4 - a_3 = 7.7 - 6.0 = 1.7, a_3 - a_2 = 6.0 - 4.3 = 1.7, 4. - a_1 = 4.3 - 2.6 = 1.7$. This sequence is arithmetic with common difference 1.7.

23. $a_1 = 4 + 7(1) = 11, a_2 = 4 + 7(2) = 18, a_3 = 4 + 7(3) = 25, a_4 = 4 + 7(4) = 32, a_5 = 4 + 7(5) = 39$. This sequence is arithmetic, the common difference is $d = 7$ and $a_n = 4 + 7n = 4 + 7n - 7 + 7 = 11 + 7(n - 1)$.

25. $a_1 = \frac{1}{1 + 2(1)} = \frac{1}{3}, a_2 = \frac{1}{1 + 2(2)} = \frac{1}{5}, a_3 = \frac{1}{1 + 2(3)} = \frac{1}{7}, a_4 = \frac{1}{1 + 2(4)} = \frac{1}{9}, a_5 = \frac{1}{1 + 2(5)} = \frac{1}{11}$. Since $a_4 - a_3 = \frac{1}{9} - \frac{1}{7} = -\frac{2}{63}$ and $a_3 - a_2 = \frac{1}{7} - \frac{1}{3} = -\frac{2}{21}$, the terms of the sequence do not have a common difference. This sequence is not arithmetic.

27. $a_1 = 6(1) - 10 = -4, a_2 = 6(2) - 10 = 2, a_3 = 6(3) - 10 = 8, a_4 = 6(4) - 10 = 14, a_5 = 6(5) - 10 = 20$. This sequence is arithmetic, the common difference is $d = 6$ and $a_n = 6n - 10 = 6n - 6 + 6 - 10 = -4 + 6(n - 1)$.

29. $2, 5, 8, 11, \dots$. Then $d = a_2 - a_1 = 5 - 2 = 3, a_5 = a_4 + 3 = 11 + 3 = 14, a_n = 2 + 3(n - 1)$, and $a_{100} = 2 + 3(99) = 299$.

31. $21, 13, 5, -3, \dots$. Then $d = a_2 - a_1 = 13 - 21 = -8, a_5 = a_4 - 8 = -3 - 8 = -11, a_n = 21 - 8(n - 1)$, and $a_{100} = 21 - 8(99) = -771$.

33. $4, 9, 14, 19, \dots$. Then $d = a_2 - a_1 = 9 - 4 = 5, a_5 = a_4 + 5 = 19 + 5 = 24, a_n = 4 + 5(n - 1)$, and $a_{100} = 4 + 5(99) = 499$.

35. $-12, -8, -4, 0, \dots$. Then $d = a_2 - a_1 = -8 - (-12) = 4, a_5 = a_4 + 4 = 0 + 4 = 4, a_n = -12 + 4(n - 1)$, and $a_{100} = -12 + 4(99) = 384$.

37. $25, 26.5, 28, 29.5, \dots$. Then $d = a_2 - a_1 = 26.5 - 25 = 1.5, a_5 = a_4 + 1.5 = 29.5 + 1.5 = 31, a_n = 25 + 1.5(n - 1)$, and $a_{100} = 25 + 1.5(99) = 173.5$.

39. $2, 2 + s, 2 + 2s, 2 + 3s, \dots$. Then $d = a_2 - a_1 = 2 + s - 2 = s, a_5 = a_4 + s = 2 + 3s + s = 2 + 4s, a_n = 2 + (n - 1)s$, and $a_{100} = 2 + 99s$.

41. $a_{10} = \frac{55}{2}, a_2 = \frac{7}{2}$, and $a_n = a + d(n - 1)$. Then $a_2 = a + d = \frac{7}{2} \Leftrightarrow d = \frac{7}{2} - a$. Substituting into $a_{10} = a + 9d = \frac{55}{2}$ gives $a + 9\left(\frac{7}{2} - a\right) = \frac{55}{2} \Leftrightarrow a = \frac{1}{2}$. Thus, the first term is $a_1 = \frac{1}{2}$.

43. $a_{100} = 98$ and $d = 2$. Note that $a_{100} = a + 99d = a + 99(2) = a + 198$. Since $a_{100} = 98$, we have $a + 198 = a_{100} = 98 \Leftrightarrow a = -100$. Hence, $a_1 = -100, a_2 = -100 + 2 = -98$, and $a_3 = -100 + 4 = -96$.

45. The arithmetic sequence is $1, 4, 7, \dots$. So $d = 4 - 1 = 3$ and $a_n = 1 + 3(n - 1)$. Then $a_n = 88 \Leftrightarrow 1 + 3(n - 1) = 88 \Leftrightarrow 3(n - 1) = 87 \Leftrightarrow n - 1 = 29 \Leftrightarrow n = 30$. So 88 is the 30th term.

47. $a = 1, d = 2, n = 10$. Then $S_{10} = \frac{10}{2}[2a + (10 - 1)d] = 5(2 \cdot 1 + 9 \cdot 2) = 100$.

49. $a = 5, d = -4, n = 20$. Then $s_{20} = \frac{20}{2}[2a + (20 - 1)d] = 10(2 \cdot 5 - 19 \cdot 4) = -660$.

51. $a_1 = 55, d = 12, n = 10$. Then $S_{10} = \frac{10}{2}[2a + (10 - 1)d] = 5(2 \cdot 55 + 9 \cdot 12) = 1090$.

53. $1 + 5 + 9 + \dots + 401$ is a partial sum of an arithmetic series with $a = 1$ and $d = 5 - 1 = 4$. The last term is $401 = a_n = 1 + 4(n - 1)$, so $n - 1 = 100 \Leftrightarrow n = 101$. So the partial sum is $S_{101} = \frac{101}{2}(1 + 401) = 101 \cdot 201 = 20{,}301$.

55. $250 + 233 + 216 + \dots + 97$ is a partial sum of an arithmetic sequence with $a = 250$ and $d = 233 - 250 = -17$. The last term is $97 = a_n = 250 - 17(n - 1)$, so $n - 1 = \frac{97 - 250}{-17} = 9 \Leftrightarrow n = 10$. So the partial sum is $S_{10} = \frac{10}{2}(250 + 97) = 1735$.

57. $0.7 + 2.7 + 4.7 + \cdots + 56.7$ is a partial sum of an arithmetic sequence with $a = 0.7$ and $d = 2.7 - 0.7 = 2$. The last term is $56.7 = a_n = 0.7 + 2(n - 1) \Leftrightarrow 28 = n - 1 \Leftrightarrow n = 29$. So the partial sum is $S_{29} = \frac{29}{2}(0.7 + 56.7) = 832.3$.

59. $\sum_{k=0}^{10}(3 + 0.25k)$ is a partial sum of an arithmetic sequence with $a = 3 + 0.25 \cdot 0 = 3$ and $d = 0.25$. The last term is $a_{11} = 3 + 0.25 \cdot 10 = 5.5$. So the partial sum is $S_{11} = \frac{11}{2}(3 + 5.5) = 46.75$.

61. Let x denote the length of the side between the length of the other two sides. Then the lengths of the three sides of the triangle are $x - a$, x, and $x + a$, for some $a > 0$. Since $x + a$ is the longest side, it is the hypotenuse, and by the Pythagorean Theorem, we know that $(x - a)^2 + x^2 = (x + a)^2 \Leftrightarrow x^2 - 2ax + a^2 + x^2 = x^2 + 2ax + a^2 \Leftrightarrow x^2 - 4ax = 0 \Leftrightarrow x(x - 4a) = 0 \Rightarrow x = 4a$ ($x = 0$ is not a possible solution). Thus, the lengths of the three sides are $x - a = 4a - a = 3a$, $x = 4a$, and $x + a = 4a + a = 5a$. The lengths $3a$, $4a$, $5a$ are proportional to 3, 4, 5, and so the triangle is similar to a 3-4-5 triangle.

63. The sequence $1, \frac{3}{5}, \frac{3}{7}, \frac{1}{3}, \ldots$ is harmonic if $1, \frac{5}{3}, \frac{7}{3}, 3, \ldots$ forms an arithmetic sequence. Since $\frac{5}{3} - 1 = \frac{7}{3} - \frac{5}{3} = 3 - \frac{7}{3} = \frac{2}{3}$, the sequence of reciprocals is arithmetic and thus the original sequence is harmonic.

65. We have an arithmetic sequence with $a = 5$ and $d = 2$. We seek n such that $2700 = S_n = \frac{n}{2}[2a + (n - 1)d]$. Solving for n, we have $2700 = \frac{n}{2}[10 + 2(n - 1)] \Leftrightarrow 5400 = 10n + 2n^2 - 2n \Leftrightarrow n^2 + 4n - 2700 = 0 \Leftrightarrow (n - 50)(n + 54) = 0 \Leftrightarrow n = 50$ or $n = -54$. Since n is a positive integer, 50 terms of the sequence must be added to get 2700.

67. The diminishing values of the computer form an arithmetic sequence with $a_1 = 12{,}500$ and common difference $d = -1875$. Thus the value of the computer after 6 years is $a_7 = 12{,}500 + (7 - 1)(-1875) = \$1250$.

69. The increasing values of the man's salary form an arithmetic sequence with $a_1 = 30{,}000$ and common difference $d = 2300$. Then his total earnings for a ten-year period are $S_{10} = \frac{10}{2}[2(30{,}000) + 9(2300)] = 403{,}500$. Thus his total earnings for the 10 year period are \$403,500.

71. The number of seats in the nth row is given by the nth term of an arithmetic sequence with $a_1 = 15$ and common difference $d = 3$. We need to find n such that $S_n = 870$. So we solve $870 = S_n = \frac{n}{2}[2(15) + (n - 1)3]$ for n. We have $870 = \frac{n}{2}(27 + 3n) \Leftrightarrow 1740 = 3n^2 + 27n \Leftrightarrow 3n^2 + 27n - 1740 = 0 \Leftrightarrow n^2 + 9n - 580 = 0 \Leftrightarrow (x - 20)(x + 29) = 0 \Rightarrow n = 20$ or $n = -29$. Since the number of rows is positive, the theater must have 20 rows.

73. The number of gifts on the 12th day is $1 + 2 + 3 + 4 + \cdots + 12$. Since $a_2 - a_1 = a_3 - a_2 = a_4 - a_3 = \cdots = 1$, the number of gifts on the 12th day is the partial sum of an arithmetic sequence with $a = 1$ and $d = 1$. So the sum is $S_{12} = 12\left(\frac{1 + 12}{2}\right) = 6 \cdot 13 = 78$.

8.3   GEOMETRIC SEQUENCES

1. A geometric sequence is a sequence where the *ratio* between successive terms is constant.

3. True. If we know the first and second terms of a geometric sequence then we can find all other terms.

5. (a) $a_1 = 5\,(2)^0 = 5$, $a_2 = 5\,(2)^1 = 10$,

$a_3 = 5\,(2)^2 = 20$, $a_4 = 5\,(2)^3 = 40$,

$a_5 = 5\,(2)^4 = 80$

(b) The common ratio is 2.

(c)

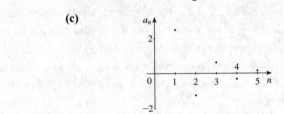

7. (a) $a_1 = \frac{5}{2}\left(-\frac{1}{2}\right)^0 = \frac{5}{2}$, $a_2 = \frac{5}{2}\left(-\frac{1}{2}\right)^1 = -\frac{5}{4}$,

$a_3 = \frac{5}{2}\left(-\frac{1}{2}\right)^2 = \frac{5}{8}$, $a_4 = \frac{5}{2}\left(-\frac{1}{2}\right)^3 = -\frac{5}{16}$,

$a_5 = \frac{5}{2}\left(-\frac{1}{2}\right)^4 = \frac{5}{32}$

(b) The common ratio is $-\frac{1}{2}$.

(c)

9. $a = 3$, $r = 5$. So $a_n = ar^{n-1} = 3\,(5)^{n-1}$ and $a_4 = 3 \cdot 5^3 = 375$.

11. $a = \frac{5}{2}$, $r = -\frac{1}{2}$. So $a_n = ar^{n-1} = \frac{5}{2}\left(-\frac{1}{2}\right)^{n-1}$ and $a_4 = \frac{5}{2} \cdot \left(-\frac{1}{2}\right)^3 = -\frac{5}{16}$.

13. $\dfrac{a_2}{a_1} = \dfrac{4}{2} = 2$, $\dfrac{a_3}{a_2} = \dfrac{8}{4} = 2$, and $\dfrac{a_4}{a_3} = \dfrac{16}{8} = 2$. Since these ratios are the same, the sequence is geometric with the common ratio 2.

15. $\dfrac{a_2}{a_1} = \dfrac{96}{144} = \dfrac{2}{3}$, $\dfrac{a_3}{a_2} = \dfrac{64}{96} = \dfrac{2}{3}$, and $\dfrac{a_4}{a_3} = \dfrac{128/3}{64} = \dfrac{2}{3}$. Since these ratios are the same, the sequence is geometric with the common ratio $\frac{2}{3}$.

17. $\dfrac{a_2}{a_1} = \dfrac{3/2}{3} = \dfrac{1}{2}$, $\dfrac{a_3}{a_2} = \dfrac{3/4}{3/2} = \dfrac{1}{2}$, and $\dfrac{a_4}{a_3} = \dfrac{3/8}{3/4} = \dfrac{1}{2}$. Since these ratios are the same, the sequence is geometric with the common ratio $\frac{1}{2}$.

19. $\dfrac{a_2}{a_1} = \dfrac{1/3}{1/2} = \dfrac{2}{3}$ and $\dfrac{a_4}{a_3} = \dfrac{1/5}{1/4} = \dfrac{4}{5}$. Since these ratios are not the same, this is not a geometric sequence.

21. $\dfrac{a_2}{a_1} = \dfrac{1.1}{1.0} = 1.1$, $\dfrac{a_3}{a_2} = \dfrac{1.21}{1.1} = 1.1$, and $\dfrac{a_4}{a_3} = \dfrac{1.331}{1.21} = 1.1$. Since these ratios are the same, the sequence is geometric with the common ratio 1.1.

23. $a_1 = 2\,(3)^1 = 6$, $a_2 = 2\,(3)^2 = 18$, $a_3 = 2\,(3)^3 = 54$, $a_4 = 2\,(3)^4 = 162$, and $a_5 = 2\,(3)^5 = 486$. This sequence is geometric, the common ratio is $r = 3$, and $a_n = a_1 r^{n-1} = 6\,(3)^{n-1}$.

25. $a_1 = \dfrac{1}{4}$, $a_2 = \dfrac{1}{4^2} = \dfrac{1}{16}$, $a_3 = \dfrac{1}{4^3} = \dfrac{1}{64}$, $a_4 = \dfrac{1}{4^4} = \dfrac{1}{256}$, and $a_5 = \dfrac{1}{4^5} = \dfrac{1}{1024}$. This sequence is geometric, the common ratio is $r = \frac{1}{4}$ and $a_n = a_1 r^{n-1} = \dfrac{1}{4}\left(\dfrac{1}{4}\right)^{n-1}$.

27. Since $\ln a^b = b \ln a$, we have $a_1 = \ln\left(5^0\right) = \ln 1 = 0$, $a_2 = \ln\left(5^1\right) = \ln 5$, $a_3 = \ln\left(5^2\right) = 2 \ln 5$, $a_4 = \ln\left(5^3\right) = 3 \ln 5$, $a_5 = \ln\left(5^4\right) = 4 \ln 5$. Since $a_1 = 0$ and $a_2 \neq 0$, this sequence is not geometric.

29. $2, 6, 18, 54, \ldots$. Then $r = \dfrac{a_2}{a_1} = \dfrac{6}{2} = 3$, $a_5 = a_4 \cdot 3 = 54\,(3) = 162$, and $a_n = 2 \cdot 3^{n-1}$.

31. $0.3, -0.09, 0.027, -0.0081, \ldots$. Then $r = \dfrac{a_2}{a_1} = \dfrac{-0.09}{0.3} = -0.3$, $a_5 = a_4 \cdot (-0.3) = -0.0081\,(-0.3) = 0.00243$, and $a_n = 0.3\,(-0.3)^{n-1}$.

33. $144, -12, 1, -\frac{1}{12}, \ldots$. Then $r = \dfrac{a_2}{a_1} = \dfrac{-12}{144} = -\dfrac{1}{12}$, $a_5 = a_4 \cdot \left(-\dfrac{1}{12}\right) = -\dfrac{1}{12}\left(-\dfrac{1}{12}\right) = \dfrac{1}{144}$, $a_n = 144\left(-\dfrac{1}{12}\right)^{n-1}$.

35. $3, 3^{5/3}, 3^{7/3}, 27, \ldots$ Then $r = \dfrac{a_2}{a_1} = \dfrac{3^{5/3}}{3} = 3^{2/3}$, $a_5 = a_4 \cdot \left(3^{2/3}\right) = 27 \cdot 3^{2/3} = 3^{11/3}$, and

$a_n = 3\left(3^{2/3}\right)^{n-1} = 3 \cdot 3^{(2n-2)/3} = 3^{(2n+1)/3}$.

37. $1, s^{2/7}, s^{4/7}, s^{6/7}, \ldots$ Then $r = \dfrac{a_2}{a_1} = \dfrac{s^{2/7}}{1} = s^{2/7}$, $a_5 = a_4 \cdot s^{2/7} = s^{6/7} \cdot s^{2/7} = s^{8/7}$, and $a_n = \left(s^{2/7}\right)^{n-1} = s^{(2n-2)/7}$.

39. $a_1 = 8, a_2 = 4$. Thus $r = \dfrac{a_2}{a_1} = \dfrac{4}{8} = \dfrac{1}{2}$ and $a_5 = a_1 r^{5-1} = 8\left(\dfrac{1}{2}\right)^4 = \dfrac{8}{16} = \dfrac{1}{2}$.

41. $a_3 = \dfrac{100}{9}$ and $a_6 = \dfrac{800}{243}$. Thus, $\dfrac{a_6}{a_3} = \dfrac{a_1 r^5}{a_1 r^2} = r^3 \Leftrightarrow r^3 = \dfrac{800/243}{100/9} = \dfrac{8}{27}$, so $r = \sqrt[3]{\dfrac{8}{27}} = \dfrac{2}{3}$. Therefore, $a_3 = r^2 a_1 \Leftrightarrow$

$a_1 = \dfrac{a_3}{r^2} = \dfrac{100/9}{(2/3)^2} = \dfrac{100/9}{4/9} = 25$. The second term is $a_2 = a_1 r = 25\left(\dfrac{2}{3}\right) = \dfrac{50}{3}$.

43. $a_8 = 640$ and $a_3 = 20$. Thus, $r^5 = \dfrac{a_8}{a_3} \Leftrightarrow r^5 = \dfrac{640}{20} = 32$, so $r = 2$. Therefore, $a_1 = \dfrac{a_3}{r^2} = \dfrac{20}{4} = 5$ and $a_n = 5 \cdot 2^{n-1}$.

45. $r = \dfrac{2}{5}, a_4 = \dfrac{5}{2}$. Since $r = \dfrac{a_4}{a_3}$, we have $a_3 = \dfrac{a_4}{r} = \dfrac{5/2}{2/5} = \dfrac{25}{4}$.

47. The geometric sequence is $2, 6, 18, \ldots$. Thus $r = \dfrac{a_2}{a_1} = \dfrac{6}{2} = 3$. We need to find n so that $a_n = 2 \cdot 3^{n-1} = 118{,}098 \Leftrightarrow$

$3^{n-1} = 59{,}049 \Leftrightarrow n - 1 = \log_3 59{,}049 = 10 \Leftrightarrow n = 11$. Therefore, 118,098 is the 11th term of the geometric sequence.

49. $a = 5, r = 2, n = 6$. Then $S_6 = 5\dfrac{1 - 2^6}{1 - 2} = (-5)(-63) = 315$.

51. $a_3 = 28, a_6 = 224, n = 6$. So $\dfrac{a_6}{a_3} = \dfrac{ar^5}{ar^2} = r^3$. So we have $r^3 = \dfrac{a_6}{a_3} = \dfrac{224}{28} = 8$, and hence $r = 2$. Since $a_3 = a \cdot r^2$, we

get $a = \dfrac{a_3}{r^2} = \dfrac{28}{2^2} = 7$. So $S_6 = 7\dfrac{1 - 2^6}{1 - 2} = (-7)(-63) = 441$.

53. $1 + 3 + 9 + \cdots + 2187$ is a partial sum of a geometric sequence, where $a = 1$ and $r = \dfrac{a_2}{a_1} = \dfrac{3}{1} = 3$. Then the last term is

$2187 = a_n = 1 \cdot 3^{n-1} \Leftrightarrow n - 1 = \log_3 2187 = 7 \Leftrightarrow n = 8$. So the partial sum is $S_8 = (1)\dfrac{1 - 3^8}{1 - 3} = 3280$.

55. $\sum_{k=0}^{10} 3\left(\dfrac{1}{2}\right)^k$ is a partial sum of a geometric sequence, where $a = 3$, $r = \dfrac{1}{2}$, and $n = 11$. So the partial sum is

$S_{11} = (3)\dfrac{1 - \left(\frac{1}{2}\right)^{11}}{1 - \left(\frac{1}{2}\right)} = 6\left[1 - \left(\dfrac{1}{2}\right)^{11}\right] = \dfrac{6141}{1024} \approx 5.997070313$.

57. $1 + \dfrac{1}{3} + \dfrac{1}{9} + \dfrac{1}{27} + \cdots$ is an infinite geometric series with $a = 1$ and $r = \dfrac{1}{3}$. Therefore, it is convergent with sum

$S = \dfrac{a}{1 - r} = \dfrac{1}{1 - \left(\frac{1}{3}\right)} = \dfrac{3}{2}$.

59. $1 - \dfrac{1}{3} + \dfrac{1}{9} - \dfrac{1}{27} + \cdots$ is an infinite geometric series with $a = 1$ and $r = -\dfrac{1}{3}$. Therefore, it is convergent with sum

$S = \dfrac{a}{1 - r} = \dfrac{1}{1 - \left(-\frac{1}{3}\right)} = \dfrac{3}{4}$.

61. $1 + \dfrac{3}{2} + \left(\dfrac{3}{2}\right)^2 + \left(\dfrac{3}{2}\right)^3 + \cdots$ is an infinite geometric series with $a = 1$ and $r = \dfrac{3}{2} > 1$. Therefore, the series diverges.

63. $3 - \dfrac{3}{2} + \dfrac{3}{4} - \dfrac{3}{8} + \cdots$ is an infinite geometric series with $a = 3$ and $r = -\dfrac{1}{2}$. Therefore, it is convergent with sum

$S = \dfrac{3}{1 - \left(-\frac{1}{2}\right)} = 2$.

65. $3 - 3(1.1) + 3(1.1)^2 - 3(1.1)^3 + \cdots$ is an infinite geometric series with $a = 3$ and $r = 1.1 > 1$. Therefore, the series diverges.

67. $\frac{1}{\sqrt{2}} + \frac{1}{2} + \frac{1}{2\sqrt{2}} + \frac{1}{4} + \cdots$ is an infinite geometric series with $a = \frac{1}{\sqrt{2}}$ and $r = \frac{1}{\sqrt{2}}$. Therefore, the sum of the series is

$$S = \frac{\frac{1}{\sqrt{2}}}{1 - \frac{1}{\sqrt{2}}} = \frac{1}{\sqrt{2} - 1} = \sqrt{2} + 1.$$

69. $0.777\ldots = \frac{7}{10} + \frac{7}{100} + \frac{7}{1000} + \cdots$ is an infinite geometric series with $a = \frac{7}{10}$ and $r = \frac{1}{10}$. Thus

$$0.777\ldots = \frac{a}{1 - r} = \frac{\frac{7}{10}}{1 - \frac{1}{10}} = \frac{7}{9}.$$

71. $0.030303\ldots = \frac{3}{100} + \frac{3}{10,000} + \frac{3}{1,000,000} + \cdots$ is an infinite geometric series with $a = \frac{3}{100}$ and $r = \frac{1}{100}$. Thus

$$0.030303\ldots = \frac{a}{1 - r} = \frac{\frac{3}{100}}{1 - \frac{1}{100}} = \frac{3}{99} = \frac{1}{33}.$$

73. $0.\overline{112} = 0.112112112\ldots = \frac{112}{1000} + \frac{112}{1,000,000} + \frac{112}{1,000,000,000} + \cdots$ is an infinite geometric series with $a = \frac{112}{1000}$ and

$r = \frac{1}{1000}$. Thus $0.112112112\ldots = \frac{a}{1 - r} = \frac{\frac{112}{1000}}{1 - \frac{1}{1000}} = \frac{112}{999}$.

75. Since we have 5 terms, let us denote $a_1 = 5$ and $a_5 = 80$. Also, $\frac{a_5}{a_1} = r^4$ because the sequence is geometric, and so

$r^4 = \frac{80}{5} = 16 \Leftrightarrow r = \pm 2$. If $r = 2$, the three geometric means are $a_2 = 10$, $a_3 = 20$, and $a_4 = 40$. (If $r = -2$, the three geometric means are $a_2 = -10$, $a_3 = 20$, and $a_4 = -40$, but these are not between 5 and 80.)

77. (a) The value at the end of the year is equal to the value at beginning less the depreciation, so
$V_n = V_{n-1} - 0.2 V_{n-1} = 0.8 V_{n-1}$ with $V_1 = 160,000$. Thus $V_n = 160,000 \cdot 0.8^{n-1}$.

(b) $V_n < 100,000 \Leftrightarrow 0.8^{n-1} \cdot 160,000 < 100,000 \Leftrightarrow 0.8^{n-1} < 0.625 \Leftrightarrow (n-1) \log 0.8 < \log 0.625 \Leftrightarrow$
$n - 1 > \frac{\log 0.625}{\log 0.8} = 2.11$. Thus it will depreciate to below \$100,000 during the fourth year.

79. Since the ball is dropped from a height of 80 feet, $a = 80$. Also since the ball rebounds three-fourths of the distance fallen, $r = \frac{3}{4}$. So on the nth bounce, the ball attains a height of $a_n = 80 \left(\frac{3}{4}\right)^n$. Hence, on the fifth bounce, the ball goes

$a_5 = 80 \left(\frac{3}{4}\right)^5 = \frac{80 \cdot 243}{1024} \approx 19$ ft high.

81. Let a_n be the amount of water remaining at the nth stage. We start with 5 gallons, so $a = 5$. When 1 gallon (that is, $\frac{1}{5}$ of the mixture) is removed, $\frac{4}{5}$ of the mixture (and hence $\frac{4}{5}$ of the water in the mixture) remains. Thus, $a_1 = 5 \cdot \frac{4}{5}$, $a_2 = 5 \cdot \frac{4}{5} \cdot \frac{4}{5}, \ldots$,

and in general, $a_n = 5 \left(\frac{4}{5}\right)^n$. The amount of water remaining after 3 repetitions is $a_3 = 5 \left(\frac{4}{5}\right)^3 = \frac{64}{25}$, and after

5 repetitions it is $a_5 = 5 \left(\frac{4}{5}\right)^5 = \frac{1024}{625}$.

83. Let a_n be the height the ball reaches on the nth bounce. From the given information, a_n is the geometric sequence
$a_n = 9 \cdot \left(\frac{1}{3}\right)^n$. (Notice that the ball hits the ground for the fifth time after the fourth bounce.)
(a) $a_0 = 9$, $a_1 = 9 \cdot \frac{1}{3} = 3$, $a_2 = 9 \cdot \left(\frac{1}{3}\right)^2 = 1$, $a_3 = 9 \cdot \left(\frac{1}{3}\right)^3 = \frac{1}{3}$, and $a_4 = 9 \cdot \left(\frac{1}{3}\right)^4 = \frac{1}{9}$. The total distance traveled is
$a_0 + 2a_1 + 2a_2 + 2a_3 + 2a_4 = 9 + 2 \cdot 3 + 2 \cdot 1 + 2 \cdot \frac{1}{3} + 2 \cdot \frac{1}{9} = \frac{161}{9} = 17\frac{8}{9}$ ft.

(b) The total distance traveled at the instant the ball hits the ground for the nth time is

$$
\begin{aligned}
D_n &= 9 + 2 \cdot 9 \cdot \tfrac{1}{3} + 2 \cdot 9 \cdot \left(\tfrac{1}{3}\right)^2 + 2 \cdot 9 \cdot \left(\tfrac{1}{3}\right)^3 + 2 \cdot 9 \cdot \left(\tfrac{1}{3}\right)^4 + \cdots + 2 \cdot 9 \cdot \left(\tfrac{1}{3}\right)^{n-1} \\
&= 2\left[9 + 9 \cdot \tfrac{1}{3} + 9 \cdot \left(\tfrac{1}{3}\right)^2 + 9 \cdot \left(\tfrac{1}{3}\right)^3 + 9 \cdot \left(\tfrac{1}{3}\right)^4 + \cdots + 9 \cdot \left(\tfrac{1}{3}\right)^{n-1}\right] - 9 \\
&= 2\left[9 \cdot \frac{1 - \left(\tfrac{1}{3}\right)^n}{1 - \tfrac{1}{3}}\right] - 9 = 27\left[1 - \left(\tfrac{1}{3}\right)^n\right] - 9 = 18 - \left(\tfrac{1}{3}\right)^{n-3}
\end{aligned}
$$

85. Let $a_1 = 1$ be the man with 7 wives. Also, let $a_2 = 7$ (the wives), $a_3 = 7a_2 = 7^2$ (the sacks), $a_4 = 7a_3 = 7^3$ (the cats), and $a_5 = 7a_4 = 7^4$ (the kits). The total is $a_1 + a_2 + a_3 + a_4 + a_5 = 1 + 7 + 7^2 + 7^3 + 7^4$, which is a partial sum of a geometric sequence with $a = 1$ and $r = 7$. Thus, the number in the party is $S_5 = 1 \cdot \dfrac{1 - 7^5}{1 - 7} = 2801$.

87. Let a_n be the height the ball reaches on the nth bounce. We have $a_0 = 1$ and $a_n = \frac{1}{2}a_{n-1}$. Since the total distance d traveled includes the bounce up as well and the distance down, we have

$$
\begin{aligned}
d &= a_0 + 2 \cdot a_1 + 2 \cdot a_2 + \cdots = 1 + 2\left(\tfrac{1}{2}\right) + 2\left(\tfrac{1}{2}\right)^2 + 2\left(\tfrac{1}{2}\right)^3 + 2\left(\tfrac{1}{2}\right)^4 + \cdots \\
&= 1 + 1 + \tfrac{1}{2} + \left(\tfrac{1}{2}\right)^2 + \left(\tfrac{1}{2}\right)^3 + \cdots = 1 + \sum_{i=0}^{\infty}\left(\tfrac{1}{2}\right)^i = 1 + \frac{1}{1 - \tfrac{1}{2}} = 3
\end{aligned}
$$

Thus the total distance traveled is about 3 m.

89. (a) If a square has side x, then by the Pythagorean Theorem the length of the side of the square formed by joining the midpoints is, $\sqrt{\left(\dfrac{x}{2}\right)^2 + \left(\dfrac{x}{2}\right)^2} = \sqrt{\dfrac{x^2}{4} + \dfrac{x^2}{4}} = \dfrac{x}{\sqrt{2}}$. In our case, $x = 1$ and the side of the first inscribed square is $\dfrac{1}{\sqrt{2}}$, the side of the second inscribed square is $\dfrac{1}{\sqrt{2}} \cdot \dfrac{1}{\sqrt{2}} = \left(\dfrac{1}{\sqrt{2}}\right)^2$, the side of the third inscribed square is $\left(\dfrac{1}{\sqrt{2}}\right)^3$, and so on. Since this pattern continues, the total area of all the squares is

$$
A = 1^2 + \left(\frac{1}{\sqrt{2}}\right)^2 + \left(\frac{1}{\sqrt{2}}\right)^4 + \left(\frac{1}{\sqrt{2}}\right)^6 + \cdots = 1 + \frac{1}{2} + \left(\frac{1}{2}\right)^2 + \left(\frac{1}{2}\right)^3 + \cdots = \frac{1}{1 - \frac{1}{2}} = 2.
$$

(b) As in part (a), the sides of the squares are $1, \dfrac{1}{\sqrt{2}}, \left(\dfrac{1}{\sqrt{2}}\right)^2, \left(\dfrac{1}{\sqrt{2}}\right)^3, \ldots$. Thus the sum of the perimeters is

$S = 4 \cdot 1 + 4 \cdot \dfrac{1}{\sqrt{2}} + 4 \cdot \left(\dfrac{1}{\sqrt{2}}\right)^2 + 4 \cdot \left(\dfrac{1}{\sqrt{2}}\right)^3 + \cdots$, which is an infinite geometric series with $a = 4$ and $r = \dfrac{1}{\sqrt{2}}$. Thus the sum of the perimeters is $S = \dfrac{4}{1 - \frac{1}{\sqrt{2}}} = \dfrac{4\sqrt{2}}{\sqrt{2} - 1} = \dfrac{4\sqrt{2}}{\sqrt{2} - 1} \cdot \dfrac{\sqrt{2} + 1}{\sqrt{2} + 1} = \dfrac{4 \cdot 2 + 4\sqrt{2}}{2 - 1} = 8 + 4\sqrt{2}$.

91. Let a_n denote the area colored blue at nth stage. Since only the middle squares are colored blue, $a_n = \frac{1}{9} \times$ (area remaining yellow at the $(n-1)$th stage). Also, the area remaining yellow at the nth stage is $\frac{8}{9}$ of the area remaining yellow at the preceding stage. So $a_1 = \frac{1}{9}$, $a_2 = \frac{1}{9}\left(\frac{8}{9}\right)$, $a_3 = \frac{1}{9}\left(\frac{8}{9}\right)^2$, $a_4 = \frac{1}{9}\left(\frac{8}{9}\right)^3, \ldots$. Thus the total area colored blue $A = \frac{1}{9} + \frac{1}{9}\left(\frac{8}{9}\right) + \frac{1}{9}\left(\frac{8}{9}\right)^2 + \frac{1}{9}\left(\frac{8}{9}\right)^3 + \cdots$ is an infinite geometric series with $a = \frac{1}{9}$ and $r = \frac{8}{9}$. So the total area is $A = \dfrac{\frac{1}{9}}{1 - \frac{8}{9}} = 1$.

93. Let $a_1, a_2, a_3, \ldots$ be a geometric sequence with common ratio r. Thus $a_2 = a_1 r$, $a_3 = a_1 \cdot r^2, \ldots, a_n = a_1 \cdot r^{n-1}$. Hence,

$$\frac{1}{a_2} = \frac{1}{a_1 \cdot r} = \frac{1}{a_1} \cdot \frac{1}{r}, \frac{1}{a_3} = \frac{1}{a_1 \cdot r^2} = \frac{1}{a_1} \cdot \frac{1}{r^2} = \frac{1}{a_1} \cdot \left(\frac{1}{r}\right)^2, \ldots \frac{1}{a_n} = \frac{1}{a_1 \cdot r^{n-1}} = \frac{1}{a_1} \cdot \frac{1}{r^{n-1}} = \frac{1}{a_1} \left(\frac{1}{r}\right)^{n-1}, \text{ and so}$$

$\frac{1}{a_1}, \frac{1}{a_2}, \frac{1}{a_3}, \ldots$ is a geometric sequence with common ratio $\frac{1}{r}$.

95. Since $a_1, a_2, a_3, \ldots$ is an arithmetic sequence with common difference d, the terms can be expressed as $a_2 = a_1 + d$,

$a_3 = a_1 + 2d, \ldots, a_n = a_1 + (n-1)d$. So $10^{a_2} = 10^{a_1+d} = 10^{a_1} \cdot 10^d$, $10^{a_3} = 10^{a_1+2d} = 10^{a_1} \cdot \left(10^d\right)^2, \ldots,$

$10^{a_n} = 10^{a_1+(n-1)d} = 10^{a_1} \cdot \left(10^d\right)^{n-1}$, and so $10^{a_1}, 10^{a_2}, 10^{a_3}, \ldots$ is a geometric sequence with common ratio $r = 10^d$.

8.4   MATHEMATICS OF FINANCE

1. An annuity is a sum of money that is paid in regular equal payments. The *amount* of an annuity is the sum of all the individual payments together with all the interest.

3. $n = 10$, $R = \$1000$, $i = 0.06$. So $A_f = R\dfrac{(1+i)^n - 1}{i} = 1000\dfrac{(1+0.06)^{10} - 1}{0.06} = \$13{,}180.79.$

5. $n = 20$, $R = \$5000$, $i = 0.12$. So $A_f = R\dfrac{(1+i)^n - 1}{i} = 5000\dfrac{(1+0.12)^{20} - 1}{0.12} = \$360{,}262.21.$

7. $n = 16$, $R = \$300$, $i = \dfrac{0.08}{4} = 0.02$. So $A_f = R\dfrac{(1+i)^n - 1}{i} = 300\dfrac{(1+0.02)^{16} - 1}{0.02} = \$5{,}591.79.$

9. $A_f = 5000$, $n = 4 \cdot 2 = 8$, $i = \dfrac{0.10}{4} = 0.025$. So $R = \dfrac{iA_f}{(1+i)^n - 1} = \dfrac{(0.025)(5000)}{(1.025)^8 - 1} = \$572.34.$

11. $R = 1000$, $n = 20$, $i = \dfrac{0.09}{2} = 0.045$. So $A_p = R\dfrac{1 - (1+i)^{-n}}{i} = 1000\dfrac{1 - (1.045)^{-20}}{0.045} = \$13{,}007.94.$

13. $R = \$200$, $n = 20$, $i = \dfrac{0.09}{2} = 0.045$. So $A_p = R\dfrac{1 - (1+i)^{-n}}{i} = (200)\dfrac{1 - (1+0.045)^{-20}}{0.045} = \$2601.59.$

15. $A_p = \$12{,}000$, $i = \dfrac{0.105}{12} = 0.00875$, $n = 48$. Then $R = \dfrac{iA_p}{1 - (1+i)^{-n}} = \dfrac{(0.00875)(12000)}{1 - (1+0.00875)^{-48}} = \$307.24.$

17. $A_p = \$100{,}000$, $i = \dfrac{0.08}{12} \approx 0.006667$, $n = 360$. Then $R = \dfrac{iA_p}{1 - (1+i)^{-n}} = \dfrac{(0.006667)(100{,}000)}{1 - (1+0.006667)^{-360}} = \$733.76.$

Therefore, the total amount paid on this loan over the 30 year period is $(360)(733.76) = \$264{,}153.60.$

19. $R = 3500$, $n = 12(30) = 360$, $i = \dfrac{0.06}{12} = 0.005$. So $A_p = R\dfrac{1 - (1+i)^{-n}}{i} = 3500\dfrac{1 - (1.005)^{-360}}{0.005} = \$583{,}770.65.$

Therefore, Dr. Gupta can afford a loan of $583,770.65.

21. $R = 220$, $n = 12(3) = 36$, $i = \dfrac{0.08}{12} \approx 0.00667$. The amount borrowed is

$A_p = R\dfrac{1 - (1+i)^{-n}}{i} = 220\dfrac{1 - (1.00667)^{-36}}{0.00667} = \$7{,}020.60$. So she purchased the car for

$7,020.60 + $2000 = $9020.60.

23. $A_p = 100{,}000$, $n = 360$, $i = \dfrac{0.0975}{12} = 0.008125$.

(a) $R = \dfrac{iA_p}{1 - (1+i)^{-n}} = \dfrac{(0.008125)(100{,}000)}{1 - (1+0.008125)^{-360}} = \$859.15.$

(b) The total amount that will be paid over the 30 year period is $(360)(859.15) = \$309{,}294.00.$

(c) $R = \$859.15$, $i = \dfrac{0.0975}{12} = 0.008125$, $n = 360$. So $A_f = 859.15\dfrac{(1+0.008125)^{360} - 1}{0.008125} = \$1{,}841{,}519.29.$

25. $A_p = \$640$, $R = \$32$, $n = 24$. We want to solve the equation $R = \dfrac{i A_p}{1 - (1+i)^n}$

for the interest rate i. Let x be the interest rate, then $i = \dfrac{x}{12}$. So we can express R

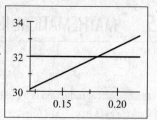

as a function of x by $R(x) = \dfrac{\dfrac{x}{12} \cdot 640}{1 - \left(1 + \dfrac{x}{12}\right)^{-24}}$. We graph $R(x)$ and $y = 32$ in

the rectangle $[0.12, 0.22] \times [30, 34]$. The x-coordinate of the intersection is about

0.1816, which corresponds to an interest rate of 18.16%.

27. $A_p = \$189.99$, $R = \$10.50$, $n = 20$. We want to solve the equation

$R = \dfrac{i A_p}{1 - (1+i)^n}$ for the interest rate i. Let x be the interest rate, then $i = \dfrac{x}{12}$. So

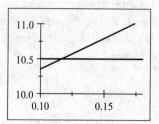

we can express R as a function of x by $R(x) = \dfrac{\dfrac{x}{12} \cdot 189.99}{1 - \left(1 + \dfrac{x}{12}\right)^{-20}}$. We graph

$R(x)$ and $y = 10.50$ in the rectangle $[0.10, 0.18] \times [10, 11]$. The x-coordinate of

the intersection is about 0.1168, which corresponds to an interest rate of 11.68%.

29. (a) The present value of the kth payment is $PV = R(1+i)^{-k} = \dfrac{R}{(1+i)^k}$. The present value of an annuity is the sum of

the present values of each of the payments of R dollars, as shown in the time line.

Time	1	2	3	4		$n-1$	n
Payment	R	R	R	R	$\cdots$	R	R

Present value

$R/(1+i)$
$R/(1+i)^2$
$R/(1+i)^3$
$R/(1+i)^4$
$\vdots$
$R/(1+i)^{n-1}$
$R/(1+i)^n$

(b) $A_p = \dfrac{R}{1+i} + \dfrac{R}{(1+i)^2} + \dfrac{R}{(1+i)^3} + \cdots + \dfrac{R}{(1+i)^n}$

$= \dfrac{R}{1+i} + \left(\dfrac{R}{1+i}\right)\left(\dfrac{1}{1+i}\right) + \left(\dfrac{R}{1+i}\right)\left(\dfrac{1}{1+i}\right)^2 + \cdots + \left(\dfrac{R}{1+i}\right)\left(\dfrac{1}{1+i}\right)^{n-1}$

This is a geometric series with $a = \dfrac{R}{1+i}$ and $r = \dfrac{1}{1+i}$. Since $S_n = a\dfrac{1-r^n}{1-r}$, we have

$A_p = \left(\dfrac{R}{1+i}\right)\dfrac{1 - \left[\dfrac{1}{(1+i)}\right]^n}{1 - \left(\dfrac{1}{1+i}\right)} = R\dfrac{1 - (1+i)^{-n}}{(1+i)\left[1 - \left(\dfrac{1}{1+i}\right)\right]} = R\dfrac{1 - (1+i)^{-n}}{(1+i) - 1} = R\dfrac{1 - (1+i)^{-n}}{i}$.

31. (a) Using the hint, we calculate the present value of the remaining 240 payments with $R = 724.17$, $i = 0.0075$, and

$n = 240$. Since $A_p = R\dfrac{1 - (1+i)^{-n}}{i} = (724.17)\dfrac{1 - (1.0075)^{-240}}{0.0075} = 80{,}487.84$, they still owe \$80,487.84 on their

mortgage.

(b) On their next payment, $0.0075\,(80{,}487.84) = \$603.66$ is interest and $\$724.17 - 603.66 = \$120.51$ goes toward the

principal.

8.5 MATHEMATICAL INDUCTION

1. Mathematical induction is a method of proving that a statement $P(n)$ is true for all *natural* numbers n. In Step 1 we prove that $P(1)$ is true.

3. Let $P(n)$ denote the statement $2 + 4 + 6 + \cdots + 2n = n(n+1)$.

Step 1: $P(1)$ is the statement that $2 = 1(1+1)$, which is true.

Step 2: Assume that $P(k)$ is true; that is, $2 + 4 + 6 + \cdots + 2k = k(k+1)$. We want to use this to show that $P(k+1)$ is true. Now

$$
\begin{aligned}
2 + 4 + 6 + \cdots + 2k + 2(k+1) &= k(k+1) + 2(k+1) \qquad \text{induction hypothesis} \\
&= (k+1)(k+2) = (k+1)[(k+1)+1]
\end{aligned}
$$

Thus, $P(k+1)$ follows from $P(k)$. So by the Principle of Mathematical Induction, $P(n)$ is true for all n.

5. Let $P(n)$ denote the statement $5 + 8 + 11 + \cdots + (3n+2) = \dfrac{n(3n+7)}{2}$.

Step 1: We need to show that $P(1)$ is true. But $P(1)$ says that $5 = \dfrac{1 \cdot (3 \cdot 1 + 7)}{2}$, which is true.

Step 2: Assume that $P(k)$ is true; that is, $5 + 8 + 11 + \cdots + (3k+2) = \dfrac{k(3k+7)}{2}$. We want to use this to show that $P(k+1)$ is true. Now

$$
\begin{aligned}
5 + 8 + 11 + \cdots + (3k+2) + [3(k+1)+2] &= \frac{k(3k+7)}{2} + (3k+5) \qquad \text{induction hypothesis} \\
&= \frac{3k^2 + 7k}{2} + \frac{6k+10}{2} = \frac{3k^2 + 13k + 10}{2} \\
&= \frac{(3k+10)(k+1)}{2} = \frac{(k+1)[3(k+1)+7]}{2}
\end{aligned}
$$

Thus, $P(k+1)$ follows from $P(k)$. So by the Principle of Mathematical Induction, $P(n)$ is true for all n.

7. Let $P(n)$ denote the statement $1 \cdot 2 + 2 \cdot 3 + 3 \cdot 4 + \cdots + n(n+1) = \dfrac{n(n+1)(n+2)}{3}$.

Step 1: $P(1)$ is the statement that $1 \cdot 2 = \dfrac{1 \cdot (1+1) \cdot (1+2)}{3}$, which is true.

Step 2: Assume that $P(k)$ is true; that is, $1 \cdot 2 + 2 \cdot 3 + 3 \cdot 4 + \cdots + k(k+1) = \dfrac{k(k+1)(k+2)}{3}$. We want to use this to show that $P(k+1)$ is true. Now

$$
\begin{aligned}
1 \cdot 2 + 2 \cdot 3 + 3 \cdot 4 + \cdots &+ k(k+1) + (k+1)[(k+1)+1] \\
&= \frac{k(k+1)(k+2)}{3} + (k+1)(k+2) \qquad \text{induction hypothesis} \\
&= \frac{k(k+1)(k+2)}{3} + \frac{3(k+1)(k+2)}{3} = \frac{(k+1)(k+2)(k+3)}{3}
\end{aligned}
$$

Thus, $P(k+1)$ follows from $P(k)$. So by the Principle of Mathematical Induction, $P(n)$ is true for all n.

9. Let $P(n)$ denote the statement $1^3 + 2^3 + 3^3 + \cdots + n^3 = \dfrac{n^2(n+1)^2}{4}$.

Step 1: $P(1)$ is the statement that $1^3 = \dfrac{1^2 \cdot (1+1)^2}{4}$, which is clearly true.

Step 2: Assume that $P(k)$ is true; that is, $1^3 + 2^3 + 3^3 + \cdots + k^3 = \dfrac{k^2(k+1)^2}{4}$. We want to use this to show that $P(k+1)$ is true. Now

$$
\begin{aligned}
1^3 + 2^3 + 3^3 + \cdots + k^3 + (k+1)^3 &= \frac{k^2(k+1)^2}{4} + (k+1)^3 \qquad \text{induction hypothesis}\\
&= \frac{(k+1)^2\left[k^2 + 4(k+1)\right]}{4} = \frac{(k+1)^2\left[k^2 + 4k + 4\right]}{4}\\
&= \frac{(k+1)^2(k+2)^2}{4} = \frac{(k+1)^2\left[(k+1)+1\right]^2}{4}
\end{aligned}
$$

Thus, $P(k+1)$ follows from $P(k)$. So by the Principle of Mathematical Induction, $P(n)$ is true for all n.

11. Let $P(n)$ denote the statement $2^3 + 4^3 + 6^3 + \cdots + (2n)^3 = 2n^2(n+1)^2$.

Step 1: $P(1)$ is true since $2^3 = 2(1)^2(1+1)^2 = 2 \cdot 4 = 8$.

Step 2: Assume that $P(k)$ is true; that is, $2^3 + 4^3 + 6^3 + \cdots + (2k)^3 = 2k^2(k+1)^2$. We want to use this to show that $P(k+1)$ is true. Now

$$
\begin{aligned}
2^3 + 4^3 + 6^3 + \cdots + (2k)^3 + [2(k+1)]^3 &= 2k^2(k+1)^2 + [2(k+1)]^3 \qquad \text{induction hypothesis}\\
&= 2k^2(k+1)^2 + 8(k+1)(k+1)^2 = (k+1)^2\left(2k^2 + 8k + 8\right)\\
&= 2(k+1)^2(k+2)^2 = 2(k+1)^2\left[(k+1)+1\right]^2
\end{aligned}
$$

Thus, $P(k+1)$ follows from $P(k)$. So by the Principle of Mathematical Induction, $P(n)$ is true for all n.

13. Let $P(n)$ denote the statement $1 \cdot 2 + 2 \cdot 2^2 + 3 \cdot 2^3 + 4 \cdot 2^4 + \cdots + n \cdot 2^n = 2\left[1 + (n-1)2^n\right]$.

Step 1: $P(1)$ is the statement that $1 \cdot 2 = 2[1 + 0]$, which is clearly true.

Step 2: Assume that $P(k)$ is true; that is, $1 \cdot 2 + 2 \cdot 2^2 + 3 \cdot 2^3 + 4 \cdot 2^4 + \cdots + k \cdot 2^k = 2\left[1 + (k-1)2^k\right]$. We want to use this to show that $P(k+1)$ is true. Now

$$
1 \cdot 2 + 2 \cdot 2^2 + 3 \cdot 2^3 + 4 \cdot 2^4 + \cdots + k \cdot 2^k + (k+1) \cdot 2^{(k+1)}
$$
$$
\begin{aligned}
&= 2\left[1 + (k-1)2^k\right] + (k+1) \cdot 2^{k+1} \qquad \text{induction hypothesis}\\
&= 2\left[1 + (k-1) \cdot 2^k + (k+1) \cdot 2^k\right] = 2\left[1 + 2k \cdot 2^k\right]\\
&= 2\left[1 + k \cdot 2^{k+1}\right] = 2\left\{1 + [(k+1) - 1]2^{k+1}\right\}
\end{aligned}
$$

Thus $P(k+1)$ follows from $P(k)$. So by the Principle of Mathematical Induction, $P(n)$ is true for all n.

15. Let $P(n)$ denote the statement $n^2 + n$ is divisible by 2.

Step 1: $P(1)$ is the statement that $1^2 + 1 = 2$ is divisible by 2, which is clearly true.

Step 2: Assume that $P(k)$ is true; that is, $k^2 + k$ is divisible by 2. Now

$(k+1)^2 + (k+1) = k^2 + 2k + 1 + k + 1 = \left(k^2 + k\right) + 2k + 2 = \left(k^2 + k\right) + 2(k+1)$. By the induction hypothesis,

$k^2 + k$ is divisible by 2, and clearly $2(k+1)$ is divisible by 2. Thus, the sum is divisible by 2, so $P(k+1)$ is true.

Therefore, $P(k+1)$ follows from $P(k)$. So by the Principle of Mathematical Induction, $P(n)$ is true for all n.

17. Let $P(n)$ denote the statement that $n^2 - n + 41$ is odd.

Step 1: $P(1)$ is the statement that $1^2 - 1 + 41 = 41$ is odd, which is clearly true.

Step 2: Assume that $P(k)$ is true; that is, $k^2 - k + 41$ is odd. We want to use this to show that $P(k+1)$ is true. Now,

$(k+1)^2 - (k+1) + 41 = k^2 + 2k + 1 - k - 1 + 41 = \left(k^2 - k + 41\right) + 2k$, which is also odd because $k^2 - k + 41$ is

odd by the induction hypothesis, $2k$ is always even, and an odd number plus an even number is always odd. Therefore, $P(k+1)$ follows from $P(k)$. So by the Principle of Mathematical Induction, $P(n)$ is true for all n.

19. Let $P(n)$ denote the statement that $8^n - 3^n$ is divisible by 5.

Step 1: $P(1)$ is the statement that $8^1 - 3^1 = 5$ is divisible by 5, which is clearly true.

Step 2: Assume that $P(k)$ is true; that is, $8^k - 3^k$ is divisible by 5. We want to use this to show that $P(k+1)$ is true. Now,

$8^{k+1} - 3^{k+1} = 8 \cdot 8^k - 3 \cdot 3^k = 8 \cdot 8^k - (8-5) \cdot 3^k = 8 \cdot \left(8^k - 3^k\right) + 5 \cdot 3^k$, which is divisible by 5 because $8^k - 3^k$

is divisible by 5 by our induction hypothesis, and $5 \cdot 3^k$ is divisible by 5. Thus $P(k+1)$ follows from $P(k)$. So by the Principle of Mathematical Induction, $P(n)$ is true for all n.

21. Let $P(n)$ denote the statement $n < 2^n$.

Step 1: $P(1)$ is the statement that $1 < 2^1 = 2$, which is clearly true.

Step 2: Assume that $P(k)$ is true; that is, $k < 2^k$. We want to use this to show that $P(k+1)$ is true. Adding 1 to both sides of $P(k)$ we have $k + 1 < 2^k + 1$. Since $1 < 2^k$ for $k \geq 1$, we have $2^k + 1 < 2^k + 2^k = 2 \cdot 2^k = 2^{k+1}$. Thus $k + 1 < 2^{k+1}$, which is exactly $P(k+1)$. Therefore, $P(k+1)$ follows from $P(k)$. So by the Principle of Mathematical Induction, $P(n)$ is true for all n.

23. Let $P(n)$ denote the statement $(1+x)^n \geq 1 + nx$, if $x > -1$.

Step 1: $P(1)$ is the statement that $(1+x)^1 \geq 1 + 1x$, which is clearly true.

Step 2: Assume that $P(k)$ is true; that is, $(1+x)^k \geq 1 + kx$. Now, $(1+x)^{k+1} = (1+x)(1+x)^k \geq (1+x)(1+kx)$, by the induction hypothesis. Since $(1+x)(1+kx) = 1 + (k+1)x + kx^2 \geq 1 + (k+1)x$ (since $kx^2 \geq 0$), we have $(1+x)^{k+1} \geq 1 + (k+1)x$, which is $P(k+1)$. Thus $P(k+1)$ follows from $P(k)$. So the Principle of Mathematical Induction, $P(n)$ is true for all n.

25. Let $P(n)$ be the statement that $a_n = 5 \cdot 3^{n-1}$.

Step 1: $P(1)$ is the statement that $a_1 = 5 \cdot 3^0 = 5$, which is true.

Step 2: Assume that $P(k)$ is true; that is, $a_k = 5 \cdot 3^{k-1}$. We want to use this to show that $P(k+1)$ is true. Now, $a_{k+1} = 3a_k = 3 \cdot \left(5 \cdot 3^{k-1}\right)$, by the induction hypothesis. Therefore, $a_{k+1} = 3 \cdot \left(5 \cdot 3^{k-1}\right) = 5 \cdot 3^k$, which is exactly $P(k+1)$. Thus, $P(k+1)$ follows from $P(k)$. So by the Principle of Mathematical Induction, $P(n)$ is true for all n.

27. Let $P(n)$ be the statement that $x - y$ is a factor of $x^n - y^n$ for all natural numbers n.

Step 1: $P(1)$ is the statement that $x - y$ is a factor of $x^1 - y^1$, which is clearly true.

Step 2: Assume that $P(k)$ is true; that is, $x - y$ is a factor of $x^k - y^k$. We want to use this to show that $P(k+1)$ is true. Now, $x^{k+1} - y^{k+1} = x^{k+1} - x^k y + x^k y - y^{k+1} = x^k(x-y) + \left(x^k - y^k\right)y$, for which $x - y$ is a factor because $x - y$ is a factor of $x^k(x-y)$, and $x - y$ is a factor of $\left(x^k - y^k\right)y$, by the induction hypothesis. Thus $P(k+1)$ follows from $P(k)$. So by the Principle of Mathematical Induction, $P(n)$ is true for all n.

29. Let $P(n)$ denote the statement that F_{3n} is even for all natural numbers n.

Step 1: $P(1)$ is the statement that F_3 is even. Since $F_3 = F_2 + F_1 = 1 + 1 = 2$, this statement is true.

Step 2: Assume that $P(k)$ is true; that is, F_{3k} is even. We want to use this to show that $P(k+1)$ is true. Now,

$F_{3(k+1)} = F_{3k+3} = F_{3k+2} + F_{3k+1} = F_{3k+1} + F_{3k} + F_{3k+1} = F_{3k} + 2 \cdot F_{3k+1}$, which is even because F_{3k} is even by

the induction hypothesis, and $2 \cdot F_{3k+1}$ is even. Thus $P(k+1)$ follows from $P(k)$. So by the Principle of Mathematical

Induction, $P(n)$ is true for all n.

31. Let $P(n)$ denote the statement that $F_1^2 + F_2^2 + F_3^2 + \cdots + F_n^2 = F_n \cdot F_{n+1}$.

Step 1: $P(1)$ is the statement that $F_1^2 = F_1 \cdot F_2$ or $1^2 = 1 \cdot 1$, which is true.

Step 2: Assume that $P(k)$ is true, that is, $F_1^2 + F_2^2 + F_3^2 + \cdots + F_k^2 = F_k \cdot F_{k+1}$. We want to use this to show that

$P(k+1)$ is true. Now

$$
\begin{aligned}
F_1^2 + F_2^2 + F_3^2 + \cdots + F_k^2 + F_{k+1}^2 &= F_k \cdot F_{k+1} + F_{k+1}^2 &&\text{induction hypothesis}\\
&= F_{k+1}\left(F_k + F_{k+1}\right)\\
&= F_{k+1} \cdot F_{k+2} &&\text{by definition of the Fibonacci sequence}
\end{aligned}
$$

Thus $P(k+1)$ follows from $P(k)$. So by the Principle of Mathematical Induction, $P(n)$ is true for all n.

33. Let $P(n)$ denote the statement $\begin{bmatrix} 1 & 1 \\ 1 & 0 \end{bmatrix}^n = \begin{bmatrix} F_{n+1} & F_n \\ F_n & F_{n-1} \end{bmatrix}$.

Step 1: Since $\begin{bmatrix} 1 & 1 \\ 1 & 0 \end{bmatrix}^2 = \begin{bmatrix} 1 & 1 \\ 1 & 0 \end{bmatrix}\begin{bmatrix} 1 & 1 \\ 1 & 0 \end{bmatrix} = \begin{bmatrix} 2 & 1 \\ 1 & 1 \end{bmatrix} = \begin{bmatrix} F_3 & F_2 \\ F_2 & F_1 \end{bmatrix}$, it follows that $P(2)$ is true.

Step 2: Assume that $P(k)$ is true; that is, $\begin{bmatrix} 1 & 1 \\ 1 & 0 \end{bmatrix}^k = \begin{bmatrix} F_{k+1} & F_k \\ F_k & F_{k-1} \end{bmatrix}$. We show that $P(k+1)$ follows from this. Now,

$$
\begin{aligned}
\begin{bmatrix} 1 & 1 \\ 1 & 0 \end{bmatrix}^{k+1} &= \begin{bmatrix} 1 & 1 \\ 1 & 0 \end{bmatrix}^k \begin{bmatrix} 1 & 1 \\ 1 & 0 \end{bmatrix} = \begin{bmatrix} F_{k+1} & F_k \\ F_k & F_{k-1} \end{bmatrix}\begin{bmatrix} 1 & 1 \\ 1 & 0 \end{bmatrix} &&\text{induction hypothesis}\\
&= \begin{bmatrix} F_{k+1}+F_k & F_{k+1} \\ F_k+F_{k-1} & F_k \end{bmatrix} = \begin{bmatrix} F_{k+2} & F_{k+1} \\ F_{k+1} & F_k \end{bmatrix} &&\text{by definition of the Fibonacci sequence}
\end{aligned}
$$

Thus $P(k+1)$ follows from $P(k)$. So by the Principle of Mathematical Induction, $P(n)$ is true for all $n \geq 2$.

35. Since $F_1 = 1$, $F_2 = 1$, $F_3 = 2$, $F_4 = 3$, $F_5 = 5$, $F_6 = 8$, $F_7 = 13, \ldots$ our conjecture is that $F_n \geq n$, for all $n \geq 5$. Let
$P(n)$ denote the statement that $F_n \geq n$.

Step 1: $P(5)$ is the statement that $F_5 = 5 \geq 5$, which is clearly true.

Step 2: Assume that $P(k)$ is true; that is, $F_k \geq k$, for some $k \geq 5$. We want to use this to show that $P(k+1)$ is true. Now,
$F_{k+1} = F_k + F_{k-1} \geq k + F_{k-1}$ (by the induction hypothesis)$\geq k + 1$ (because $F_{k-1} \geq 1$). Thus $P(k+1)$ follows from
$P(k)$. So by the Principle of Mathematical Induction, $P(n)$ is true for all $n \geq 5$.

37. (a) $P(n) = n^2 - n + 11$ is prime for all n. This is false as the case for $n = 11$ demonstrates: $P(11) = 11^2 - 11 + 11 = 121$, which is not prime since $11^2 = 121$.

(b) $n^2 > n$, for all $n \geq 2$. This is true. Let $P(n)$ denote the statement that $n^2 > n$.

Step 1: $P(2)$ is the statement that $2^2 = 4 > 2$, which is clearly true.

Step 2: Assume that $P(k)$ is true; that is, $k^2 > k$. We want to use this to show that $P(k+1)$ is true. Now $(k+1)^2 = k^2 + 2k + 1$. Using the induction hypothesis (to replace k^2), we have $k^2 + 2k + 1 > k + 2k + 1 = 3k + 1 > k + 1$, since $k \geq 2$. Therefore, $(k+1)^2 > k + 1$, which is exactly $P(k+1)$. Thus $P(k+1)$ follows from $P(k)$. So by the Principle of Mathematical Induction, $P(n)$ is true for all n.

(c) $2^{2n+1} + 1$ is divisible by 3, for all $n \geq 1$. This is true. Let $P(n)$ denote the statement that $2^{2n+1} + 1$ is divisible by 3.

Step 1: $P(1)$ is the statement that $2^3 + 1 = 9$ is divisible by 3, which is clearly true.

Step 2: Assume that $P(k)$ is true; that is, $2^{2k+1} + 1$ is divisible by 3. We want to use this to show that $P(k+1)$ is true. Now, $2^{2(k+1)+1} + 1 = 2^{2k+3} + 1 = 4 \cdot 2^{2k+1} + 1 = (3+1)2^{2k+1} + 1 = 3 \cdot 2^{2k+1} + \left(2^{2k+1} + 1\right)$, which is divisible by 3 since $2^{2k+1} + 1$ is divisible by 3 by the induction hypothesis, and $3 \cdot 2^{2k+1}$ is clearly divisible by 3. Thus $P(k+1)$ follows from $P(k)$. So by the Principle of Mathematical Induction, $P(n)$ is true for all n.

(d) The statement $n^3 \geq (n+1)^2$ for all $n \geq 2$ is false. The statement fails when $n = 2$: $2^3 = 8 < (2+1)^2 = 9$.

(e) $n^3 - n$ is divisible by 3, for all $n \geq 2$. This is true. Let $P(n)$ denote the statement that $n^3 - n$ is divisible by 3.

Step 1: $P(2)$ is the statement that $2^3 - 2 = 6$ is divisible by 3, which is clearly true.

Step 2: Assume that $P(k)$ is true; that is, $k^3 - k$ is divisible by 3. We want to use this to show that $P(k+1)$ is true. Now
$$(k+1)^3 - (k+1) = k^3 + 3k^2 + 3k + 1 - (k+1) = k^3 + 3k^2 + 2k = k^3 - k + 3k^2 + 2k + k = \left(k^3 - k\right) + 3\left(k^2 + k\right).$$
The term $k^3 - k$ is divisible by 3 by our induction hypothesis, and the term $3\left(k^2 + k\right)$ is clearly divisible by 3. Thus $(k+1)^3 - (k+1)$ is divisible by 3, which is exactly $P(k+1)$. So by the Principle of Mathematical Induction, $P(n)$ is true for all n.

(f) $n^3 - 6n^2 + 11n$ is divisible by 6, for all $n \geq 1$. This is true. Let $P(n)$ denote the statement that $n^3 - 6n^2 + 11n$ is divisible by 6.

Step 1: $P(1)$ is the statement that $(1)^3 - 6(1)^2 + 11(1) = 6$ is divisible by 6, which is clearly true.

Step 2: Assume that $P(k)$ is true; that is, $k^3 - 6k^2 + 11k$ is divisible by 6. We show that $P(k+1)$ is then also true. Now

$$\begin{aligned}
(k+1)^3 - 6(k+1)^2 + 11(k+1) &= k^3 + 3k^2 + 3k + 1 - 6k^2 - 12k - 6 + 11k + 11 \\
&= k^3 - 3k^2 + 2k + 6 = k^3 - 6k^2 + 11k + \left(3k^2 - 9k + 6\right) \\
&= \left(k^3 - 6k^2 + 11k\right) + 3\left(k^2 - 3k + 2\right) = \left(k^3 - 6k^2 + 11k\right) + 3(k-1)(k-2)
\end{aligned}$$

In this last expression, the first term is divisible by 6 by our induction hypothesis. The second term is also divisible by 6. To see this, notice that $k - 1$ and $k - 2$ are consecutive natural numbers, and so one of them must be even (divisible by 2). Since 3 also appears in this second term, it follows that this term is divisible by 2 and 3 and so is divisible by 6. Thus $P(k+1)$ follows from $P(k)$. So by the Principle of Mathematical Induction, $P(n)$ is true for all n.

8.6 THE BINOMIAL THEOREM

1. An algebraic expression of the form $a + b$, which consists of a sum of two terms, is called a *binomial*.

3. The binomial coefficients can be calculated directly using the formula $\binom{n}{k} = \dfrac{n!}{k!\,(n-k)!}$. So $\binom{4}{3} = \dfrac{4!}{3!\,1!} = 4$.

5. $(x + y)^6 = x^6 + 6x^5y + 15x^4y^2 + 20x^3y^3 + 15x^2y^4 + 6xy^5 + y^6$

7. $\left(x + \dfrac{1}{x}\right)^4 = x^4 + 4x^3 \cdot \dfrac{1}{x} + 6x^2 \left(\dfrac{1}{x}\right)^2 + 4x \left(\dfrac{1}{x}\right)^3 + \left(\dfrac{1}{x}\right)^4 = x^4 + 4x^2 + 6 + \dfrac{4}{x^2} + \dfrac{1}{x^4}$

9. $(x - 1)^5 = x^5 - 5x^4 + 10x^3 - 10x^2 + 5x - 1$

11. $\left(x^2y - 1\right)^5 = \left(x^2y\right)^5 - 5\left(x^2y\right)^4 + 10\left(x^2y\right)^3 - 10\left(x^2y\right)^2 + 5x^2y - 1 = x^{10}y^5 - 5x^8y^4 + 10x^6y^3 - 10x^4y^2 + 5x^2y - 1$

13. $(2x - 3y)^3 = (2x)^3 - 3(2x)^2 3y + 3 \cdot 2x (3y)^2 - (3y)^3 = 8x^3 - 36x^2y + 54xy^2 - 27y^3$

15. $\left(\dfrac{1}{x} - \sqrt{x}\right)^5 = \left(\dfrac{1}{x}\right)^5 - 5\left(\dfrac{1}{x}\right)^4 \sqrt{x} + 10\left(\dfrac{1}{x}\right)^3 x - 10\left(\dfrac{1}{x}\right)^2 x\sqrt{x} + 5\left(\dfrac{1}{x}\right) x^2 - x^2\sqrt{x}$

$\qquad = \dfrac{1}{x^5} - \dfrac{5}{x^{7/2}} + \dfrac{10}{x^2} - \dfrac{10}{x^{1/2}} + 5x - x^{5/2}$

17. $\dbinom{6}{4} = \dfrac{6!}{4!\,2!} = \dfrac{6 \cdot 5 \cdot 4!}{2 \cdot 1 \cdot 4!} = 15$

19. $\dbinom{100}{98} = \dfrac{100!}{98!\,2!} = \dfrac{100 \cdot 99 \cdot 98!}{98! \cdot 2 \cdot 1} = 4950$

21. $\dbinom{3}{1}\dbinom{4}{2} = \dfrac{3!}{1!\,2!}\dfrac{4!}{2!\,2!} = \dfrac{3 \cdot 2! \cdot 4 \cdot 3 \cdot 2!}{1 \cdot 2! \cdot 2 \cdot 1 \cdot 2!} = 18$

23. $\dbinom{5}{0} + \dbinom{5}{1} + \dbinom{5}{2} + \dbinom{5}{3} + \dbinom{5}{4} + \dbinom{5}{5} = (1 + 1)^5 = 2^5 = 32$

25. $(x + 2y)^4 = \dbinom{4}{0}x^4 + \dbinom{4}{1}x^3 \cdot 2y + \dbinom{4}{2}x^2 \cdot 4y^2 + \dbinom{4}{3}x \cdot 8y^3 + \dbinom{4}{4}16y^4 = x^4 + 8x^3y + 24x^2y^2 + 32xy^3 + 16y^4$

27. $\left(1 + \dfrac{1}{x}\right)^6 = \dbinom{6}{0}1^6 + \dbinom{6}{1}1^5\left(\dfrac{1}{x}\right) + \dbinom{6}{2}1^4\left(\dfrac{1}{x}\right)^2 + \dbinom{6}{3}1^3\left(\dfrac{1}{x}\right)^3 + \dbinom{6}{4}1^2\left(\dfrac{1}{x}\right)^4 + \dbinom{6}{5}1\left(\dfrac{1}{x}\right)^5 + \dbinom{6}{6}\left(\dfrac{1}{x}\right)^6$

$\qquad = 1 + \dfrac{6}{x} + \dfrac{15}{x^2} + \dfrac{20}{x^3} + \dfrac{15}{x^4} + \dfrac{6}{x^5} + \dfrac{1}{x^6}$

29. The first three terms in the expansion of $(x + 2y)^{20}$ are $\dbinom{20}{0}x^{20} = x^{20}$, $\dbinom{20}{1}x^{19} \cdot 2y = 40x^{19}y$, and $\dbinom{20}{2}x^{18} \cdot (2y)^2 = 760x^{18}y^2$.

31. The last two terms in the expansion of $\left(a^{2/3} + a^{1/3}\right)^{25}$ are $\dbinom{25}{24}a^{2/3} \cdot \left(a^{1/3}\right)^{24} = 25a^{26/3}$, and $\dbinom{25}{25}a^{25/3} = a^{25/3}$.

33. The middle term in the expansion of $\left(x^2 + 1\right)^{18}$ occurs when both terms are raised to the 9th power. So this term is $\dbinom{18}{9}\left(x^2\right)^9 1^9 = 48{,}620x^{18}$.

35. The 24th term in the expansion of $(a + b)^{25}$ is $\dbinom{25}{23}a^2b^{23} = 300a^2b^{23}$.

37. The 100th term in the expansion of $(1 + y)^{100}$ is $\dbinom{100}{99}1^1 \cdot y^{99} = 100y^{99}$.

39. The term that contains x^4 in the expansion of $(x + 2y)^{10}$ has exponent $r = 4$. So this term is $\dbinom{10}{4}x^4 \cdot (2y)^{10-4} = 13{,}440x^4y^6$.

41. The rth term is $\dbinom{12}{r}a^r\left(b^2\right)^{12-r} = \dbinom{12}{r}a^r b^{24-2r}$. Thus the term that contains b^8 occurs where $24 - 2r = 8 \Leftrightarrow r = 8$. So the term is $\dbinom{12}{8}a^8b^8 = 495a^8b^8$.

43. $x^4 + 4x^3y + 6x^2y^2 + 4xy^3 + y^4 = (x + y)^4$

45. $8a^3 + 12a^2b + 6ab^2 + b^3 = \dbinom{3}{0}(2a)^3 + \dbinom{3}{1}(2a)^2 b + \dbinom{3}{2}2ab^2 + \dbinom{3}{3}b^3 = (2a + b)^3$

47. $\dfrac{(x + h)^3 - x^3}{h} = \dfrac{x^3 + 3x^2h + 3xh^2 + h^3 - x^3}{h} = \dfrac{3x^2h + 3xh^2 + h^3}{h} = \dfrac{h\left(3x^2 + 3xh + h^2\right)}{h} = 3x^2 + 3xh + h^2$

49. $(1.01)^{100} = (1 + 0.01)^{100}$. Now the first term in the expansion is $\binom{100}{0} 1^{100} = 1$, the second term is $\binom{100}{1} 1^{99} (0.01) = 1$, and the third term is $\binom{100}{2} 1^{98} (0.01)^2 = 0.495$. Now each term is nonnegative, so $(1.01)^{100} = (1 + 0.01)^{100} > 1 + 1 + .0.495 > 2$. Thus $(1.01)^{100} > 2$.

51. $\binom{n}{1} = \dfrac{n!}{1! \, (n-1)!} = \dfrac{n \, (n-1)!}{1 \, (n-1)!} = \dfrac{n}{1} = n$. $\binom{n}{n-1} = \dfrac{n!}{(n-1)! \, 1!} = \dfrac{n \, (n-1)!}{(n-1)! \, 1} = n$. Therefore, $\binom{n}{1} = \binom{n}{n-1} = n$.

53. **(a)** $\binom{n}{r-1} + \binom{n}{r} = \dfrac{n!}{(r-1)! \, [n - (r-1)]!} + \dfrac{n!}{r! \, (n-r)!}$.

(b) $\dfrac{n!}{(r-1)! \, [n - (r-1)]!} + \dfrac{n!}{r! \, (n-r)!} = \dfrac{r \cdot n!}{r \cdot (r-1)! \, (n-r+1)!} + \dfrac{(n-r+1) \cdot n!}{r! \, (n-r+1) \, (n-r)!}$

$= \dfrac{r \cdot n!}{r! \, (n-r+1)!} + \dfrac{(n-r+1) \cdot n!}{r! \, (n-r+1)!}$

Thus a common denominator is $r! \, (n-r+1)!$.

(c) Therefore, using the results of parts (a) and (b),

$$\binom{n}{r-1} + \binom{n}{r} = \dfrac{n!}{(r-1)! \, [n - (r-1)]!} + \dfrac{n!}{r! \, (n-r)!} = \dfrac{r \cdot n!}{r! \, (n-r+1)!} + \dfrac{(n-r+1) \cdot n!}{r! \, (n-r+1)!}$$

$$= \dfrac{r \cdot n! + (n-r+1) \cdot n!}{r! \, (n-r+1)!} = \dfrac{n! \, (r+n-r+1)}{r! \, (n-r+1)!} = \dfrac{n! \, (n+1)}{r! \, (n+1-r)!} = \dfrac{(n+1)!}{r! \, (n+1-r)!} = \binom{n+1}{r}$$

55. By the Binomial Theorem, the volume of a cube of side $x + 2$ inches is

$(x+2)^3 = \binom{3}{0} x^3 + \binom{3}{1} x^2 (2) + \binom{3}{2} x (2)^2 + \binom{3}{3} 2^3 = x^3 + 3 \cdot 2x^2 + 3 \cdot 4x + 8 = x^3 + 6x^2 + 12x + 8$. The volume of a cube of side x inches is x^3, so the difference in volumes is $x^3 + 6x^2 + 12x + 8 - x^3 = 6x^2 + 12x + 8$ cubic inches.

57. Notice that $(100!)^{101} = (100!)^{100} \cdot 100!$ and $(101!)^{100} = (101 \cdot 100!)^{100} = 101^{100} \cdot (100!)^{100}$. Now $100! = 1 \cdot 2 \cdot 3 \cdot 4 \cdots 99 \cdot 100$ and $101^{100} = 101 \cdot 101 \cdot 101 \cdots 101$. Thus each of these last two expressions consists of 100 factors multiplied together, and since each factor in the product for 101^{100} is larger than each factor in the product for $100!$, it follows that $100! < 101^{100}$. Thus $(100!)^{100} \cdot 100! < (100!)^{100} \cdot 101^{100}$. So $(100!)^{101} < (101!)^{100}$.

59. $0 = 0^n = (-1 + 1)^n = \binom{n}{0} (-1)^0 (1)^n + \binom{n}{1} (-1)^1 (1)^{n-1} + \binom{n}{2} (-1)^2 (1)^{n-2} + \cdots + \binom{n}{n} (-1)^n (1)^0$

$= \binom{n}{0} - \binom{n}{1} + \binom{n}{2} - \cdots + (-1)^k \binom{n}{k} + \cdots + (-1)^n \binom{n}{n}$

CHAPTER 8 REVIEW

1. $a_n = \dfrac{n^2}{n+1}$. Then $a_1 = \dfrac{1^2}{1+1} = \dfrac{1}{2}$, $a_2 = \dfrac{2^2}{2+1} = \dfrac{4}{3}$, $a_3 = \dfrac{3^2}{3+1} = \dfrac{9}{4}$, $a_4 = \dfrac{4^2}{4+1} = \dfrac{16}{5}$, and $a_{10} = \dfrac{10^2}{10+1} = \dfrac{100}{11}$.

3. $a_n = \dfrac{(-1)^n + 1}{n^3}$. Then $a_1 = \dfrac{(-1)^1 + 1}{1^3} = 0$, $a_2 = \dfrac{(-1)^2 + 1}{2^3} = \dfrac{2}{8} = \dfrac{1}{4}$, $a_3 = \dfrac{(-1)^3 + 1}{3^3} = 0$,

$a_4 = \dfrac{(-1)^4 + 1}{4^3} = \dfrac{2}{64} = \dfrac{1}{32}$, and $a_{10} = \dfrac{(-1)^{10} + 1}{10^3} = \dfrac{1}{500}$.

5. $a_n = \dfrac{(2n)!}{2^n n!}$. Then $a_1 = \dfrac{(2 \cdot 1)!}{2^1 \cdot 1!} = 1$, $a_2 = \dfrac{(2 \cdot 2)!}{2^2 \cdot 2!} = 3$, $a_3 = \dfrac{(2 \cdot 3)!}{2^3 \cdot 3!} = \dfrac{6 \cdot 5 \cdot 4}{8} = 15$,

$a_4 = \dfrac{(2 \cdot 4)!}{2^4 \cdot 4!} = \dfrac{8 \cdot 7 \cdot 6 \cdot 5}{16} = 105$, and $a_{10} = \dfrac{(2 \cdot 10)!}{2^{10} \cdot 10!} = 654{,}729{,}075$.

7. $a_n = a_{n-1} + 2n - 1$ and $a_1 = 1$. Then $a_2 = a_1 + 4 - 1 = 4$, $a_3 = a_2 + 6 - 1 = 9$, $a_4 = a_3 + 8 - 1 = 16$, $a_5 = a_4 + 10 - 1 = 25$, $a_6 = a_5 + 12 - 1 = 36$, and $a_7 = a_6 + 14 - 1 = 49$.

9. $a_n = a_{n-1} + 2a_{n-2}$, $a_1 = 1$ and $a_2 = 3$. Then $a_3 = a_2 + 2a_1 = 5$, $a_4 = a_3 + 2a_2 = 11$, $a_5 = a_4 + 2a_3 = 21$, $a_6 = a_5 + 2a_4 = 43$, and $a_7 = a_6 + 2a_5 = 85$.

11. (a) $a_1 = 2(1) + 5 = 7$, $a_2 = 2(2) + 5 = 9$,

$a_3 = 2(3) + 5 = 11$, $a_4 = 2(4) + 5 = 13$,

$a_5 = 2(5) + 5 = 15$

(b)

(c) $S_5 = 7 + 9 + 11 + 13 + 15 = 55$

(d) This sequence is arithmetic with common difference 2.

13. (a) $a_1 = \dfrac{3^1}{2^2} = \dfrac{3}{4}$, $a_2 = \dfrac{3^2}{2^3} = \dfrac{9}{8}$, $a_3 = \dfrac{3^3}{2^4} = \dfrac{27}{16}$,

$a_4 = \dfrac{3^4}{2^5} = \dfrac{81}{32}$, $a_5 = \dfrac{3^5}{2^6} = \dfrac{243}{64}$

(b)

(c) $S_5 = \dfrac{3}{4} + \dfrac{9}{8} + \dfrac{27}{16} + \dfrac{81}{32} + \dfrac{243}{64} = \dfrac{633}{64}$

(d) This sequence is geometric with common ratio $\frac{3}{2}$.

15. $5, 5.5, 6, 6.5, \ldots$. Since $5.5 - 5 = 6 - 5.5 = 6.5 - 6 = 0.5$, this is an arithmetic sequence with $a_1 = 5$ and $d = 0.5$. Then $a_5 = a_4 + 0.5 = 7$.

17. $t - 3, t - 2, t - 1, t, \ldots$. Since $(t - 2) - (t - 3) = (t - 1) - (t - 2) = t - (t - 1) = 1$, this is an arithmetic sequence with $a_1 = t - 3$ and $d = 1$. Then $a_5 = a_4 + 1 = t + 1$.

19. $t^3, t^2, t, 1, \ldots$. Since $\dfrac{t^2}{t^3} = \dfrac{t}{t^2} = \dfrac{1}{t}$, this is a geometric sequence with $a_1 = t^3$ and $r = \dfrac{1}{t}$. Then $a_5 = a_4 \cdot r = 1 \cdot \dfrac{1}{t} = \dfrac{1}{t}$.

21. $\dfrac{3}{4}, \dfrac{1}{2}, \dfrac{1}{3}, \dfrac{2}{9}, \ldots$. Since $\dfrac{\frac{1}{2}}{\frac{3}{4}} = \dfrac{\frac{1}{3}}{\frac{1}{2}} = \dfrac{\frac{2}{9}}{\frac{1}{3}} = \dfrac{2}{3}$, this is a geometric sequence with $a_1 = \dfrac{3}{4}$ and $r = \dfrac{2}{3}$. Then

$a_5 = a_4 \cdot r = \dfrac{2}{9} \cdot \dfrac{2}{3} = \dfrac{4}{27}$.

23. $3, 6i, -12, -24i, \ldots$. Since $\dfrac{6i}{3} = 2i$, $\dfrac{-12}{6i} = \dfrac{-2}{i} = \dfrac{-2i}{i^2} = 2i$, $\dfrac{-24i}{-12} = 2i$, this is a geometric sequence with common ratio $r = 2i$.

25. $a_6 = 17 = a + 5d$ and $a_4 = 11 = a + 3d$. Then, $a_6 - a_4 = 17 - 11 \Leftrightarrow (a + 5d) - (a + 3d) = 6 \Leftrightarrow 6 = 2d \Leftrightarrow d = 3$. Substituting into $11 = a + 3d$ gives $11 = a + 3 \cdot 3$, and so $a = 2$. Thus $a_2 = a + (2 - 1)d = 2 + 3 = 5$.

27. $a_3 = 9$ and $r = \dfrac{3}{2}$. Then $a_5 = a_3 \cdot r^2 = 9 \cdot \left(\dfrac{3}{2}\right)^2 = \dfrac{81}{4}$.

29. (a) $A_n = 32{,}000 \cdot 1.05^{n-1}$

(b) $A_1 = \$32{,}000$, $A_2 = 32{,}000 \cdot 1.05^1 = \$33{,}600$, $A_3 = 32{,}000 \cdot 1.05^2 = \$35{,}280$, $A_4 = 32{,}000 \cdot 1.05^3 = \$37{,}044$,

$A_5 = 32{,}000 \cdot 1.05^4 = \$38{,}896.20$, $A_6 = 32{,}000 \cdot 1.05^5 = \$40{,}841.01$, $A_7 = 32{,}000 \cdot 1.05^6 = \$42{,}883.06$,

$A_8 = 32{,}000 \cdot 1.05^7 = \$45{,}027.21$

31. Let a_n be the number of bacteria in the dish at the end of $5n$ seconds. So $a_0 = 3$, $a_1 = 3 \cdot 2$, $a_2 = 3 \cdot 2^2$, $a_3 = 3 \cdot 2^3, \ldots$. Then, clearly, a_n is a geometric sequence with $r = 2$ and $a = 3$. Thus at the end of $60 = 5(12)$ seconds, the number of bacteria is $a_{12} = 3 \cdot 2^{12} = 12{,}288$.

33. Suppose that the common ratio in the sequence $a_1, a_2, a_3, \ldots$ is r. Also, suppose that the common ratio in the sequence $b_1, b_2, b_3, \ldots$ is s. Then $a_n = a_1 r^{n-1}$ and $b_n = b_1 s^{n-1}$, $n = 1, 2, 3, \ldots$. Thus $a_n b_n = a_1 r^{n-1} \cdot b_1 s^{n-1} = (a_1 b_1)(rs)^{n-1}$. So the sequence $a_1 b_1, a_2 b_2, a_3 b_3, \ldots$ is geometric with first term $a_1 b_1$ and common ratio rs.

35. (a) $6, x, 12, \ldots$ is arithmetic if $x - 6 = 12 - x \Leftrightarrow 2x = 18 \Leftrightarrow x = 9$.

(b) $6, x, 12, \ldots$ is geometric if $\dfrac{x}{6} = \dfrac{12}{x} \Leftrightarrow x^2 = 72 \Leftrightarrow x = \pm 6\sqrt{2}$.

37. $\sum_{k=3}^{6} (k+1)^2 = (3+1)^2 + (4+1)^2 + (5+1)^2 + (6+1)^2 = 16 + 25 + 36 + 49 = 126$

39. $\sum_{k=1}^{6} (k+1) \, 2^{k-1} = 2 \cdot 2^0 + 3 \cdot 2^1 + 4 \cdot 2^2 + 5 \cdot 2^3 + 6 \cdot 2^4 + 7 \cdot 2^5 = 2 + 6 + 16 + 40 + 96 + 224 = 384$

41. $\sum_{k=1}^{10} (k-1)^2 = 0^2 + 1^2 + 2^2 + 3^2 + 4^2 + 5^2 + 6^2 + 7^2 + 8^2 + 9^2$

43. $\sum_{k=1}^{50} \dfrac{3^k}{2^{k+1}} = \dfrac{3}{2^2} + \dfrac{3^2}{2^3} + \dfrac{3^3}{2^4} + \dfrac{3^4}{2^5} + \cdots + \dfrac{3^{49}}{2^{50}} + \dfrac{3^{50}}{2^{51}}$

45. $3 + 6 + 9 + 12 + \cdots + 99 = 3\,(1) + 3\,(2) + 3\,(3) + \cdots + 3\,(33) = \sum_{k=1}^{33} 3k$

47. $1 \cdot 2^3 + 2 \cdot 2^4 + 3 \cdot 2^5 + 4 \cdot 2^6 + \cdots + 100 \cdot 2^{102}$

$$= (1)\,2^{(1)+2} + (2)\,2^{(2)+2} + (3)\,2^{(3)+2} + (4)\,2^{(4)+2} + \cdots + (100)\,2^{(100)+2}$$
$$= \sum_{k=1}^{100} k \cdot 2^{k+2}$$

49. $1 + 0.9 + (0.9)^2 + \cdots + (0.9)^5$ is a geometric series with $a = 1$ and $r = \dfrac{0.9}{1} = 0.9$. Thus, the sum of the series is

$$S_6 = \dfrac{1 - (0.9)^6}{1 - 0.9} = \dfrac{1 - 0.531441}{0.1} = 4.68559.$$

51. $\sqrt{5} + 2\sqrt{5} + 3\sqrt{5} + \cdots + 100\sqrt{5}$ is an arithmetic series with $a = \sqrt{5}$ and $d = \sqrt{5}$. Then $100\sqrt{5} = a_n = \sqrt{5} + \sqrt{5}\,(n-1)$

$\Leftrightarrow n = 100$. So the sum is $S_{100} = \dfrac{100}{2} \left(\sqrt{5} + 100\sqrt{5} \right) = 50 \left(101\sqrt{5} \right) = 5050\sqrt{5}$.

53. $\sum_{n=0}^{6} 3 \cdot (-4)^n$ is a geometric series with $a = 3$, $r = -4$, and $n = 7$. Therefore, the sum of the series is

$$S_7 = 3 \cdot \dfrac{1 - (-4)^7}{1 - (-4)} = \tfrac{3}{5} \left(1 + 4^7 \right) = 9831.$$

55. $1 - \dfrac{2}{5} + \dfrac{4}{25} - \dfrac{8}{125} + \cdots$ is a geometric series with $a = 1$ and $r = -\dfrac{2}{5}$. Therefore, it is convergent with sum

$$S = \dfrac{a}{1 - r} = \dfrac{1}{1 - \left(-\frac{2}{5} \right)} = \dfrac{5}{7}.$$

57. $5 - 5\,(1.01) + 5\,(1.01)^2 - 5\,(1.01)^3 + \cdots$ is an infinite geometric series with $a = 5$ and $r = 1.01$. Because $|r| = 1.01 > 1$, the series diverges.

59. $-1 + \dfrac{9}{8} - \left(\dfrac{9}{8} \right)^2 + \left(\dfrac{9}{8} \right)^3 - \cdots$ is an infinite geometric series with $a = -1$ and $r = -\dfrac{9}{8}$. Because $|r| = \dfrac{9}{8} > 1$, the series diverges.

61. We have an arithmetic sequence with $a = 7$ and $d = 3$. Then $S_n = 325 = \dfrac{n}{2}\,[2a + (n-1)\,d] = \dfrac{n}{2}\,[14 + 3\,(n-1)] = \dfrac{n}{2}\,(11 + 3n) \Leftrightarrow 650 = 3n^2 + 11n \Leftrightarrow (3n + 50)\,(n - 13) = 0 \Leftrightarrow n = 13$ (because $n = -\dfrac{50}{3}$ is inadmissible). Thus, 13 terms must be added.

63. This is a geometric sequence with $a = 2$ and $r = 2$. Then $S_{15} = 2 \cdot \dfrac{1 - 2^{15}}{1 - 2} = 2 \left(2^{15} - 1 \right) = 65{,}534$, and so the total number of ancestors is 65,534.

65. $A = 10{,}000$, $i = 0.03$, and $n = 4$. Thus, $10{,}000 = R\,\dfrac{(1.03)^4 - 1}{0.03} \Leftrightarrow R = \dfrac{10{,}000 \cdot 0.03}{(1.03)^4 - 1} = \$2390.27$.

67. Let $P(n)$ denote the statement that $1 + 4 + 7 + \cdots + (3n - 2) = \dfrac{n(3n-1)}{2}$.

Step 1: $P(1)$ is the statement that $1 = \dfrac{1[3(1)-1]}{2} = \dfrac{1 \cdot 2}{2}$, which is true.

Step 2: Assume that $P(k)$ is true; that is, $1 + 4 + 7 + \cdots + (3k - 2) = \dfrac{k(3k-1)}{2}$. We want to use this to show that $P(k+1)$ is true. Now

$$
\begin{aligned}
1 + 4 + 7 + 10 + \cdots + (3k - 2) + [3(k+1) - 2] &= \frac{k(3k-1)}{2} + 3k + 1 \quad \text{induction hypothesis} \\
&= \frac{k(3k-1)}{2} + \frac{6k+2}{2} = \frac{3k^2 - k + 6k + 2}{2} \\
&= \frac{3k^2 + 5k + 2}{2} = \frac{(k+1)(3k+2)}{2} \\
&= \frac{(k+1)[3(k+1) - 1]}{2}
\end{aligned}
$$

Thus, $P(k+1)$ follows from $P(k)$. So by the Principle of Mathematical Induction, $P(n)$ is true for all n.

69. Let $P(n)$ denote the statement that $\left(1 + \dfrac{1}{1}\right)\left(1 + \dfrac{1}{2}\right)\left(1 + \dfrac{1}{3}\right) \cdot\ \cdots\ \cdot \left(1 + \dfrac{1}{n}\right) = n + 1$.

Step 1: $P(1)$ is the statement that $1 + \frac{1}{1} = 1 + 1$, which is clearly true.

Step 2: Assume that $P(k)$ is true; that is, $\left(1 + \dfrac{1}{1}\right)\left(1 + \dfrac{1}{2}\right)\left(1 + \dfrac{1}{3}\right) \cdot\ \cdots\ \cdot \left(1 + \dfrac{1}{k}\right) = k + 1$. We want to use this to show that $P(k+1)$ is true. Now

$$
\begin{aligned}
\left(1 + \frac{1}{1}\right)&\left(1 + \frac{1}{2}\right)\left(1 + \frac{1}{3}\right) \cdot\ \cdots\ \cdot \left(1 + \frac{1}{k}\right)\left(1 + \frac{1}{k+1}\right) \\
&= \left[\left(1 + \frac{1}{1}\right)\left(1 + \frac{1}{2}\right)\left(1 + \frac{1}{3}\right) \cdot\ \cdots\ \cdot \left(1 + \frac{1}{k}\right)\right]\left(1 + \frac{1}{k+1}\right) \\
&= (k+1)\left(1 + \frac{1}{k+1}\right) \quad \text{induction hypothesis} \\
&= (k+1) + 1
\end{aligned}
$$

Thus, $P(k+1)$ follows from $P(k)$. So by the Principle of Mathematical Induction, $P(n)$ is true for all n.

71. $a_{n+1} = 3a_n + 4$ and $a_1 = 4$. Let $P(n)$ denote the statement that $a_n = 2 \cdot 3^n - 2$.

Step 1: $P(1)$ is the statement that $a_1 = 2 \cdot 3^1 - 2 = 4$, which is clearly true.

Step 2: Assume that $P(k)$ is true; that is, $a_k = 2 \cdot 3^k - 2$. We want to use this to show that $P(k+1)$ is true. Now

$$
\begin{aligned}
a_{k+1} &= 3a_k + 4 \quad \text{definition of } a_{k+1} \\
&= 3\left(2 \cdot 3^k - 2\right) + 4 \quad \text{induction hypothesis} \\
&= 2 \cdot 3^{k+1} - 6 + 4 = 2 \cdot 3^{k+1} - 2
\end{aligned}
$$

Thus $P(k+1)$ follows from $P(k)$. So by the Principle of Mathematical Induction, $P(n)$ is true for all n.

73. $\dbinom{5}{2}\dbinom{5}{3} = \dfrac{5!}{2!\,3!} \cdot \dfrac{5!}{3!\,2!} = \dfrac{5 \cdot 4}{2} \cdot \dfrac{5 \cdot 4}{2} = 10 \cdot 10 = 100$

75. $\sum_{k=0}^{5} \binom{5}{k} = \binom{5}{0} + \binom{5}{1} + \binom{5}{2} + \binom{5}{3} + \binom{5}{4} + \binom{5}{5} = 2\left(\frac{5!}{0!\,5!} + \frac{5!}{1!\,4!} + \frac{5!}{2!\,3!}\right) = 2(1 + 5 + 10) = 32$

77. $(A - B)^3 = \binom{3}{0}A^3 - \binom{3}{1}A^2 B + \binom{3}{2}AB^2 - \binom{3}{3}B^3 = A^3 - 3A^2 B + 3AB^2 - B^3$

79. $\left(1 - x^2\right)^6 = \binom{6}{0}1^6 - \binom{6}{1}1^5x^2 + \binom{6}{2}1^4x^4 - \binom{6}{3}1^3x^6 + \binom{6}{4}1^2x^8 - \binom{6}{5}x^{10} + \binom{6}{6}x^{12}$

$$= 1 - 6x^2 + 15x^4 - 20x^6 + 15x^8 - 6x^{10} + x^{12}$$

81. The 20th term is $\binom{22}{19}a^3b^{19} = 1540a^3b^{19}$.

83. The rth term in the expansion of $(A + 3B)^{10}$ is $\binom{10}{r}A^r(3B)^{10-r}$. The term that contains A^6 occurs when $r = 6$. Thus, the term is $\binom{10}{6}A^6(3B)^4 = 210A^6 81B^4 = 17{,}010A^6B^4$.

CHAPTER 8 TEST

1. $a_n = 2n^2 - n \Rightarrow a_1 = 1, a_2 = 6, a_3 = 15, a_4 = 28, a_5 = 45, a_6 = 66$, and $S_6 = 1 + 6 + 15 + 28 + 45 + 66 = 161$.

3. (a) The common difference is $d = 5 - 2 = 3$.

(b) $a_n = 2 + (n - 1)3$

(c) $a_{35} = 2 + 3(35 - 1) = 104$

5. (a) $a_1 = 25, a_4 = \frac{1}{5}$. Then $r^3 = \dfrac{\frac{1}{5}}{25} = \dfrac{1}{125} \Leftrightarrow r = \frac{1}{5}$, so $a_5 = ra_4 = \frac{1}{25}$.

(b) $S_8 = 25\dfrac{1 - \left(\frac{1}{5}\right)^8}{1 - \frac{1}{5}} = \dfrac{5^8 - 1}{12{,}500} = \dfrac{97{,}656}{3125}$

7. Let the common ratio for the geometric series $a_1, a_2, a_3, \ldots$ be r, so that $a_n = a_1 r^{n-1}$, $n = 1, 2, 3, \ldots$. Then $a_n^2 = \left(a_1 r^{n-1}\right)^2 = \left(a_1^2\right)\left(r^2\right)^{n-1}$. Therefore, the sequence $a_1^2, a_2^2, a_3^2, \ldots$ is geometric with common ratio r^2.

9. (a) The geometric sum $\frac{1}{3} + \frac{2}{3^2} + \frac{2^2}{3^3} + \frac{2^3}{3^4} + \cdots + \frac{2^9}{3^{10}}$ has $a = \frac{1}{3}$, $r = \frac{2}{3}$, and $n = 10$. So

$$S_{10} = \frac{1}{3} \cdot \frac{1 - (2/3)^{10}}{1 - (2/3)} = \frac{1}{3} \cdot 3\left(1 - \frac{1024}{59{,}049}\right) = \frac{58{,}025}{59{,}049}.$$

(b) The infinite geometric series $1 + \frac{1}{2^{1/2}} + \frac{1}{2} + \frac{1}{2^{3/2}} + \cdots$ has $a = 1$ and $r = 2^{-1/2} = \frac{1}{\sqrt{2}}$. Thus,

$$S = \frac{1}{1 - 1/\sqrt{2}} = \frac{\sqrt{2}}{\sqrt{2} - 1} = \frac{\sqrt{2}}{\sqrt{2} - 1} \cdot \frac{\sqrt{2} + 1}{\sqrt{2} + 1} = 2 + \sqrt{2}.$$

11. $\left(2x + y^2\right)^5 = \binom{5}{0}(2x)^5 + \binom{5}{1}(2x)^4 y^2 + \binom{5}{2}(2x)^3\left(y^2\right)^2 + \binom{5}{3}(2x)^2\left(y^2\right)^3 + \binom{5}{4}(2x)\left(y^2\right)^4 + \binom{5}{5}\left(y^2\right)^5$

$$= 32x^5 + 80x^4y^2 + 80x^3y^4 + 40x^2y^6 + 10xy^8 + y^{10}$$

13. (a) Each week he gains 24% in weight, that is, $0.24a_n$. Thus, $a_{n+1} = a_n + 0.24a_n = 1.24a_n$ for $n \geq 1$. a_0 is given to be 0.85 lb. Then $a_0 = 0.85$, $a_1 = 1.24(0.85)$, $a_2 = 1.24(1.24(0.85)) = 1.24^2(0.85)$, $a_3 = 1.24\left(1.24^2(0.85)\right) = 1.24^3(0.85)$, and so on. So we can see that $a_n = 0.85(1.24)^n$.

(b) $a_6 = 1.24a_5 = 1.24(1.24a_4) = \cdots = 1.24^6 a_0 = 1.24^6(0.85) \approx 3.1$ lb

(c) The sequence $a_1, a_2, a_3, \ldots$ is geometric with common ratio 1.24.

FOCUS ON MODELING   Modeling with Recursive Sequences

1. (a) Since there are 365 days in a year, the interest earned per day is $\dfrac{0.0365}{365} = 0.0001$. Thus the amount in the account at the end of the nth day is $A_n = 1.0001A_{n-1}$ with $A_0 = \$275{,}000$.

(b) $A_0 = \$275,000$, $A_1 = 1.0001 A_0 = 1.0001 \cdot 275,000 = \$275,027.50$,

$A_2 = 1.0001 A_1 = 1.0001 (1.0001 A_0) = 1.0001^2 A_0 = \$275,055.00$,

$A_3 = 1.0001 A_2 = 1.0001^3 A_0 = \$275,082.51$, $A_4 = 1.0001^4 A_0 = \$275,110.02$, $A_5 = 1.0001^5 A_0 = \$275,137.53$,

$A_6 = 1.0001^6 A_0 = \$275,165.04$, $A_7 = 1.0001^7 A_0 = \$275,192.56$

(c) $A_n = 1.0001^n \cdot 275,000$

3. (a) Since there are 12 months in a year, the interest earned per day is $\dfrac{0.03}{12} = 0.0025$. Thus the amount in the account at the end of the nth month is $A_n = 1.0025 A_{n-1} + 100$ with $A_0 = \$100$.

(b) $A_0 = \$100$, $A_1 = 1.0025 A_0 + 100 = 1.0025 \cdot 100 + 100 = \$200.25$,

$A_2 = 1.0025 A_1 + 100 = 1.0025 (1.0025 \cdot 100 + 100) + 100 = 1.0025^2 \cdot 100 + 1.0025 \cdot 100 + 100 = \$300.75$,

$A_3 = 1.0025 A_2 + 100 = 1.0025 \left(1.0025^2 \cdot 100 + 1.0025 \cdot 100 + 100\right) + 100$

$= 1.0025^3 \cdot 100 + 1.0025^2 \cdot 100 + 1.0025 \cdot 100 + 100 = \$401.50$,

$A_4 = 1.0025 A_3 + 100 = 1.0025 \left(1.0025^3 \cdot 100 + 1.0025^2 \cdot 100 + 1.0025 \cdot 100 + 100\right) + 100$

$= 1.0025^4 \cdot 100 + 1.0025^3 \cdot 100 + 1.0025^2 \cdot 100 + 1.0025 \cdot 100 + 100 = \$502.51$

(c) $A_n = 1.0025^n \cdot 100 + \cdots + 1.0025^2 \cdot 100 + 1.0025 \cdot 100 + 100$, the partial sum of a geometric series, so

$A_n = 100 \cdot \dfrac{1 - 1.0025^{n+1}}{1 - 1.0025} = 100 \cdot \dfrac{1.0025^{n+1} - 1}{0.0025}$.

(d) Since 5 years is 60 months, we have $A_{60} = 100 \cdot \dfrac{1.0025^{61} - 1}{0.0025} \approx \$6580.83$.

5. (a) The amount A_n of pollutants in the lake in the nth year is 30% of the amount from the preceding year ($0.30 A_{n-1}$) plus the amount discharged that year (2400 tons). Thus $A_n = 0.30 A_{n-1} + 2400$.

(b) $A_0 = 2400$, $A_1 = 0.30 (2400) + 2400 = 3120$,

$A_2 = 0.30 [0.30 (2400) + 2400] + 2400 = 0.30^2 (2400) + 2400 (2400) + 2400 = 3336$,

$A_3 = 0.30 \left[0.30^2 (2400) + 2400 (2400) + 2400\right] + 2400$

$= 0.03^3 (2400) + 0.30^2 (2400) + 2400 (2400) + 2400 = 3400.8$,

$A_4 = 0.30 \left[0.03^3 (2400) + 0.30^2 (2400) + 2400 (2400) + 2400\right] + 2400$

$= 0.03^4 (2400) + 0.03^3 (2400) + 0.30^2 (2400) + 2400 (2400) + 2400 = 3420.2$

(c) A_n is the partial sum of a geometric series, so

$A_n = 2400 \cdot \dfrac{1 - 0.30^{n+1}}{1 - 0.30} = 2400 \cdot \dfrac{1 - 0.30^{n+1}}{0.70}$

$\approx 3428.6 \left(1 - 0.30^{n+1}\right)$

(e)

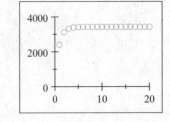

(d) $A_6 = 2400 \cdot \dfrac{1 - 0.30^7}{0.70} = 3427.8$ tons. The sum of a geometric

series, is $A = 2400 \cdot \dfrac{1}{0.70} = 3428.6$ tons.

7. (a) In the nth year since Victoria's initial deposit the amount V_n in her CD is the amount from the preceding year (V_{n-1}), plus the 5% interest earned on that amount ($0.05 V_{n-1}$), plus \$500 times the number of years since her initial deposit ($500n$). Thus $V_n = 1.05 V_{n-1} + 500n$.

(b) Ursula's savings surpass Victoria's savings in the 35th year.

9 PROBABILITY AND STATISTICS

9.1 COUNTING

1. The Fundamental Counting Principle says that if one event can occur in m ways and a second event can occurs in n ways, then the two events can occur in order in $m \times n$ ways. So if you have two choices for shoes and three choices for hats, then the number of different shoe-hat combinations you can wear is $2 \times 3 = 6$.

3. The number of ways of choosing r objects from n objects is called the number of *combinations* of n objects taken r at a time, and is given by the formula $C(n, r) = \dfrac{n!}{r!\,(n-r)!}$.

5. $P(8, 3) = \dfrac{8!}{(8-3)!} = \dfrac{8!}{5!} = 8 \cdot 7 \cdot 6 = 336$

7. $P(11, 4) = \dfrac{11!}{(11-4)!} = \dfrac{11!}{7!} = 11 \cdot 10 \cdot 9 \cdot 8 = 7920$

9. $P(100, 1) = \dfrac{100!}{(100-1)!} = \dfrac{100!}{99!} = 100$

11. $C(8, 3) = \dfrac{8!}{3!\,(8-3)!} = \dfrac{8!}{3!\,5!} = \dfrac{8 \cdot 7 \cdot 6}{3 \cdot 2 \cdot 1} = 56$

13. $C(11, 4) = \dfrac{11!}{4!\,7!} = \dfrac{11 \cdot 10 \cdot 9 \cdot 8}{4 \cdot 3 \cdot 2 \cdot 1} = 330$

15. $C(100, 1) = \dfrac{100!}{1!\,99!} = \dfrac{100}{1} = 100$

17. By the Fundamental Counting Principle, the number of possible single-scoop ice cream cones is
$$\begin{pmatrix} \text{number of ways to} \\ \text{choose the flavor} \end{pmatrix} \cdot \begin{pmatrix} \text{number of ways to} \\ \text{choose the type of cone} \end{pmatrix} = 4 \cdot 3 = 12.$$

19. **(a)** By the Fundamental Counting Principle, the possible number of ways 8 horses can complete a race, assuming no ties in any position, is
$$\begin{pmatrix} \text{number of ways to} \\ \text{choose the 1st finisher} \end{pmatrix} \cdot \begin{pmatrix} \text{number of ways to} \\ \text{choose the 2nd finisher} \end{pmatrix} \cdots \cdots \begin{pmatrix} \text{number of ways to} \\ \text{choose the 8th finisher} \end{pmatrix} = 8 \cdot 7 \cdot 6 \cdot 5 \cdot 4 \cdot 3 \cdot 2 \cdot 1$$
$$= 8! = 40{,}320$$

(b) By the Fundamental Counting Principle, the possible number of ways the first, second, and third place can be decided, assuming no ties, is $\begin{pmatrix} \text{number of ways to} \\ \text{choose the 1st finisher} \end{pmatrix} \cdot \begin{pmatrix} \text{number of ways to} \\ \text{choose the 2nd finisher} \end{pmatrix} \cdot \begin{pmatrix} \text{number of ways to} \\ \text{choose the 3th finisher} \end{pmatrix} = 8 \cdot 7 \cdot 6 = 336$.

21. The number of possible seven-digit phone numbers is
$\begin{pmatrix} \text{number of ways to} \\ \text{choose the 1st digit} \end{pmatrix} \cdot \begin{pmatrix} \text{number of ways to} \\ \text{choose the 2nd digit} \end{pmatrix} \cdots \cdots \begin{pmatrix} \text{number of ways to} \\ \text{choose the 7th digit} \end{pmatrix}$. Since the first digit cannot be a 0 or a 1, there are only 8 digits to choose from, while there are 10 digits to choose from for the other six digits in the phone number. Thus the number of possible seven-digit phone numbers is $8 \cdot 10 \cdot 10 \cdot 10 \cdot 10 \cdot 10 \cdot 10 = 8{,}000{,}000$.

23. Since there are 4 main courses, there are 6 ways to choose a main course. Likewise, there are 5 drinks and 3 desserts so there are 5 ways to choose a drink and 3 ways to choose a dessert. So the number of different meals consisting of a main course, a drink, and a dessert is $\begin{pmatrix} \text{number of ways to} \\ \text{choose the main course} \end{pmatrix} \cdot \begin{pmatrix} \text{number of ways to} \\ \text{choose a drink} \end{pmatrix} \cdot \begin{pmatrix} \text{number of ways to} \\ \text{choose a dessert} \end{pmatrix} = (4)(5)(3) = 60$.

25. The number of possible sequences of heads and tails when a coin is flipped 5 times is
$\begin{pmatrix} \text{number of possible} \\ \text{outcomes on the 1st flip} \end{pmatrix} \cdot \begin{pmatrix} \text{number of possible} \\ \text{outcomes on the 2nd flip} \end{pmatrix} \cdots \cdots \begin{pmatrix} \text{number of possible} \\ \text{outcomes on the 5th flip} \end{pmatrix} = (2)(2)(2)(2)(2)$
$$= 2^5 = 32$$

Here there are only two choices, heads or tails, for each flip.

27. Since there are six different faces on each die, the number of possible outcomes when a red die and a blue die and a white die are rolled is

$$\begin{pmatrix} \text{number of possible} \\ \text{outcomes on the red die} \end{pmatrix} \cdot \begin{pmatrix} \text{number of possible} \\ \text{outcomes on the blue die} \end{pmatrix} \cdot \begin{pmatrix} \text{number of possible} \\ \text{outcomes on the white die} \end{pmatrix} = (6)\,(6)\,(6) = 6^3 = 216.$$

29. The number of different California license plates possible is

$$\begin{pmatrix} \text{number of ways to} \\ \text{choose a nonzero digit} \end{pmatrix} \cdot \begin{pmatrix} \text{number of ways} \\ \text{to choose 3 letters} \end{pmatrix} \cdot \begin{pmatrix} \text{number of ways} \\ \text{to choose 3 digits} \end{pmatrix} = (9)\left(26^3\right)\left(10^3\right) = 158{,}184{,}000.$$

31. Since successive numbers cannot be the same, the number of possible choices for the second number in the combination is only 59. The third number in the combination cannot be the same as the second in the combination, but it can be the same as the first number, so the number of possible choices for the third number in the combination is also 59. So the number of possible combinations consisting of a number in the clockwise direction, a number in the counterclockwise direction, and then a number in the clockwise direction is $(60)\,(59)\,(59) = 208{,}860.$

33. Since a student can hold only one office, the number of ways that a president, a vice-president and a secretary can be chosen from a class of 30 students is

$$\begin{pmatrix} \text{number of ways} \\ \text{to choose a president} \end{pmatrix} \cdot \begin{pmatrix} \text{number of ways to} \\ \text{choose a vice-president} \end{pmatrix} \cdot \begin{pmatrix} \text{number of ways} \\ \text{to choose a secretary} \end{pmatrix} = (30)\,(29)\,(28) = 24{,}360.$$

35. We have seven choices for the first digit and 10 choices for each of the other 8 digits. Thus, the number of Social Security numbers is $7 \cdot 10^8 = 700{,}000{,}000.$

37. **(a)** The number of ways to select 5 of the 8 objects is $C\,(8, 5) = \dfrac{8!}{5!\,3!} = 56.$

(b) A set with 8 elements has $2^8 = 256$ subsets.

39. Each subset of toppings constitutes a different way a hamburger can be ordered. Since a set with 10 elements has $2^{10} = 1024$ subsets, there are 1024 different ways to order a hamburger.

41. **(a)** The number of ways to seat ten people in a row of ten chairs is $10! = 3{,}628{,}800.$

(b) The number of ways to choose six out of ten people and seat them in six chairs is $C\,(10, 6) \cdot 6! = \dfrac{10!}{6!\,4!} \cdot 6! = \dfrac{10!}{4!} = 151{,}200.$

43. In selecting these officers, order is important and repetition is not allowed, so the number of ways of choosing 3 officers from 15 students is $P\,(15, 3) = 2730.$

45. Since the order of finish is important, we want the number of permutations of 8 objects (the contestants) taken three at a time, which is $P\,(8, 3) = \dfrac{8!}{(8-3)!} = \dfrac{8!}{5!} = 8 \cdot 7 \cdot 6 = 336.$

47. The number of ways of ordering 9 distinct objects (the contestants) is $P\,(9, 9) = 9! = 362{,}880.$ Here a runner cannot finish more than once, so no repetitions are allowed, and order is important.

49. The number of ways of ordering 1000 distinct objects (the contestants) taking 3 at a time is $P\,(1000, 3) = 1000 \cdot 999 \cdot 998 = 997{,}002{,}000.$ We are assuming that a person cannot win more than once, that is, there are no repetitions.

51. We first place Jack in the first seat, and then seat the remaining four students. Thus the number of these arrangements is

$$\begin{pmatrix} \text{number of ways to} \\ \text{seat Jack in the first seat} \end{pmatrix} \cdot \begin{pmatrix} \text{number of ways to seat} \\ \text{the remaining four students} \end{pmatrix} = P\,(1, 1) \cdot P\,(4, 4) = 1!\,4! = 24.$$

53. Here we have 6 objects, of which 2 are blue marbles and 4 are red marbles. Thus the number of distinguishable permutations is $\dfrac{6!}{2!\,4!} = \dfrac{6 \cdot 5 \cdot 4!}{2 \cdot 4!} = 15.$

55. The number of distinguishable permutations of 12 objects (the 12 coins), from like groups of size 4 (the pennies), of size 3 (the nickels), of size 2 (the dimes) and of size 3 (the quarters) is $\dfrac{12!}{4!\,3!\,2!\,3!} = 277{,}200.$

57. The number of distinguishable permutations of 12 objects (the 12 ice cream cones) from like groups of size 3 (the vanilla cones), of size 2 (the chocolate cones), of size 4 (the strawberry cones), and of size 5 (the butterscotch cones) is $\dfrac{14!}{3!\,2!\,4!\,5!} = 2{,}522{,}520.$

59. The number of distinguishable permutations of 8 objects (the 8 cleaning tasks) from like groups of size 5, 2, and 1 workers, respectively is $\dfrac{8!}{5!\,2!\,1!} = 168.$

61. Here we are interested in the number of ways of choosing three objects (the three members of the committee) from a set of 25 objects (the 25 members). The number of combinations of 25 objects taken three at a time is $C\,(25,\,3) = \dfrac{25!}{3!\,22!} = 2300.$

63. We want the number of ways of choosing a group of three from a group of 12. This number is $C\,(12,\,3) = \dfrac{12!}{3!\,9!} = 220.$

65. We want the number of ways of choosing a group (the 5-card hand) where order of selection is not important. The number of combinations of 52 objects (the 52 cards) taken 5 at a time is $C\,(52,\,5) = \dfrac{52!}{5!\,47!} = 2{,}598{,}960.$

67. The order of selection is not important, hence we must calculate the number of combinations of 10 objects (the 10 questions) taken 7 at a time, this gives $C\,(10,\,7) = \dfrac{10!}{7!\,3!} = 120.$

69. We assume that the order in which he plays the pieces in the recital is not important, so the number of combinations of 12 objects (the 12 pieces) taken 8 at a time is $C\,(12,\,8) = \dfrac{12!}{8!\,4!} = 495.$

71. The order in which the pants are selected is not important and no pair of pants is repeated, so the number of combinations of ten pairs of pants taken three at a time is $C\,(10,\,3) = \dfrac{10!}{3!\,7!} = 120.$

73. Since the order in which the numbers are selected is not important, the number of combinations of 49 numbers taken 6 at a time is $C\,(49,\,6) = \dfrac{49!}{6!\,43!} = 13{,}983{,}816.$

75. **(a)** The number of ways of choosing 5 students from the 20 students is $C\,(20,5) = \dfrac{20!}{5!\,15!} = 15{,}504.$

   (b) The number of ways of choosing 5 students for the committee from the 12 females is $C\,(12,\,5) = \dfrac{12!}{5!\,7!} = 792.$

   (c) We use the Fundamental Counting Principle to count the number of possible committees with 3 females and 2 males. Thus, we get
   $$\binom{\text{number of ways to choose the}}{\text{3 females from the 12 females}} \cdot \binom{\text{number of ways to choose the}}{\text{2 males from the 8 males}} = C\,(12,\,3) \cdot C\,(8,\,2) = (220)\,(28) = 6160.$$

77. The number of ways the committee can be chosen is
   $$\binom{\text{number of ways to}}{\text{choose 1 president}} \cdot \binom{\text{number of ways to}}{\text{choose 1 vice-president}} \cdot \binom{\text{number of ways to}}{\text{choose 4 other members}} = C\,(20,\,1) \cdot C\,(19,\,1) \cdot C\,(18,\,4)$$
   $$= 20 \cdot 19 \cdot 3060 = 1{,}162{,}800$$

79. The number of ways the committee can be chosen is
   $$\binom{\text{number of ways to}}{\text{choose 2 of 6 freshmen}} \cdot \binom{\text{number of ways to}}{\text{choose 3 of 8 sophomores}} \cdot \binom{\text{number of ways to}}{\text{choose 4 of 12 juniors}} \cdot \binom{\text{number of ways to}}{\text{choose 5 of 10 seniors}}$$
   $$= C\,(6,\,2) \cdot C\,(8,\,3) \cdot C\,(12,\,4) \cdot C\,(10,\,5) = 15 \cdot 56 \cdot 495 \cdot 252 = 104{,}781{,}600$$

81. We choose 3 forwards from the forwards, 2 defensemen from the defensemen, and the goalie from the two goalies. Thus the number of ways to pick the 6 starting players is
   $$\binom{\text{number of ways to}}{\text{pick 3 of 12 forwards}} \cdot \binom{\text{number of ways to}}{\text{pick 2 of 6 defensemen}} \cdot \binom{\text{number of ways to}}{\text{pick 1 of 2 goalies}} = C\,(12,\,3) \cdot C\,(6,\,2) \cdot C\,(2,\,1)$$
   $$= (220)\,(15)\,(2) = 6600$$

83. We count the total number of committees and subtract the number that contain both Barry and Harry. The total number of committees possible is $C(10, 4)$ and the number that contain both Barry and Harry is $C(8, 2)$, so the number of possible committees is $C(10, 4) - C(8, 2) = 210 - 28 = 182$.

85. Since the two algebra books must be next to each other, we first consider them as one object. So we now have four objects to arrange and there are 4! ways to arrange these four objects. Now there are 2 ways to arrange the two algebra books. Thus the number of ways that 5 mathematics books may be placed on a shelf if the two algebra books are to be next to each other is $2 \cdot 4! = 48$.

87. (a) To find the number of ways the men and women can be seated we first select and place a man in the first seat and then arrange the other 7 people. Thus we get $\begin{pmatrix} \text{select 1 of} \\ \text{the 4 men} \end{pmatrix} \cdot \begin{pmatrix} \text{arrange the} \\ \text{remaining 7 people} \end{pmatrix} = C(4, 1) \cdot P(7, 7) = 4 \cdot 7! = 20{,}160$.

   (b) To find the number of ways the men and women can be seated we first select and place a woman in the first and last seats and then arrange the other 6 people. Thus we get

$$\begin{pmatrix} \text{arrange 2 of} \\ \text{the 4 women} \end{pmatrix} \cdot \begin{pmatrix} \text{arrange the} \\ \text{remaining 6 people} \end{pmatrix} = P(4, 2) \cdot P(6, 6) = 12 \cdot 6! = 8{,}640.$$

89. The number of ways the top finalist can be chosen is

$$\begin{pmatrix} \text{number of ways} \\ \text{to choose the 6} \\ \text{semifinalists from the 30} \end{pmatrix} \cdot \begin{pmatrix} \text{number of ways} \\ \text{to choose the 2} \\ \text{finalists from the 6} \end{pmatrix} \cdot \begin{pmatrix} \text{number of ways} \\ \text{to choose the winner} \\ \text{from the 2 finalists} \end{pmatrix} = C(30, 6) \cdot C(6, 2) \cdot C(2, 1)$$

$$= (593{,}775)\,(15)\,(2) = 17{,}813{,}250$$

91. Since there are 26 letters, the possible number of combinations of the first and the last initials is $(26)(26) = 676$. Since $677 > 676$, there must be at least two people that have the same first and last initials in any group of 677 people.

93. We are only interested in selecting a set of three marbles to give to Luke and a set of two marbles to give to Mark, not the order in which we hand out the marbles. Since both $C(10, 3) \cdot C(7, 2)$ and $C(10, 2) \cdot C(8, 3)$ count the number of ways this can be done, these numbers must be equal. (Calculating these values shows that they are indeed equal.) In general, if we wish to find two distinct sets of k and r objects selected from n objects ($k + r \le n$), then we can either first select the k objects from the n objects and then select the r objects from the $n - k$ remaining objects, or we can first select the r objects from the n objects and then the k objects from the $n - r$ remaining objects. Thus $\binom{n}{r} \cdot \binom{n-r}{k} = \binom{n}{k} \cdot \binom{n-k}{r}$.

9.2   PROBABILITY

1. The set of all possible outcomes of an experiment is called the *sample space*. A subset of the sample space is called an *event*. The sample space for the experiment of tossing two coins is $S = \{HH, HT, TH, TT\}$, and the event "getting at least one head" is $E = \{HH, HT, TH\}$. The probability of getting at least one head is $P(E) = \dfrac{n(E)}{n(S)} = \dfrac{3}{4}$.

3. The conditional probability of E given that F occurs is $P(E \mid F) = \dfrac{n(E \cap F)}{n(F)}$. So in tossing a die, the conditional probability of the event E "getting a six" given that that the event F "getting an even number" has occurred is $P(E \mid F) = \dfrac{1}{3}$.

5. (a) $S = \{1, 2, 3, 4, 5, 6\}$

   (b) $E = \{2, 4, 6\}$

   (c) $E = \{5, 6\}$

7. Let H stand for head and T for tails.

   (a) The sample space is $S = \{HH, HT, TH, TT\}$.

   (b) Let E be the event of getting exactly two heads, so $E = \{HH\}$. Then $P(E) = \frac{n(E)}{n(S)} = \frac{1}{4}$.

   (c) Let F be the event of getting at least one head. Then $F = \{HH, HT, TH\}$, and $P(F) = \frac{n(F)}{n(S)} = \frac{3}{4}$.

   (d) Let G be the event of getting exactly one head, that is, $G = \{HT, TH\}$. Then $P(G) = \frac{n(G)}{n(S)} = \frac{2}{4} = \frac{1}{2}$.

9. (a) Let E be the event of rolling a six. Then $P(E) = \frac{n(E)}{n(S)} = \frac{1}{6}$.

   (b) Let F be the event of rolling an even number. Then $F = \{2, 4, 6\}$. So $P(F) = \frac{n(F)}{n(S)} = \frac{3}{6} = \frac{1}{2}$.

   (c) Let G be the event of rolling a number greater than 5. Since 6 is the only face greater than 5, $P(G) = \frac{n(G)}{n(S)} = \frac{1}{6}$.

11. (a) Let E be the event of choosing a king. Since a deck has four kings, $P(E) = \frac{n(E)}{n(S)} = \frac{4}{52} = \frac{1}{13}$.

   (b) Let F be the event of choosing a face card. Since there are three face cards per suit and four suits,

$$P(F) = \frac{n(F)}{n(S)} = \frac{12}{52} = \frac{3}{13}.$$

   (c) Let F be the event of choosing a face card. Then $P(F') = 1 - P(F) = 1 - \frac{3}{13} = \frac{10}{13}$.

13. (a) Let E be the event of selecting a red ball. Since the jar contains five red balls, $P(E) = \frac{n(E)}{n(S)} = \frac{5}{8}$.

   (b) Let F be the event of selecting a yellow ball. Since there is only one yellow ball,

$$P(F') = 1 - P(F) = 1 - \frac{n(F)}{n(S)} = 1 - \frac{1}{8} = \frac{7}{8}.$$

   (c) Let G be the event of selecting a black ball. Since there are no black balls in the jar, $P(G) = \frac{n(G)}{n(S)} = \frac{0}{8} = 0$.

15. (a) Let E be the event of dealing five hearts. Since there are 13 hearts, $P(E) = \frac{C(13, 5)}{C(52, 5)} = \frac{1287}{2{,}598{,}960} \approx 0.000495$.

   (b) Let E be the event of choosing five cards of the same suit. Since there are four suits and 13 cards in each suit, $n(E) = 4 \cdot C(13, 5)$. Also, $n(S) = C(52, 5)$. Therefore, $P(E) = \frac{4 \cdot C(13, 5)}{C(52, 5)} = \frac{5{,}148}{2{,}598{,}960} \approx 0.00198$.

   (c) Let E be the event of dealing five face cards. Since there are 3 face cards for each suit and 4 suits,

$$P(E) = \frac{C(12, 5)}{C(52, 5)} = \frac{792}{2{,}598{,}960} \approx 0.000305.$$

   (d) Let E be the event of dealing a royal flush (ace, king, queen, jack, and 10 of the same suit). Since there is only one such sequence for each suit, there are only 4 royal flushes, so $P(E) = \frac{4}{C(52, 5)} = \frac{4}{2{,}598{,}960} \approx 1.53908 \times 10^{-6}$.

17. (a) Let E be the event of choosing two red balls. Since there are three red balls, $n(E) = C(3, 2)$. Also, $n(S) = C(8, 2)$. Therefore, $P(E) = \frac{n(E)}{n(S)} = \frac{C(3, 2)}{C(8, 2)} = \frac{3}{28} \approx 0.107$.

   (b) Let E be the event of choosing two white balls. Since there are five white balls, $n(E) = C(5, 2)$, so

$$P(E) = \frac{n(E)}{n(S)} = \frac{C(5, 2)}{C(8, 2)} = \frac{10}{28} = \frac{5}{14} \approx 0.357.$$

19. (a) Let E be the probability that at least one card is a spade. The number of hands that do not contain a spade is the number of possible 5-card hands using the other three suits, that is, $C(39, 5)$. Thus, $P(E) = 1 - \frac{C(39, 5)}{C(52, 5)} = \frac{7411}{9520} \approx 0.778$.

   (b) Let E be the probability that at least one card is a face card. The number of hands that *do not* contain a face card is the number of possible 5-card hands using the cards of the deck that are not face cards, that is, $C(40, 5)$. Thus,

$$P(E) = 1 - \frac{C(40, 5)}{C(52, 5)} = \frac{6221}{8330} \approx 0.747.$$

21. (a) Let E be the event that the spinner stops on red. Since 12 of the regions are red, $P(E) = \frac{12}{16} = \frac{3}{4}$.

(b) Let F be the event that the spinner stops on an even number. Since 8 of the regions are even-numbered, $P(F) = \frac{8}{16} = \frac{1}{2}$.

(c) Since 4 of the even-numbered regions are red, $P(E \cup F) = P(E) + P(F) - P(E \cap F) = \frac{3}{4} + \frac{1}{2} - \frac{4}{16} = 1$.

23. (a) Yes, the events are mutually exclusive since the number cannot be both even and odd. So
$$P(E \cup F) = P(E) + P(F) = \frac{3}{6} + \frac{3}{6} = 1.$$

(b) No, the events are not mutually exclusive since 6 is both even and greater than 4. So
$$P(E \cup F) = P(E) + P(F) - P(E \cap F) = \frac{3}{6} + \frac{2}{6} - \frac{1}{6} = \frac{2}{3}.$$

25. (a) No, the events E and F are not mutually exclusive since the jack, queen, and king of spades are both face cards and spades. So $P(E \cup F) = P(E) + P(F) - P(E \cap F) = \frac{13}{52} + \frac{12}{52} - \frac{3}{52} = \frac{11}{26}$.

(b) Yes, the events E and F are mutually exclusive since the card cannot be both a heart and a spade. So
$$P(E \cup F) = P(E) + P(F) = \frac{13}{52} + \frac{13}{52} = \frac{1}{2}.$$

27. (a) Let E denote a roll of five and F a roll greater than three. Using the formula for conditional probability, we have
$$P(E \mid F) = \frac{n(E \cap F)}{n(F)} = \frac{1}{3}.$$

(b) Let E denote a roll of three and F an odd roll. Using the formula for conditional probability, we have
$$P(E \mid F) = \frac{n(E \cap F)}{n(F)} = \frac{1}{3}.$$

29. Let E denote the spinner stopping on an even number and F the spinner stopping on red. Then
$$P(E \mid F) = \frac{n(E \cap F)}{n(F)} = \frac{4}{12} = \frac{1}{3}.$$

31. (a) There is only one red ball numbered 3 and one green ball numbered 3. If the ball drawn is numbered 3, then the probability it is red is $\frac{1}{2}$.

(b) There is only one ball numbered 7 and it is green, so if the ball drawn is numbered 7, then the probability it is green is 1.

(c) There are two even-numbered red balls and three even-numbered green balls, so if the ball is even-numbered, then the probability it is a red ball is $\frac{2}{5}$.

(d) There are five red balls, and two are even-numbered, so if the ball drawn is red, then the probability it is even-numbered is $\frac{2}{5}$.

33. (a) Let E be the event of drawing a black ball first. Because the jar contains seven black balls and three white balls, the probability of the first ball being black is $P(E) = \frac{7}{10}$. The probability of the second ball being white is $\frac{3}{9} = \frac{1}{3}$, so the probability of the intersection is $\frac{7}{10} \cdot \frac{1}{3} = \frac{7}{30}$.

(b) Here the probability of the second ball being black is $\frac{6}{9} = \frac{2}{3}$, so the probability of the intersection is $\frac{7}{10} \cdot \frac{2}{3} = \frac{7}{15}$.

35. (a) The probability of the first card being an ace is $\frac{4}{52}$ and the probability of the second being a king is $\frac{4}{51}$, so the probability of the intersection is $\frac{4}{52} \cdot \frac{4}{51} = \frac{4}{663}$.

(b) The probability of the first card being an ace is $\frac{4}{52}$ and the probability of the second being an ace is $\frac{3}{51}$, so the probability of the intersection is $\frac{4}{52} \cdot \frac{3}{51} = \frac{1}{221}$.

37. Let E be the event of getting a 1 on the first roll, and let F be the event of getting an even number on the second roll. Since these events are independent, $P(E \cap F) = P(E) \cdot P(F) = \frac{1}{6} \cdot \frac{3}{6} = \frac{1}{12}$.

39. (a) Yes. What happens on spinner A does not influence what happens on spinner B.

(b) The probability that A stops on red and B stops on yellow is $P(E \cap F) = P(E) \cdot P(F) = \left(\frac{2}{4}\right)\left(\frac{2}{8}\right) = \frac{1}{8}$.

41. (a) Let B and G stand for "boy"and "girl". Then

$$S = \{BBBB, GBBB, BGBB, BBGB, BBBG, GGBB, GBGB, GBBG,$$
$$BGGB, BGBG, BBGG, BGGG, GBGG, GGBG, GGGB, GGGG\}$$

(b) Let E be the event that the couple has only boys. Then $E = \{BBBB\}$ and $P(E) = \frac{1}{16}$.

(c) Let F be the event that the couple has 2 boys and 2 girls. Then

$$F = \{GGBB, GBGB, GBBG, BGGB, BGBG, BBGG\}, \text{ so } P(F) = \frac{6}{16} = \frac{3}{8}.$$

(d) Let G be the event that the couple has 4 children of the same sex. Then $G = \{BBBB, GGGG\}$, and $P(G) = \frac{2}{16} = \frac{1}{8}$.

(e) Let H be the event that the couple has at least 2 girls. Then H' is the event that the couple has fewer than two girls. Thus, $H' = \{BBBB, GBBB, BGBB, BBGB, BBBG\}$, so $n(H') = 5$, and $P(H) = 1 - P(H') = 1 - \frac{5}{16} = \frac{11}{16}$.

43. Let E be the event that the ball lands in an odd numbered slot. Since there are 18 odd numbers between 1 and 36, $P(E) = \frac{18}{38} = \frac{9}{19}$.

45. Let E be the event of picking the 6 winning numbers. Since there is only one way to pick these,

$$P(E) = \frac{1}{C(49, 6)} = \frac{1}{13,983,816} \approx 7.15 \times 10^{-8}.$$

47. The sample space consist of all possible True-False combinations, so $n(S) = 2^{10}$. Let E be the event that the student answers all 10 questions correctly. Since there is only one way to answer all 10 questions correctly, $P(E) = \frac{1}{2^{10}} = \frac{1}{1024}$.

49. (a) Let E be the event that the monkey types "Hamlet" as his first word. Since "Hamlet" contains 6 letters and there are 48 typewriter keys, $P(E) = \frac{1}{48^6} \approx 8.18 \times 10^{-11}$.

(b) Let F be the event that the monkey types "to be or not to be" as his first words. Since this phrase has 18 characters (including the blanks), $P(F) = \frac{1}{48^{18}} \approx 5.47 \times 10^{-31}$.

51. Let E be the event that the monkey will arrange the 11 blocks to spell *PROBABILITY* as his first word. The number of ways of arranging these blocks is the number of distinguishable permutations of 11 blocks. Since there are two blocks labeled B and two blocks labeled I, the number of distinguishable permutations is $\frac{11!}{2!\,2!}$. Only one of these arrangements spells the word *PROBABILITY*. Thus $P(E) = \frac{1}{\frac{11!}{2!\,2!}} = \frac{2!\,2!}{11!} \approx 1.00 \times 10^{-7}$.

53. (a) Let E be the event that the pea is tall. Since tall is dominant, $E = \{TT, Tt, tT\}$. So $P(E) = \frac{3}{4}$.

(b) E' is the event that the pea is short. So $P(E') = 1 - P(E) = 1 - \frac{3}{4} = \frac{1}{4}$.

55. Let E be the event that the player wins on spin 1, and let F be the event that the player wins on spin 2. What happens on the first spin does not influence what happens on the second spin, so the events are independent. Thus, $P(E \cap F) = P(E) \cdot P(F) = \frac{1}{38} \cdot \frac{1}{38} = \frac{1}{1444}$.

57. Let E, F and G denote the events of rolling two ones on the first, second, and third rolls, respectively, of a pair of dice. The events are independent, so $P(E \cap F \cap G) = P(E) \cdot P(F) \cdot P(G) = \frac{1}{36} \cdot \frac{1}{36} \cdot \frac{1}{36} = \frac{1}{36^3} \approx 2.14 \times 10^{-5}$.

59. Let E be the event that the marble is red and F be the event that the number is odd-numbered. Then E' is the event that the marble is blue, and F' is the event that the marble is even-numbered.

(a) $P(E) = \frac{6}{16} = \frac{3}{8}$

(b) $P(F) = \frac{8}{16} = \frac{1}{2}$

(c) $P(E \cup F) = P(E) + P(F) - P(E \cap F) = \frac{6}{16} + \frac{8}{16} - \frac{3}{16} = \frac{11}{16}$

(d) $P(E' \cup F') = P(E') + P(F') - P(E' \cap F') = \frac{10}{16} + \frac{8}{16} - \frac{5}{16} = \frac{13}{16}$.

61. The probability of getting 2 red balls by picking from jar B is $\left(\frac{5}{7}\right)\left(\frac{4}{6}\right) = \frac{10}{21}$. The probability of getting 2 red balls by picking one ball from each jar is $\left(\frac{3}{7}\right)\left(\frac{5}{7}\right) = \frac{15}{49}$. The probability of getting 2 red balls after putting all balls in one jar is $\left(\frac{8}{14}\right)\left(\frac{7}{13}\right) = \frac{4}{13}$. Hence, picking both balls from jar B gives the greatest probability.

63. Let E be the event that she opens the lock within an hour. The number of combinations she can try in one hour is $10 \cdot 60 = 600$. The number of possible combinations is $P(40, 3)$ assuming that no number can be repeated. Thus

$$P(E) = \frac{600}{P(40, 3)} = \frac{600}{59,280} = \frac{5}{494} \approx 0.010 \,.$$

65. Let E be the event that Paul stands next to Phyllis. To find $n(E)$ we treat Paul and Phyllis as one object and find the number of ways to arrange the 19 objects and then multiply the result by the number of ways to arrange Paul and Phyllis. So $n(E) = 19! \cdot 2!$. The sample space is all the ways that 20 people can be arranged. Thus $P(E) = \dfrac{19! \cdot 2!}{20!} = \dfrac{2}{20} = 0.10$.

67. Let E be the event that the monkey arranges the 11 blocks to spell *PROBABILITY*. The number of ways of arranging these blocks is the number of distinguishable permutations of 11 blocks. Since there are two blocks labeled B and two blocks labeled I, the number of distinguishable permutations is $\dfrac{11!}{2! \, 2!}$. Only one of these arrangements spells the word *PROBABILITY*. Thus $P(E) = \dfrac{1}{\dfrac{11!}{2! \, 2!}} = \dfrac{2! \, 2!}{11!} = \dfrac{1}{9,979,200}$.

9.3 BINOMIAL PROBABILITY

1. A binomial experiment is one in which there are exactly *two* outcomes. One outcome is called *success* and the other is called *failure*.

3. $P(2 \text{ successes in } 5) = C(5, 2) \cdot \left(0.7^2\right)\left(0.3^3\right) = 0.13230$

5. $P(0 \text{ success in } 5) = C(5, 0) \cdot \left(0.7^0\right)\left(0.3^5\right) = 0.00243$

7. $P(1 \text{ success in } 5) = C(5, 1) \cdot \left(0.7^1\right)\left(0.3^4\right) = 0.02835$

9. $P(\text{at least 4 successes}) = P(4 \text{ successes}) + P(5 \text{ successes}) = 0.36015 + 0.16807 = 0.52822$

11. $P(\text{at most 1 failure}) = P(0 \text{ failure}) + P(1 \text{ failure}) = P(5 \text{ successes}) + P(4 \text{ successes}) = 0.52822$

13. $P(\text{at least 2 successes}) = P(2 \text{ successes}) + P(3 \text{ successes}) + P(4 \text{ successes}) + P(5 \text{ successes})$
$= 0.13230 + 0.30870 + 0.36015 + 0.16807 = 0.96922$

15. (a)

Outcome	Probability
1	0.2
2	0.2
3	0.2
4	0.2
5	0.2

(b)

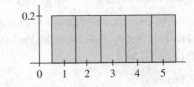

17. (a)

r	Probability
0	$\frac{1}{16}$
1	$\frac{1}{4}$
2	$\frac{3}{8}$
3	$\frac{1}{4}$
4	$\frac{1}{16}$

(b)

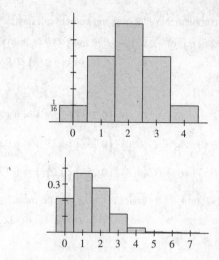

19. (a)

r	Probability
0	0.2097
1	0.3670
2	0.2753
3	0.1147
4	0.0287
5	0.0043
6	0.00036
7	0.000013

(b)

21. Here "success" is "face is 4" and P (face is 4) $= \frac{1}{6}$. Then P (2 successes in 6) $= C(6, 2) \cdot \left(\frac{1}{6}\right)^2 \left(\frac{5}{6}\right)^4 = 0.20094$.

23. P (4 successes in 10) $= C(10, 4) \cdot \left(0.4^4\right)\left(0.6^6\right) \approx 0.25082$

25. (a) P (5 in 10) $= C(10, 5) \cdot \left(0.45^5\right)\left(0.55^5\right) \approx 0.23403$

(b) P (at least 3) $= 1 - P$ (at most 2) $= P$ (0 in 10) $+ P$ (1 in 10) $+ P$ (2 in 10)
$$= 1 - \left[C(10, 0) \cdot \left(0.45^0\right)\left(0.55^{10}\right) + C(10, 1) \cdot \left(0.45^1\right)\left(0.55^9\right) + C(10, 2) \cdot \left(0.45^2\right)\left(0.55^8\right)\right]$$
$$\approx 0.90044$$

27. (a) The complement of at least 1 germinating is no seed germinating, so
$$P \text{ (at least 1 germinates)} = 1 - P \text{ (0 germinates)} = 1 - C(4, 0) \cdot \left(0.75^0\right)\left(0.25^4\right) \approx 0.99609.$$

(b) P (at least 2 germinate) $= P$ (2 germinates) $+ P$ (3 germinates) $+ P$ (4 germinates)
$$= C(4, 2) \cdot \left(0.75^2\right)\left(0.25^2\right) + C(4, 3) \cdot \left(0.75^3\right)\left(0.25^1\right) + C(4, 4) \cdot \left(0.75^4\right)\left(0.25^0\right)$$
$$\approx 0.94922$$

(c) P (4 germinates) $= C(4, 4) \cdot \left(0.75^4\right)\left(0.25^0\right) \approx 0.31641$

29. (a) P (all 10 are boys) $= C(10, 10) \cdot \left(0.52^{10}\right)\left(0.48^0\right) \approx 0.0014456$.

(b) P (all 10 are girls) $= C(10, 0) \cdot \left(0.52^0\right)\left(0.48^{10}\right) \approx 0.00064925$.

(c) P (5 in 10 are boys) $= C(10, 5) \cdot \left(0.52^5\right)\left(0.48^5\right) \approx 0.24413$.

31. (a) P (3 in 3) $= C(3, 3) \cdot \left(0.005^3\right)\left(0.995^0\right) \approx 0.000000125$.

(b) The complement of "one or more bulbs is defective" is "none of the bulbs is defective." So
$$P \text{ (at least 1 defective)} = 1 - P \text{ (none is defective)} = 1 - C(3, 0) \cdot \left(0.005^0\right)\left(0.995^3\right) \approx 0.014925.$$

33. The complement of "2 or more workers call in sick" is "0 or 1 worker calls in sick." So

$$P \text{ (2 or more)} \quad = 1 - [P \text{ (0 in 8)} + P \text{ (1 in 8)}]$$
$$= 1 - \left[C \text{ (8, 0)} \cdot \left(0.04^0 \right) \left(0.96^8 \right) + C \text{ (8, 1)} \cdot \left(0.04^1 \right) \left(0.96^7 \right) \right] \approx 0.038147$$

35. (a) $P \text{ (6 in 6)} = C \text{ (6, 6)} \cdot \left(0.75^6 \right) \left(0.25^0 \right) \approx 0.17798$

   (b) $P \text{ (0 in 6)} = C \text{ (6, 0)} \cdot \left(0.75^0 \right) \left(0.25^6 \right) \approx 0.00024414$

   (c) $P \text{ (3 in 6)} = C \text{ (6, 3)} \cdot \left(0.75^3 \right) \left(0.25^3 \right) \approx 0.13184$

   (d) $P \text{ (at least 2 seasick)} \quad = 1 - P \text{ (at most 1 seasick)} = 1 - [P \text{ (6 in 6 OK)} + P \text{ (5 in 6 OK)}]$
$$= 1 - \left[C \text{ (6, 6)} \cdot \left(0.75^6 \right) \left(0.25^0 \right) + C \text{ (6, 5)} \cdot \left(0.75^5 \right) \left(0.25^1 \right) \right] \approx 0.46606$$

37. (a) The complement of "at least one gets the disease" is "none gets the disease." Then

$$P \text{ (at least 1 gets the disease)} = 1 - P \text{ (0 gets the disease)} = 1 - C \text{ (4, 0)} \cdot \left(0.25^0 \right) \left(0.75^4 \right) \approx 0.68359.$$

   (b) $P \text{ (at least 3 get the disease)} \quad = P \text{ (3 get the disease)} + P \text{ (4 get the disease)}$
$$= C \text{ (4, 3)} \cdot \left(0.25^3 \right) \left(0.75^1 \right) + C \text{ (4, 4)} \cdot \left(0.25^4 \right) \left(0.75^0 \right) \approx 0.05078$$

39. Fred (a nonsmoker) is already in the room, so this exercise concerns the remaining 4 participants assigned to the room.

   (a) $P \text{ (1 in 4 is a smoker)} = C \text{ (4, 1)} \cdot \left(0.3^1 \right) \left(0.7^3 \right) = 0.4116$

   (b) $P \text{ (at least 1 smoker)} = 1 - P \text{ (0 in 4 is a smoker)} = 1 - C \text{ (4, 0)} \cdot \left(0.3^0 \right) \left(0.7^4 \right) = 0.7599$

41. (a) $P \text{ (8 or more recover)} = P \text{ (0 dies)} + P \text{ (1 dies)} + P \text{ (2 die)}$

$$= C \text{ (10, 0)} (0.6)^0 (0.4)^{10} + C \text{ (10, 1)} (0.6)^1 (0.4)^9 + C \text{ (10, 2)} (0.6)^2 (0.4)^8 \approx 0.0123$$

   (b) Yes, the drug appears to be effective.

43. (a)

Number of heads	Probability
0	0.003906
1	0.03125
2	0.109375
3	0.21875
4	0.273475
5	0.21875
6	0.109375
7	0.03125
8	0.003906

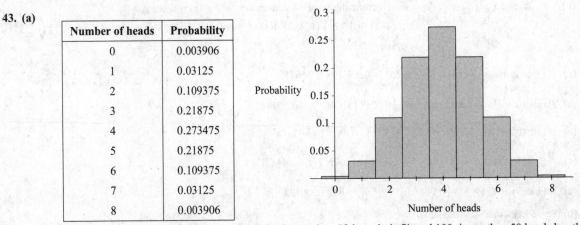

If $n = 8$, then 4 heads has the greatest probability of occurring. If the coin is flipped 100 times, then 50 heads has the greatest probability of occurring.

(b)

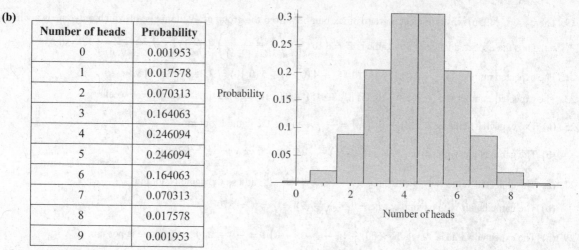

Number of heads	Probability
0	0.001953
1	0.017578
2	0.070313
3	0.164063
4	0.246094
5	0.246094
6	0.164063
7	0.070313
8	0.017578
9	0.001953

If $n = 9$, then 4 and 5 heads are the most likely outcomes. If the coin is flipped 101 times, then 50 and 51 heads are the most likely outcomes.

9.4 EXPECTED VALUE

1. If a game gives payoffs of $10 and $100 with probabilities 0.9 and 0.1, respectively, then the expected value of this game is $E = 10 \times 0.9 + 100 \times 0.1 = \$19$.

3. Mike gets $2 with probability $\frac{1}{2}$ and $1 with probability $\frac{1}{2}$. Thus, $E = (2)\left(\frac{1}{2}\right) + (1)\left(\frac{1}{2}\right) = 1.5$, and so his expected winnings are $1.50 per game.

5. Since the probability of drawing the ace of spades is $\frac{1}{52}$, the expected value of this game is $E = (100)\left(\frac{1}{52}\right) + (-1)\left(\frac{51}{52}\right) = \frac{49}{52} \approx 0.94$. So your expected winnings are $0.94 per game.

7. Since the probability that Carol rolls a six is $\frac{1}{6}$, the expected value of this game is $E = (3)\left(\frac{1}{6}\right) + (0.50)\left(\frac{5}{6}\right) = \frac{5.5}{6} \approx 0.9167$. So Carol expects to win $0.92 per game.

9. Since the probability that the die shows an even number equals the probability that that die shows an odd number, the expected value of this game is $E = (2)\left(\frac{1}{2}\right) + (-2)\left(\frac{1}{2}\right) = 0$. So Tom should expect to break even after playing this game many times.

11. Since it costs $0.50 to play, if you get a silver dollar, you win only $1 - 0.50 = \$0.50$. Thus the expected value of this game is $E = (0.50)\left(\frac{2}{10}\right) + (-0.50)\left(\frac{8}{10}\right) = -0.30$. So your expected winnings are −$0.30 per game. In other words, you should expect to lose $0.30 per game.

13. You can either win $35 or lose $1, so the expected value of this game is $E = (35)\left(\frac{1}{38}\right) + (-1)\left(\frac{37}{38}\right) = -\frac{2}{38} = -0.0526$. Thus the expected value is −$0.0526 per game.

15. By the rules of the game, a player can win $10 or $5, break even, or lose $100. Thus the expected value of this game is $E = (10)\left(\frac{10}{100}\right) + (5)\left(\frac{10}{100}\right) + (-100)\left(\frac{2}{100}\right) + (0)\left(\frac{78}{100}\right) = -0.50$. So the expected winnings per game are −$0.50.

17. If the stock goes up to $20, she expects to make $20 - \$5 = \$15$. And if the stock falls to $1, then she has lost $5 - \$1 = \$4$. So the expected value of her profit is $E = (15)(0.1) - (4)(0.9) = -2.1$. Thus, her expected profit per share is −$2.10, that is, she should expect to lose $2.10 per share. She did not make a wise investment.

19. There are $C(49, 6)$ ways to select a group of six numbers from the group of 49 numbers, of which only one is a winning set. Thus the expected value of this game is $E = \left(10^6 - 1\right)\left(\dfrac{1}{C(49, 6)}\right) + (-1)\left(1 - \dfrac{1}{C(49, 6)}\right) \approx -\$0.93$.

21. The expected number is $E = 2(0.15) + 3(0.45) + 4(0.30) + 5(0.10) = 3.35$ hours of TV.

23. The expected number is $3(0.30) + 2(0.45) + 1(0.15) + 0(0.10) = 1.95$ times in any given week.

25. **(a)** The expected value is $\frac{1}{52}(12) - \frac{51}{52}\left(\frac{1}{2}\right) = -\frac{27}{104}$, and so the game is not fair.

(b) The game is fair with payout x, where $\frac{1}{52}(x) - \frac{51}{52}\left(\frac{1}{2}\right) = 0 \Leftrightarrow x = \frac{51}{2} = \$25.50$.

27. **(a)** The expected value is $\frac{1}{6} \cdot \frac{1}{6}(30) - \frac{35}{36}(2) = \frac{30-70}{36} = -\frac{10}{9}$, and so the game is not fair.

(b) The game is fair with payout x, where $\frac{1}{36}x - \frac{35}{36}(2) = 0 \Leftrightarrow x = \$70$.

29. **(a)** The expected value is $\frac{1}{52} \cdot \frac{1}{6} \cdot \frac{1}{2}(600) - \left(1 - \frac{1}{52} \cdot \frac{1}{6} \cdot \frac{1}{2}\right)1 = -\frac{23}{624}$, and so the game is not fair.

(b) The game is fair with payout x, where $\frac{1}{52} \cdot \frac{1}{6} \cdot \frac{1}{2}x - \left(1 - \frac{1}{52} \cdot \frac{1}{6} \cdot \frac{1}{2}\right) = 0 \Leftrightarrow x = \$623$.

31. If you win, you win \$1 million minus the price of the stamp. If you lose, you lose only the price of the stamp (currently 44 cents). So the expected value of this game is $(999{,}999.56) \cdot \dfrac{1}{20 \times 10^6} + (-0.44) \cdot \dfrac{20 \times 10^6 - 1}{20 \times 10^6} = -0.39$. Thus, you expect to lose 39 cents on each entry, and so it's not worth it.

CHAPTER 9 REVIEW

1. The number of possible outcomes is
$$\begin{pmatrix}\text{number of outcomes} \\ \text{when a coin is tossed}\end{pmatrix} \cdot \begin{pmatrix}\text{number of outcomes} \\ \text{a die is rolled}\end{pmatrix} \cdot \begin{pmatrix}\text{number of ways} \\ \text{to draw a card}\end{pmatrix} = (2)(6)(52) = 624.$$

3. **(a)** Order is not important, and there are no repetitions, so the number of different two-element subsets is
$$C(5, 2) = \frac{5!}{2!\,3!} = \frac{5 \cdot 4}{2} = 10.$$

(b) Order is important, and there are no repetitions, so the number of different two-letter words is $P(5, 2) = \dfrac{5!}{3!} = 20$.

5. You earn a score of 70% by answering exactly 7 of the 10 questions correctly. The number of different ways to answer the questions correctly is $C(10, 7) = \dfrac{10!}{7!\,3!} = 120$.

7. You must choose two of the ten questions to omit, and the number of ways of choosing these two questions is
$$C(10, 2) = \frac{10!}{2!\,8!} = 45.$$

9. The maximum number of employees using this security system is
$$\begin{pmatrix}\text{number of choices} \\ \text{for the first letter}\end{pmatrix} \cdot \begin{pmatrix}\text{number of choices} \\ \text{for the second letter}\end{pmatrix} \cdot \begin{pmatrix}\text{number of choices} \\ \text{for the third letter}\end{pmatrix} = (26)(26)(26) = 17{,}576.$$

11. We could count the number of ways of choosing 7 of the flips to be heads; equivalently we could count the number of ways of choosing 3 of the flips to be tails. Thus, the number of different ways this can occur is $C(10, 7) = C(10, 3) = \dfrac{10!}{3!\,7!} = 120$.

13. Let x be the number of people in the group. Then $C(x, 2) = 10 \Leftrightarrow \dfrac{x!}{2!\,(x-2)!} = 10 \Leftrightarrow \dfrac{x!}{(x-2)!} = 20 \Leftrightarrow x(x-1) = 20$ $\Leftrightarrow x^2 - x - 20 = 0 \Leftrightarrow (x-5)(x+4) = 0 \Leftrightarrow x = 5$ or $x = -4$. So there are 5 people in this group.

15. A letter can be represented by a sequence of length 1, a sequence of length 2, or a sequence of length 3. Since each symbol is either a dot or a dash, the possible number of letters is

$$\begin{pmatrix} \text{number of letters} \\ \text{using 3 symbols} \end{pmatrix} + \begin{pmatrix} \text{number of letters} \\ \text{using 2 symbols} \end{pmatrix} + \begin{pmatrix} \text{number of letters} \\ \text{using 1 symbol} \end{pmatrix} = 2^3 + 2^2 + 2 = 14.$$

17. (a) Since we cannot choose a major and a minor in the same subject, the number of ways a student can select a major and a minor is $P(16, 2) = 16 \cdot 15 = 240$.

 (b) Again, since we cannot have repetitions and the order of selection is important, the number of ways to select a major, a first minor, and a second minor is $P(16, 3) = 16 \cdot 15 \cdot 14 = 3360$.

 (c) When we select a major and 2 minors, the order in which we choose the minors is not important. Thus the number of ways to select a major and 2 minors is $\begin{pmatrix} \text{number of ways} \\ \text{to select a major} \end{pmatrix} \cdot \begin{pmatrix} \text{number of ways to} \\ \text{select two minors} \end{pmatrix} = 16 \cdot C(15, 2) = 16 \cdot 105 = 1680.$

19. Because the letters are distinct, the number of anagrams of the word *RANDOM* is $6! = 720$.

21. Because three letters are the same, the number of anagrams of the word *BUBBLE* is $\dfrac{6!}{3!} = 120$.

23. (a) The possible number of committees is $C(18, 7) = 31{,}824$.

 (b) Since we must select the 4 men from the group of 10 men and the 3 women from the group of 8 women, the possible number of committees is $\begin{pmatrix} \text{number of ways to} \\ \text{choose 4 of 10 men} \end{pmatrix} \cdot \begin{pmatrix} \text{number of ways to} \\ \text{choose 3 of 8 women} \end{pmatrix} = C(10, 4) \cdot C(8, 3) = 210 \cdot 56 = 11{,}760.$

 (c) We remove Susie from the group of 18, so the possible number of committees is $C(17, 7) = 19{,}448$.

 (d) The possible number of committees is

$$\begin{pmatrix} \text{possible number of} \\ \text{committees with 5 women} \end{pmatrix} + \begin{pmatrix} \text{possible number of} \\ \text{committees with 6 women} \end{pmatrix} + \begin{pmatrix} \text{possible number of} \\ \text{committees with 7 women} \end{pmatrix}$$
$$= C(8, 5) \cdot C(10, 2) + C(8, 6) \cdot C(10, 1) + C(8, 7) \cdot C(10, 0) = 56 \cdot 45 + 28 \cdot 10 + 8 \cdot 1 = 2808$$

 (e) Since the committee is to have 7 members, "at most two men" is the same as "at least five women," which we found in part (d). So the number is also 2808.

 (f) We select the specific offices first, then complete the committee from the remaining members of the group. So the number of possible committees is

$$\begin{pmatrix} \text{number of ways to choose} \\ \text{a chairman, a vice-chairman,} \\ \text{and a secretary} \end{pmatrix} \cdot \begin{pmatrix} \text{number of ways to choose} \\ \text{4 other members} \end{pmatrix} = P(18, 3) \cdot C(15, 4) = 4896 \cdot 1365 = 6{,}683{,}040.$$

25. (a) The probability that the ball is red is $\dfrac{10}{15} = \dfrac{2}{3}$.

 (b) The probability that the ball is even numbered is $\dfrac{8}{15}$.

 (c) The probability that the ball is white and an odd number is $\dfrac{2}{15}$.

 (d) The probability that the ball is red or odd numbered is $P(\text{red}) + P(\text{odd}) - P(\text{red} \cap \text{odd}) = \dfrac{10}{15} + \dfrac{7}{15} - \dfrac{5}{15} = \dfrac{12}{15} = \dfrac{4}{5}$.

27. (a) $S = \{HHH, HHT, HTH, HTT, THH, THT, TTH, TTT\}$.

 (b) $P(HHH) = \dfrac{1}{8}$

 (c) $P(\text{2 or more heads}) = P(\text{exactly 2 heads}) + P(\text{3 heads}) = \dfrac{3}{8} + \dfrac{1}{8} = \dfrac{4}{8} = \dfrac{1}{2}$

 (d) $P(\text{tails on the first toss}) = \dfrac{4}{8} = \dfrac{1}{2}$

29. Since rolling a die and selecting a card is independent,

$P(\text{both show a six}) = P(\text{die shows a six}) \cdot P(\text{card is a six}) = \dfrac{1}{6} \cdot \dfrac{4}{52} = \dfrac{1}{78}$.

31. (a) Since these events are independent, the probability of getting the ace of spades, a six, and a head is $\dfrac{1}{52} \cdot \dfrac{1}{6} \cdot \dfrac{1}{2} = \dfrac{1}{624}$.

 (b) The probability of getting a spade, a six, and a head is $\dfrac{13}{52} \cdot \dfrac{1}{6} \cdot \dfrac{1}{2} = \dfrac{1}{48}$.

(c) The probability of getting a face card, a number greater than 3, and a head is $\frac{12}{52} \cdot \frac{3}{6} \cdot \frac{1}{2} = \frac{3}{52}$.

33. (a) Since there are four kings in a standard deck, $P\,(4\text{ kings}) = \dfrac{C\,(4, 4)}{C\,(52, 4)} = \dfrac{1}{\frac{52\cdot51\cdot50\cdot49}{4\cdot3\cdot2\cdot1}} = \dfrac{1}{270{,}725} \approx 3.69 \times 10^{-6}$.

(b) Since there are 13 spades in a standard deck, $P\,(4\text{ spades}) = \dfrac{C\,(13, 4)}{C\,(52, 4)} = \dfrac{\frac{13\cdot12\cdot11\cdot10}{4\cdot3\cdot2\cdot1}}{\frac{52\cdot51\cdot50\cdot49}{4\cdot3\cdot2\cdot1}} = \dfrac{11}{4165} \approx 0.00264$.

(c) Since there are 26 red cards and 26 black cards, $P\,(\text{all same color}) = \dfrac{2 \cdot C\,(26, 4)}{C\,(52, 4)} = \dfrac{2 \cdot \frac{26\cdot25\cdot24\cdot23}{4\cdot3\cdot2\cdot1}}{\frac{52\cdot51\cdot50\cdot49}{4\cdot3\cdot2\cdot1}} = \dfrac{92}{833} \approx 0.11044$.

35. She knows the first digit and must arrange the other four digits. Since only one of the $P\,(4, 4) = 24$ arrangements is correct, the probability that she guesses correctly is $\frac{1}{24}$.

37. (a) Since there are only two colors of socks, any 3 socks must contain a matching pair.

(b) *Method 1:* If the two socks drawn form a matching pair then they are either both red or both blue. So

$$P\left(\begin{array}{c}\text{choosing a}\\\text{matching pair}\end{array}\right) = P\left(\begin{array}{c}\text{both red or}\\\text{both blue}\end{array}\right) = P\,(\text{both red}) + P\,(\text{both blue}) = \frac{C\,(20, 2)}{C\,(50, 2)} + \frac{C\,(30, 2)}{C\,(50, 2)} \approx 0.51.$$

Method 2: The complement of choosing a matching pair is choosing one sock of each color. So

$$P\,(\text{choosing a matching pair}) = 1 - P\,(\text{different colors}) = 1 - \frac{C\,(20, 1) \cdot C\,(30, 1)}{C\,(50, 2)} \approx 1 - 0.49 = 0.51.$$

39. (a) Order is important, and repeats are possible. Thus there are 10 choices for each digit. So the number of different Zip+4 codes is $10 \cdot 10 \cdot \cdots \cdot 10 = 10^9$.

(b) If a Zip+4 code is to be a palindrome, the first 5 digits can be chosen arbitrarily. But once chosen, the last 4 digits are determined. Since there are 10 ways to choose each of the first 5 digits, there are 10^5 palindromes.

(c) By parts (a) and (b), the probability that a randomly chosen Zip+4 code is a palindrome is $\frac{10^5}{10^9} = 10^{-4}$.

41. (a) $P\,(\text{king}) = \frac{4}{52} = \frac{1}{13}$

(b) $P\,(\text{king or ace}) = \frac{8}{52} = \frac{2}{13}$

(c) The probability that the card is a king given that it is a face card is $\dfrac{\text{number of kings}}{\text{number of face cards}} = \dfrac{4}{12} = \dfrac{1}{3}$.

(d) The probability that the card is a king given that it is not an ace is $\dfrac{\text{number of kings}}{\text{number of non-aces}} = \dfrac{4}{48} = \dfrac{1}{12}$.

43. (a) $P\,(5\text{ in }5\text{ are white flesh}) = C\,(5, 5) \cdot \left(0.3^5\right)\left(0.7^0\right) = 0.00243$

(b) $P\,(0\text{ in }5\text{ are white flesh}) = C\,(5, 0) \cdot \left(0.3^0\right)\left(0.7^5\right) \approx 0.16807$

(c) $P\,(2\text{ in }5\text{ are white flesh}) = C\,(5, 2) \cdot \left(0.3^2\right)\left(0.7^3\right) = 0.3087$

(d) $P\,(3\text{ or more are red flesh}) = P\,(3\text{ in }5\text{ are red flesh}) + P\,(4\text{ in }5\text{ are red flesh}) + P\,(5\text{ in }5\text{ are red flesh})$

$$= C\,(5, 3) \cdot \left(0.3^2\right)\left(0.7^3\right) + C\,(5, 4) \cdot \left(0.3^1\right)\left(0.7^4\right) + C\,(5, 5) \cdot \left(0.3^0\right)\left(0.7^5\right)$$

$$= 0.83692$$

45. The probabilities are as follows:

Zero heads: $C\,(4, 0)\,(0.7)^0\,(0.3)^4 = 0.0081$

One heads: $C\,(4, 1)\,(0.7)^1\,(0.3)^3 = 0.0756$

Two heads: $C\,(4, 2)\,(0.7)^2\,(0.3)^2 = 0.2646$

Three heads: $C\,(4, 3)\,(0.7)^3\,(0.3)^1 = 0.4116$

Four heads: $C\,(4, 4)\,(0.7)^4\,(0.3)^0 = 0.2401$

Outcome (heads)	Probability
0	0.0081
1	0.0756
2	0.2646
3	0.4116
4	0.2401

47. Since Mary makes a guess as to the order of ratification of the 13 original states, the number of such guesses is $P(13, 13) = 13!$, while the probability that she guesses the correct order is $\frac{1}{13!}$. Thus the expected value is

$$E = (1,000,000)\left(\frac{1}{13!}\right) + (0)\left(\frac{13! - 1}{13!}\right) = 0.00016.$$ So Mary's expected winnings are $0.00016.

CHAPTER 9 TEST

1. The order is fixed, but for each grandchild they have three choices of pictures. Thus, the number of possibilities is $3 \times 3 \times 3 \times 3 = 81$.

3. (a) If repetition is allowed, then each letter can be chosen in 26 ways and each digit in 10 ways, so the number of possible passwords is $26^4 \cdot 10^3 = 456,976,000$.

(b) If repetition is not allowed, then the first letter can be chosen in 26 ways, the second in 25 ways, the third in 24 ways, and the fourth in 23 ways. The first digit can be chosen in 10 ways, the second in 9 ways, and the third in 8 ways. Thus, in this case the total number of possible passwords is $26 \cdot 25 \cdot 24 \cdot 23 \cdot 10 \cdot 9 \cdot 8 = 258,336,000$.

5. There are two choices to be made: choose a road to travel from Ajax to Barrie, and then choose a different road from Barrie to Ajax. Since there are 4 roads joining the two cities, we need the number of permutations of 4 objects (the roads) taken 2 at a time (the road there and the road back). This number is $P(4, 2) = 4 \cdot 3 = 12$.

7. (a) We want the number of ways of arranging 4 distinct objects (the letters L, O, V, E). This is the number of permutations of 4 objects taken 4 at a time. Therefore, the number of anagrams of the word LOVE is $P(4, 4) = 4! = 24$.

(b) We want the number of distinguishable permutations of 6 objects (the letters K, I, S, S, E, S) consisting of three like groups of size 1 and a like group of size 3 (the S's). Therefore, the number of different anagrams of the word KISSES is $\frac{6!}{1!\,1!\,1!\,3!} = \frac{6!}{3!} = 120$.

9. One card is drawn from a deck.

(a) Since there are 26 red cards, the probability that the card is red is $\frac{26}{52} = \frac{1}{2}$.

(b) Since there are 4 kings, the probability that the card is a king is $\frac{4}{52} = \frac{1}{13}$.

(c) Since there are 2 red kings, the probability that the card is a red king is $\frac{2}{52} = \frac{1}{26}$.

11. Let E be the event of choosing 3 men. Then $P(E) = \frac{n(E)}{n(S)} = \frac{\text{number of ways to choose 3 men}}{\text{number of ways to choose 3 people}} = \frac{C(5, 3)}{C(15, 3)} \approx 0.022$.

13. There are 4 students and 12 astrological signs. Let E be the event that at least 2 have the same astrological sign. Then E' is the event that no 2 have the same astrological sign. It is easier to find E'. So

$$P(E') = \frac{\text{number of ways to assign 4 different astrological signs}}{\text{number of ways to assign 4 astrological signs}} = \frac{P(12, 4)}{12^4} = \frac{12 \cdot 11 \cdot 10 \cdot 9}{12 \cdot 12 \cdot 12 \cdot 12} = \frac{55}{96}.$$

Therefore, $P(E) = 1 - P(E') = 1 - \frac{55}{96} = \frac{41}{96} \approx 0.427$.

15. A deck of cards contains 4 aces, 12 face cards, and 36 other cards. So the probability of an ace is $\frac{4}{52} = \frac{1}{13}$, the probability of a face card is $\frac{12}{52} = \frac{3}{13}$, and the probability of a non-ace, non-face card is $\frac{36}{52} = \frac{9}{13}$. Thus the expected value of this game is $E = (10)\left(\frac{1}{13}\right) + (1)\left(\frac{3}{13}\right) + (-.5)\left(\frac{9}{13}\right) = \frac{8.5}{13} \approx 0.654$, that is, about $0.65.

FOCUS ON MODELING   The Monte Carlo Method

1. **(a)** You should find that with the switching strategy, you win about 90% of the time. The more games you play, the closer to 90% your winning ratio will be.

   (b) The probability that the contestant has selected the winning door to begin with is $\frac{1}{10}$, since there are ten doors and only one is a winner. So the probability that he has selected a losing door is $\frac{9}{10}$. If the contestant switches, he exchanges a losing door for a winning door (and vice versa), so the probability that he loses is now $\frac{1}{10}$, and the probability that he wins is now $\frac{9}{10}$.

3. **(a)** You should find that player A wins about $\frac{7}{8}$ of the time. That is, if you play this game 80 times, player A should win approximately 70 times.

   (b) The game will end when either player A gets one more head or player B gets three more tails. Each toss is independent, and both heads and tails have probability $\frac{1}{2}$, so we obtain the following probabilities.

Outcome	Probability
H	$\frac{1}{2}$
TH	$\frac{1}{2} \cdot \frac{1}{2} = \frac{1}{4}$
TTH	$\frac{1}{2} \cdot \frac{1}{2} \cdot \frac{1}{2} = \frac{1}{8}$
TTT	$\frac{1}{2} \cdot \frac{1}{2} \cdot \frac{1}{2} = \frac{1}{8}$

   Since Player A wins for any outcome that ends in heads, the probability that he wins is $\frac{1}{2} + \frac{1}{4} + \frac{1}{8} = \frac{7}{8}$.

5. With 1000 trials, you are likely to obtain an estimate for π that is between 3.1 and 3.2.

7. **(a)** We can use the following TI-83 program to model this experiment. It is a minor modification of the one given in Problem 5.

```
PROGRAM:PROB7
:0→P
:For(N,1,1000)
:rand→X:rand→Y
:P+((X+Y)<1)→P
:End
:Disp "PROBABILITY IS APPROX",P/1000
```

   You should find that the probability is very close to $\frac{1}{2}$.

   (b) Following the hint, the points in the square for which $x + y < 1$ are the ones that lie below the line $x + y = 1$. This triangle has area $\frac{1}{2}$ (it takes up half the square), so the probability that $x + y < 1$ is $\frac{1}{2}$.

CUMULATIVE REVIEW TEST: CHAPTERS 8 and 9

1. (a) $a_n = \dfrac{n}{2n+1}$, so $a_7 = \dfrac{7}{2(7)+1} = \dfrac{7}{15}$ and $a_{20} = \dfrac{20}{2(20)+1} = \dfrac{20}{41}$.

(b) $a_7 = \dfrac{2 \cdot 7^2 + 1}{7^3 - 7 + 4} = \dfrac{99}{340}$ and $a_{20} = \dfrac{2 \cdot 20^2 + 1}{20^3 - 20 + 4} = \dfrac{801}{7984}$.

(c) $a_n = \frac{1}{2} + 3\,(n-1)$, so $a_7 = \frac{1}{2} + 3\,(6) = \frac{37}{2}$ and $a_{20} = \frac{1}{2} + 3\,(19) = \frac{115}{2}$.

(d) $a_n = 12\left(\frac{5}{6}\right)^{n-1}$, so $a_7 = 12\left(\frac{5}{6}\right)^6 = \dfrac{15,625}{3888}$ and $a_{20} = 12\left(\frac{5}{6}\right)^{19} = \dfrac{19,073,486,328,125}{50,779,978,334,208}$.

(e) $a_1 = 0.01$ and $a_n = -2a_{n-1}$ can be expressed without recursion as a geometric series with initial term 0.01 and common ratio $r = -2$. Thus, $a_n = 0.01\,(-2)^{n-1}$, $a_7 = 0.01\,(-2)^6 = 0.64$, and $a_{20} = 0.01\,(-2)^{19} = -5242.88$.

3. The amount of the mortgage is $A_p = 350,000 - 35,000 = 315,000$, $n = 15\,(12) = 180$, and $i = \dfrac{0.06}{12} = 0.005$, so

$$R = \frac{iA_p}{1 - (1+i)^{-n}} = \frac{0.005\,(315,000)}{1 - (1.005)^{-180}} = \$2658.15.$$

5. (a) $\left(2x - \frac{1}{2}\right)^5 = \binom{5}{0}(2x)^5\left(\frac{1}{2}\right)^0 - \binom{5}{1}(2x)^4\left(\frac{1}{2}\right)^1 + \binom{5}{2}(2x)^3\left(\frac{1}{2}\right)^2 - \binom{5}{3}(2x)^2\left(\frac{1}{2}\right)^3 + \binom{5}{4}(2x)^1\left(\frac{1}{2}\right)^4 - \binom{5}{5}(2x)^0\left(\frac{1}{2}\right)^5$

$$= 32x^5 - 40x^4 + 20x^3 - 5x^2 + \tfrac{5}{8}x - \tfrac{1}{32}$$

(b) The term containing x^4 is $\binom{12}{8}(2x)^4\left(\frac{1}{2}\right)^8 = \dfrac{495}{16}x^4$.

7. (a) The probability that all three dice show a particular number is $\left(\frac{1}{6}\right)^3$, so the probability that all three dice show the same number (any number) is $6 \cdot \left(\frac{1}{6}\right)^3 = \frac{1}{36}$.

(b) The probability that all three dice show an even number is $\left(\frac{1}{2}\right)^3 = \frac{1}{8}$.

(c) The events for which the sum is 15 are $(3,6,6)$, $(6,3,6)$, $(6,6,3)$, $(4,5,6)$, $(4,6,5)$, $(5,4,6)$, $(5,6,4)$, $(6,4,5)$, $(6,5,4)$, and $(5,5,5)$. So the probability of one of these 10 events is $\dfrac{10}{n\,(S)} = \dfrac{10}{6^3} = \dfrac{5}{108}$.

9. (a) $(p+q)^5 = \binom{5}{0}p^5q^0 + \binom{5}{1}p^4q^1 + \binom{5}{2}p^3q^2 + \binom{5}{3}p^2q^3 + \binom{5}{4}p^1q^4 + \binom{5}{5}p^0q^5$. The event whose probability is $\binom{5}{2}p^3q^2$ is the probability of getting 3 heads and 2 tails.

(b) The probability that there are 2 heads and 3 tails is $\binom{5}{3}p^2q^3 = 10\left(\frac{2}{3}\right)^2\left(\frac{1}{3}\right)^3 = \frac{40}{243} \approx 0.16$.

APPENDIXES

A   CALCULATIONS AND SIGNIFICANT FIGURES

1. $3.27 - 0.1834 \approx 3.09$

3. $28.36 \times 501.375 \approx 14{,}220$

5. $(1.36)^3 \approx 2.52$

7. $3.3\,(642.75 + 66.787) \approx 3.3\,(709.54) \approx 2300$

9. $\left(5.10 \times 10^{-3}\right)\left(12.4 \times 10^{7}\right)\left(6.007 \times 10^{-6}\right) \approx 3.80$

11. The circumference is $2\pi r \approx 2\pi\,(5.27) \approx 33.1$ ft and the area is $\pi r^2 \approx \pi\,(5.27)^2 \approx 87.3$ ft^2.

13. The force is $F = G\dfrac{m_1 m_2}{r^2} \approx 6.67428 \times 10^{-11} \cdot \dfrac{(11{,}426)^2}{(57{,}200)^2} \approx 2.66 \times 10^{-12}$ N.

B   GRAPHING WITH A GRAPHING CALCULATOR

1. $y = x^4 + 2$

   (a) $[-2, 2]$ by $[-2, 2]$

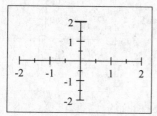

   (b) $[0, 4]$ by $[0, 4]$

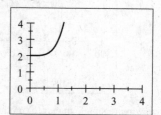

   (c) $[-8, 8]$ by $[-4, 40]$

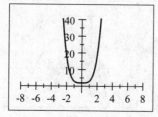

   (d) $[-40, 40]$ by $[-80, 800]$

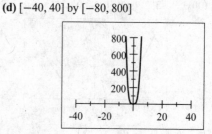

   The viewing rectangle in part (c) produces the most appropriate graph of the equation.

3. $y = 100 - x^2$

   (a) $[-4, 4]$ by $[-4, 4]$

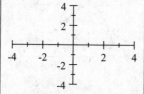

   (b) $[-10, 10]$ by $[-10, 10]$

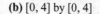

359

(c) $[-15, 15]$ by $[-30, 110]$

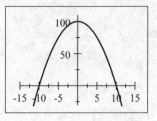

(d) $[-4, 4]$ by $[-30, 110]$

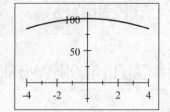

The viewing rectangle in part (c) produces the most appropriate graph of the equation.

5. $y = 10 + 25x - x^3$

(a) $[-4, 4]$ by $[-4, 4]$

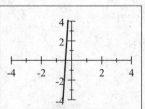

(b) $[-10, 10]$ by $[-10, 10]$

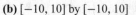
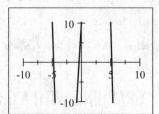

(c) $[-20, 20]$ by $[-100, 100]$

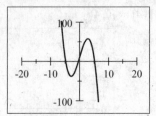

(d) $[-100, 100]$ by $[-200, 200]$

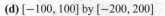
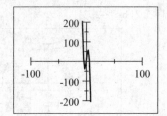

The viewing rectangle in part (c) produces the most appropriate graph of the equation.

7. $y = 100x^2$, $[-2, 2]$ by $[-10, 400]$

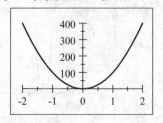

9. $y = 4 + 6x - x^2$, $[-4, 10]$ by $[-10, 20]$

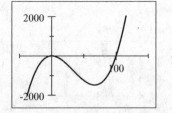

11. $y = \sqrt[4]{256 - x^2}$. We require that $256 - x^2 \geq 0 \Rightarrow$ $-16 \leq x \leq 16$, so we graph $y = \sqrt[4]{256 - x^2}$ in the viewing rectangle $[-20, 20]$ by $[-1, 5]$.

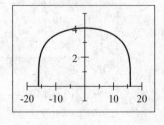

13. $y = 0.01x^3 - x^2 + 5$, $[-50, 150]$ by $[-2000, 2000]$

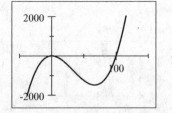

15. $y = \dfrac{1}{x^2 - 2x}$, $[-2, 4]$ by $[-8, 8]$

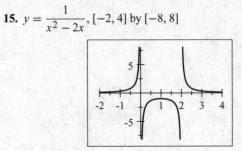

17. $y = 1 + |x - 1|$, $[-3, 5]$ by $[-1, 5]$

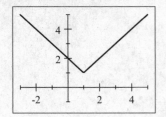

19. Although the graphs of $y = -3x^2 + 6x - \frac{1}{2}$ and

$y = \sqrt{7 - \frac{7}{12}x^2}$ appear to intersect in the viewing

rectangle $[-4, 4]$ by $[-1, 3]$, there is no point of

intersection. You can verify this by zooming in.

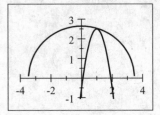

21. The graphs of $y = 6 - 4x - x^2$ and $y = 3x + 18$ appear to

have two points of intersection in the viewing rectangle

$[-6, 2]$ by $[-5, 20]$. You can verify that $x = -4$ and

$x = -3$ are exact solutions.

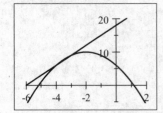

23. $x^2 + y^2 = 9 \Leftrightarrow y^2 = 9 - x^2 \Rightarrow y = \pm\sqrt{9 - x^2}$. So we

graph the functions $y_1 = \sqrt{9 - x^2}$ and $y_2 = -\sqrt{9 - x^2}$ in

the viewing rectangle $[-6, 6]$ by $[-4, 4]$.

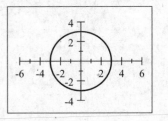

25. $4x^2 + 2y^2 = 1 \Leftrightarrow 2y^2 = 1 - 4x^2 \Leftrightarrow y^2 = \dfrac{1 - 4x^2}{2} \Rightarrow$

$y = \pm\sqrt{\dfrac{1 - 4x^2}{2}}$. So we graph the functions

$y_1 = \sqrt{\dfrac{1 - 4x^2}{2}}$ and $y_2 = -\sqrt{\dfrac{1 - 4x^2}{2}}$ in the viewing

rectangle $[-1.2, 1.2]$ by $[-0.8, 0.8]$.

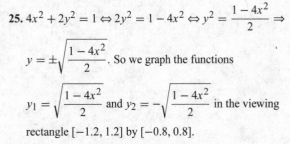